STORM FORCE

BRITAIN'S WILDEST WEATHER

Michael Fish MBE, Ian McCaskill
and Paul Hudson

GREAT NORTHERN

Lightning. Exeter 12th June 2006. (Matt Clark, IJMet/TORRO)

STORM FORCE

Michael Fish MBE, Ian McCaskill
and Paul Hudson

Great Northern Books
PO Box 213, Ilkley, LS29 9WS
www.greatnorthernbooks.co.uk

First published 2007

ISBN: 978 1 905080 32 8

Design and layout: David Burrill

Printed in Spain

CIP Data
A catalogue for this book is available from the British Library

Many thanks to The International Journal of Meteorology (IJMet)
and TORRO (The Tornado and Storm Research Organisation)
for their photographic contribution.
www.ijmet.org   www.torro.org.uk

Photographs except where otherwise
credited are courtesy Yorkshire Post Newspapers

CONTENTS

*Running before the storm – a natural instinct for hundreds of years,
as seen here on a deserted beach.*

Prologue 1

I never wanted to be a TV Weatherman: my aim was to be a forecaster or even a researcher into adverse conditions. Looking back, the most likely 'triggers' for my interest in meteorology were the Great Smog of December 1952 and the floods of 1953. I was lucky to have a geography master at Prep school and a physics master at 'Big' school who were both excellent and enthusiastic teachers, and very fortunate that, having decided at the age of ten that I wanted to be a weatherman, I was able to get the right qualifications (maths and physics) and then end up a few weeks after leaving school in a career that would last 42 years. Ironically my careers master had said to me "Whatever you do, DON'T join the Civil Service, the pay is awful". He was right; my early salary was £25 per month and it cost me £28 a month just to get to work! For non-shift workers at the Met Office, it is not much better now.

1962 was a memorable year in the history of the Met Office for two reasons. First, Her Majesty the Queen opened the new headquarters in Bracknell, enabling the first ever numerical model of the atmosphere to be produced on a vast new computer. From that time on the days of human beings were numbered as it was soon apparent that computers could do a better job. Second, and far more important, I joined!

My first winter with the Met Office was the severe winter of 1962-63, and I was regularly battling the elements as I tried to drive from my home in Eastbourne to Gatwick Airport where I was based. I spent about two years at Gatwick and Elmdon (Birmingham International) airports making tea, being a dogsbody, plotting charts and making observations. In 1965 I was promoted and posted to Bracknell as part of the numerical forecasting team. The boss only had women working for him, until I arrived, and it wasn't too long before I tried to escape, this time to do a Sandwich Course in Applied Physics at the City University, London. While at City I was sent for work experience to the London Weather Centre. I had a posting notice that said "detached for a period of 6 months, expected to return to permanent station, Bracknell". I was there until I retired 38 years later.

Whilst there, in 1971, Bert Foord and Graham Parker suddenly gained promotion and Civil Service red tape being what it was, were no longer allowed to appear on television as it was regarded as being too menial a task for men of their grade. The Met Office decided on a complete new image of first a female (not me, but Barbara Edwards) and second a young, handsome, sophisticated, well dressed much younger man (me). Try as they might over the years neither the Met Office nor the BBC ever managed to get rid of me until I was compulsorily retired in 2004.

I have vivid memories of the snow in June 1975 and the drought summer of 1976. It was highly amusing to know that the drought would break in three or four days time just as a Minister for Drought was appointed! There was also a memorable occasion when I 'lost' thousands of racing pigeons in unexpected fog, and the very mild Christmas forecast for 1970 that ended up with deep snow!

However, the weather event with which I am inextricably linked in the public mind is the Great Storm of 1987. If only I had a penny for each time the infamous 'Hurricane' clip has been broadcast, I would be a multimillionaire. As it is I think I will have it engraved on my tombstone, when the time comes!

"Earlier on today apparently a lady rang the BBC and said she heard that there was a hurricane on the way. Well don't worry if you're watching, there isn't." It is even more unfortunate as my remark had NOTHING to do with the storm, was NOT made the evening before and NO woman rang the BBC. All these things were pointed out at the time but, as is often the case with the Press, if the facts get in the way of a good story then they are not printed.

Bill Giles was actually the forecaster on duty just before the event and it was he who just said, "It will be breezy up the Channel", and focused on the expected heavy rain instead. He kept quiet about this until he had collected his OBE and retired! Of course he was not alone, as he was only interpreting the advice given by the Chief Forecaster at the Met Office headquarters which, at that time, was in Bracknell. He, in turn, was hampered by a lack of information from the Bay of Biscay, from where the storm originated, due to industrial action in France. It was then left up to poor old Ian McCaskill, who was on the night shift (and, as it turned out, a day shift as well) to pick up the pieces. My broadcast the morning before had actually said "batten down the hatches, there's some very stormy weather on the way" – pretty good, I would say!

Even though it was made clear to the media at the time that the famous phone call to the BBC about a hurricane had never been made, one paper, the Daily Mail, even managed to run a full-page exclusive interview with this non-existent person on the tenth anniversary of the storm. The truth is that a call was made by a colleague FROM the BBC to his mother, who was about to leave for Florida on holiday, to reassure her that the hurricane in the Caribbean, featured on the day's news bulletins, would not affect her plans. I turned it around to protect him, so that it did not appear as if he had made a long distance call from work, which was strictly against BBC regulations.

The evening forecast that was put out in Britain just before the storm was an error but it was nothing like as bad as was claimed at the time. As early as the previous weekend the Met Office had been forecasting severe weather. As Thursday approached, however, the data we were receiving and the computer models based on it became more equivocal, suggesting that the gales might only affect the Channel and France. As it turned out the gale warnings for sea areas were both timely and adequate. However, by the time most British people had gone to bed that night, no warnings of exceptional weather on land had been issued, although that was rectified in the early hours as new data was received.

This so-called "gigantic cock-up" also must be put into context. Because all weather systems are interconnected and cannot be treated in isolation, the Met Office computer produces a Global Model of the weather. On a global scale, therefore, a small difference of

Aftermath of the Great Storm of 16 October 1987, which caused exceptional damage to trees with some fifteen million felled within a few hours. The devastation is epitomised by this scene at the Royal Botanic Gardens at Kew, where a third of the specimen trees were lost. (PA)

track of a weather system from up the Channel to across south-east England is a minute occurrence but, of course, it makes a very big difference to the weather in that area.

By daybreak on Friday 16 October 1987 there was chaos across the South and East of England and I struggled into work at 4.30 in the morning having spent the hours before propping up trees and fences. I was not aware of just how severe the storm had been for some time, because, if windy, the drive to work seemed normal and on arriving at the BBC most communications were out because of unknown power cuts. Bracknell had also lost its phones.

A lot of the damage, especially to trees, was because of unfortunate timing; had the storm broken a few weeks later, many of the trees would have survived. It had been a mild and very wet autumn and the trees were still in full leaf, unpruned and standing in soggy ground. The wind battered them from the south and south-east, directions from which they were not braced (because the prevailing winds in Britain are from the west) and with roots in wet mud they just couldn't hold on.

By lunchtime the media had begun to pick up on my remarks and, naturally, preferred to ignore the explanation for them. For the next few days the Director-General of the Met Office, John Houghton, and myself were besieged at home by hundreds of reporters and cameramen. Saturday's papers even called for his resignation as it was alleged that other Met Offices had got the forecast right. This was yet another fallacy; none of the meteorological services on the continent had produced forecasts of severe weather over southern England. In any case, even if warnings had been given, it would have probably led to a greater loss of life as people ventured outside to try and prevent storm damage to their property – many of the casualties of the storm were caused in this way – so it was probably better that people remained safely tucked up in bed.

Only time will tell whether the storm was a truly exceptional, once in a lifetime event, or the first sign that the predictions for Global Warming were correct and that we must brace ourselves for ever more frequent and devastating storms? Perhaps one day a true hurricane will occur – 1987 certainly was not! Throughout history we have had similar storms but they do already appear to be becoming more frequent and more vicious.

Much good also came out of it in terms of improving our information gathering and systems to ensure swifter and more accurate forecasting of extreme weather in future. More powerful and efficient computers were installed, steps were taken to obtain more observations of developing weather systems, leading to improved computer modelling, and a new procedure was introduced for the dissemination of severe weather warnings. It seems to have worked; we have not been "caught out" again and the storm that devastated much of France a few years later was forecast by us, but not the French.

Despite, or possibly even because of 1987, my career as a TV weatherman brought me honorary Doctorates from the City University, London, and Exeter University, the Freedom of the City of London in 1997, and the award of an MBE in the Queen's Birthday Honours List in 2004. Less seriously, I've also been declared a "National Treasure" by the Press, voted both the "Worst Dressed" and "Best Dressed" man on television and made "Tie-man of the Year" four times, in recognition of the fish motif ties that I always wear.

My career opened many other doors to me, most of which I ran through at top speed. I've travelled widely in Africa, training television weathermen there, and I've guested on television programmes covering the full spectrum from the sublime to the ridiculous, including Basil Brush, Blue Peter, The Sky At Night and Eastenders. I also tried my hand at full-blown acting and recently toured the country with "The Play What I Wrote". I've even had a chart career, albeit by proxy, with punk group Rachel and Nicki's 1985 release "I wish, I wish, he was like Michael Fish", and A Tribe of Toff's 1988 Top Ten single "John Kettley is a Weatherman".

In retirement I'm Patron of several charities, including Age Concern, and I pass the rest of my time giving talks and lectures and writing articles and, since travel is my hobby, if this book sells well you won't see me for dust!

Michael Fish MBE

Prologue 2

8.30pm on the eve of the Great Storm of 1987. The nightshift is starting at the London Weather Centre, the busy heart of weather forecasting for London and the South of England. It all takes place in a dusty and surprisingly anonymous room in High Holborn, in the heart of London. We are all there to receive an amazingly anodyne report on the upcoming weather. For five days we have been watching an active hurricane travelling from the Caribbean towards Britain; not that we have seen much of it (these are the old days, after all!). Our shipping forecasts have been devastatingly accurate and, as a result, no merchant ship has tried to cross the line of our forecast track of the hurricane.

When an observation from a ship or land station supports our forecast – or doesn't – we call it "surface truth" and, either way, it is invaluable. We bang on about how we rely on ship reports and every digit is recorded – probably in stone – to keep those precious records together for ever. Everything is noted: wave height, wave direction (even in mid-Atlantic for goodness sake), sea temperature, weather and atmospheric pressure. It doesn't matter if some are missing. We can interpolate from the others, if we're lucky. But not this time. There are no reports, nothing at all – niente, nada. All we have had are the images looking down from the satellite, showing a vague, leaf-shaped cloud, tracking across the Atlantic towards us.

The 6pm computer run came in. Perhaps unsurprisingly, given the same lack of solid information, it merely echoed the midday computer run. It showed an active but small "Low", running up the English Channel, producing gales but little else. So this was what the dayshift hands over: not much.

Then, shortly after we arrived for the nightshift, a report from a Channel Light Vessel came in. We really, truly, listen to them. Fifty knot winds it said, not a gust to 50 knots, but a steady wind of fifty knots – fifty-seven miles an hour. We took notice; that is to say, we panicked. Fifty knots in the relatively sheltered waters of the English Channel really meant something. We immediately went back to our charts, filling in the new information. It soon became clear that the storm-centre was not likely to move, relatively harmlessly, through the English Channel, but would soon start to turn left towards England – believe me, this is what they do!

By 10pm the nightshift was fully involved in amending the forecasts left for us by the day shift. All the time, the wind-noise outside the office was increasing. No one present had ever heard a loud and sustained roar like it. We didn't go outside to check it; the idea of a nineteenth century Welsh slate flying from the roof on the sixth floor focussed our minds wonderfully. For the rest of the night, the senior forecaster (me) turned my attention to the Indian subcontinent, Africa, Asia and Australia.

When daybreak arrived, the phones started ringing in earnest and the reports of damage started to come in. Kew Gardens had been flattened, Sevenoaks was now Oneoak. You can guess the rest. At 8am the swing doors should have opened to herald the arrival of the dayshift. No one came! We all did eighteen-hour shifts instead before our relief eventually arrived.

One thing on which we all agreed was that it had been so fortunate that the storm took place at night, because the number of casualties was mercifully small, much fewer, for example, than the toll during a similar storm three years later (the Burns' Night Storm), which broke during the day and claimed many lives. My daughter had learned her lesson well and hid my car keys to prevent me from driving until the storm had passed.

What lessons have we learned from the Great Storm of 1987? Global warming is playing its part, and for a hurricane to maintain its identity for five days across the Atlantic must mean something, and the frequency and intensity of such storms must give us pause to reflect. My sincere apologies to you all; it will be better next time!

Ian McCaskill

Lines written to commemorate the Great Storm of 16th October 1987
After William Rees McGonagall

'Twas in October of the year nineteen-hundred-and-eighty-seven
That an almighty storm came down from heaven.
The hurricane came completely out of the blue,
And the fact that it was coming nobody knew.

The whole population was tucked up in bed
When suddenly the tempest roared over their head.
Like a raging lion, it roared and roared
While most of the people just slept and snored.

But when they woke, they saw scenes of devastation
Which has laid waste almost the whole of the nation.
At least this was true in Sussex and Kent
When many a greenhouse got rather bent.

Huge trees were uprooted and fell over the roads.
Lorries on the M25 were reported to have shed their loads.
Meanwhile many a home was deprived of power
Which made it in every sense Britain's darkest hour.

Millions were unable to get to work
And this was not because they wished to shirk,
But when they set out to catch the 8.23
They found the way barred by an enormous tree.

Then out came the chainsaws and the strong cups of char
As neighbours spoke to each other for the first time since the War.
They all pulled together as they had done in the Blitz,
Instead of sitting around looking at pictures of Page 3 bosoms.

And everywhere there was only one topic of conversation
Among every section of the population,
As they began to search for someone to blame
They all soon agreed it was the Met men's night of shame.

"Why weren't we told?" went up the cry,
(Though how this would have helped, no one could descry).
And at last the British people had only one universal wish,
The public execution of Messrs McCaskill and Fish.'

With grateful acknowledgements to Private Eye, 30 October 1987.

Churches and cathedrals have been prey to damage by storms and lightning ever since records began. One of the most spectacular – and costly – incidents occurred when a lightning bolt hit the south transept of York Minster during the early hours of 9 July 1984.

Down the Centuries

Since our prehistoric ancestors took one look out of the cave opening and decided to put the sabre tooth tiger hunt on hold until the Ice Age was over, Britain's inhabitants have been grumbling about the weather. Global warming may be making the modern climate increasingly tempestuous and unpredictable but storms, tornados, lightning strikes, floods and the rest of the wild weather in Mother Nature's formidable armoury have been savaging Britain throughout our history.

Caesar's first attempt to invade Britain in 55 BC was abandoned after his cavalry ships were forced back to Gaul by a ferocious storm in the Channel; the invasion eventually took place the following year 54 BC. In 566 AD a great storm battered the coasts of Kent, Sussex, and Hampshire and in 1014 AD "on the eve of St. Michael's Day (29 September), came the great sea flood, which spread wide over the land, and ran so far up as it never did before, overwhelming many towns and an innumerable multitude of people."

On 23 October 1091, Britain's first known tornado and one of the two strongest ever known in the UK struck and severely damaged the church of St Mary le Bow in London. The wind speed was not recorded – the anemometer was not invented for several more centuries but it is estimated to have been 210-240 miles per hour, based on the force needed to pile-drive four rafters – each 26 feet long – so far into the ground that only four feet remained protruding above the surface. Other churches in the area were flattened, along with over 600 houses.

In the same turbulent decade a great snowstorm struck Ireland in 1095 "killing many" and in 1099 a violent North Sea storm-surge flooded huge areas of the English and Dutch coasts, killing thousands of people. As the Anglo-Saxon Chronicle recorded, "On the festival of St Martin (11 November), the sea flood sprung up to such a height and did so much harm as no man remembered that it ever did before".

In February 1216 King John, perhaps venting his anger at being forced to sign Magna Carta the previous year, invaded Scotland with terrible ferocity. King Alexander II of Scotland then launched an equally savage reprisal raid on Cumberland. He led his men back to Scotland, laden with booty, but as they forded the River Eden across the Solway Firth, the main medieval route between the two countries, they were surprised by a storm-surge, driven on by a south-westerly gale. The ford was treacherous at the best of times. The safest route was often hard to discern, quicksands awaited those who strayed too far from the path, and the tide came in at such speed that a seventh century document said that "if the best steed in Saxonland (i.e. England) ridden by the best horseman, were to start from the edge when the tide begins to flow, he could only bring his rider ashore by swimming, so extensive is the

Ouse Bridge in York has witnessed many a dramatic flood down the centuries, including the occasion in 1564 when it was washed away. Floodwaters were even more unwelcome than usual in March 1963 as the bitter winter weather had created masses of jagged ice.

strand and so impetuous the tide."

With a gale and a storm-surge behind it, the tidal rip became a terrifying wall of foam and water. The first warning to Alexander and his heavy laden men would have been a distant, sullen roar, swelling in sound from moment to moment. Then came a great, white-crested wave, spanning the Solway from shore to shore and rumbling onwards at terrifying speed with an angry roar like surf breaking on a rocky coast. That ferocious wave engulfed Alexander's army under thousands of tons of water and swept 1,900 men of "that plundering crew" to their deaths.

The earliest known British waterspouts, also the earliest recorded in Europe, occurred in the English Channel in June 1233, a sight that must have terrified the medieval fishermen and seamen who witnessed it.

Churches and great cathedrals, sometimes taking more than a century to construct, dominated the medieval landscape. Their towering spires pointed the way to heaven and their massive size and apparent indestructibility offered a reassuring image of stability and permanence in a turbulent, often terrifying age. It was sometimes illusory. Between 1271 and 1279 AD the west gable end of St Andrews Cathedral was blown down in a savage storm and in 1409 the south transept followed it to destruction in another storm. Norwich Cathedral spire was also destroyed by a ferocious gale in 1362. In itself, such damage or destruction in "great storms" could have a shattering psychological impact on the population, but it was also often taken as a portent of some looming disaster.

Flooding was an even greater peril than storm winds, particularly in England's low-lying eastern counties. In December 1287, almost 200 people were drowned at Hickling in Norfolk alone, when a tidal surge inundated the land and "suffocated or drowned men and women sleeping in their beds, with infants in their cradles, and all kinds of cattle and freshwater fishes. Many, when surrounded by the waters, sought a place of refuge by mounting into trees, but benumbed by the cold, they were overtaken by the water and fell into it and were drowned." In 1421 another tidal surge – "St Elizabeth's Flood" – inundated huge areas of eastern England and claimed 10,000 lives in the Netherlands as seventy-two villages were drowned.

In the sixteenth century, Old St Paul's in London lost its spire in a lightning strike. It was still awaiting rebuilding a century later when the entire cathedral was destroyed in the Great Fire of London.

In 1564 the Ouse Bridge in York was washed away by a flood that also carried off the dozen houses built on the bridge, drowning all the inhabitants. The sixteenth century also saw a great storm on All Saints' Day (1 November) 1570 that was known for generations after as the "All Saints' Flood". A North Sea tidal surge, it deluged Britain's East Coast but also breached the Dutch sea defences, leaving large parts of Rotterdam and other Dutch cities under water, and almost obliterating Amsterdam. Estimates of the total dead across Europe ranged from 100,000 people to as many as half a million.

In 1588, the people of England were for once grateful for a storm, as the early autumn gales of that year, hailed as "a Protestant Wind", completed the destruction of the ships of the Spanish Armada that had been begun by the English fleet in a series of battles in the Channel. Driven at first by southerly gales, the Armada tried to return to Spain by the northern route around Scotland, but ferocious Atlantic storms "with the waves reaching the sky" then battered the already crippled Spanish ships and drove them "like flocks of

starlings" towards the unforgiving coasts of Western Ireland. Twenty-six ships were wrecked there "with all the chivalry and flower of the Armada." Those who survived the rocks and the pounding surf were often stripped and robbed by the Irish, and many of the survivors were hunted down and killed by English troops garrisoned in Ireland.

Storms such as that which ravaged the remnants of the Armada were rarely so welcomed by Britons, particularly those living on the coast and exposed to the full ferocity of wind and wild weather. Storms often destroyed harbour defences, sea-cliffs, roads and rows of houses, but that same storm that had so battered the Armada, even swallowed an entire village – Singleton Thorp on the Fylde coast of Lancashire – which sank beneath the waves, never to reappear.

On 4 January 1601 another whole English village – Eccles by the Sea, Norfolk – disappeared when it too was engulfed by the seas. When the storm abated only the tower of the village church remained, standing on the beach. Everything else - houses and inhabitants - had disappeared forever. The church tower remained as a forlorn monument for almost two more centuries before it too was claimed by the sea.

Six years after Eccles by the Sea disappeared, the once great Welsh port of Kenfig was also lost to a storm. A thriving medieval port, Kenfig had been under a death sentence for centuries as the winter storms steadily piled up mountainous sand dunes that choked the life from the port. In 1445, Leland noted that "there is a village and a castell on the east side of Kenfig, both in ruins and almost choked and devoured with the sands that the Severn Sea castith up." The city with its navigable river, port, hospital, law courts, a church said to date from 520 AD, and a castle with a moat 45 feet wide and 15 feet deep, was slowly suffocated by the sands. The final death knell came in the great storm of 1607 which buried it forever. The lost city still lies beneath the sprawling sand dunes of the South Wales coast near Porthcawl. In the previous year, 1606, storms and flooding had caused the death of 2,000 people in the Severn Estuary, the worst confirmed flood death toll in UK history.

On 3 September 1658 England's first, and so far only experiment with republicanism, ended with the death of Oliver Cromwell. A violent storm was raging at the time and many saw it as further proof of God's wrath or God's sorrow depending on which side of the monarchist/republican divide they were sitting at the time. The precocious Isaac Newton, a mere sapling of fifteen at the time, did not concern himself with interpreting the storm but instead measured its strength by first jumping as far as he could with the wind at his back and then jumping back with the wind in his face. He measured the distance in both directions, compared it with the length he could jump on a still day, and computed the strength of the wind – possibly the first man in history to jump to a conclusion!

(Opposite) Anticipating a gathering storm has exercised mankind throughout history and the ominous warning of what we now call cumulonimbus cloud soon came to be recognised. What could not be seen until recent times was an aerial view, as in the top photograph taken on 17 August 2006. It shows large cumulonimbus over the West Country, creating a storm that formed over the Blackdown Hills and moved north-east towards Bristol. (Matt Clark, IJMet/TORRO)

Nor in the past was it possible to attempt a 'storm chase' by car, as was done to obtain the lower photo taken at Frodsham, near Warrington, in July 2005. Even today it is not always wholly successful. The photographer comments: "It was a really photogenic cumulonimbus and was teasing us in that every time we tried to get closer it seemed to die, but when we ignored it, it seemed to fire up again." (Samantha Hall, IJMet/TORRO)

Descriptions of the 'greatest storm' of 1703 refer to a sky 'full of meteors and vaporous fires'. The phraseology may be slightly exaggerated, but there can be little doubt of the awe in which a largely uneducated populace held a storm such as the one depicted here. This modern-day example of lightning and associated thunderclouds was photographed just after 11.07pm on 26 July 2006 using a seven-second time exposure. The location is near Stamford, Cambridgeshire. (Steve Warren, IJMet/TORRO)

1703 –
The Greatest Storm

There had been storms aplenty in the preceding centuries but on 26 November 1703 (the date is "Old Style", using the Julian calendar that was not replaced by the Gregorian calendar in Britain until 1752; the equivalent modern date is 7 December), a storm broke that dwarfed every other one ever seen in Britain. It was then and remains the worst storm in British history and it claimed 8,000 lives.

The great storm was witnessed by Daniel Defoe whose account of it, written in the immediate aftermath, made his reputation and is still in print to this day. There had been severe gales for the preceding fortnight of that November, and "several stacks of chimneys were blown down and several ships were lost", but this was a mere prelude to the great storm. It was not a true hurricane, but it was a storm of unparalleled ferocity, probably originating far out in the Atlantic and feeding its formidable energy on its three or four day journey across the ocean in this unusually warm and wet year. On 26 November it struck the British coast, carving a swathe of devastation "like an Army of Terror in its furious March", ravaging the whole of Britain south of a line from North Wales to the Humber.

It raged all afternoon and all night, increasing in strength until it "blew with the greatest violence" around dawn. Even in daylight, it was "excessively dark" throughout the duration of the storm, with incessant peals of thunder and bolts of lightning stabbing down from the dark skies. "The air was seen full of meteors and vapourous fires and in some places both thundrings and unusual flashes of lightning, to the great terror of the inhabitants." It was even claimed that the storm had provoked an earthquake, with one man in Nottingham claiming to have felt "three shakes, like the three rocks of a cradle, to and again."

As the shrieking wind increased still further, "most people expected the fall of their houses" but "nobody durst quit their tottering habitations, for the bricks, tiles and stones from the tops of the houses, flew with such force and so thick in the streets that no one thought fit to venture out, though their houses were near demolish'd within."

In Oxfordshire a tornado tore across the fields "a spout marching directly with the wind and I can think of nothing I can compare it to better than the trunk of an elephant, which it resembled, only much bigger. It extended to a great length and swept the ground as it went, leaving a mark behind. Meeting with an oak that stood in the middle of the field, it snapped the body of it asunder... It sucked up the water that was in the cart ruts and coming to an old barn, tumbled it down."

Men and animals were swept off their feet by the storm winds and carried for yards through the air before being dumped back to earth, and even huge rocks, trees and sheets of lead were tossed about like straws. In Dorset a stone of "near 4 hundredweight... though it lay so long as to be fixed in the ground and was as much out of the wind as could be... was carried from the place where it lay into a hollow way (sunken path) at least seven yards from the place." In the neighbouring county of Hampshire, "some stones of a vast weight" estimated at 2–3 hundredweight, were blown from the tower of Fareham church a distance of at least fifty feet. In Gloucester "26 sheets of lead hanging all together", each one of which weighed 350 pounds, took off like a sail from "the middle aisle of our church and were carried over the north aisle, which is a very large one, without touching it and into the church yard ten yards distant."

In Northamptonshire "a very great headed elm was blown over the park wall into the road, and yet never touched the wall." At Slimbridge near the River Severn, "a vast tree, thought to be the most large and flourishing elm in the land, was torn up by the roots, some of which are really bigger than one's middle and several than a man's thigh... and yet thrown up near perpendicular. The trunk together with the loaden roots is well judged to be thirteen ton at least."

Even more curiously, after the main branches had been cut off a huge walnut tree near Shaftesbury in Dorset that had been uprooted in the storm, it "replanted itself", sliding back into the hole torn in the earth when it was uprooted "and now stands in the same place and posture it stood in before it was blown down."

Roof tiles were "blown from a house above thirty or forty yards and stuck from five to eight inches into the solid earth." The lead from one hundred church roofs was "roll'd up like a Roll of Parchment and blown in some places clear off from the buildings", and in Chatham, "the lead of the church rolled up together and blown off from the church above 20 rod (300 feet) distance, and being taken up afterwards and weighed, it appeared to weigh over 26 hundred weight."

The winds turned the high tides into storm-surges, "the water rising six or eight foot higher than it was ever known to do in the Memory of Man". Near Bristol "the waters broke with such violence that they came six miles into the country, drowning much cattle... and 15,000 sheep in one level", and destroying 800 houses. The flooding at Bristol "spoiled or damnified 1500 hogsheads (large barrels holding over fifty gallons) of sugar and tobacco, besides great quantities of other goods. The covering of the land with salt water is a damage (that) cannot well be estimated".

The storm ravaged Royal Navy and merchant ships and fishing vessels as they desperately sought shelter. One small boat was lifted from the sea at Whitstable in Kent "by the violence of the wind" and driven through the air, turning over and over, until it hit rising ground some 750 feet inland. As it landed, it struck a man "who was in the way and broke his knee to pieces." Another boat, the Association, somehow remaining afloat, was driven by the wind all the way from the Thames to the coast of Norway, and a ship loaded with tin at Helford in Cornwall, was "blown from her anchors with only one man and two boys on board, without anchor, cable or boat." The ship was eventually driven aground on the Isle of Wight and although carrying not a shred of sail, she had covered the distance of "80 leagues" (c240

'Violence of the sea', as experienced by mariners during the 'Greatest Storm' of 1703, can be awe-inspiring. A photographer in the right place at the right time captured this superb image of waves foaming fifty feet into the air above a promenade after a February 1954 storm at sea.

miles) in just eight hours, an average speed of thirty miles an hour.

At the height of the Great Storm, many of the ships trying to shelter in the Downs off the east coast of Kent or in harbours from Cornwall to the Wash "had not a mast standing nor an anchor or cable left them, (and) went out to sea wherever the winds drove them." According to one sailor, there were "above forty merchant ships cast away and sunk. To see Admiral Beaumont and all the rest of his men, how they climbed up the main mast, hundreds at a time, crying out for help and thinking to save their lives, and in the twinkling of an eye were drowned."

The Sterling Castle, "a ship of eighty guns and about 600 men... had cut away all her masts. The men were all in the confusion of death and despair; she had neither anchor, not cable, nor boat to help her, the sea breaking over her in a terrible manner that sometimes she seemed all under water and they knew as well as we that saw her, that they (were driven) by the Tempest directly for the Goodwin (Sands) where they could expect nothing but destruction. The cries of the men and the firing of their guns, one by one, every half minute for help, terrified us in such a manner that I think we were half-dead with the horror of it."

Some seamen were offered the cruel illusion of safety when, "having hung upon the masts and rigging of the ships, or floated upon the broken pieces of the wrecks", they were washed up on the Goodwin Sands when the tide was low. They had a reprieve of only a few hours for at high tide the sands are completely covered by the sea. Although they signalled for help, were visible from the shore and boats even went "very near them in search of booty", only one man came to their aid. The Mayor of Deal was horrified at his fellow citizens' indifference, and when the Customs House officers refused him either their men or their boats, he paid out of his own pocket any that would volunteer to sail with him and with their aid took the boats by force at his own personal and financial risk. By this means he was reckoned to have saved 200 of the men on the Goodwin Sands, before the tide rose and drowned the unfortunate remainder. When the Queen's agents refused to provide money for the distressed seamen that he had rescued, the Mayor once more used his own funds to provide food and clothing for them. His name was Thomas Powell and for his courage and humanity among the indifference and venality of his fellow citizens, he deserves to be remembered with honour in any account of the Great Storm.

Another sailor was stranded on the Goodwin Sands twice during the storm, yet somehow survived both times. His ship, the Mary, sank after running aground on the Sands and he was the only survivor, clinging on to a piece of wreckage until he was picked up by the Northumberland. However, that ship then suffered the same fate as the Mary and "coming ashore upon the same Sand, was split to pieces by the violence of the sea." Yet by "a singular providence" the man was one of the few to be rescued, "all the rest perishing." A sailor from Brighthelmston was another fortunate survivor who "was taken up after he had hung by his hands and feet on the top of a mast 48 hours, the sea raging so high that no boat durst go near him."

Milford Haven in West Wales had been exposed to the full force of the storm. The captain of HMS Dolphin heard "guns firing from one ship or other all night for help, though it was impossible to assist each other, the sea was so high and the darkness of the night such... When daylight appeared it was a dismal sight to behold the ships driving up and down, one foul of another, without masts, some sunk and others upon the rocks... some split in pieces, the men all drowned... nigh 30 merchant ships and vessels without masts are lost and what men are lost is not known."

On this never to be forgotten night, not a harbour anywhere on the coast could offer a safe haven. Off the east coast, the Russia fleet, 100-strong, "was absolutely dispersed and scattered." Hundreds of vessels were lost, including four men-of-war, and thousands of sailors perished.

Henry Winstanley, the designer of the wooden lighthouse on the Eddystone Rocks off Plymouth, had the misfortune to be inside the lighthouse as the storm broke. By first light not a trace of man or lighthouse remained, both had vanished into the boiling seas forever, "a person whose loss is very much regretted by such as knew him as a very useful man to his country." By one of those remarkable twists of fate, the model of the lighthouse in Mr Winstanley's house at Littlebury in Essex, "fell down and was broken to pieces" at the same time as the actual lighthouse was destroyed. And in a particularly cruel irony, soon after it collapsed and disappeared beneath the waves, the Winchelsea, "a homeward bound Virginiaman, was split upon the rock, where that building (had) stood, and most of her men were drowned."

One of the greatest scenes of destruction of shipping was in the Port of London where Defoe himself saw some 700 ships on the Thames between Radcliffe Cross and Limehouse all piled together. "Some vessels lay heeling off with the bow of another ship over her waist and the stern of another upon her forecastle, the bowsprits of some driven into the cabin windows of others. Some lay with their sterns tossed up so high that the tide flowed into their forecastles before they could come to rights. Some lay so leaning upon others that the undermost vessels would sink before the others could float; the numbers of masts, bowsprits and yards split and broke, the staving (of) the heads and sterns and carved work, the tearing and destruction of rigging and the squeezing of boats to pieces between the ships is not to be reckoned. There was hardly a vessel to be seen that had not suffered some damage or other."

In addition to the larger ships, some 500 wherries, 300 ships' boats, and at least 100 barges and lighters were sunk in the Thames alone. Further inland, the "water of the River Thames and other places was in a very strange manner blown up into the air" and the fish in a pond in St James' Park "to the number of at least 200, were blown out and lay by the bankside". In the midst of the maelstrom of wind and water, 400 windmills were "overset and broken to pieces" or destroyed by fire after the friction from their wildly-whirling sails caused them to burst into flames.

Westminster Abbey had the lead stripped from its roof, the cathedrals of Worcester, Gloucester and Ely were badly damaged and the stone pinnacles were toppled and much of the medieval stained glass smashed at King's College Chapel in Cambridge. The royal palace at Whitehall lost its great weathervane and the roof of the guard house was blown off.

In the Parish of St Cray in Kent, "a great long stable in the town, near the church, was blown off the foundations entirely at one sudden blast... and cast out onto the highway, over the heads of five horses and a carter feeding them, and not one of them hurt, nor the rack or manger touched, which are yet standing to the admiration of all beholders." In Northamptonshire "an honest yeoman being upon a ladder to save his hovel, was blown off and fell upon a plough, died outright and never spoke word more".

Even when the worst of the storm had passed, the winds barely abated throughout Sunday and Monday and on Tuesday night they increased again, bringing fresh gales which,

Even in daylight hours, the great storm of 1703 brought 'unusual flashes of lightning to the great terror of the inhabitants'. Such events still cause consternation today, as instanced on 19 June 2005 when Ian Loxley, Gallery Archivist for the Cloud Appreciation Society, took this photograph near Gainsborough. He comments: "It was a very hot and humid Father's Day. A cricket match had just finished on the playing field – the storm had been rumbling about during the match causing the umpires to glance nervously skywards. I was teetering on some steps photographing the storm and nearly fell off when the flash occurred."
(Ian Loxley, IJMet/TORRO)

though not as strong as at the peak of the storm, wreaked further havoc on already damaged shipping and buildings. "Several ships which escaped the Great Storm perished this night and several people who had repaired their houses had them untiled again." The winds continued so fierce "for near a fortnight, that no ship stirred out of harbour and all the vessels, great or small, that were at sea made for some port or other for shelter... When the storm was over and the winds began to be tolerable, almost all the shipping in England was more or less out of repair, for there was very little shipping in the nation but what had received some damage or other."

When Defoe and the other shaken survivors of the storm emerged from their houses, they found "the Havock the Storm had made" had left London looking like a war zone, with every street "covered with Tyle-sherds (shards) and Heaps of Rubbish from the Tops of the Houses, lying almost at every Door." Proving the truth of the adage about an ill wind, the price of tiles rose from 21 shillings per thousand to 6 pounds a thousand and reeds for thatching were so scarce that the price rose threefold and many people turned to "bean, helm (heather) and furse (gorse)... things never known to be put to such use before."

Hundreds of thousands of trees had been laid flat by the storm. Twenty-five large estates each lost over 1000 trees, and "450 Parks and Groves" had from 200 to 1000 large trees uprooted. But as one melancholic farmer in Somerset noted, "our loss in the apple trees is the greatest because we shall want liquor to make our hearts merry."

As much as 25 miles from the coast, the torrential rain that had fallen had carried the salt tang of the sea and countless witnesses talked afterwards of trees and brushes rimed with salt, land poisoned and crops stunted and killed by it. "The grass was so salt that the cattle would not eat for several days... the sheep on the Downs in the morning would not eat till hunger compelled them and afterwards drank like fishes."

Defoe drew comparisons with the Great Fire of London, "an exceeding loss... yet that desolation was confined to a small space (and) the loss fell on the wealthiest part of the people, whereas this loss is universal and its extent general: not a house, not a family that had anything to lose, but have lost something by this storm."

Even making full allowance for the understandable hyperbole and exaggerations of Defoe and the other eye- and hearsay-witnesses in that distant age, it is clear that the Great Storm of 1703 was the most violent, sustained and destructive that these shores have ever seen: "I can give you no account but this; but sure such a Tempest never was in the world. ... No Pen can describe it, no Tongue can express it, no Thought conceive it unless by one in the extremity of it."

****

In 1748 a single bolt of lightning was claimed to have killed a Scottish farmer's son and no less than 320 ewes.

In October 1755 "red rain" – coloured by fine sand carried north by the winds from storms raging in the Sahara – fell over large parts of Europe. On 14 October nine inches of red rain fell on Italy and as the storm clouds moved north and cooled as they were driven over the Alps, the rain turned to snow and the upper slopes and snow covered summits of the mountains were coloured a vivid red.

The Lynmouth flood devastation of 1770, when the river 'rose to such a degree as was never known by the memory of any man' and brought down great rocks, proved to be a foretaste of tragic events in the summer of 1952. As can be seen, the later flood again moved huge rocks and swept away everything in its path.

In 1770 Lynmouth in North Devon was devastated by a flood as "the river by the late rain rose to such a degree as was never known by the memory of any man now living, which brought down great rocks of several tons each, and choked up the harbour, and also carried away the foundation under the Kay (quay) on that side of the river six foot down and ninety foot long, and some places two foot under the Kay, which stands now in great danger of falling."

In 1771, there was "great loss of life" throughout Northern England during the worst floods ever seen there. When the Tees burst its banks, half of the town of Yarm was swept away.

On September 22, 1810, a tornado roared across Hampshire from Old Portsmouth to Southsea Common and caused tremendous damage. Houses were flattened and many others so badly damaged that they had to be demolished; chimneys were blown down and the lead on the roof of a bank was "rolled up like a piece of canvas and blown from its situation."

On Christmas Day 1836 the UK's heaviest ever Christmas snowstorm killed many people and left drifts more than 25 feet high in its wake. In Lewes, Sussex, the snow formed a cornice many feet thick on the lip of a steep cliff. The next day, Boxing Day, the cornice collapsed, triggering an avalanche of wet snow that tore down the cliff face and devastated the row of cottages at the bottom. Eight of the inhabitants were killed. A pub with the tasteless name of "The Snowdrop" now stands on the site.

William Tattershall of Silkstone in South Yorkshire lost 30,000 rose trees in the "Great Storm" of 1838, but a far more catastrophic loss was suffered in the same storm. Torrential rain falling for two hours led to floods that cut off the village from the outside world, and one of the swollen streams that burst its banks inundated the mineshaft of the Huskar pit, drowning twenty-six children aged between seven and seventeen, who were working underground.

On 6 to 7 January 1839 "The Night of the Big Wind" struck Ireland, killing over four hundred people.

1859 – The Royal Charter Storm

On 25 October 1859 a massive storm broke over the Irish Sea, lashing the British coast and battering hundreds of ships at sea. Among them was the Royal Charter, "The Gold Ship", a 2,700 ton, iron-hulled, triple-masted, steam- and sail-powered clipper. One of the fastest ships afloat, she could carry 600 passengers and make the 10,000 mile passage between Australia and England nearly a month faster than any of her competitors.

Under Captain Thomas Taylor, the Royal Charter had made the outbound voyage from Liverpool to Melbourne in a record 59 days. She began the return voyage on 26 August 1859 with a full complement of one hundred and twelve officers and crew, and three hundred and seventy one passengers, whose personal effects, carried in the strong room below decks, included £150,000 in gold sovereigns and £322,440 in gold bullion, mined from the booming goldfields of Victoria.

The Royal Charter sailed west across the Indian Ocean, rounded the Cape of Good Hope, and by 24 October was off Cork on the Irish Coast. Despite a flat calm that night, Captain Taylor boasted to his first class passengers at a farewell dinner that his ship would reach Liverpool by the following evening. By the next morning, the Royal Charter was within sight of the coast of North Wales and less than a hundred miles from her destination, but the captain noted that the barometer was falling rapidly and told the crew that he could see storm clouds approaching over the mountains of Snowdonia. Some passengers later claimed that he was urged to shelter from the storm in Holyhead Harbour on Anglesey, but preferred to run for Liverpool to preserve the ship's reputation for speed.

The storm that he had seen approaching had already ravaged the coasts of Devon and Cornwall, causing great damage and loss of life. In Pembrokeshire, West Wales, the storm winds created a tidal surge that swept the ancient church of St Brynach at Cwm-yr-Eglwys into the sea. At New Quay, the storm "wrought more damage on lives and goods belonging to this town, both here and elsewhere, than any previous storm in history. The wind on Tuesday night was so strong that it was dangerous to venture outside. Things were even worse amongst the ships, most of which were totally out of control, having broken their chains and become mixed up in a chaotic mess. The lighthouse and a large section of the sea-wall had been washed away by the power of the sea."

That storm now reached the coast of North Wales, and around Caernarfon and Bangor "one of the most awful storms ever seen here" tore vessels from their anchors and hurled them against the rocks and "the effect of the storm on the eastern side of Port Penrhyn was truly frightening." The howling north-easterlies sent waves "like mountains" roaring into Llandudno Bay. As the breakers smashed into the shore, flying spume and spray filled the air and reduced visibility to a few feet. Even the oldest inhabitants had never seen the seas rage with such violence before. A row of thatched cottages was at once overwhelmed by the waves, forcing the inhabitants to flee for their lives. A newly-built pier was reduced to matchwood and strewn over the beach, and the toll booth from the pier was picked up on the crest of a huge wave and dumped at the eastern end of the bay.

As the Royal Charter rounded the north coast of Anglesey, the gales shifted to northerly and increased still further. Less than an hour after passing the Skerries, the storms had reached Force 10 on the Beaufort scale and the ebbing tide conspired with the wind to drive the ship towards the shore. The storm continued to intensify as the evening wore on. By midnight, the winds had reached Hurricane Force 12 and they continued at that strength until the following afternoon. Even in the relatively sheltered estuary of the River Mersey, the winds were peaking at well over 80 miles an hour.

Probably regretting his earlier decision to make for Liverpool, Captain Taylor had tried to summon a pilot off the Skerries but no pilot there would now brave the fury of the storm, and even though he displayed a blue distress light off Point Lynas, the pilot boats there were also unable or unwilling to put to sea into those terrifying waves.

By eleven-thirty that night, the Royal Charter was no longer answering its helm and was being pounded by waves sending thousands of tons of water coursing over the decks. Taylor decided to square the sails, shut down the engines and try to ride out the storm. He dropped anchor four miles off the Anglesey coast, but his chosen anchorage was exposed to the full force of the winds and the surge sweeping down the Irish Sea on the now rising tide.

The anchors held for a while but the storm winds, the towering seas and the tidal surge were too powerful to resist and at around 1.30am, the port anchor cable snapped with a sound like a cannon shot. The starboard cable held for another hour and then it too snapped. Even though Taylor ordered the masts cut down to reduce the wind drag and ran the steam engines at full power, the Royal Charter could neither make headway nor even hold station against the storm and was driven inexorably toward the Anglesey shore.

The stricken ship first grounded on a sandbank but was then lifted off by the next mountainous wave and hurled on to the rocks of Moelfre Bay. The impact broke the Royal Charter's back and split the ship in two. Some people were killed instantly, thrown onto the rocks by the force of the impact, and the ferocious waves breaking over the wreck swept almost every other man, woman and child on board into the sea. Some passengers, returning home having made their fortunes in the Australian goldfields, tried to reach the shore still carrying their gold, but it only hastened their end as the weight dragged them under.

Guze Ruggier (a Maltese seaman whose name was anglicised in contemporary accounts to "Joe Rodgers"), and one of the few sailors aboard who could swim, dived overboard and despite the pounding surf, dragged a hawser from the ship to the shore. At the fourth attempt, he succeeded in fixing the rope to a rock and with the help of a couple of dozen local men who had rushed down to the shore to help, he formed a human chain that rescued a few of those drowning in the sea or still clinging to the wreckage, but not one of the women or children aboard nor any of the ship's officers survived. In total, although the wreck had happened just fifty yards from the shore, only 41 of the 483 aboard lived. Ruggier – "Rodgers of the Royal Charter" – was subsequently honoured and rewarded by the Royal National Life Boat Institution and the Board of Trade for his heroism.

The storm continued to track northwards, scything across Scotland and the North Sea, and leaving further havoc in its wake. In total more than 200 vessels were lost with over 800 lives. Of the dead, more than half had been aboard the Royal Charter. Several other savage gales struck Britain over the next few days, battering already storm-damaged vessels and by 9 November the storms had caused a total of three hundred and twenty five shipwrecks.

Bodies continued to wash up on the beaches surrounding Moelfre for weeks after the tragedy. Among them was Isaac Griffiths, a sailor aboard the Royal Charter, who had been born in Moelfre and so drowned within half a mile of his birthplace. The bodies were housed in a temporary mortuary at the church where, according to Charles Dickens, "forty four shipwrecked men and women lay here at one time, awaiting burial".

Along with the bodies, large amounts of gold washed up along the coast near Moelfre, including gold sovereigns "scattered far and wide over the beach, like seashells." Beachcombers and divers were said to have recovered all but £30,000 of the gold bullion from the wreck and many local families became rich.

Dickens included the story of the disaster and its aftermath in his book The Uncommercial Traveller: "O reader, haply turning this page by the fireside at Home, and hearing the night wind rumble in the chimney, that slight obstruction was the uppermost fragment of the Wreck of the Royal Charter, Australian trader and passenger ship, Homeward bound, that struck here on the terrible morning of the twenty-sixth of this October, broke into three parts, went down with her treasure of at least five hundred human lives, and has never stirred since!"

A memorial still stands on the clifftop above the wreck site, commemorating the place "where the Royal Charter met its end, and the memory of those who died." An official enquiry into the storm recommended that the Meteorological Department of the Board of Trade (what we now call the "Met Office"), established only five years previously to encourage meteorological observations on land and at sea and collate them to aid the understanding of weather patterns, should in future use the new telegraph system to spread storm warnings to British coastal towns and ships at sea. The government also agreed to distribute "storm glasses", to small fishing communities in the hope of preventing or at least minimising further catastrophes.

Captain (later Admiral) Robert Fitzroy, Director of the Department, analysed the Royal Charter storm from the limited data available and was the first to raise the idea of using a synoptic chart to foretell the weather and the first to describe that process as a "weather forecast". Fitzroy, who had captained the Beagle on its famous voyage and befriended the then unknown Charles Darwin, was no mean scientist in his own right. A keen naturalist, he also compiled coastal charts of southern South America and the Falklands Islands that were of such accuracy that they were only recently superseded. By meticulous analysis of wind and weather patterns, Fitzroy also invented that saviour of seamen's lives and subject of a billion British conversations, the weather forecast. He was indirectly upstaged by his former passenger when Fitzroy's paper on "British Storms" was read at the same meeting of the

British Association as the legendary debate about Darwin's theories between Samuel Wilberforce, Bishop of Oxford, and Thomas Henry Huxley. Nonetheless, an official storm warning system was initiated on 6 June 1860 and the first storm warning was issued on 6 February 1861 – the first weather forecast ever produced.

****

On 24 March 1878 the Naval Training Ship, HMS Eurydice sank in a squall off the Isle of Wight as a "polar low" swept down from the North.

The first Tay Bridge, the longest in the world when it was opened in February 1878. The 'high girders' at the centre collapsed during a great gale on 28 December 1879, taking with them a train and its 75 passengers.

1879 –
Tay Bridge Disaster

Twenty years after the Royal Charter Storm, another terrible storm and its attendant waterspouts caused the most notorious engineering disaster in British history. On 28 December 1879 a gale estimated at Force 10 or 11 on the Beaufort scale – Storm to Violent Storm Force – was raging in the North Sea and pounding the Scottish coast. The high tower at Kilchurn Castle on Loch Awe was blown down in the gales, many Scottish houses lost their roofs and telegraph wires came down right across the country. On the Firth of Tay, the gales were blowing straight along the estuary, shrieking around the iron girders of the railway bridge. Opened only nineteen months before in February 1878, the Tay Bridge had been hailed as an engineering masterpiece. Queen Victoria travelled over it soon after its opening and the designer, Thomas Bouch, was knighted for his work. A very experienced engineer, Bouch had full responsibility for the design, construction and maintenance of the bridge.

It was almost two miles in length – the longest bridge in the world at the time. Most of Bouch's previous bridges were built from a lattice of iron girders supported on slender cast iron columns and braced with wrought iron struts and ties. The Tay Bridge was a new departure for him. Supported on masonry piers, 72 of the 85 spans carrying the single-track railway over the river were supported from below by iron girders set beneath the track, but the thirteen spans in the centre of the bridge ran through a tunnel of girders raised above the the track. These 'high girders' were 27 foot tall and had an 88 foot clearance above the high water mark of the estuary.

At 7.14 on that wild evening of 28 December 1879, a train from Edinburgh, bound for Dundee and carrying seventy-five passengers and crew, thundered onto the southern end of the bridge. As it did so, the spray whipped by the winds from the surface of the Firth seemed to coalesce with the driving rain, forming first one and then two whirling grey-black columns: waterspouts. Lit by the fitful moonlight that pierced the winter darkness, these waterborne tornados grew with frightening speed, twin towering columns bearing tons of water, roaring towards the bridge.

As the train approached the 'high girders' at the centre of the bridge, the waterspouts engulfed the structure. When they passed on, the central spans of the bridge and the train had disappeared. The bridge had suffered a catastrophic failure, plunging the train and its seventy-five passengers and crew into the frigid, storm-tossed waters of the Firth of Tay.

In the darkness and wild weather, few had witnessed the incident and even they were uncertain exactly what they had seen. One witness spoke of seeing the train running along the rails, "and then suddenly was observed a flash of fire." He believed that the train had left the rails and gone over the bridge. Another rushed to inform the station master at the Tay Bridge station in Dundee who immediately contacted the signal boxes at either end of the bridge. The signalman on the Fife shore confirmed that the train had entered the bridge at 7.14, but it had not passed the signal box at the north end of the bridge, and though the signalman there had seen nothing untoward, when he tested the telegraph wires that ran across the bridge, the line was dead.

"Mr. Smith, the stationmaster and Mr. Roberts, locomotive superintendent, determined, notwithstanding the fierce gale, to walk across the bridge as far as possible from the north side, with the view of ascertaining the extent of the disaster." The first thing that caught their eye was water spurting from a severed main that had been laid across the bridge to supply water from the Dundee reservoirs to the village of Newport on the south bank of the Firth. "Going a little further, they could distinctly see by the aid of the strong moonlight that there was a large gap in the bridge caused by the fall, so far as they could discern, of two or three of the largest spars. They thought, however, that they observed a red light on the south part of the bridge and were of the opinion that the train had been brought to a standstill on the driver noticing the accident."

They retreated to the north shore and because of the darkness and the continuing wild weather, it was several hours before several bags of mail carried by the train were found washed up on the shore at Broughty Ferry, four miles downstream, confirming that the train and its six carriages had actually plunged into the Tay along with the midsection of the bridge.

Around ten o'clock that night, as the gale began to moderate, the Provost of Dundee "and a number of leading citizens" set out in a steamboat for the bridge. Meanwhile dense crowds of people had gathered around the Tay Bridge station in the city, hoping for news of friends and relatives, "strong men and women wringing their hands in despair... the return of the steamboat is anxiously awaited."

By first light, the full extent of the tragedy was revealed, as the jagged gap torn in the bridge and the shattered carriages and wreckage in the swirling waters below told their own story. Boats patrolled the Firth all that day and search parties combed both shores of the Firth, but no survivors were found and only forty-six of the seventy-five bodies were ever recovered. The engine that had pulled the fateful train was recovered from the bed of the Tay and put back into service. Nicknamed "The Diver" in the gallows humour of the railway workers, it remained in service with the North British Railway for another thirty years.

The collapse of the bridge after just nineteen months horrified the nation and sent shock waves through the Victorian engineering industry. A Court of Inquiry set up to ascertain the cause, heard that the bridge had been built to a very tight budget and, unable to buy prefabricated iron sections from established suppliers, Bouch had decided to set up his own foundry. Some of the iron from this foundry proved to be of very poor quality. On 2 October 1877, while the bridge was still being constructed, "a high girder on a barge was lost to storms for a night" during a similar gale, "but then only one of the workmen lost his life." Another section was dropped to the seabed whilst being lifted into place. These sections were still used even though the dropped section had been slightly twisted out of true. Engine drivers subsequently reported "a slight change of direction" as they entered that section.

After taking evidence from eyewitnesses, railwaymen, engineers and experts from the Board of Trade, the Court of Inquiry concluded that the iron superstructure was of poor quality and had been badly maintained. Even worse, insufficient allowance had been made for wind pressure in the design of the bridge: "The fall of the bridge was occasioned by the insufficiency of the cross bracing and its fastenings to sustain the force of the gale."

Sir Thomas Bouch had based his design on a maximum wind pressure of ten pounds per square foot, even though his proposed design for a bridge over the Firth of Forth – on which he was working at the time of the Tay Bridge collapse – allowed for three times as much wind-loading. Bouch vehemently denied the charges of negligence, but his career was over. The contract for the design of the Forth Bridge was taken from Bouch and given to Benjamin Baker and Sir John Fowler, whose "robust" design, still in use today, erred massively on the side of caution The same men were also awarded the redesign of the Tay Bridge.

Disgraced and broken-hearted, Bouch died just ten months after the fall of his monument, the Tay Bridge. The surviving wrought iron girders from the original bridge were used in the construction of its replacement and are still in use. Visible from the new bridge, the masonry piers that once supported the iron columns of Bouch's bridge also still stand in the riverbed to this day, a mute testament to the horror of that long ago December night and the power and fury of wild weather.

* * * *

Temperatures plunged around the world in 1883 after Krakatoa, a volcano off the east coast of Java, exploded with such force that the sound was heard in Australia. Dust and sulphur dioxide was blasted high into the atmosphere and circled the globe, obscuring the sun and bringing colder, wetter weather to Britain and many other countries for twelve months afterwards.

On 8 July 1890 – high summer even in the Scottish Highlands – the observatory on Ben Nevis reported that eight inches of snow had fallen.

On 10 March 1891, at a time when many hoped the worst of winter was already over, Cornwall and Devon suffered their worst ever blizzard. Storm force winds whipped up drifts of twenty feet or more and on Dartmoor a gorge 300 feet deep was said to have been completely filled with snow. The storms sank sixty-five ships at sea, stranded fourteen trains – one took four days to travel from Paddington to Plymouth – and killed 220 people and over 6,000 sheep.

Northerly gales and a Spring Tide combined to breach flood defences in East Anglia on 29 November 1897, inundating thousands of acres of farmland and leaving the entire coast from Harwich to Southend under water. The sea defences on the north bank of the Thames estuary were also breached in a number of places, causing localised flooding. Only the fact that the height of the surge came in the middle of the day, rather than at night, prevented scores of fatalities.

The blizzard of March 1891 was exceptional in its severity but less extreme occurrences were once almost routine. Storm force winds and drifting snow regularly closed roads over high ground, as shown in this 1955 scene of lorries stranded on Woodhead pass, east of Manchester.
.

On Friday 12 July 1900, a ferocious thunderstorm struck Ilkley in Yorkshire. 3.75 inches of rain fell in 75 minutes and flood waters tearing down the steep hillsides above the town washed away roads and bridges and dumped huge boulders and hundreds of tons of debris in the main streets.

On 13 June 1903 the longest continuous period of rain ever recorded began in London. The capital suffered an unbroken downpour that continued for 59 hours. The same amount of continuous rain was also recorded in Leeds, 200 miles to the north.

In November 1905, a railway steamer, the Hilda, was wrecked off St Malo during a storm and went down with the loss of 128 lives.

The town of Broadstairs in Kent was completely flooded on 26 October 1909 after six inches of rain fell over the previous three days.

On 15 April 1912, the Titanic went down in the North Atlantic with the loss of over 1,500 lives, after hitting an iceberg in poor visibility.

On St Swithin's Day – 15 July – in 1913 three inches of rain fell on Mayfield in Sussex. Despite the rhyme:

> *St Swithin's Day, if it does rain*
> *Full forty days, it will remain*
> *St Swithin's Day, if it be fair*
> *For forty days, t'will rain no more...*
> *the next day was fine.*

On 27 October 1913, one of Britain's worst ever tornado disasters killed six people at Edwardsville, near Cardiff.

On 28 June 1917, 9.56 inches of rain were recorded in a 24 hour period at Bruton in Somerset.

A 227 gram hailstone fell at Plumstead in London in 1925, the heaviest hailstone then recorded in the UK.

On the night of 6-7 January 1928, a northerly gale combined with a high tide pushed water levels in the Thames estuary so high that at several places in the City, Southwark, Westminster and Hammersmith, the embankments were over-topped and the low-lying riverside districts were flooded. When a section of the embankment near Lambeth Bridge collapsed, water flooded into the basements of nearby houses so quickly that people were unable to escape and fourteen were drowned.

On 22 October 1928 a tornado toured some of London's famous landmarks, damaging buildings at Victoria Station, Piccadilly Circus, Oxford Circus and Euston Station on its passage through the city.

The miserable summer of 1930 reached its nadir on 22 July when, in the course of four sodden days, 11.97 inches of rain fell in Castleton, Derbyshire.

Ferocious thunderstorms breaking over the West Midlands on 14 June 1931 generated a tornado that wreaked havoc along a twelve mile path through Birmingham and caused two and a half inches of rain to fall on Cannock in less than 40 minutes. At least 4 inches of rain also fell on the hills of the North-West, causing floods that destroyed roads and bridges and drowned hundreds of cows and sheep.

On 15 August 1931, large hailstones falling during a violent storm at Southwold caused cattle to stampede and bruised and cut bathers on the beach. On the same day, sleeping inhabitants of Great Yarmouth and Lowestoft were pitched out of bed at 1.30 in the morning by an earth tremor.

In 1932 the wettest May for 160 years saw extensive flooding in the Midlands and North with the Trent and Don Valleys particularly badly affected.

On Thursday 9 July 1936 a hail storm in Norfolk caused extensive damage in Norwich and killed 300 ducks near Thetford.

An ice storm or 'glazed frost' at the end of January 1940, the longest and most severe ever recorded in Britain, caused chaos and massive transport disruption, and damaged tens of thousands of trees. The supercooled mist and rain froze on contact with twigs, leaves, branches, paths and roads and every other surface.

The ice storm continued for two days in places; Cirencester had 48 hours of continuous freezing rain in temperatures that never rose above -2°C. Telegraph poles and wires bowed and then snapped, tree branches were ripped off by the enormous weight of ice and birds were unable to fly because of the accumulations of ice on their wings. A number of ponies were found frozen to death, their bodies completely encased in ice.

Travel was almost impossible by any means. Trains were massively disrupted by frozen points, signals and rolling stock, roads were like ice rinks and even the gentlest gradient proved impossible to climb. Hospital casualty departments were inundated with people who had broken bones in falls on ice covered pavements.

At Rugby, the familiar local landmark of the 'radio aerial' – the dozen 820 feet masts of the radio transmission system – was fatally overloaded by the weight of ice. The release gear at the mastheads should have lowered the aerial to protect it but the automatic slipping gear had also frozen solid and the aerial finally collapsed under the strain.

On 17 March 1947, fourteen people were killed by gales that swept Britain. Gusts peaked at 98 mph in Suffolk and entire houses were demolished. The gales accelerated the thaw of that year's record snowfall, bringing warm Atlantic air and heavy rainfall that combined with the melting snow to produce some of the most extensive floods in history.

On 12 August 1948 the greatest floods ever known in Scotland created havoc throughout Lothian and the Borders. Four inches of rain fell that day alone over an area stretching from Edinburgh to the English border. In Lammermuir, East Lothian, which had already been deluged with 5.12 inches that week, a further 5.48 inches fell in 24 hours. A new lake three-quarters of a mile wide appeared at Eyemouth, the River Tweed rose seventeen feet above normal and some of the more remote valleys were flooded to a depth of forty feet. Roads and forty bridges were swept away by the floods, which rose with such terrifying speed that thousands of farm animals drowned before they could be rescued. Many farmers, still struggling to recover from their losses of livestock and crops in the terrible winter of 1947, were wiped out by this new calamity.

Tornados usually last a few minutes at most, but at four o'clock on the afternoon of 21 May

Water-borne transport can come into its own in times of extreme flood with the Army and Police often paddling to the rescue.

Hail falling near Tiverton in Devon in April 2006 created an angry sky, although the consequences were not as dire as the Norfolk hailstorm of seventy years earlier, which caused extensive damage and killed 300 ducks. (Matt Clark, IJMet/TORRO)

1950, the UK's most prolonged tornado ever touched-down at Little London, near Wendover in Bucks. It lasted for four hours, devastating parts of Wendover, Aston Clinton, Puttenham, Linslade, Lidlington, Bedford and Blakeney, as it tracked 70 miles across Eastern England to Coveney in Cambridgeshire. There it parted company with the ground, but continued as a funnel cloud for another thirty miles, crossing the Norfolk coast at Shipham about 8 o'clock that evening and was last seen still moving east across the North Sea.

On 10 July 1950 a hailstorm killed the astonishing total of 3,000 ducks at Illington in Norfolk.

1952 – Lynmouth Flood Disaster

In the late afternoon of 14 August 1952 dense black clouds piling up over the moors brought a premature dusk to the West Country and torrential rain began to fall. It continued all night and throughout the next day. The downpour was at its worst over Exmoor and in the course of those 24 hours nine inches of rain was recorded at Longstone Barrow on the heights of the moor.

August was already a prodigiously wet month – the rainfall over North Devon was 250 times the normal monthly figure – and the whole of Exmoor was affected by floods, causing considerable damage on the Barle, Exe, Heddon and Bray rivers. At the village of Exbridge, to the south of Dulverton, huge trees from a sawmill a quarter of a mile upstream were hurled against the bridge, damaging its stone piers. At Dulverton itself, a garage beside the river Exe had its walls ripped open by the surging storm waters and a dozen cars disappeared into the swirling flood. When the waters receded, great baulks of timber remained littering the ground, marking the course of the flood like the "great gaps ripped in the river banks, uprooted trees stranded in fields, and where the banks are wooded, deeply cut ditches beside the nettled highway."

Three boy scouts were swept away when floods ripped through their camp at Filleigh, near South Molton. The Reverend NC Wieland got most of the remaining eighteen boys in his charge – all from Moss Side in Manchester – to the shelter of a nearby bungalow, but then "heard a call for help. He went back to the camp, wading and swimming several yards to rescue 12 year-old Robert Kennedy, who was clinging to a tree." The search for the missing boys went on all night with the aid of searchlights and flares, but their drowned bodies were found the following morning.

The worst effects of the flooding were on the East and West Lyn rivers, already swollen to near-flood levels even before the catastrophic storm. Those rivers drained the north part of Exmoor, merging in a steep, narrow and rocky valley just four miles long, running down through Lynton and Lynmouth to the sea. The river had once taken a more gentle route to the sea, but coastal erosion during the Ice Age had torn a hole in the wall of that valley and the river now followed a plunging, breakneck course through rocky crags and oak woods clinging to the steep slopes, and dropping 1,500 feet from the heights of the moor to the sea in little over four miles. In its lower reaches the valley was a V-shaped gorge, so narrow at the bottom that there was not even sufficient level ground for a road beside the river.

As the sun began to set on 15 August 1952, every watercourse, stream and river, and the two small reservoirs at Woolhanger and North Furzehill, both on tributaries of the West Lyn, burst their banks. Writer S H Burton described what happened next. "Down every gully and natural depression, and northward running combe, the thousands of tons of water flowed into the East and West Lyn rivers. Farley Water and Hoaroak Water joined the already swollen East Lyn at Watersmeet. Half a dozen streams converging at the head waters of the West Lyn brought the deluge from the western Chains, and at Barbrook Mill another influx from Woolhanger Common joined the raging torrent, sweeping bridges and houses away before starting the last deadly descent into Lynmouth." An avalanche of 90 million tonnes of water came surging off the moor and blasted through that precipitous, narrow valley, destroying everything in its path.

Debris and uprooted trees blocked bridges and formed temporary dams that then collapsed, sending fresh walls of water carrying whole trees and boulders the size of cars, crashing and smashing down through the valley. In their path stood the villages of Lynton and, 600 feet lower down the valley, Lynmouth. During Victorian times, the picturesque twin villages were known as "Little Switzerland", and they were still a popular holiday destination, swelling the summer population.

Every one of those inhabitants – resident and visitor alike – was now in peril as the flood burst upon the lower valley with explosive force. The torrential rain, thunder and lightning piercing the gathering darkness added to the drama and the horror of the scene. The four main road bridges were swept away one by one, scores of buildings were flattened or badly damaged and every boat in the harbour was torn from its moorings and washed out to sea.

In Victorian times, the River Lyn had been diverted into a confined channel and many houses and shops had been built along its former course. The river now reverted to part of its former, more direct, course, destroying many of these buildings and the streets and bridges that served them. A fisherman, Ken Oxenholme, battling his way uphill from the devastated harbour, found Lynmouth High Street a raging torrent and, frantic, had to run up through the woods to reach his wife and child in their caravan at the top of the village. "As we watched, we saw a row of cottages near the river, in the flashes of lightning because it was dark by this time, fold up like a pack of cards and they were swept out by the river with the agonising screams of some of the local inhabitants who I knew very well."

Tom Denham, the owner of the Lyndale Hotel, had seen his cellars flood before and was not unduly worried at first when the waters began to rise, but "about half-past nine there was a tremendous roar. The West Lyn had broken its banks and pushed against the side of the hotel, bringing with it thousands of tons of rocks and debris in its course. It carried away the chapel opposite and a fruit shop. Three people in the fruit shop were swept against the lounge windows of the hotel. We managed to pull them through in the nick of time. In all we had sixty people in the hotel all night." Arthur Brooks of Croydon saw those "three people being washed out to sea. We managed to get a hold of them and brought them through the window. By the morning boulders were piled twenty feet high outside that window."

The 'unimaginable devastation' at Lynmouth on 16 August 1952 after the floodwaters had subsided. On the right is the Lyndale Hotel, where terrified guests spent the night on the top floor after the West Lyn broke its banks and pounded against the side of the building. The 90 million tonnes of water that surged through the village claimed thirty-four lives.

Among the other hotel guests were Mr HL Watson of Catford and his wife and family. He described the waters rising rapidly from seven that evening, then at nine o'clock, "it was just like an avalanche coming through our hotel, bringing down boulders from the hills and breaking down walls, doors and windows. Within half an hour the guests had evacuated the ground floor. In another ten minutes the second floor was covered, and then we made for the top floor where we spent the night."

As many residents and visitors retreated to the upper floors and roof spaces of their houses, others, including the local policeman, Derek Harper, who had only recently completed his training, were battling the raging waters to save others who were in danger of being swept away. He was later awarded the George Medal for the part he played in rescuing people from the flood. Thirteen other local people received lesser awards for bravery.

The next morning, as the floods subsided as quickly as they had come, the full extent of the destruction was visible for the first time. "Along the tragic streets of desolation are the boulders and trees, like spent ammunition that ancient gods might have hurled down from aloft. It was a force against which the little shops and houses of the narrow streets, renowned as one of the most quietly picturesque even in North Devon, could not stand." The Western Morning News spoke of "deaths on a wartime scale, destruction... worse than in the heaviest blitz, hundreds of residents and visitors personally ruined and destitute – the story stuns the human mind."

Troops and council workers arriving to clear the aftermath of the storm found a scene of unimaginable devastation. The path of the flood was marked by flattened and tottering buildings. Uprooted trees completely stripped of their bark by the force of the water, and the mangled remnants of houses, cars and other human debris, were embedded in enormous piles of boulders. The largest boulder moved by the flood measured around 350 cubic feet. River banks had been gouged back to the bedrock, walls and hedges torn apart, and deep potholes and pits as much as twelve feet deep had been carved out of the ground and the road surfaces, leaving bewildered trout stranded in pools in the middle of the High Street.

In places, the course of the river had been permanently altered, as the floods had sliced through the old meanders and formed a new channel, bypassing the old which was now blocked by a layer of debris – tree trunks, boulders and mounds of stone torn from the river bed – six to ten feet deep. The harbour was completely choked with boulders, rubble and the detritus of the flood. Hundreds of villagers had been left homeless and those houses that still stood were filled to first floor level with debris and stinking mud. There were no mains services, as water and gas mains and electricity power lines had all been severed. Twelve bodies had already been recovered and twenty-four people were still missing, feared dead. The eventual death toll was thirty-four.

Although the local papers carried reports of the tragedy, The Times that Saturday morning, 16 August 1952, had no word of the disaster. A short article headlined "Traffic delayed by floods: 2.81 inches of rain at Plymouth" instead concentrated on traffic problems in London and across the South and the lightning strikes that had hit "many houses" in London, killed a 22 year-old woman from Buckinghamshire as she was walking down a Middlesex street and badly injuring a young boy near the Middlesex Hospital. After detailing the delays to Underground services in the capital, The Times then turned its attention to the South-West but concentrated on the flooding in Plymouth where "in several low-lying areas, water was in places more than a foot deep. The city fire services received fifty five emergency

Police organising the evacuation of the entire population of Lynmouth – an operation completed within two days of the flood disaster. Such was the national sympathy that a relief fund raised the then huge sum of £300,000 within two weeks.

calls from flooded buildings within half an hour, lorries were pressed into service to tow stranded cars to safety; other cars were marooned. In one street, in which drains became blocked, water ran in through the back doors of houses and out through the front doors. At the head post office, clerks worked knee-deep in water."

The newspaper then moved on to reports that "a performance of Ben Jonson's Volpone at the Shakespeare Memorial Theatre in Stratford-upon-Avon had been interrupted for fifteen minutes when a storm caused the main electricity supply to fail. Heavy machinery had to be moved by hand instead of by electrically operated lifts and slides."

Only on the Monday morning, 18 August, three days after the cataclysmic floods in Lynmouth, did news of the area's plight begin to dominate the national news agenda. By then, during the course of Sunday 17 August, the entire population of Lynmouth, both the tourists and temporary visitors and the permanent inhabitants, had been evacuated. With the bridge to Lynton destroyed, the evacuation was to the east and the majority of holidaymakers were given temporary accommodation in "halls at Minehead, farther along the coast."

With the village deserted, Army engineers began installing a Bailey bridge across the river at Barbrook and "military detachments" started working alongside workmen from the county and local authorities to clear the rubble and debris, a first step towards restoring the village to normality. However, the chairwoman of the local council, Mrs S Slater, issued a warning to inhabitants that "the whole job will probably take six months: that is, building an additional bridge, diverting the stream approximately to its original position, rebuilding the bridge destroyed in the main street in Lynmouth and building retaining walls." Electricity and water supplies would be restored as soon as possible and "emergency chlorination" carried out to prevent the risk of disease. Lady Fortescue, wife of the Lord Lieutenant of the county and a pillar of the Red Cross, also reported that the organisation was sending supplies of mattresses, blankets and pillows and clothes for men, women and children.

Elsewhere on Exmoor, villagers attempted to repair their homes and lives and mop up after the flood. At Dulverton, next door to the ruined garage, the village inn was "in scarcely more happy plight, with people still retrieving furniture and household goods or putting out to dry in the fresh air things that on Friday night were submerged for several hours." At Simonsbath, the road had been torn up by the storm and the bridge across the river Barle, though still intact, "had its approaches on either side badly damaged. The village presented the same unhappy picture of devastation, worst nearest the river." When the moor road reached the southern entrance to Watersmeet, two and a half miles from Lynmouth, it came "to a sudden stop, broken as cleanly as if it had been cut from the river bank by a guillotine." Beyond it lay a valley of desolation, extending all the way to the sea.

By the next day "across the chasm that remains where part of the high street ran, there is now a rough footbridge. The people using the footbridge have been almost exclusively those concerned with the task of making Lynmouth accessible to the stricken folk whose homes still stand there in greater or less degree of ruin. During much of the day the tragic scene has been made more sombre by steady downpours of rain. From time to time, evacuated inhabitants of the village have come along to gather a suitcase full of such household goods as it has been possible for them to retrieve from the ruins of their homes. They leave the scene again in vans and lorries which await them at the foot of the hills on either side of the village. One of the improvised pathways now most frequently used for crossing the village beyond the footbridge towards the foot of the Countisbury Hill road passes through the Lyndale Hotel. It is a melancholy route, beginning with stepping stones that lead through a broken window frame, across a room strewn with rubble, along passages where the strips of carpet are sodden pulp and beyond still more heaps of debris, to a doorway on a final pile of rocks and fallen masonry."

Understandably the inhabitants of the area remained very jittery and each fresh fall of rain brought with it a rash of rumours. Half the 150 inhabitants of Challacombe, twelve miles from Lynmouth, fled in terror to the neighbouring hills in pouring rain after hearing rumours that the reservoir above the village had burst its banks. Because of "confusion over names" another forty people from Parracombe also fled their homes. "Some carried furniture out of their houses before leaving." The village constable told villagers that it was a false alarm, but not all of them were persuaded.

Ilfracombe Urban District Council was in the middle of its monthly meeting when "someone burst in and said the bank had given way. The chairman, three members and council officials went fifteen miles in cars to see what had happened. Police and road patrol men gave warning that the valley might be inundated. Fire brigades were warned and an ambulance and council workmen in lorries were sent to the spot."

However, having examined the reservoir, the Ilfracombe Water Board Inspector, Mr WA Lewis, pronounced it perfectly safe. The water was a few feet above normal but the overflow was only two inches. A "haystack" had apparently floated into the outflow, blocking it for a while and this had led to the surge in water that had triggered the alarm. "We cannot stress too strongly that there is no reason to panic," he said. "It would take another cloudburst to do any real harm." The Chairman of Ilfracombe Council, Mr MC Meredith, said it was "a great pity that there was this scare but I do not really blame the man who started it. None of us could take a chance after what has already happened."

An hour after the alarm, the villagers were told it was safe to go home. Others had never left, having taken a more phlegmatic view of the panic and remained in their homes. Mr J Potter, the Challacombe postmaster who had lived in the village for 35 years, felt that "even if the reservoir burst, it could hardly be worse than what we had to put up with on Friday when the floods washed right up our walls." No sooner had the panic died down in Challacombe than a rumour swept Lynton that "additional damage" being done by wind, rain and water in Lynmouth had undermined the cliff-side, which was "about to fall". Police and staff from the town information centre were forced to tour the streets issuing denials of the rumour.

However, the greatest worry for many owners of flood-ravaged buildings in Lynmouth and elsewhere was whether their insurance policies covered them against their losses. While the "risk of loss or damage by storm or tempest" was routinely covered in comprehensive insurance policies, another clause just as routinely excluded "risk of destruction or damage by flood, subsidence or landslip. This risk may be insured against if the cover is specially desired, and premium rates are quoted which take into account particular circumstances including the character of the neighbourhood." Lynmouth residents, reading between the lines of small-print legalese, would not have been reassured.

The Queen and Queen Mary, the Queen Mother, sent messages of sympathy to the people of Lynmouth and the surrounding area, which were read out at a packed meeting in the town hall. A flood relief fund for the victims of the disaster was established by the Mayor of Plymouth, who set the ball rolling with a donation of £1,000. The Duchy of Cornwall contributed £100 – not overgenerous in comparison, but a more substantial sum in those far-off days than it appears today, though it was dwarfed by a contribution of £500 from an anonymous resident of Barnstaple, and holidaymakers at Butlins raised a similar amount. The Lord Lieutenants of Devon and Somerset, Lord Fortescue and Lord Hylton, made a joint appeal for donations to the fund. "We invite not only the people of the West Country, but everyone who has known and loved Lynmouth and the quiet villages of north Devon and west Somerset which have suffered so grievously in this disaster, to contribute to a fund for the relief of all those who have suffered."

The disaster had touched hearts throughout Britain and donations poured in from every quarter. A cinema owner in Manchester donated one night's takings, the London Fire Brigade sent £100 from its welfare fund, Barnstaple Town FC played a friendly match against Bristol City with all proceeds going to the fund, the southern area of the British Legion sent £200, and the Postmaster General announced that all gifts of clothing and other items would be delivered post free to Lynmouth. Such was the national sympathy aroused by the devastation of these tiny communities that the flood relief fund had raised more than £300,000 by the end of August.

Sympathy for the victims was also mingled with curiosity about the cause of this terrible disaster. Such a prolonged and massive downpour in such a confined area was almost unprecedented and rumours soon began to circulate that the flash floods had been caused by Ministry of Defence experiments in rain making. Military theories suggested that seeding clouds with dry ice could trigger heavy storms over battlefields, hampering the movement of enemy troops, armoured vehicles and supplies.

Survivors of the disaster spoke of the air smelling of sulphur on the afternoon before the floods, and said that the rainfall was so hard and furious that it hurt people's faces. They made repeated calls for a public inquiry but were ignored and the MoD issued blanket denials that it had been involved in any rain-making experiments.

A BBC investigation around the fiftieth anniversary of the disaster revealed that classified documents relating to the secret experiments had "disappeared", but BBC researchers did unearth fresh information including RAF logbooks and personal testimony, suggesting that rain making experiments had taken place in some areas. The experiments were allegedly code named "Operation Cumulus," though some dubbed them "Operation Witch Doctor." A glider pilot, Alan Yates, described a flight over Bedfordshire in which he sprayed "salt crystals" into the air, and claimed that he was later told that it had triggered a devastating downpour in Staines, 50 miles away.

However the MOD has continued to maintain its denials that any "cloud seeding" experiments took place in August 1952. Michael Fish concludes: "Met Office forecasters are also employees of the Ministry of Defence and neither I nor any of my colleagues have never seen or heard anything that would substantiate the claims of cloud-seeding. Even if it had been taking place – and in my view, the categorical answer is that it didn't – it would not have had the effects claimed for it. 55 years later, a few of the dwindling number of survivors and the descendants of those who died in the floods are still seeking a definitive answer to the question 'What caused the Lynmouth Disaster?' The short answer is that – as has also happened at Lynmouth in previous centuries, and at many other places in Britain and abroad – a combination of steep, narrow valleys, ferocious rainfall in a concentrated area and human folly in seeking to divert rivers and build houses on their former courses, can combine to create floods that will have devastating consequences.

1953 –
The Great Storm-Surge

Intense floods like that in Lynmouth are often very localised, but less than six months after the Lynmouth disaster, on Saturday 31 January 1953, one of the worst floods ever seen in Britain engulfed the whole of the North Sea coast from the Orkneys right down to Dungeness in Kent.

The storm that broke that day was ferocious enough, but what made it catastrophic was the tidal "storm-surge" that accompanied it – the worst ever recorded in these islands. Throughout 31 January, the northerly winds blew at Storm Force 10 or 11, generating waves well over 25 feet high and driving the water towards the coast. The surge was amplified as it travelled southwards and the high tide advancing down the coast brought with it catastrophic flooding.

The storm that was to devastate vast areas of the country had begun on Friday 30 January 1953 as an unremarkable low-pressure zone off the southern coast of Iceland. During that day, as it moved eastwards, it deepened rapidly and, by noon the next day, 31 January, it was centred over the North Sea midway between North-East Scotland and southern Norway. The atmospheric pressure at the heart of the depression was well over 60 millibars lower than the ridge of high pressure that had built up over the Atlantic Ocean and the steep "pressure gradient" between them was a guarantee of storm winds.

Sweeping down from the north, the storms first battered the Orkneys. That Saturday morning had dawned still and calm, with a sharp frost, but within a few hours the worst storm in the islands' history was raging. A "sustained wind – not a gust – of 125 mph" was recorded at Costa Head in Evie on the West Mainland during the storm – the highest wind speed ever seen in Britain.

The storm tore away parts of the coastline, flattened buildings and reduced roads to rubble. The Orcadian newspaper reported that "gigantic seas in the harbour tore open the sea front and in little over three hours, crumpled the sea wall and washed away the roadway along the entire length of the Ayre Road." Peter Baikie, a builder on the islands, saw "three tremendous seas" strike the sea wall, which disappeared "in a smother of foam". When the waters ebbed away, almost the whole of the sea defences had disappeared along a 300 yard stretch, exposing the Ayre Hotel and the surrounding buildings to the full force of the storm and leaving the road "reduced to a pile of debris." The sea wall also gave way on Shore Street and the waves scoured a huge hole ten feet deep out of the road, exposing water pipes, electric cables and gas mains. Kirkwall's water main was severed and a twenty-yard length was lost to the waves without trace, cutting off the supply to the town and further flooding Junction Road. Four hundred yards of the main road from Kirkwall to Stromness was also washed away.

A Dutch tanker, sheltering in Kirkwall, added to the damage when it was driven stern-first across the harbour and "knocked down about fifty feet of the basin wall", and another ship, St Magnus, had to put to sea and ride out the storm after her cables snapped and she was driven from her anchorage by the winds. Many small boats also sank or broke adrift; phone lines were ripped down and thousands of chickens were killed as their henhouses were overturned and smashed, or blown out to sea; fisherman heard squawking from one henhouse as it was carried off by the waves. Dead birds were strewn everywhere and on Rousay they were "picked up by the barrow-load."

Shipping, air and bus services were brought to a standstill, the West Mainland Mart building in Stromness was "reduced to matchwood" and many people who braved the storm suffered bruises and broken bones as the shrieking winds blew them off their feet. As soon as the weather eased, the inhabitants of the houses along the seafront rushed outside once more to try and barricade their broken windows against the next high tide.

The storms had changed the face of the Orkneys forever but, by a miracle, no one had been killed. The inhabitants of mainland Britain were not to be so lucky. Although the East Coast bore the main brunt of the winds and the storm-surge driven before them, the West Coast did not escape unscathed. Coastal and inland communities suffered severe storm damage and the winds blew down more Scottish trees in one night than the forestry industry felled in an entire year. At Southampton, Cunard took the decision to postpone the sailing of the Queen Mary, due to leave for New York at 11.30 that morning and, because of the risk from the gales, the Cunard liner Franconia could not be moved to the Prince's Landing Stage in Liverpool to embark passengers, but other ships sailed on schedule and for those already at sea there was no escaping the storms.

A 228 ton Fleetwood trawler, Michael Griffith, vanished without trace in the towering seas off Barra Head in the Hebrides, the first of ten ships to sink during the storm. Other trawlers, lifeboats, a warship and RAF aircraft braved the conditions to search the area but the ship's fifteen-man crew were all drowned. The Islay lifeboat, which had kept up the search throughout the night, had to make for the island of Colonsay after two members of the crew were overcome with exhaustion. Sixty-six members of the crew of the steamship Clan Macquarrie were rescued by breeches buoy after the ship was driven onto the rocks near Borve on the Isle of Lewis and holed beneath the waterline. Twenty-seven herring drifters were driven ashore and badly damaged at Ullapool, Ross and Cromarty.

A car ferry, the 2,700 ton Princess Victoria, left Stranraer in Scotland at 7.45 on Saturday morning on its regular crossing to Larne in Northern Ireland. With 125 passengers and forty-nine crew aboard, it steamed slowly up the relatively sheltered waters of Loch Ryan, but as it rounded the head of the loch, it was hit by the full force of the gales and raging seas. The stern car-loading doors burst open under the impact and were buckled by the mountainous waves, preventing them being re-closed, and the car-decks at once began to fill with water. The cargo also shifted and the ferry developed an increasing list to starboard.

At 8.46 that morning the captain radioed "Not under command, car deck flooded, need a tug" and, as his ship wallowed ever lower in the waves, he sent an SOS 45 minutes later. However, the ferocious gales were driving the ship off its usual course and in the appalling weather and visibility prevailing, none of the ships and aircraft sent to bring aid to the stricken ferry could locate it. At 2 o'clock that afternoon, still not spotted by any of the search and rescue craft, the Princess Victoria finally sank four miles east of the Copeland Islands, off the coast of County Down in Northern Ireland.

One of the few survivors, Captain James Kerr, the master of a collier who was returning to his home in Belfast said that "before the ship foundered, floats were dropped into the sea. There was no panic. I never saw a crowd of people acting so coolly. Some of the boats were smashed as they were being lowered, but two were not. I was in one of these. The other was crowded and when the ship rolled over, was swept across her keel. It capsized and I don't think any of those in it could have been saved."

In all 133 lives were lost in the disaster, including all the women and children on board, who were flung into the boiling sea when their lifeboat capsized. Only forty-one of the passengers and crew survived, most of them rescued by the Donaghdee lifeboat. Aircraft from RAF Aldergrove dropped rescue equipment into the sea and searched the area in vain for other survivors and the lifeboat, the destroyer HMS Contest and the Liverpool-Belfast steamer also made searches of the area but found no sign of life.

On land, there was a trail of devastation across the north of Scotland. Roads and rail lines were blocked by fallen trees, power supplies failed in Caithness and Sutherland as pylons and telegraph poles were flattened, and stacks of hay and straw were scattered. The streets of Wick, Thurso, Peterhead and scores of other Scottish towns and cities were littered with fragments of shattered slates, tiles, masonry and chimney pots. A three hundredweight stone pinnacle, blown from a church in Thurso, smashed straight through the roof of a garage, causing extensive damage, and elsewhere in the town temporary housing at a transit camp was demolished by the winds, leaving several families homeless.

As the storm swept south, havoc and death travelled with it and coastal communities suffered the worst of it, with mountainous seas and the deadly tidal surge accompanying the gales. In mid-afternoon, still an hour short of high tide, the River Tees burst its banks. The massive North Bay sea wall at Scarborough was breached in several places by the overpowering force of the waves, and the Scarborough lifeboat was washed from its boathouse. Inland, a one-ton stone pinnacle came crashing down from York Minster, the worst fall of masonry at the cathedral for 200 years. It fell into the road below, fortunately at a time when no one was passing. A little later that wild afternoon, the shingle spit of Spurn Head on the Humber estuary was breached and, soon after nightfall, the exposed coast of Lincolnshire took the full brunt of the storm-surge.

The sea defences were breached at intervals all the way from Barton-upon-Humber to the Wash, and almost the entire defences on the 34 mile stretch between Cleethorpes and Skegness were destroyed. The "stepped sea wall, backed with sand" proved woefully inadequate against the brutal force of the tidal surge. Millions of tons of sand was stripped from beaches, dunes buttressed with timber piles were breached, embankments and concrete sea walls collapsed, the promenades of Mablethorpe and Sutton-on-Sea were wrecked and with the coastal defences in ruins, sea water inundated huge tracts of prime agricultural land. Virtually every street in Mablethorpe and Sutton was flooded by water that was twenty feet deep in places and twelve people were drowned by the floods at the little

Damage and destruction when Britain's worst ever tidal 'storm-surge' swept down the East Coast on 31 January 1953.
1. Remnants of chalets and uprooted lampposts on Scarborough's Marine Drive after the massive North Bay sea wall was breached in many places.

2. More solid brick-built houses at Spurn Point fared little better.

3. The shattered promenade at Mablethorpe in Lincolnshire.

village of Sutton alone.

Unaware of the disaster about to engulf them, most of the people of East Anglia were still getting ready to enjoy their Saturday night. The Lowestoft Choral Society was holding its annual dinner at the Suffolk Hotel, the villagers of Sea Palling had organised a whist drive and countless people were listening to the radio – few had television in those days – getting ready to head for the pub or preparing for dinner with friends.

On the other side of the North Sea, the Dutch had issued a warning of "rather high tides" at 10 o'clock on Saturday morning and revised it to issue an alert about "dangerously high tides" at 4.45 that afternoon. Yet in Britain, although the Met Office had reported that "exceptionally strong north-west to north winds are becoming established over the North Sea" and strengthened the warning in a later forecast with a prediction of "northerly gales of exceptional severity", this fell some way short of a warning of a tidal surge. When it arrived, its scale and extent were completely unexpected and even as the flood waters were bursting into people's homes, the Met Office was still only warning of gales and heavy seas.

Part of the problem was that responsibilities had been divided for bureaucratic convenience, with the Met Office responsible for anything to do with atmospheric conditions and the Admiralty's Hydrographic Department responsible for reporting on the sea. Its work was mainly involved in preparing tide tables a year in advance and even if it had had a means of predicting the storm-surge that night, it had no way of communicating that knowledge to the outside world, least of all on a Saturday night when few, if any of its staff were at work.

Even worse, so many telephone lines were brought down by the storm in Lincolnshire and Norfolk that virtually no warnings of its severity were passed to counties farther south until it was too late. As a Ministry of Agriculture report later noted, "Farmers had little or no warning, other than from those few who were outside during the evening and noticed that inland drains suddenly began to flow upstream."

As the surge reached the southern shores of The Wash, embankments were overtopped with fatal consequences. Michael Pollard, author of a fine account of the floods, North Sea Surge, interviewed a King's Lynn ambulance driver, Bertie Hart, who had "lived by and watched the river Great Ouse for sixty years and got used to its moods and changes." For the previous three days he had noticed that "outgoing tides were not getting away as they should have done. There was a lot of water still in the river when the incoming tides came." His unease was increased that afternoon when he went to a football match and talked to a friend, a river pilot, who told him "the motion of the sea had been strange, like something he had never experienced before." Nonetheless, though Mr Hart was "very much concerned about the wind, never did we think about floods."

The damage and death toll along the coast was made worse because many of the seaside developments, from Norfolk all the way to Canvey Island in Essex, were flimsy wood-framed bungalows and prefabs, never intended for year-round occupation. Nonetheless, many of them had permanent residents and when the sea defences gave way, these fragile homes and their occupants suffered as a result, with the very young and the very old by far the most vulnerable.

The River Ouse had never flooded at King's Lynn before; it did so now and the flood waters surging through the town "in a great wave seven feet high" were a foot higher than had ever been recorded. "Nobody knew it was coming, for whatever sound the water made was hidden by the noise of the wind." Some 1,800 houses were evacuated, 700 of the occupants were housed at Gaywood Park School, but fifteen people drowned – all elderly people who could not get out of their ground floor rooms in time – and another sixty-five died in the low-lying villages between King's Lynn and Hunstanton. Police, servicemen and local volunteers worked all night "often wading in icy water up to their armpits to try to reach other old people who had been trapped."

Ambulanceman Bertie Hart went on duty at 6.30 that evening and was sent to collect an old army lorry that had a high enough wheelbase to get through at least some of the flood water. He passed a scene of devastation: flattened trees and houses, and fire arcing from a flooded electricity substation. "On the Saddlebar Road there were three cars upside down and forty-gallon drums of oil had been rolled along by the force of the water, knocking down the side of a house before falling into the River Nar... the most terrible scene of the disaster was south of Hunstanton where the natural bank, though strengthened, was overwhelmed. The majority of the dwellings built on it, including some permanently occupied, were destroyed or damaged and sixty-one people lost their lives."

Salt water penetrated the water mains at Hunstanton, adding to the distress of the inhabitants by rendering the water supply undrinkable. Tankers bringing drinking water to the stricken area were unable to get through until the flood waters receded. Many American servicemen from the nearby USAF bases lived on this coast and many of them threw themselves into rescue efforts alongside the British residents. One of them, 22 year-old USAF Corporal Reis Leming, later became the first non-Briton ever to be awarded the George Medal for bravery after single-handedly rescuing twenty-seven people from the South Beach area of Hunstanton, manoeuvring his dinghy among the remains of the beachside bungalows for four hours, often up to his neck in water as he clambered out of the dinghy to search buildings for survivors. He stopped only when he collapsed from exposure after four hours and his feat was all the more remarkable because he couldn't swim and could easily have drowned. Several of his countrymen living in seafront bungalows were among the dead.

Peter Beckerton of Snettisham was awarded a posthumous Albert Medal for his bravery after being swept away and drowned while making a vain attempt to rescue a sixty year-old invalid and his wife from a neighbouring bungalow. Unaware of his fate, his parents with their other two children and two friends of theirs, narrowly escaped with their lives in their own flooded bungalow. They used their bedsheets to lash their thirty year-old lugsail boat to the front of their bungalow, and stood up to their necks in water for seven hours holding on to it, while the children baled it out with cake tins. Baulks of wood and pieces of furniture battered them as they were swept past by the raging waters. A bicycle, somehow floating, was hurled into the boat by the flood waters and just as quickly thrown out again by Mrs Beckerton. Most astonishingly of all they watched a bungalow, appearing complete and untouched, whirl past them on the flood and disappear. They were eventually rescued, bruised, battered and alive, but only learned of the tragic fate of their eldest son the next day.

The bungalow they had seen was probably the same one that then collided with the 7.27 pm train from Hunstanton to King's Lynn. Halfway between Hunstanton and Heacham it was crossing an embankment over the marshes when the train driver saw the scarcely credible sight of a huge wave approaching with a bungalow riding on its crest. It struck the steam engine "squarely on the smoke-box" and damaged the brakes, leaving the train stranded. As the waters rose higher, the engine fire was extinguished, the train lights went out, and the passengers had to stand on the seats of the darkened train to keep clear of the water. "For six hours the engine-men and guard kept up the morale of the passengers and at length succeeded in effecting temporary repairs to the brakes and, by using the floorboards of the tender as fuel, in raising sufficient steam to propel the train slowly back to Hunstanton."

Two seventy year-old women died in the first rush of waters into north Norfolk. Mrs M Middleton, 73, of Salthouse, was swept out of the kitchen of her house by the flood and later found dead in her garden, and Mrs Edie Dix who lived alone at Willow Cottage, Wiverton, was found dead in the ground floor of her house when rescuers reached it the next day. At Cromer, the pier, pavilion and the shed housing the lifeboat were "severely battered and the lifeboat, Henry Bloggs, lies on her side in the shattered boathouse." At Wells-next-the-Sea, a 160-ton motor torpedo boat was picked up by the storm-surge and dumped high and dry on the quayside. Wells butcher Charles Ramm had closed for the night when someone raced up the street and shouted, "There's a wall of water coming through on the marsh." Mr Ramm had to leave his seventy cattle and two hundred sheep in fields between Wells and Holkham to their fate. "We couldn't get anywhere near. About all we could do that night was to get the dog out. We found him in the stable, standing in the crib with his front paws up on the wall to keep his nose out of the water. We got him out about 11 o'clock."

At Burnham Norton a farmer put his sow and her new litter of piglets onto the sofa floating in his living room before he and his family retreated upstairs. The pigs were found safe and well, and still reclining on the sofa when the waters went down again. Another pig found its own food and accommodation by taking up residence in a potato clamp, where it was found hale and hearty when the farm, abandoned to the floods, was reoccupied. Another pig was not so lucky. In King's Lynn, one old man asked the ambulancemen if they could remove a dead pig that had floated in through his living room window on the flood, as the sight of it was upsetting his wife. Unfortunately, it was too heavy for the ambulancemen to move and it had to be left until more help arrived.

For every story of human or animal survival there was a tale of woe. In total one hundred people died in the area around Heacham, among them two American children found on the beach; their parents were missing, believed drowned. An hour before nightfall Ivor Stuart had walked down to the sea shore because "it had been blowing hard and I felt disturbed, so I went to have a look at the tide coming in. It was a strange sight; the previous tide had not gone out, it had been penned in the Wash by the wind. It was as if we were looking uphill, the new tide was coming in on top of the old. At 6 pm the water was rushing up from the sea. I told the wife to gather up our foster son, Christopher. The water was coming up through the floorboards. I told the American couple next door to get out quick. The girl was pregnant and she had the baby next day; no doubt it was the shock."

Colin Turner, a 19 year-old national serviceman based at RAF Swanton Morley in Norfolk, was one of those scrambled to aid in the makeshift rescue effort. "Eight of us were dropped in Heacham and told to do what we could. It was 4 am, pitch dark, you could hear the waves crashing. It was terrifying and bitterly cold, with the wind blowing a gale. We could hear people crying for help. We tied ropes round our waists and waded in. Some were in upstairs rooms, balanced on the furniture, some were on the roofs of bungalows. I have no idea how many we pulled out. The sea just came up and caught them. We plunged into the water up to our necks, soaked, but we kept on all night. In the morning we waded out to see if there was anyone left alive and we could see the drowned people floating in the bungalows. It was not something a nineteen year-old is prepared for."

At the villages of Horsey and Eccles, waves bursting over the concrete sea walls tore away the dunes beyond. A mile and a half of the sea wall at Eccles was destroyed and there and elsewhere many people were swept to their deaths before they even realised that a flood was upon them. Seven were drowned in the neighbouring village of Sea Palling, including three children, one aged just six months, when a number of houses collapsed. Villagers at their whist drive had to drop their cards and rush to save their own lives and those of their families and friends. Two 70 year-old ladies somehow managed to climb to the relative safety of the roof of their bungalow, but they were soaking wet and exposed to the full force of the gales and one then died later that night from hypothermia.

The landlord of the Life Boat Inn drowned as he tried to reach a rescue boat, and a family lost two children, a daughter who was on her way home after completing her paper round but was then swept away and drowned, and a baby girl who was seized by the flood waters and torn from her father's back as he tried to carry her through the floods to safety.

The Ridley family were luckier. They wedged a ladder so that they could reach their roof when "just then our nearest neighbours came through the back door, an old lady about eighty and her daughter who was in her forties. With the doors open front and back, the water just roared through, catching up and sweeping away all the furniture. My poor Mum just said one sentence, 'My poor home. All my home.' After that she was silent as though numb.

"Dad and I got the older two women up the ladder, Dad going up first to help them up. I pushed them up the ladder with my head. It was pretty hard as the wind was tearing past at a great rate and the sea was rising rapidly all the time. Then we got Eva (a family friend) up, which was difficult because she was in a plaster cast owing to a slipped disc. Then I followed. We got ourselves all huddled by the chimney stack, holding one another on, as the roof was wet, slippery and I suspect a bit icy.

"I watched the water still rising and was in fear of us all being swept off the roof. We saw animals, trees and large objects float by. At last, after what seemed a lifetime, the water receded and we could see the roof of a house appear in the distance. Dad decided our best plan was to get down and inside; if only to get out of the driving rain and wind." Their ordeal on the rooftop had lasted five and a half hours.

Massive damage was done to Yarmouth seafront with parts of the sea wall destroyed and shelters, huts and railings "torn to pieces", while the banks of Breydon Water and the River Bure at the northern end of the town both collapsed. Nine people drowned and 10,000 had to be evacuated from their homes. In Lowestoft, a party of forty children and a few adult helpers were trapped in St John's Church. Two police detectives and an uniformed constable went to the rescue. They had to wade through chest-deep water and the constable had to dive under it to reach the door handle. With the aid of a local boatman, who rowed his dinghy up the aisle of the church, Detective Constable Allenby Sparkes carried the children in turn from the altar where they were huddled to the boat and they were then ferried to safety in small groups. When he came to carry one of the helpers, a woman of substance, to the boat, she gave Sparkes "explicit instructions not to get her fur coat wet. Had I not been in a House of God, I would probably have dropped her in the water and made her walk to the boat."

The darkness hid the scale of the tragedy and a Norfolk farmer's wife was not the only one who "thought we were the only ones". Officials were also struggling to comprehend

The disastrous storm-surge of 31 January 1953 worked its way remorselessly down the East Coast, quickly reaching Yarmouth where the depth of water brought rail travel to a halt. (PA Archive/PA Photos)

what was happening. As one said, "During the night such alarming reports kept coming in that many thought that darkness and uncertainty had magnified the disaster. However, when daylight came it was found that darkness had in fact hidden much of the damage and that things were far worse than could have been imagined."

The Sorick family, Americans serving at the USAF Bentwaters base, were settling down to supper with friends at their Southwold home when the surge hit the area. As their bungalow was picked up by the flood and swept 400 yards across the marshes, the six adults, with the Sorick's seven month-old baby, managed to clamber through a skylight onto the roof. They were eventually rescued unharmed after several hours by a 74 year-old lifeboatman "Mobbs" Mayhew, who also retrieved four adults and a child from the top of a drifting beach hut. He was later awarded the British Empire Medal for his rescue efforts, but five residents of the devastated Ferry Road area of Southwold were drowned. Among the dead were a mother and her four year-old son.

By midnight, Felixstowe, Harwich and Maldon were also flooded, with much loss of life. At Felixstowe the River Orwell burst its banks and every single house on a development of prefabs was swept from its site by the floods. Three hundred people were left homeless and twenty-one bodies had already been recovered by the following day. William Leggett awoke to the sound of his house creaking, groaning and rocking on its foundations. As the floodwaters overwhelmed his house, he swam for his life. "I half-swam and was half-taken by the current, I could hear my neighbours shouting and screaming but there was nothing I could do. When I looked round, my house was following me down the road. I've been under machine gun fire during the war, but this was far more frightening."

Neighbours in stronger houses on Langer Road watched, horrified, as the prefabs were swept away. "As one floated past, we saw a woman washed off it." She was one of thirty-nine people from the town who died. Among them were several entire families, including the Bushnalls and their two year-old son and six-month old daughter; the Salmons, who had an eight year-old son; and the Pettitts with their six year-old son and two year-old daughter.

There were thirty breaks in the banks of the River Deben between Felixstowe and Woodbridge and the course of the river was "altered entirely". At Harwich, where the housing stock was more solidly constructed, only eight people died despite "a wall of water rolling towards us that seemed as high as a three-storeyed house", but 3,000 people were made homeless. The town was left without gas or electricity and all but one of the pumping stations were out of order, leading to warnings that "drastic steps must be taken to deal with sewage." Late that night a rescuer in Harwich saw the bizarre sight of the husband of the local RSPCA representative rowing a dinghy "loaded with cats at the stern, dogs in the middle and a most magnificent parrot perched on the stem and shouting its head off: 'Get out of here, you buggers'."

Just after midnight, as the surge peaked on the Essex coast, the sea engulfed a colony of flimsy holiday chalets and retirement homes at Jaywick, near Clacton. About 1,500 were only occupied in summer, but 250 were year-round homes to retired people and thirty-seven of them were drowned as their houses collapsed around them. Many old people had to smash their way through their ceilings to reach the roof void. One elderly woman spent 31 hours in darkness in her roof space, with only her cat for company, before her weakening cries for help were heard by rescuers. But the roof voids of their houses proved an illusory refuge for some old people as many chalets were swept off on the flood waters. A man tried to scramble to safety through the rising flood with his semi-invalid wife and their three year-old grandchild. "As we pushed open the door, the full force of the water hit us. I was up to my neck. My wife just disappeared." She was drowned, but her husband saved himself and their grandchild by clinging to a barbed wire fence until rescuers reached him.

The sea walls on Canvey Island collapsed, effectively "drowning" the island and its ramshackle settlements. Once more, many of the prefabs and jerry-built houses intended as holiday homes had been turned into permanent homes either by retired couples, or people working locally or commuting into London. As in Jaywick, there were no records of which houses were permanently occupied and in the early hours, few of those were showing any lights, giving rescuers a near-impossible task of identifying the houses in which people might be trapped.

The Manser family and their ten children – nine boys and a girl – lived three and four to a room in a cramped single-storey prefab that had gas but no electricity and its only water supply was a pump at the end of the road. Christopher, thirteen years old, had been given a new coat and a pair of trousers that evening, the first clothes he had ever had that were not second-hand. They lay, unworn, at the end of the bed as he went to sleep that night.

About half past midnight their fox terrier dog woke him with its frenzied barking. "Straightaway we were up to our waists in ice cold water. It was pitch-dark and by now the other children were crying out in their bedrooms. Somehow we found our parents and gathered in the central room. I was clutching one youngster, and Ian (his fifteen year-old elder brother) and my mother had the two babies. The water was rising by the second. My father found his lighter and lit the gas mantle; it flickered for about four minutes and then went out, then all was dark again."

Christopher was now up to his chest and the other children were standing on furniture to keep themselves above the rising water. It was nearly five feet deep in the house, which was itself raised on stilts, two foot six inches above the ground. An elevated path ran about 500 yards behind the bungalow. Ian decided to try and swim to it and summon help. He disappeared into the flood while his family waited in the dark, hearing the water swirling ever higher around them. Mrs Manser was huddled on a window sill in water up to her chest, holding the two babies. Christopher tried to keep himself and two other children afloat, while his father cared for the others.

Christopher now "started to get very tired and could not get the children's faces out of the water. We waited for help but it never came and we thought Ian must have been lost. My mother made us all shout for help; we found out later that a neighbour in the two-storey house next door had tried to reach us but couldn't make it. The water continued to rise and was now up to my bottom lip. My mother made us shout and sing. We sang hymns. I remember praying to God and promising to go to Sunday School every week if He would only help."

Mr Manser then thought to smash the asbestos ceiling panels above them and "standing on the remains of a submerged table, between us we passed up the older children so that they sat straddled across the rafters, out of the water." As Christopher climbed up to join them, the table broke and was carried away by the flood, leaving his mother still stranded on the window ledge with the babies. One of the children on the rafters then fell off into the water. Christopher "jumped in after him and kept him afloat. There was no means of getting back as all the furniture had gone. I was back where I'd started and twice as cold and sleepy."

He looked at the peaceful faces of the child he was supporting and the two babies his mother was holding. Only later did he realise that all three were already dead. The family remained there all night and it was well after dawn when a man in a kayak heard their shouts and rescued them, transferring them one at a time to the elevated path behind the house. By the time Christopher was rescued he had been in the water for ten and a half hours and was far gone in hypothermia. He was taken to a house on high ground above the flood, where "a woman wrapped me in blankets, hot water bottles all round, and rubbed and rubbed and rubbed until I came round. How I blessed her." His brother Ian and the rest of the older members of the family also survived, but the three youngest children were all dead.

The surge had swept on, over Tilbury and into London's docklands, flooding refineries, pumping stations, warehouses, factories, and putting gasworks and electricity generating stations out of commission. The ensuing power and gas cuts only added to the mounting panic of the population. Almost the entire district of Tilbury was flooded. Over 3000 people

A policeman carries a small child to safety on Canvey Island in the Thames Estuary. It was dark by the time the storm-surge reached the island, giving rescuers little chance of finding those trapped in prefabs and jerry-built houses. Ten thousand people were evacuated in the next 48 hours, but help came too late for the fifty-eight who died. (PA Archive/PA Photos)

were rescued from the stricken areas of Tilbury and Purfleet – where one breach of the sea-wall extended 200 feet and in places the floods were fifteen feet deep – and many of them spent the night on the floors of commandeered schools in the neighbouring districts. The floods also put the Thurrock sewage works out of action, leading to calls from police for residents of Grays and other districts "not to use lavatories, baths or sinks any more than absolutely necessary". Since many of them were already under water, the advice could have been seen as superfluous.

In the East End of London, more than a hundred yards of embankments and sea walls collapsed, flooding over 1,000 houses and turning the streets of Silvertown and West Ham into rivers. Between 200 and 300 caravan dwellers could only watch as their homes were washed away when the Thames breached the river wall in the Belvedere Marshes near Woolwich. Within minutes 700 acres of the marshes were under three feet of water. Several Thameside factories were also flooded and the nightwatchman at one was drowned. Eight homeless Barking families were sheltered for the night in the Barking police station. Rotherhithe was under four feet of water, the wall of the Surrey Docks was breached at Osprey Street and the Grand Surrey Canal overflowed its banks at Peckham and Camberwell.

The floodwaters also hit the naval dockyard at Sheerness on the Isle of Sheppey, displacing the main lock gates, swamping the 715 ton submarine Sirdar, which was in dry-dock, and capsizing the 1,060 ton frigate Berkeley Castle. The ship was "on her beam ends with her mast snapped off", surrounded by mangled scaffolding and timbers. The Captain-in-Charge at Sheerness said that they knew that "the biggest catastrophe that has ever befallen a dockyard in peacetime was about to happen an hour before midnight when the tide passed the danger mark, but there was nothing we could do. The sea swept across the dockyard like a tidal wave." It was thought the frigate would have to be broken up where she lay.

Further upstream, thousands of people gathered on the Embankment to watch the tide on the Thames. At high tide, the water level was almost level with the roadway at Chelsea and it overflowed at Putney where swans could be seen swimming along the submerged road. Wandsworth, Kew Bridge, Mortlake, Richmond and Barnes were also flooded and there was a fatality at Barnes where a 73 year-old man, Herbert Haines, who was living in an underground air raid shelter in the back garden of a house in Rectory Road, was drowned by flood waters that inundated his refuge as he was lying in bed.

In Whitstable in Kent, the surge overwhelmed the town's new sea wall and swept through the West Beach district. Hundreds of families were marooned in their houses and 500 were eventually rescued by boat, including an elderly couple who were found standing on the draining board of their kitchen sink, up to their chests in water.

By now, almost 100,000 hectares of eastern England were flooded and 307 people had died. There were 280 breaches of the sea walls in Essex alone. Evacuation of people from the stricken areas went on all night and throughout the next day with boats, punts, amphibious army vehicles, lorries and helicopters ferrying people to hastily established emergency reception centres, from where they were directed to temporary accommodation in hotels, schools, village halls and private houses. The entire populations of Mablethorpe and Sutton on Sea – four to five thousand people in all – were evacuated on Sunday. "Some waded out up their waists in water and others were rescued from bedroom windows by improvised ropes." By 10 o'clock on Sunday night, when the evacuations stopped for the night, only 300 people remained to be rescued from Mablethorpe and Sutton. Hundreds of

No sign of life at the partially destroyed Beach Hotel at Sutton on Sea, Lincolnshire, where virtually the entire population was evacuated on the day following the storm-surge.

It was a similar story at Mablethorpe, a total evacuation of over four thousand people leaving deserted streets and water everywhere. Signs such as 'Fried Fish Saloon' and 'Snack Bar' had a distinctly hollow ring!

people were given shelter in nearby houses or with relatives in the surrounding areas, the remainder were received at emergency centres in Louth and Alford and then distributed to temporary accommodation in the area.

A mass evacuation by Army and Naval detachments, firemen and police, also removed the entire population of Canvey Island – 10,000 people – in the space of the following 48 hours, but for fifty-eight inhabitants, rescue came too late. Detachments of Civil Defence workers and WVS volunteers helped to care for the bedraggled survivors and a convoy of buses from Southend moved them to temporary accommodation in schools at Benfleet, Rayleigh, Billericay and Southend. As the evacuation went on, many rescue trucks and cars were caught by the flood-tide on the Sunday afternoon and had to be abandoned "with water lapping over their bonnets."

Human tragedies continued. A local taxi driver was caught by the floods and swam back to his house where his wife and three children were marooned. Just as he reached the house, neighbours saw it collapse on top of him. Perhaps no incident was more distressing than when a seventeen year-old Army cadet was trying to rescue a stranded family by boat. As he was doing so, he saw a baby floating by on a door which had been wrenched from its hinges. He dived out of the boat to attempt a rescue but driftwood struck him in the back. He was taken to hospital, badly injured, and the baby was later found dead.

The flooding and the death toll in London and the South-East would have been far worse had the wind not swung more into the west, driving the storm-surge across the North Sea towards the Netherlands. England's gain proved to be a catastrophic loss for the Dutch as the sea burst through fifty dykes and inundated over 322,500 acres of reclaimed polder land, leaving up to one sixth of the whole country underwater and 1,835 people drowned. The gales and tidal surge also left shipping stranded around Rotterdam, one of the world's busiest ports, and a Swedish steamer ran aground near the island of Texel, a German ship was stranded near Ymuiden, and Finnish and Belgian steamers grounded near Flushing.

On land, the scale of the disaster was immense. The maritime provinces of North and South Brabant and Zeeland were almost totally submerged. A belt of flooded land stretched continuously along the coast and up the Maas estuary and the islands of Zeeland were "divided into fragments." The Beverland peninsula was inundated and cut off from the mainland and the Ramekens dyke, bombed and rebuilt in the Second World War, was once more destroyed, flooding the town of Flushing. Dordrecht was totally cut off, Rotterdam so badly flooded that it was approachable only from the north, and the main road and rail links to Belgium were cut both sides of the Moerdijk Bridge. In South Holland, Gravendeel was marooned and 600 people were crowded into the upper storey of a secondary school. Large parts of many other Dutch towns and cities were also under water and the Dutch government declared a nationwide State of Emergency.

Despite their own flood problems, the British government at once offered the Dutch "500 men with picks and shovels, 30 three-ton trucks, more than eighty digging and earth-moving appliances and 50,000 sandbags, all at 24 hours notice." British troops stationed at the Hook of Holland transit camp were also "hard at work all day, helping with the removal of refugees." All Dutch servicemen were ordered to report at once to help with flood relief work and in many towns the burgomasters ordered all the men in their areas to "report with spades and working clothes to strengthen the dykes." It took six months to repair the breaches in the dykes and pump out the last of the floodwater.

By a cruel irony, the surge occurred only a few days after a Dutch Ministry had published a policy document outlining the "Delta Project", a plan to prevent a flood disaster by damming all the tidal inlets and estuaries in the provinces of Zeeland and South Holland. After the disaster, urgent if belated action was taken to implement this plan. Belgium was also badly hit by flooding with Antwerp's docks and the port city of Ostend, the worst affected areas.

The embankments breached by the storms in East Anglia proved to be a mixed blessing in the aftermath of the flood. Designed for defence against the sea and often enclosing low-lying land originally reclaimed from the sea, few had viable means of draining off water that had collected on the wrong side of them, and one person noted that around Wells and Hunstanton in Norfolk, "a great deal of the water on reclaimed land was still there in late February, and it had thus almost returned to the condition of the old coastline with lagoons and a long offshore bar."

The scene in the flooded areas almost defied description: a wasteland of water in which the already bloating bodies of thousands of dead cows, pigs, sheep and hens, floated in a sea of debris, much of which had once been people's homes. At Snettisham only twenty out of one hundred bungalows still stood along the edge of the beach and even they were all "in ruins". Some bungalows had been carried bodily by the flood "under the line of overhead electric cables, across the big pitholes dug out for the shingle works, across the shunting yards, over a high hedge, over a line of bungalows at a lower level and finally landed on top of the flood bank, at least 800 or 900 yards from the original site and still remained more or less whole, with most of the china intact inside. Others were completely smashed to firewood and not a sign of them is left."

At Holkham the tideline was marked by the corpses of sheep, wildfowl that had been unable to take flight before the storm-surge was upon them, and dead pheasants that festooned tree branches. Although the worst of the flood had subsided, it was only a few hours before another high tide and desperate work to repair the sea defences went on in the teeth of the gale that still blew with savage force. All that Sunday "you couldn't hear yourself speak, you could hardly stand against it."

Tens of thousands of people the length of England and Scotland were in temporary accommodation, housed in schools, police stations, town halls and other public buildings. There they were fed and cared for by local volunteers, the Red Cross, the St John's Ambulance Brigade, the WVS, the Salvation Army and Ministry of Food convoys "cooking food and distributing tea and sandwiches."

Astonishingly, just as in the Lynmouth flood disaster, despite the death and devastation the floods had created, the national media were extremely slow to react. The lead story that morning was the sinking of the Princess Victoria, but where they were mentioned at all, the floods that had ravaged the whole of the East Coast were dismissed in a few paragraphs. Most of the British population remained ignorant of the catastrophe that had engulfed their fellow-countrymen until Monday morning when the papers at last began to carry stories revealing the full extent of the disaster.

That morning, 240 stranded inhabitants of Foulness Island were finally rescued by motorboats, dinghies and military landing craft which "set out from inlets and creeks to cross the mile wide stretch of water." RAF helicopters also ferried some people out. Casualties suffering from shock and exposure were taken to hospitals in Southend, Chelmsford,

Billericay and Braintree. As the inhabitants were being taken out, a dozen farmers and two RSPCA officials were arriving there to see what, if anything, could be done to save the stranded cattle on Foulness.

On Monday afternoon the Queen toured flooded areas within range of her Sandringham residence, though newspapers reported that when she "drove three miles down a country land to reach Snettisham beach, because of a hailstorm the Queen remained in the car and looked across the devastated area with its 300 residents busily saving what they could from the wreckage of their homes."

Further south, 36 hours after the inundation of Canvey Island, "the sombre and chilly exodus was still continuing and the tragic roll of dead and missing still cannot be accurately stated". Eighty bodies had already been found but 400 or 500 people still remained unaccounted for, though many of them had simply fled their homes for safer areas. More corpses were being found all the time. In the Newlands area on the north of the island, "the body of a woman could be seen hanging from a tree and another body could be seen hanging from a shattered roof." In every part of the East Coast, there were still pockets of deep floodwaters and men, women and children still marooned. All of them had to be brought to safety, fed and housed and, meanwhile, the grim task of finding and retrieving the dead continued.

Occasionally, unexpected survivors turned up as well; a 70 year-old woman was found in a house at Newlands, Canvey Island, three days after the flood, and helicoptered to hospital in Southend. She had been sitting on a table in her flooded living room since early on the Sunday morning. She had "passed the two days and nights without food or drink, without a clock, without knowing what had befallen her neighbours or when she would be found. She told police that she had begun to lose her sense of time." In the same area rescuers also made the discovery of "an eight week-old baby in a floating pram, adrift in a house porch." The baby was safe and well, but it faced an uncertain future; "the father was dead and the mother believed to be dead."

Until the breaches in the sea defences were repaired, each high tide brought yet more water flooding back inland, triggering terrifying memories for many of the inhabitants. In many areas, including around King's Lynn, "police loudspeaker cars toured the streets to announce the passing of danger after each high tide." The most frightening thought of all was that another Spring Tide was due in a fortnight and this one was predicted to be two feet higher than the one they had just endured.

A tug service ferried thousands of sandbags between Chatham and the Isle of Grain to plug the gaps in the sea defences – one 175 yards long – and allow work on restoring the new oil refinery there to begin. The refinery had only just opened and "had begun to deliver finished products on Saturday for the first time." The night shift managed to shut it down within fifteen minutes once the sea flooded in.

Elsewhere, tens of thousands of soldiers, farmers and volunteers of all sizes and ages laboured unceasingly over the following days to fill the breaches. Landing craft were used to dump boulders, rubble and slag from power stations onto the seaward side of broken embankments, endless human chains passed sandbags along the line to fill smaller gaps and in some places "repairs were carried out by rowing sand-filled bags out over the marshes to the breaches in many scores of small boats." As the days passed, contractors with heavy plant – bulldozers, draglines, cranes, industrial pumps and other equipment – arrived in increasing numbers to speed the work. It was largely successful, although some of the larger breaches were still only partially repaired, the Spring Tide of the night of 13-14 February passed off with only minor flooding.

One of the most delicate tasks was the removal and disposal of hundreds of tons of explosives from a munitions factory on Bramble Island, which had been rendered dangerously unstable by their immersion in seawater. Harwich police had earlier warned that "many tons of explosives were drifting and might be dangerous if tampered with." Those explosives that could be retrieved and all those damaged by seawater still stored within the factory were eventually loaded onto a concrete barge, towed out to sea and sunk, with the site marked by a Trinity House buoy to warn off shipping. However, ten tins of hydrogen cyanide, washed away from Associated Fumigators Ltd, of Canning Town, when the flood waters swept in, were never found.

Like the permanent repairs to the sea defences, the clean-up after the floods continued for months. Even those houses that had survived intact were often filled with stinking mud and slime to first floor level, in which was usually embedded the reeking corpses of dead animals and fish. The clean-up also revealed one even more macabre sight. A couple from Harwich had abandoned their house as the flood waters rose. The person with whom they eventually took refuge noticed that the woman in particular "seemed very nervous, constantly drumming her fingers on the table." That nervousness increased still further when her husband was hospitalised with typhoid, and the following Saturday the woman simply disappeared, going out of the door and not coming back. Her host put her behaviour down to the shock and stress of the floods, but when an electrician went up into the loft of the couple's house to rewire it prior to restoring the power, he discovered a suitcase that contained "several tiny baby skeletons".

The woman's husband knew nothing of the skeletons, and, reading the story in his hospital bed, complained "if only she'd told me, I'd have put the suitcase out of the window" from where the floods would have carried it off. The woman was eventually tracked down, tried and sentenced to gaol for infanticide. She died just over two years later.

Agriculture had suffered a devastating blow. Some of the most fertile lands in England were now contaminated by salt and the Minister for Agriculture warned that "crops on land under salt water can be written off as far as this year's harvest is concerned and any land similarly placed which was going to be ploughed and sown for this year's harvest can also be written off." Families only just accustoming themselves to an adequate food supply after years of rationing that had continued all through the war years and for another five peacetime years after that, now found their diet under renewed threat.

In the face of all their hardships, many in the flooded areas put on a resolutely brave face, recalling the "Dunkirk Spirit" that had helped them survive the horrors of the war. Anecdotes and jokes were exchanged among the survivors – the unhappy old man who had kept his life savings, £900, down the back of his sofa and had to endure the sight of it floating away on the flood, never to be seen again; the happier pensioner who had pinned up his sodden life savings, £160 in one pound notes, on his washing line to dry them out; and the Suffolk man who came down from his upstairs refuge after the floods to find two pigs in his front room, seated at either end of his sofa – but for many there were no smiles.

As the floods subsided, people the length of the East Coast had to come to terms with the loss of homes, businesses, possessions, livestock, pets and in many cases, friends and

The Queen officially opens the Thames Flood Barrier near Woolwich. An early response to extreme weather events, it was completed in 1982 at a cost of £460 million. (PA Archive/PA Photos).

loved ones. Many never got over it. Government agencies often added to the miseries of the survivors by their bureaucratic and hardhearted attitudes – old people who had lost their pension tokens in the floods had to wait weeks for replacements; National Health teeth and spectacles were only replaced, if at all, after long delays – and insurance companies proved particularly stone-hearted. The tragic Manser family was told by their insurers that, as the deaths of their three children were caused by "an Act of God", no pay-out would be forthcoming and the Mansers had to borrow money to pay for their children's funerals.

Other organisations and individuals were much more generous. The Lord Mayor of London's Relief Fund, opened on the Monday after the disaster, eventually raised £7.25 million in donations. Much less creditably, the government of Winston Churchill then welched on a commitment to match the sum pound for pound and only £2 million of central government money was ever added.

The "great storm" and the disastrous surge of 31 January - 1 February 1953 had been the worst to hit the UK in the 20th century, killing over 300 people. Around 24,000 houses were damaged and 180,000 acres flooded. Some of England's most prime agricultural land had been rendered sterile for at least a season by salt water, thousands of cattle, sheep, pigs and poultry had been wiped out and winter crops and stores of grain and animal feed destroyed.

The height of the surge peaked at nine feet at Southend in Essex, almost ten feet at King's Lynn in Norfolk and eleven feet in the Netherlands – enough to overwhelm all but the most massive sea defences. The one feeble consolation was that the disaster could have been even worse. Although the surge had occurred on a full moon Spring Tide, the preceding and succeeding new moon Spring tides were even higher and would have led to even more devastating flooding and loss of life.

The Government committee set up to inquire into the disaster recommended the establishment of a flood-warning organisation: the Storm Tide Warning Service. A Dutch equivalent had been set up after the great surge of January 1916, when the dykes of the Zuyder Zee were breached.

That 1953 storm-surge, and other less catastrophic ones in the succeeding years, led to measures to improve the flood defences of the south-east of England in particular. Huge steel and rubber floodgates were installed in 1972, to protect the major London docks from floods. They were closed for the first time on 12 January 1978, when a surge caused flooding from Humberside to Kent. London came close to disaster; had the surge been eighteen inches higher, the capital would have been inundated. Concerns heightened over rising sea levels and the progressive subsidence of the East Coast – in the aftermath of the last Ice Age, the UK is still gradually tilting, as the north-west, free of the weight of ice, continues to slowly rise, causing the south east of the country to sink. Coupled with the increasing frequency of extreme weather events, a major flood in central London seemed ever more probable and this led to the construction of the Thames Flood Barrier near Woolwich, completed in 1982. The continuing rise of sea levels fuelled by global warming is now leading to calls for new and even higher flood defences to be constructed to protect London.

For many people the traumas of that terrible winter's night in 1953 soon faded but some never recovered from it and were haunted by memories and vivid flashbacks for the rest of their lives. One survivor wrote "People tell me I am silly and that I should forget, but how can I sweep from my mind that awful huge wave that engulfed our home? I still relive it whenever we have a heavy storm." Reminders of the disaster also turn up even today. One farmer in Norfolk, when ploughing his fields, still occasionally unearths bottles, combs and other items from the village shop that completely disintegrated as the seas roared in that night.

****

On 8 December 1954 a rash of tornados ripped across the South from Hampshire to London. They left a trail of destroyed houses, outbuildings, greenhouses, cars and caravans in their wake. In London, the Gunnersbury railway station was destroyed along with a row of neighbouring houses, shops and a factory.

In mid-July 1955, many places in the UK recorded their wettest week since records began and the record for the wettest day ever was set on Monday 18 July, when thunderstorms became stationary over Martinstown in Dorset. Eleven inches – 279 mm – of rain fell in 15 hours, causing widespread flooding in which two people were killed. A few holidaymakers in Cornwall were given a temporary distraction from the torrential rain during the weekend when a waterspout appeared in Widemouth Bay.

On 29 July 1956, storms lashed the South-West and eleven people were killed by falling trees.

On 6 February 1958, a snow and ice storm at Munich had devastating consequences for Manchester United and British football, as the aircraft bringing the United players back from a European Cup tie suffered icing to its wings and crashed on take-off, killing half the passengers including eight members of the United team and most of the coaching staff.

The heaviest hailstone ever measured in Britain fell out of the skies during a thunderstorm at Horsham in Sussex on 5 September 1958. It weighed in at five ounces.

On 3 December 1960 six inches of rain fell across parts of Wales. Streams and rivers in the Brecon Beacons burst their banks and towns and villages across Glamorgan were flooded.

Ford Zodiac in trouble in 1960 – a year that ended with many streams and rivers bursting their banks. Six inches of rain fell across parts of Wales on 3 December 1960.

1962 – The 'Storm of the Century'

On 16 February 1962, a ferocious gale in Yorkshire – 'the storm of the century' – devastated the cities of Sheffield, Leeds and Bradford. Barometric pressure at the centre of the depression dropped close to the British record low of 925 millibars and the precipitous pressure gradient produced savage winds. It was the second ferocious gale to hit the area in four days. On 12 January, 10,000 homes in Leeds had been damaged when thunder, lightning, hail, snow and torrential rain accompanied the gales and "the sound of breaking glass" continued for hours in the city centre. Rumours spread that the storms had been caused by the effects of Russian nuclear tests in the Arctic, but those stories were dismissed by experts at the Met Office.

The storm, breaking in the early hours and continuing through most of the day, ravaged the whole of Northern England. Small children were blown off their feet by the force of the storm winds as they struggled to make their way to and from school in the teeth of the gales. Thousands of car alarms were triggered as parked cars were rocked by the howling winds, tens of thousands of people were without electricity and their telephone service as phone lines and cables were severed by falling trees and collapsing poles and pylons. Two pylons carrying 12 inch high-tension electricity cables at Jarrow and Howdon on the banks of the Tyne were felled. The pylon at Howdon demolished a time office and severely damaged a warehouse. The electricity cables it had been carrying fell into the Tyne, paralysing shipping in both directions for six hours.

In Middlesbrough three brand new cranes at Dent's Wharf were blown over in a sustained gust which an eyewitness said had "lasted at least twenty minutes." One of the cranes "swallow-dived into the water and the other two crashed and locked together." At the coastguard station at Seahouses on the Northumberland coast, the wind blew the needle off the wind recording graph, showing that the gust must have been at least 90 miles an hour. Off the coast between Hull and Grimsby, two Mansfield men spent a terrifying night drifting helpless in a small yacht after their engine failed. They were eventually rescued by the Spurn lifeboat.

The 'Storm of the Century' devastated many cities, including Bradford where pylons were wrecked at Bradford Park Avenue football ground.

In Leeds, platforms at the City Station were closed because of the danger from glass falling from the roof. Outside the station, part of City Square was also cordoned off and office workers were evacuated after the central tower of the Royal Exchange building began swaying perilously in the wind. Elsewhere in the city, a foreman of a four-storey warehouse heard the walls "groaning" and refused to let the workers go inside. His caution was justified when the upper half of the building collapsed. Hundreds of workers were evacuated in the nearby city of Bradford when a mill chimney in the Manningham district began swaying ominously.

Despite the storm, racing went ahead at Catterick racecourse, but five horses were killed

Harrogate more than lived up to its reputation as 'breezy' during the ferocious storm. Caravans at Bilton on the outskirts of the town were no match for the extreme conditions.

while competing in novice hurdles and chases. Chastened officials and jockeys later admitted that "the gale was an added obstacle for inexperienced horses." In Harrogate the gable end of the Registry of Births, Marriages and Deaths collapsed. Hundreds of "stately trees" in the town were blown down and a 200 yard avenue of pines was devastated. Forestry plantations in the Washburn Valley were said to be "in havoc" and at Savile Lea Nurseries near Halifax, a 100-foot greenhouse containing 16,000 spring bulbs was completely wrecked.

At York firemen had to brave the gales and remove a weathervane from the spire of St Helen's church and, in Wakefield, the Town Hall clock was out of commission after two of its faces were blown in by the winds. In South Yorkshire, Rotherham police rescued men's suits that were being "tossed about the street outside a tailor's shop" and a policeman was reunited with his helmet. It had blown onto the roof of the parish church in another gale the previous weekend but had then been dislodged by the fresh storms and had been handed in at the Rotherham police station.

At Huddersfield Town's football ground, where two floodlight pylons had already collapsed in the previous gale, the remaining towers were swaying so dangerously that they had to be demolished and Town's match the next day was postponed. Part of Sheffield United's floodlighting was also destroyed.

The worst devastation occurred in Sheffield where a "lee-wave effect" – where turbulent air accelerates as it descends from high ground – compressed the air into a series of waves and troughs, heightening the impact of the storm winds and gusts and leaving the city "facing a problem comparable with that of a wartime blitz." Between 6 am and 12 noon there were gusts of 80 and 90 miles an hour and one was measured at 96 mph – the highest wind speed ever recorded in the city. A 100-foot tower crane on the site of Sheffield's new College of Technology twisted and then collapsed. Church Street – one of the city's main thoroughfares – was closed because of the danger from falling coping stones, uprooted trees blocked many other roads and whole streets were completely blocked by rubble, slates and loose wood.

One hundred of Sheffield's 250 schools sustained storm damage and almost two-thirds of the city's entire housing stock of 163,000 houses had some structural damage, with one hundred completely destroyed, leaving 250 people homeless. Prefab houses were inevitably the worst affected; an entire row at Arbourthorne on high ground overlooking the city was devastated, leaving thirteen completely uninhabitable. In Errington Road, terrified householders began fleeing their homes at 5.30 in the morning – half an hour before the full force of the storm struck the city. Soon afterwards, the prefabs began tumbling one after the other, "like a pack of playing cards". A 14 year-old boy, Leslie Wiestka, heard the wind and "felt the whole house rock as if it was a box, so I rushed into my grandmother's bedroom. Just then the whole roof caved in. I held it on my shoulders and Granny scrambled out underneath. When I was sure she was safe, I let the roof fall. I went down underneath it but I wasn't hurt. I crawled out." Minutes late the whole building collapsed. His grandmother, Mrs Anna Medley, whose husband was at work when it happened, said "Leslie saved my life. When the wind roared through the house, I was too frightened to move. It was the worst thing I've ever known."

Arthur Haydock, a steelworker who lived round the corner in Errington Avenue, risked his life in his frantic attempts to save his own prefab house. He lay at full length on the roof of his house, in the hope that firemen would be able to secure it with ropes. "The roof was heaving up and down, so I climbed up there and sprawled flat out. I was there several minutes and then a fireman dashed up and yelled at me to come down. I climbed off and not a moment too soon. Seconds later the roof caved in with a crash."

Arthur Twigg, a steelworker who lived in the neighbouring Northfield Road, where another dozen houses were flattened, said, "I have never seen anything like it. The wind whipped up to an unbelievable pitch and in moments the whole row of houses went down like skittles. They seemed to be made of paper." Seventeen house roofs were also blown off in the nearby town of Barnsley.

The dead included a nineteen year-old farmer's wife, Beryl Dickinson, who was killed asleep in bed, when the gable end and chimney stack of the farmhouse collapsed. Her twenty year-old husband suffered multiple injuries but survived and their two month-old baby daughter, sleeping in her cot in the same room, suffered only minor injuries. Among the other dead were a 22 year-old Sheffield man, who was walking past an already badly damaged building in the city centre when a huge coping stone dislodged by the gale fell 40 feet and struck him, and a seventeen year-old who was killed by a falling chimney stack as he slept. Rescuers including his father, a neighbour and two police officers, tried to reach him but were themselves trapped when the bedroom floor collapsed, plunging them into the room below.

Shirley Hill, the thirty year-old wife of the vicar of St Thomas' Church, Firth Park, in the Brightside district of Sheffield, also died when the chimney of the vicarage blew down, smashing through the roof and upper floor and crushing her as she sat in an armchair in her living room. Her husband, who had been sipping tea with her only seconds before, was also buried by rubble and the broken beds, furniture, floorboards and ceiling plaster brought down with it. He struggled free and climbed out of a shattered window to summon help. In shock and bleeding, he staggered next door to the home of another vicar, the Reverend Ronald Thomson, and they made frantic attempts to rescue Mrs Hill. "We burrowed with our bare hands," the Reverend Thomson said, "but it was hopeless and we were in danger of being trapped by more falling masonry. We had to wait till the fire brigade came, they soon got her out but it was too late; she was dead." The Hill's two year-old son, Simon, was in bed upstairs at the time, but escaped without injury. Another Sheffield woman, Ida Stabbs, was killed by falling masonry; her husband Eric suffered severe head injuries and was taken to hospital where he was fighting for his life, not knowing his wife's fate.

In Bradford, a 74 year-old man was blown into the path of a car and taken to hospital with head injuries. Even more tragically a 29 year-old husband and wife were also killed in the city, leaving their 4 year old daughter and 7 year old son orphans. In Yorkshire alone, thirteen people were killed in the storm and hundreds injured, the majority struck by falling masonry and flying debris. Estimates of the damage ran to tens of millions of pounds.

All that remained of two of the three massive cooling towers that collapsed at Ferrybridge 'C' Power Station during strong winds on 1 November 1965. Remarkably there was no loss of life, primarily because it was 'tea-break' time for the workforce!

1965 – Ferrybridge Tower Collapse

On 1 November 1965 storm winds with gusts of up to 100 miles an hour lashed Britain and left a trail of death and destruction behind them. A gust of 101 miles per hour was recorded at Douglas Moor in Lanarkshire and a hurricane force gust of 117 mph was recorded not far away at the Ministry of Aviation's radio station on Lowther Hill, also in Lanarkshire.

In Belfast the storms tore the 34,000 ton Cunard liner Caronia from her moorings at the Harland and Wolff shipyard, where she was undergoing a refit, and she was blown across to the opposite bank. Off Malin Head, County Donegal, five crewmen on the Norwegian naval frigate Bergen, on their way to Londonderry for anti-submarine training, were washed overboard by mountainous seas. An air-sea rescue search involving ships, aircraft and the Portrush lifeboat was mounted and the crew of a Royal Canadian Air Force aircraft spotted a raft with a man on it who waved to them. They kept it in sight for fifteen minutes but then it was overturned by a huge wave and the next time they saw the raft it was empty. None of the five men was rescued. As winds lashed Prestwick airport, an American jet, inbound from New York, overflew the airport and then carried straight on to Berlin.

Both carriageways of the Forth Road bridge were blocked after two pantechnicons were blown onto their sides and drivers of other high-sided vehicles were turned back from this and many other high level bridges. In the Lake District, Kendal police warned of dangerous conditions on the top of Shap Fell, 1,600 feet above sea level. Four drivers of removal vans who ignored the advice had their vehicles blown over. On Windermere, houseboats and pleasure craft were blown aground and badly damaged and near Coniston nine young climbers were "blown into the air" by gusts. One was taken to hospital with severe back injuries.

There were many other casualties and fatalities. An 86 year-old widow from Haslingden in Lancashire was killed when a gust of wind blew her under the wheels of a lorry. Five people were injured in Liverpool after the winds caused the collapse of scaffolding in Lime Street. A plasterer was left hanging by his fingertips for ten minutes from a ledge 35 feet above the ground until a ladder was brought. One man was killed and another seriously injured when a thirty foot wall collapsed at a Salford engineering works that was under demolition. Co-workers and firemen tore at the rubble with their bare hands to try and reach the two men.

A giant crane was blown down in Newcastle. It fell on a car-park, writing off a number of vehicles. On Teesside a giant oil-rig, the Ocean Prince, broke free of its moorings at South Bank, Middlesbrough, and drifted across the Tees "taking a runaway journey among shipping" in the gale. The massive steelwork platform, 300 feet long, 217 feet wide and 150 feet high, collided with and damaged two other vessels and narrowly missed a third which was carrying 2,000 pounds weight of gelignite. It also pulled down high-voltage cables and damaged two jetties. The river was closed to all shipping as the rig blocked the shipping lanes. Big ships could not pass and smaller ships would have been in great danger if the rig began to move again. As attempts were made to corral the runaway rig, thirty shipyard workers trapped on board, who described the experience as "a nightmare", were rescued. Twenty of the men climbed 150 feet down a narrow ladder and jumped another twenty feet to reach a rescue launch. The others were eventually taken off after nearly twelve hours. It took nine river tugs 24 hours to bring the errant rig under control.

Ten people in a rush-hour bus queue in Wilton Road by Victoria Station in London were hospitalised after being hit by corrugated iron sheets blown from the station roof, and at Rochford in Essex a man was killed after being blown off a house roof.

The most spectacular damage caused by the storms was at Ferrybridge, near Doncaster in South Yorkshire, where the gales triggered the collapse of three of eight cooling towers at the £68 million pound Ferrybridge "C" Power Station. The other towers, though remaining upright, sprang cracks and sustained severe structural damage. Europe's largest coal-fired power station, Ferrybridge C was not yet operational but was scheduled to begin generating power the following spring. When fully operational it was designed to burn five million tons of coal a year.

The towers, each 375 feet high – the tallest in Europe – and weighing 8,000 tons, had been built only three years before. They fell down "with a roar like thunder", raising a vast dust cloud as thousands of tons of concrete came crashing to the ground. Fire and ambulance crews raced to the scene but astonishingly only three people needed treatment for relatively minor injuries. One, a seventeen year-old labourer, Bert Wilkinson, was thrown out of one of the towers seconds before it collapsed. "I opened the door and was met by a cloud of dust, and then the door swung and knocked me out of the tower. Seconds later the tower collapsed inwards."

The Great British Tea Break was credited with saving hundreds more men from injury or death. Almost the entire 5,000 strong workforce, who ten minutes earlier had been working in and around the base of the towers, were on a tea break when the disaster occurred. One of the workers who saw a tower collapse said "It just disintegrated. The gale blew a great mass of debris all over the place. It all blew away in dust, like a shower of pepper." A fifteen year-old boy, Michael Law, also saw it come down. "I was looking out of the window of a hut. I saw about a dozen men run out of the cooling tower and scatter. Then the whole lot fell inwards with a tremendous roar." Soon afterwards a second tower fell and "the roof of the boilerhouse started shifting, with pieces of aluminium blowing everywhere in the terrific gale." A third tower was then seen moving on its foundations and it too collapsed.

An anxious watch was maintained on the remaining towers, one of which had a large crack running down it, and on the giant twin 650-foot chimneys. Roads around the site were closed and the owners of a farm half a mile from the site were evacuated because of fears of flying debris.

The same gales that had destroyed the cooling towers at Ferrybridge C also caused damage and injuries throughout the country, but although the high winds at Ferrybridge had triggered the collapse, the subsequent official inquiry found the cause to be faults in the design. The vast towers had greater diameters and surface area then any previous ones, but the fault lay not in their size but in the wind tunnel tests that had been used to test their resistance. The wind speeds specified in British Standards had not been used in the tests, leading to the designed wind pressures at the top of the towers being one-fifth lower than they should have been. The designers had run the tests using an average wind speed over a one minute period, but in reality the towers would be subjected to much shorter but more powerful gusts. Even worse, the wind tunnel tests had been run on a single isolated tower without any consideration of the funnelling effect on the wind of the eight towers sited together and the turbulence and local eddying created on the leeward towers – the ones that collapsed. The disgraced designers were fortunate that no one was killed or injured in the accident; the losses, though very substantial, were purely financial.

The wreck of the 61,000-ton Torrey Canyon, broken in half and slowly sinking after striking Pollard's Rock off Land's End on 18 March 1967. Thousands of tons of oil were released into the sea in what was then the world's worst environmental disaster.
(S&G and Barratts/PA Photos)

1967 – The Torrey Canyon Disaster

The supertanker Torrey Canyon struck Pollard's Rock in the Seven Stones Reef between the Scilly Isles and Land's End on 18 March 1967. At 974 feet (297 metres) long, she was the first of the big supertankers, and was carrying a cargo of 120,000 tons of oil. The crew of the Seven Stones lightship, two miles off the reef, said they realised the tanker was in danger when she was still a mile from the disaster site, but by then neither they, nor the captain, Pastrengo Rugiati, could do anything to avert the disaster.

Although the adverse weather might have been partly to blame, the principal cause of the disaster was the disastrous navigation error of the captain. Under pressure to catch the high tide at Milford Haven at 11 o'clock that night or face a wait of six days for the next one – the size of the ship made it impossible to enter the harbour on a lower tide – the captain opted to take a short cut between the Scillies and Land's End, rather than the long way round on the seaward side that would have added two hours to his voyage.

There is a navigable channel to the east of the islands, but it is narrow for a ship of that size, and when it strayed off course, it ran onto the rocks. Although the captain and three of his crew initially stayed on board, all the crew of the Liberian-registered supertanker were eventually rescued by helicopters and lifeboats. Pounded by heavy seas, their vessel remained impaled on the rocks for eleven days as its cargo of thick crude oil steadily leaked out into the water, forming a vast slick 35 miles long and up to 20 miles wide that threatened environmental disaster for the beaches of south-west Britain and north-west France.

As the Torrey Canyon began to break up, leaking even more of its deadly cargo into the sea, in desperation the Labour government of Harold Wilson called in the armed forces to bomb the tanker to sink it and ignite the oil slick floating on the ocean. To the embarrassment of both the military and the government, despite raining down 62,000 pounds of bombs, 5,200 gallons of petrol, eleven rockets and huge quantities of napalm onto the stricken ship in the course of the day, the Torrey Canyon stubbornly refused to sink.

The RAF and the Royal Navy and their bomber pilots also came in for further ridicule, because even though they were attacking a completely stationary target – and one that certainly was not firing back – no less than one in four of the bombs they dropped missed the target altogether and detonated harmlessly in the sea. To complete their humiliation, the mission then had to be called off for the day when the rising Spring Tide swept over the burning tanker and extinguished the flames coming from it.

The bombing raids resumed at first light the next morning and once more proved a popular attraction for the tourists and locals who lined the cliffs to watch the bombs. The towering column of smoke and flame rising from the ship could be seen up to 100 miles away. The Torrey Canyon finally sank that day, 30 March 1967.

The oil had polluted beaches along the entire coast of Devon and Cornwall from Hartland Point in North Devon to Start Point, south-west of Dartmouth on the Channel coast. Across the Channel in Normandy, beaches were also covered in the thick brown sludge, killing thousands of seabirds and extinguishing virtually all marine life in the affected areas.

From the start of the disaster, dozens of ships had also been spraying the oil with detergent – 10,000 gallons was shipped from Grangemouth in Scotland alone – in an unsuccessful attempt to disperse it. However, the use of detergent merely triggered an even worse environmental catastrophe. A later report into the effects of the disaster on the marine environment found that the detergent killed far more marine life than the oil. Despite a prodigious expenditure of money and effort, the oil slick was only finally dispersed by favourable weather conditions, not by any efforts of man, military or otherwise.

An inquiry in Liberia, where the ship was registered, duly found the captain negligent in taking a short cut to save time in getting to his destination of Milford Haven in Wales.
The Wilson government was also strongly criticised for its handling of the incident, the costliest shipping accident and the world's worst environmental disaster at that time.

* * * *

On 15 January 1968, storm force winds funnelling down the Forth-Clyde valley battered Glasgow. The tall chimney stacks and steeply pitched roofs of the old tenement districts proved particularly vulnerable to the winds and twenty people were killed, mainly as a result of being struck by falling slates and masonry.

In the first week of July 1968, thunderstorms broke out across Yorkshire with such savagery that there were reports of paving stones being pitted by hailstones. Drifts of hailstones five feet deep accumulated in the streets of Bradford and the entire city centre was flooded. Bulldozers were used to clear the mounds of hailstones from the roads. In Harrogate windows were smashed, crops flattened and hundreds of birds killed by hailstones "the size of golf balls." Near Leeming, almost one and a half inches of rain fell in just eight minutes.

On 15 September 1968 the south-east of England was doused by thunderstorms that dropped record rainfalls over the following 24 hours. 6 inches of rain fell across Essex, Kent and Surrey, and Guildford town centre was flooded to a depth of 8 feet, but Tilbury topped the rainfall chart with a whopping 7.93 inches in 24 hours.

The wreck of the Torrey Canyon *dominated weather events in 1967, but a more local tragedy occurred at Wray, near Lancaster, when an August cloudburst on the Bowland fells south of the village turned the normally placid River Roeburn into a raging torrent. Many cottages were demolished in a disaster that had haunting echoes of Lynmouth fifteen years earlier.*

Wallace Arnold in deep water! The dying days of 1967 saw buses about to give up the unequal struggle on the Castleford to Hook Moor road at Allerton Bywater.

1969 –
Collapse of Emley
Moor Mast

Two kinds of weather can cause thick ice to form on any exposed object. Freezing rain occurs when rain falling out of the clouds hits a frozen surface and immediately forms glazed ice, creating skating-rink conditions on roads and pavements. Rime forms when super-cooled droplets of water freeze on contact with any exposed surface, from a blade of grass to a massive steel television mast. In certain conditions, water can remain liquid at temperatures as low as -40°C, but when it comes into contact with any object it will then freeze instantly, even if the object itself was previously at a temperature above zero. Whether glazed ice or rime is responsible, however, the effects are dangerously similar, hazardous conditions for pedestrians and drivers alike, and potentially dangerous accumulations of ice on tall, exposed structures like TV masts.

On 19 March 1969, an ice storm swept in from the north-west, blowing frigid, polar air over the country. The super-cooled rain, mist and cloud streaming over the icy Pennine moors on the wind froze as they touched any surface – roads, pavements, fields, trees – and where the wind blew over exposed, open land, it formed horizontal icicles on the downwind side of any obstacle. Tree branches snapped under the weight of ice, and the accumulations of ice on power cables, phone lines, pylons and telegraph poles were so great that they caused "cables to sag within arm's reach" or snap altogether. Pylons bowed and twisted and telegraph poles snapped off at ground level. "Imagine four hundredweight sacks of coal swinging on the wires between each telegraph pole," a Yorkshire Electricity Board spokesman said, "and you'll have some idea of the weight of this ice." When the ice fell, the lines whipped back into the air, causing vast fluctuations in voltage known as "dancing conductors." Exhausted power engineers working day and night and "shod with climbing irons, wriggled up hundreds of icy poles on the moors and high ground" in an unceasing battle to reconnect electricity supplies.

At Emley Moor, 850 feet above sea level on a Pennine hilltop near Huddersfield, engineers were working as usual at the giant television mast. The 365 metre (1,265 feet) mast – a 2.75

metre diameter steel tube braced by giant steel cables two and a half inches thick and tethered to massive blocks of concrete buried deep below ground – had been erected in 1964 to broadcast television pictures to up to seven million viewers. At the time of its construction, it was one of the tallest standing structures in the world.

In winter, the cylindrical steel mast was regularly coated with ice that formed on the mast itself and on the supporting cables. Such icing occurs when a current of cold air is overlain with warmer air. Rain falling through the cold air is then cooled rapidly as it reaches the layer of colder air and immediately turns to ice when it comes into contact with any solid object, from blades of grass to the giant Emley Moor mast. A Met Office spokesman added "It does not take long to build up a thickness and ice is nine-tenths the weight of water; it soon begins to have an appreciable effect." Icing could load huge amounts of additional weight onto the Emley Moor television mast, but it had been designed to withstand the worst weather conditions ever recorded in Britain – it could sway up to seven feet from the vertical in winds of up to 110 miles per hour – and was considered absolutely safe.

The only acknowledged hazard to the safety of the inhabitants of the tiny village of Emley Moor around the foot of the mast came when huge chunks of ice fell from the mast and went crashing to the ground, and the minor roads past the site were often closed when this was occurring. An engineer estimated that a 3 foot length of the stay-cables holding the mast could carry as much as one hundredweight of ice – and there were thousands of feet of stay-cables bracing the mast. Bars of ice up to 7 feet long and 6 inches thick had fallen from the mast in the winter of 1966-67, and in the winter of 1969 the problem of falling ice was even worse.

Jeffrey Jessop, one of the inhabitants of a row of houses in Jagger Lane, Emley Moor, less than 100 yards from the mast, said that they had been "absolutely terrified this winter. We have been terrified to go out of the house and even inside we are scared stiff as we hear the whistle of pieces of ice raining down around us. It has to be experienced to be believed. The whole house shakes and the fall of ice can go on for half the day. It is not unusual to see pieces six or seven feet long come crashing down around you. I refuse to go out of the house without my old motorcycle crash helmet, which I have kept for this purpose. Even then, when it starts falling you 'run the gun' if you are anywhere near the mast." His neighbour, Brian Wharan, said that "when I am out at work, my wife and two small children go to my parents' home about a quarter of a mile away and stay there. It is not safe for them to stay in."

After fierce complaints from the local council, the body ultimately responsible for Emley Moor, the Independent Television Authority, had installed warning signs on the approach roads and placed a flashing amber warning light half way up the mast, as a danger signal when ice was falling. This was little comfort to the people living in the shadow of the mast, who claimed that pieces of ice weighing as much as 6 hundredweight had been known to fall from it, and several houses had had their roofs pierced by smaller pieces.

The Bawtry weather station had issued a warning of the danger of icing three days before - "moderate or heavy icing will occur in freezing rain" – but there was no possibility of de-icing the Emley Moor mast as no provision for this had been included in the design brief when it was built. A spokesman for the Independent Television Authority said that they had "sought advice from other countries with the same problem and done our own research, but we have not come up with a suitable preventative system. It seems we are stuck with it and all we can do is try to minimise the danger to people and property. Short of taking the mast down, there seems little else we can do." A senior civil engineer supported this view. "It's difficult to know what, if anything, could have been done. You can't chip ice off a structure like that and there is no way of melting it. When icing sets in on that scale, I don't see anyone can do much except pray."

For the previous two or three days the ice storm had been building layer upon layer of ice on the mast and the surrounding moor. Elsewhere icing on power cables had caused havoc with electricity supplies at Whitby and on high ground throughout the North, and massive accumulations of ice had already caused the collapse of an 187 foot VHF radio tower at Rivington Pike near Bolton.

As chunks of ice kept falling from the Emley Moor mast, some local people "had quit their homes and gone to stay with relatives because of the danger." Mrs Muriel Truelove, the licensee of the village pub, the Three Acres Inn, said the ice had been "terrible" and she refused to let her daughter walk from the bus stop to the inn because of the danger to her as she passed close to the foot of the mast. The shrubs and trees around the station looked like "ships in bottles, they were like bubbles of ice, with the branches lying down on the ground." Thick mist and low cloud had hidden the upper part of the mast from view for several days, but one of the engineers at the station said that the day before, the mast "had been seen to be bending under the weight of ice."

On 19 March, the temperature rose above freezing for the first time in days and the roads were closed at one o'clock that afternoon because of the renewed threat from falling ice. The duty engineer advised the police and local authorities of the danger and the town surveyor of the Denby Dale district at once put up road blocks on Jagger Lane and Common Lane, leading to Emley Moor. His opposite number in the neighbouring district of Kirkburton also closed off the roads at his end.

As they worked, the engineers at Emley Moor could hear the regular crash of lumps of ice shattering on the ground. One of them picked up a piece of ice and put it on the scales; it weighed fifteen pounds. Jeffrey Jessop, the caretaker at the Emley Moor Methodist Chapel in the shadow of the mast, also heard the "bangs and crashes as ice fell off the tower and the guy wires twanged and whistled in the wind." All day he had also heard "strange noises coming from the mast" but he was so used to "strange noises from above" by then that he ignored them.

By 3 o'clock in the afternoon it was snowing and ice was no longer falling from the stay-wires but, after discussing it with his colleagues, the chief engineer on the shift, David Lee, decided to leave the evening shift that came on duty at 5 pm to decide whether to reopen the roads to traffic. A team of Post Office engineers had to dig out their vehicle before they left the site at about 4.30, followed by the engineers being relieved by the men on the evening shift.

At three minutes past five there was a bass rumble that shook the floors of the station buildings and every house in the area and a crash that seemed to go on for ever. The first thought of some was that an aircraft had hit the TV mast – it was directly under the flight path into Manchester airport – but in fact the mast had come crashing down on its own, with a thunder that was heard for many miles around.

The massive steel structure had crumpled like paper and as they fell, the hundreds of tons of steel and ice demolished many of the station buildings. The mast came down with such force that the part of the road it fell on was pulverised and craters 10 feet deep were gouged in the nearby fields. One of the flailing stay-wires sliced straight through the chapel, cutting a vertical line through the roof, walls and pews. The nearby cottages, including the caretaker's house, were also damaged, though the top section of the mast, made of "meshed steel girders clad in fibreglass, missed two cottages narrowly as it pitched into a field a third of a mile away." Roy Simpson's farmhouse also had a narrow escape, but his farm trailer, a tree and a hedge were demolished by falling debris and blocks of ice.

The chapel caretaker, Jeffrey Jessop, and a trustee of the chapel, Mr Silverwood Burt, were inside the chapel when it was struck, inspecting the damage to the roof caused by ice falls earlier in the day. "The ice had made eleven holes in the roof in the morning," Mr Burt said, "and more holes were made in the afternoon. Some of the pieces of ice weighed three quarters of a stone. I was leaving the vestry when the mast collapsed. I felt as if the whole building was falling around me. Slates, woodwork and plaster cascaded over me. Luckily I was wearing a crash helmet I borrowed a fortnight ago because of falling ice. It probably saved my life. I was imprisoned in smashed woodwork and I had to scramble over the pews to get out of the front door."

Mr Jessop had been at the other end of the chapel. "There was a crack like thunder," he said, "and a shower of ice hit the roof. I looked up and I saw one of the mast's supporting cables come through the roof. I dived for cover under a pew because I thought the roof was coming down on top of us. I laid there until it was all over, saying a lot of prayers. We climbed out of the rubble and went outside to see if anyone was hurt." He ran to his home to find his wife and family shocked but unhurt, and his only injury was a cut to his hand, sustained when he dived under the pew. There were no other casualties but the interior of the chapel looked as if a bomb had exploded in it and 24 hours later huge blocks of ice were still littering the churchyard.

A busload of schoolchildren also had a very narrow escape. Elizabeth Walton, an 18 year old pupil, was among those going home on the bus, but they had to get off just before Emley Moor because "the police had stopped traffic from passing the mast because great lumps of ice had been showering down. We ran as fast as we could past the mast, then when we started to walk again, we heard a fantastic noise, just like a jet thundering past only ten feet above our heads. The mast had come down." The collapse left 7,000,000 television viewers with blank screens – ironically just as a programme entitled "Do Not Adjust Your Set" was about to be broadcast.

David Lee had been one of those who had left the site shortly before the collapse. As soon as he realised what had happened, he returned to Emley Moor. By the time he got there at about 6.15pm, the police had cordoned off the whole area. Security guards with dogs were already patrolling to deter people from stealing the scrap metal; in addition to the steel, the feeders – the heavy wire conductors – were solid copper.

Police, led by Assistant Chief Constable David Bradley, and fire crews searched for any sign of bodies or injured people under the crumpled steel sections of the fallen mast strewn across the site and littering the roads outside. "I remember thinking they wouldn't find anything because I left the roads closed to traffic," David Lee said. "Of course that wasn't quite true, because whilst cars wouldn't go through because of the illuminated signs closing the roads, you'd still get pedestrians who took a chance. If I had opened the roads, as was suggested during the day, the local school bus would have been parked on the corner by the chapel at one minute past five. It always waited for a few minutes from five o'clock on school days,

The shattered remains of the 1,265 foot (365 metre) Emley Moor television mast, near Huddersfield, after its spectacular collapse during an ice storm on 19 March 1969.

and it would have been there just after five when the mast came down on the corner."

Within a few hours of the collapse "breakdown crews were hauling the mangled wreckage away." Police kept the roads closed to non-essential vehicles all night to aid the search and recovery work and to keep away "sightseers who crowded in through the fog."

By a cruel irony, a meeting between the local council and the Independent Television Authority had already been scheduled for 20 March – the day after the collapse – to discuss the concerns of residents about "the safety factors involved in living in the shadow of the 'ice monster'." The meeting was to consider a proposal that those living near the mast, including eight couples in Jagger Lane, should be rehoused and the ITA had already offered to buy the affected properties. Work had also begun on a steel 'umbrella' to shield the chapel "which had often been bombarded by falling ice."

That same day the television mast at Belmont, near Louth in Lincolnshire, a "twin" of Emley Moor, caused fresh consternation when the top 150 feet of the mast was seen to be bending out of the vertical by two foot six inches to three feet because of heavy icing. An all day and all night watch was kept on the mast in fear of another collapse. The engineer in charge, James Clarke, said he was "getting in touch with our design people to get their advice. Obviously the situation is one we don't like. The icing is exceptional and it is one of those freaks that it should occur on two masts. We have to take every possible step and be extremely careful." Just as at Emley Moor, ice had been falling from the stays for several days and it was impossible to see the upper part of the mast for much of the time because of mist and low cloud. Television transmissions continued but all members of the station's staff wore safety helmets, ice warning lights flashed continuously on the mast and, having announced that it was "slightly possible" that the mast would collapse, the police sealed all approach roads to the site.

An ITA spokesman said that Emley Moor "was the first structure of its kind in the world and no-one had any idea that this danger might occur. No one had any previous experience of this type of mast and so we had no idea that this might happen. If anyone has any suggestions for a remedy to it, we would be only too pleased to hear from them."

It was assumed that the sheer weight of the massive accumulations of ice on the Emley Moor mast had caused its collapse, but the committee of enquiry that investigated the disaster instead cited an oscillation that gradually developed in the mast in the constant wind blowing over the moor on that day. The oscillation had steadily increased in strength until it had caused the tower to fail. As a result, several other masts of similar design were made safe by suspending heavy iron chains to act as dampers, but when engineers came to design a 330 metre (1.083 feet) replacement mast for Emley Moor, they opted for a stronger, curved concrete tower with a much shorter lattice-steel mast on the top. Although the chairman of the Denby Dale Housing Committee had told a local newspaper "I am absolutely shattered by the catastrophe, I shall never agree to the mast being rebuilt," the new mast was duly erected on the site, albeit to a different design, and still stands there today.

* * * *

On the night of 28-29 September 1969 a severe gale sent a storm-surge careering down the North Sea coast, causing floods that left the centre of Hull under three feet of water. As the pent-up waters receded when the gale passed on, the tide went out so far that the Humber Ferries were cancelled because of the lack of water in the Channel.

On 9 September 1970 a tornado roared across Lincolnshire, ripping roofs from houses and exploding greenhouses. A Post Office van, minus its driver who was delivering letters at the time, was picked up, carried 30 yards and then deposited again without a scratch on it. The occupants of a henhouse were less fortunate when it was whirled 50 yards by the twister and smashed to pieces.

Rawmarsh near Rotherham in South Yorkshire was hit by a tornado on 26 September 1971. It caused damage over a 12 mile track before petering out. The most surprised victim of the

Less than two weeks after destruction of the Emley Moor mast, the weather was still in angry mood. Firemen and council workers try to keep floodwaters at bay with sandbags.

twister was probably an Alsatian dog, sleeping peacefully in its kennel when the tornado picked up the kennel and sent it flying over a tall fence to land in a new site in the next door garden.

On 2 June 1975, one of the most unusual entries ever recorded was entered in the scorebook of the County Championship cricket match between Derbyshire and Lancashire at Buxton: "Snow stopped play." Two inches of snow fell and play was abandoned for the day. Sleet was also reported as far south as London and the South Coast.

Evidence of how localised torrential rain and flooding can be was given by a thunderstorm on Hampstead Heath on 14 August 1975. While the downpour dumped 170.8 mm (almost seven inches) of rain on the Heath in 155 minutes, only 5 mm fell two miles away, while 10 inches of hail fell elsewhere in London. Within five minutes of the start of the storm on the Heath, a hundred homes were flooded to a depth of over a metre. One man drowned, the Underground was flooded and manhole covers were blown off by the force of the floods, filling the streets with raw sewage.

On 22 August 1975, several waterspouts occurred off the coast of East Anglia. One at West Mersea, Essex, whipped a dinghy out of the harbour and flung it 450 feet into the air.

On 29 February 1976 the Leap Year Day weather included a tornado in Hull, accompanied by thunder, lightning and ferociously heavy rain. The twister tossed caravans around and moved a 500-tonne crane at Hull docks almost 20 yards along its track.

Ferocious thunderstorms across the South-West on 6 December 1976 generated three tornados, damaging roofs, buildings and vehicles right across Cornwall, Devon and Dorset. One massive roof beam weighing almost 500 pounds was hurled 30 yards by the twister.

The next day, 7 December, another tornado, one of the most damaging of the century, struck Wiltshire. A large number of buildings in Landford were destroyed and debris was carried as much as a mile away, but eight calves were left unhurt when the twister picked up their shed and carried it off.

On 3 January 1978 a tornado wreaked havoc at Newmarket. A flight of geese was caught by it and 136 of them were killed, plummeting to their deaths over a wide area around the town. The tornado was one of a crop that bedevilled an area stretching from East Anglia to North Yorkshire over a four-day period. Caravans in Scunthorpe and Aldbrough were hurled into the air and thirteen trees were uprooted at Wold Newton in East Yorkshire. Buildings in the coastal town of Withernsea suffered major structural damage and a large part of the pier at Hunstanton was destroyed by another twister on 7 January. Margate pier was the next to suffer, on the night of 11-12 January, when a northerly gale caused a storm-surge that destroyed the pier.

The village hall at Llandissilio in Dyfed was completely destroyed on 12 December 1978 by a tornado that tore the building from its foundations and scattered the debris for 300 yards around. Elsewhere in the town, house roofs were ripped off and windows blown out by the force of the twister.

1979 – Fastnet Yacht Race

Dozens of yachts were lost and fifteen people were killed on 14 August 1979 when a freak storm struck the Irish Sea during the Fastnet yacht race, one of the series of five races that make up the "world championship" of yacht racing, the Admiral's Cup.

The race had begun three days earlier at Cowes on the Isle of Wight in sunshine and calm waters but as the yachts raced westwards across the Irish Sea to round the Fastnet Rock off the Irish coast, the weather conditions deteriorated rapidly. Survivors described being hit by a "great fury" at sea as the storm force winds and towering waves battered the yachts.

Rescuers from both sides of the Irish Sea worked round the clock answering distress calls from the majority of the 303 yachts taking part, but many of the smaller craft had no radios and were unable to report their positions when they got into difficulties. It was also reported that some life rafts had broken up in the raging seas and a few safety harnesses had snapped; bodies could be seen floating in the sea.

Naval helicopters from Culdrose, a Dutch warship and French trawlers collaborated on the rescue operation, the largest one ever mounted in peacetime, scouring 20,000 square miles of sea. Well over one hundred people were rescued, but fifteen yachtsmen drowned or perished from hypothermia.

One of the few boats to reach the finish line back in Plymouth was Morning Cloud, skippered by former prime minister Edward Heath, who described it as "an experience that I do not think anybody would want to go through again willingly. It was a raging sea with enormous waves and one of them picked us up and laid us on our side."

The organisers of the race, the Royal Ocean Racing Club, were severely criticised at the time, but an official report later cleared the club of blame, though new regulations limited the number of competitors in future races, imposed minimum standards and compelled all yachts to carry VHF radios. Restrictions on the use of electronic navigational aids were also subsequently lifted.

****

Skies over Britain and most of Northern Europe turned bright purple on the afternoon of 27 November 1979. Commuters stopped in their tracks to gaze at the astonishing sight, which

Lightning is a frequent hazard for golfers, but footballers can suffer too. This FA Amateur Cup match at Shirley, Warwickshire, was halted when one of the players was struck by lightning.

only faded after sunset. The cause was a series of storms in the Sahara desert that had sent thousands of tons of fine dust spiralling into the atmosphere. Southerly winds had then carried it across Europe. Over the next few days the dust was washed from the skies by rain, leaving brightly coloured dust and grit covering the streets.

Corny comedians all over Britain received a boost on 6 August 1981 when Manchester, the traditional butt of bad weather jokes, was deluged by 4.3 inches of rain in the space of a few hours.

On 21 November 1981, the largest tornado outbreak ever seen in Britain saw one hundred and fifty twisters touch down across England and Wales in the space of just over five hours.

On 3 June 1982 a bolt of lightning just missed a golfer preparing for a putt during a round at a Sheffield course, but the blast threw him 6 feet into the air. After returning to earth and regaining his composure he resumed his round, but perhaps understandably, he failed to sink the putt.

Bristol was hit by a thunderstorm on 18 June 1982 that created flash floods in the city. These were then exacerbated as hailstones – yes you guessed it, "the size of golf-balls" – formed five-foot drifts that covered the roads and blocked gutters, preventing the flood waters from escaping. As the fire brigade pumped water out of flooded cellars, the inhabitants shovelled mounds of hail out of the clogged gutters, piling it up on the pavements and verges.

Later that unusually extreme month, on 23 June 1982, a middle-aged couple in a small boat off King's Lynn used their cine camera to film a tornado that veered off the land and grew into a waterspout. Unfortunately, their record of the event was then lost when their boat exploded as it was struck by lightning. They escaped with only minor injuries, but all their possessions, including their cine camera, went to the bottom of the sea.

On 21 September 1982 a rash of tornados broke out across south-eastern England. The most serious struck Bicester in Oxfordshire, injuring four people.

Two golfers playing a round at Drogheda, County Louth, on 7 April 1983 were treated for burns and shock after being struck by lightning. Even so, they fared better than a man crossing a nearby road who was knocked to the ground by another bolt of lightning and then, while lying there, was run over by a car, receiving serious injuries from both mishaps.

On 22 April 1983 tornados struck the English Midlands, tearing off roofs, overturning caravans, lifting greenhouses and sheds off their foundations and driving slates and roof-tiles deep into the ground. In Derbyshire, another tornado demolished a hay barn and hurled the corrugated iron roofing sheets 250 yards away.

Much of the Kent fruit crop was stripped from the trees on 4 June 1983, when storms sweeping across Dorset, Hampshire, Sussex and Kent produced hailstones "the size of golf-balls" (again) which also smashed windows and destroyed hundreds of greenhouses. Three days later, on 7 June, a single hailstone punched a 3 inch hole through the rear window of a car parked in Warrington, Cheshire.

Those complaining about the British summer in July 1983 were reminded that there were worse places to be. On 21 July of that year the lowest temperature ever recorded anywhere on Earth was set when a reading from a thermometer at Vostok, Antarctica, showed a temperature of -89.2∞C (-128.6∞F).

A spectacular thunderstorm burst over the Plain of York on Sunday 8 July 1984 with disastrous consequences for York Minster. After St Elmo's Fire had been seen illuminating the towers of the Minster, a bolt of lightning hit the 13th century south transept, melting the lightning conductors. A second more powerful strike then blasted into the roof, setting fire to the ancient timbers. That part of the roof was completely destroyed. Explanations for the

The winter of 1982/83 was notable for a rash of gales. The owners of the cars shown in these two photographs were left in no doubt about the fury of the elements.

disaster ranged from attack by UFOs to Divine retribution, but North Yorkshire Fire Brigade's report to the Home Office confirmed that lightning was the most likely cause. Restoration work took four years and cost two million pounds.

On 9 November 1984 a "sandstorm" covered much of Britain in a layer of gritty brown sand, carried from the Sahara on strong southerly winds.

Aberdeen was deluged by 10.5 inches of rain in just 24 hours on 30 November 1984.

On 5 June 1985 a vicious storm in Wiltshire lashed down hailstones with such ferocity that the bodywork of cars was dented and two people were admitted to hospital with facial cuts.

On 20 March 1986 the strongest gust of wind ever recorded in Great Britain – 173 miles per hour – was measured by an anemometer sited 1,245 metres above sea level on Cairngorm in the Scottish Highlands.

One of the most unusual weather-related perils arrived on our shores on 5 May 1986: a cloud bearing radioactive dust from Chernobyl. As it passed over Britain, heavy rain fell over parts of Scotland, north-west England and Wales. No fatalities have yet been attributed to the radioactive storm but the sale of livestock and crops from the affected areas was banned for some time.

One of the driest summers on record ended on August Bank Holiday 1986 when Hurricane Charley, originating in the Caribbean, crossed the Atlantic retaining much of its energy and burst upon the Irish coast. Warned in advance by Irish weather forecasters, most holidaymakers had already abandoned the beaches. Those who remained saw winds tear up trees, destroy tents and caravans and damage buildings. The winds were accompanied by torrential rain, with falls of two or three inches being recorded right across Ireland and Britain.

On 22 August 1987 the peace of a Saturday afternoon was destroyed by a devastating storm that caused millions of pounds worth of damages to buildings, cars and crops from Essex to the Midlands. Hailstones measuring an inch and half in diameter destroyed greenhouses, broke car and house windows and smashed a hundred windows at a Colchester hospital. Bean and sugar beet plants were stripped to bare stalks and apples and pears were split in half by the force of the hailstorm.

1987 –
The Great Storm

During the night of 15 to 16 October 1987 "The Great Storm" struck southern England with the force of a bomb. The worst storm to hit England since the Great Storm of 1703, it caused the deaths of twenty-three people and remains notorious because of the perceived failure of weather forecasters to predict the storm and warn the public.

The forecast for the week given out on the BBC on Sunday 11 October had predicted "high winds and heavy rain" for the end of the week, but by midweek the charts suggested that the severe weather would be confined to the Channel and coastal areas of southern England. The first gale warnings to shipping for sea areas in the English Channel were issued at 06.30 on Thursday 15 October. Four hours later, these were upgraded to warnings of "severe gales," but the noon general weather forecast that Thursday, spoke only of a "depression expected to track along the English Channel, producing fresh to strong winds (Beaufort Force 5 to 7)."

At that time, the depression, originating in the Bay of Biscay, was still centred well to the west of La Rochelle and deepening to 970 millibars. At about six that evening, the depression suddenly deepened again to 958 millibars, with a very steep pressure gradient that was a guarantee of powerful winds.

What happened next depends on whether you prefer the truth or the legend. Michael Fish states: "According to the legend, at nine-thirty that night I prefaced the BBC Television weather forecast with the soon to be infamous remarks: 'A lady has rung in to ask if there is going to be a hurricane tonight … there is not!' I then went on to give a warning that 'actually the weather will become rather windy'. Like the other radio and TV general weather forecasts of that evening, although I had mentioned the strong winds, I suggested that heavy rain would be the main feature of that night's weather. That was the legend; the truth, as I pointed out at the start of this book, was rather different.

"I did give the weather forecast the previous morning, during which I said, 'Batten down the hatches, there's some very stormy weather on the way.' I did indeed also make the remark about there being no hurricane, but it was not in connection with the impending storm in England but a different storm over the Caribbean that was alarming the mother of a colleague who was due to fly to Florida that night. Nor was the remark made in that 9.30 evening forecast - how could it have been? I wasn't even on duty – and at no stage did any woman phone the BBC seeking reassurance."

Bill Giles was the duty forecaster that evening and his comment "it will be breezy up the Channel" and his focus on the heavy rain expected, were what provoked the later furore. But Bill was merely interpreting the advice given by the Chief Forecaster at the Met Office headquarters, then at Bracknell, and in turn, due to industrial action in France, he had been hampered by a lack of data from the source of the storm, the Bay of Biscay. Poor Ian McCaskill, whose night shift extended right through the next day as well, since no one else could get in to work to relieve him, had the thankless task of updating the forecast as the true picture emerged and fielding the barrage of questions and occasionally the abuse from the press and the public.

When Bill Giles gave the forecast at 9.30 that evening, forecasters still believed that the deepening depression would pass along the English Channel. It was unfortunate for them that the storm was developing so rapidly that they were unable accurately to predict its actual track and ferocity. It was even more unfortunate that, by the time the eye of the storm – the deep depression moving up the Channel – surprised the forecasters by suddenly veering further to the north-east, it was after most BBC viewers and listeners had switched off their sets and gone to bed.

By midnight, the still deepening depression was down to 953 millibars, but few heard the midnight bulletin revealing that the depression was now "expected to move rapidly north-east," veering over the north coasts of Cornwall and Devon and across the southern Midlands to the Wash. Fewer still heard the shipping forecast, half an hour later, that gave warnings of "severe gale conditions for sea areas Thames, Dover, Wight, Portland."

Just one hour later, at 01.30 on Friday 16 October 1987, an emergency alert was flashed to police and emergency services, including the London Fire Brigade: "Extreme wind conditions expected." Five minutes later, at 01.35, with warnings of gales of Storm Force 11 now being issued, the Met Office issued a warning to the Ministry of Defence that the anticipated scale of storm damage was such that civil authorities might need to call for the help of the armed services.

During the dark hours of that autumn night the full ferocity of the storm was unleashed right across the country from west to east. The ferocious energy driving the storm was indicated by the dramatic increases in temperature that accompanied the passage of its warm front. Increases of more than 6°C an hour were recorded in many places across the south, and at South Farnborough in Hampshire the temperature rose over nine degrees – from 8.5°C to 17.6°C – in just twenty minutes. Those rapid increases were followed by almost equally steep declines. The changes in atmospheric pressure as the depression passed over were even more startling. In just three hours between three o'clock and six o'clock that morning, instruments at the Portland Royal Naval Air Station in Dorset recorded a rise in pressure of 25.5 millibars; by a considerable margin the greatest change in pressure ever recorded in such a short time period anywhere in the British Isles.

The strongest winds were recorded along the Channel coast through Hampshire, Sussex, Surrey and Kent. In the south-east, where the greatest damage occurred, gusts of 80 miles an hour and more were recorded continually for three or four consecutive hours. Damage patterns also suggested that whirlwinds accompanied the storm. The highest wind speed on either side of the Channel was a gust soon after midnight at Quimper on the coast of Brittany that the coastguard estimated to be 138 miles an hour. The highest measured speed recorded

Dramatic scenes at Dover as waves crash against the promenade. The 'Great Storm' of 1987 caused a bulk carrier in the port to turn turtle in what coastguards described as 'murderous' seas with 'huge waves, a wall of spray and lashing seas'. (PA Archive/PA Photos)

on an anemometer was a gust of 137 mph at Pointe du Roc near Granville, Normandy.

The highest hourly average wind speed recorded in the UK was 86 miles an hour at the Royal Sovereign Lighthouse, but gusts were far stronger. One at Shoreham in Sussex peaked at almost 115 mph, and at Gorlestone in Norfolk an even stronger gust was measured at 122 mph, and figures of over 100 mph were recorded at many other places along the coast. Gatwick Airport was closed as gusts hit 99 mph and even in the relatively sheltered site of the Weather Centre in central London the gusts peaked at 93mph.

Shipping in the Channel took a ferocious battering. Numerous small boats were sunk or wrecked as their anchor cables or mooring ropes snapped in the gales and they were driven aground. A British-registered bulk carrier, the Sumnea, capsized at Dover, leaving two Singaporean seamen feared drowned. Four others were rescued, one in a critical condition. The murderous gales almost smashed the carrier into the harbour wall before it "bobbed dangerously near the entrance and suddenly keeled over."

"The seas were murderous," a coastguard spokesman said. "We could hardly see anything, with huge waves, a wall of spray and lashing wind. One of the tugs which went out to the vessel said it turned turtle after listing heavily. It happened in a matter of minutes."

A huge Sealink cross channel ferry, the Hengist, was driven aground near Folkestone in Kent. There were no passengers aboard and the twenty-two unharmed crew were rescued by breeches buoy. Nine hundred seasick passengers on two other ferries were stranded in the storm-tossed seas outside Dover and Harwich for twelve hours, unable to make port until the winds abated. Two major alerts were declared in the North Sea as first an oilfield support vessel with seventy-nine people on board broke down and began drifting towards gas drilling platforms, and then a chemical tanker was torn free from its moorings in Felixstowe harbour. Emergency services "averted disaster on both occasions after lengthy and dangerous operations." At Harwich, Tamil refugees who had recently fled from the civil war in Sri Lanka, suffered further fear and misery as the ship on which they were confined, the Earl William, also broke from its moorings and was driven on to a sandbank.

On land, people who managed to sleep through the storm that night woke to a landscape that had been torn apart. To many, yawning and stretching as they pulled back their curtains, the world outside their window had changed out of all recognition. Many other people had already been awake for hours, ripped from their slumbers by the sound of howling winds, falling trees, breaking glass, cracking roofs and gables, collapsing buildings and, in cities, an endless cacophony of car and house alarms. The terror of the night was increased by widespread power blackouts, forcing frightened householders to sit out the storms in darkness. The power cuts also affected water supplies and sewage pumping stations; 15,000 villages in Kent, Sussex and Hampshire were left without water.

Although the worst of the storm had passed by daybreak – the centre of the intense depression continued to move rapidly north-east and reached the Humber estuary around dawn – southern Britain was effectively paralysed that Friday morning. Virtually every road and railway line was blocked by fallen trees – some fifteen million had been felled by the winds in that single night – the worst devastation since the outbreak of Dutch Elm Disease in the 1960s and 1970s, which destroyed 20 million trees, but that took more than a decade and this had happened in a single night. Whole forests had been destroyed and one third of the specimen trees at the Royal Botanic Gardens at Kew in London, some of them centuries old, had been uprooted. The deputy curator, Ian Beyer, called it "the worst day in the entire

The 'Tree of Heaven', one of the jewels of the Royal Botanic Gardens at Kew, fell onto King William's Temple (visible in the background) at the height of the Great Storm of October 1987. (PA)

history of Kew. It is impossible to put any kind of financial estimate of the damage, literally hundreds of trees, many of them 200 years old, have been destroyed."

One of the jewels of the collection, the 200 year-old "Tree of Heaven," had been blown down, along with a rare "headache tree", the cluster of mulberries planted by Queen Victoria and the robinias planted in the 1850s. The priceless palm collection, housed in the badly damaged Victorian Palm House, was also now at great risk. A large plane tree had smashed through the roof, shattering the glass and allowing the winds and the cold air to affect the delicate trees. Two other historic buildings within the gardens, King William's Temple and Hanover House, had also suffered major structural damage after trees fell on them. At nearby Syon Park, hundreds of rare birds, butterflies and insects had escaped when a huge tree smashed through the roof of the London Butterfly House. The flight to freedom of most of the exotic species would end in their death; they would not survive their exposure to the gales and the autumn cold.

The loss of trees would have been much less severe had the autumn leading up to the Great Storm not been so mild and wet and had the gales blown from the normal direction in Britain - the west. The gales battered the trees from the south and south-east, and the trees, top-heavy because they were still in full leaf and unpruned, and especially vulnerable to gales because their roots were in waterlogged ground, were knocked down like ninepins.

Southern Region did not run a single train that morning as the whole network was closed down. Hundreds of roads in Suffolk, Essex and Kent were also blocked by floods, further hampering already overstretched emergency services. Farmers at Stowmarket in Suffolk braved the gales all night as they worked to move cattle and sheep to higher ground, as their fields disappeared under four feet of floodwater.

Thousands of buildings had lost roofs or chimneys or suffered other severe damage, hundreds of cars had been damaged or destroyed and eighteen people had died in Britain and four more in France as a direct result of the storms – the majority crushed by falling trees or masonry. Four people were killed in a single incident in Wales when the Tywi railway bridge collapsed in floods caused by the storm just as a train was crossing. It plunged into the river, drowning four of the passengers. Had the strongest gusts occurred during daylight rather than in the early hours of the morning when most people were in bed, the death toll would have been far higher.

The trail of catastrophic damage extended all the way from the Channel Islands to the North Sea coast. In Guernsey hundreds of greenhouses and the tomato plants and the other crops they contained were destroyed, jeopardising the livelihoods of many residents. As the gales swept eastwards, there was severe damage in the West Country and police reported that virtually every road in Dorset and Hampshire was blocked by at least one fallen tree. Two Dorset firemen, Ernest Gregory and Graham White, were among the first fatalities of the storm, killed at Highcliffe when a great 80-foot oak tree crashed on to their water tender as they answered an emergency call.

In Hampshire, electricity workers discovered the body of man in a car that had been crushed by a falling tree. Mrs Patricia Bellwood, an expert on child abuse who was staying at the Garter Hotel in Windsor after giving a lecture in Reading, was killed in her bed when a chimney toppled by the gales crashed through the ceiling. A 70 year-old hotel guest also died at Hastings when part of the Queen's Hotel collapsed. Three other people died in their beds as roofs, chimneys or trees crashed down on them. In Hastings, James Read, a 49 year-old fisherman, was killed when he was hit by part of a beach hut hurled across the sea front by the winds. William Bennister, of Rottingdean in West Sussex, died as he tried to stop his garage doors blowing away and another man died at the wheel of his car in Bromley in Kent. In London, police were struggling to identify the body of a tramp who was killed when his cardboard "house" in Lincoln Inn's Fields was crushed by falling bricks.

Property damage was on an almost unimaginable scale. The National Farmers' Union reported that farmhouses, grain stores, barns, livestock units and glasshouses had all suffered extensive damage. At one poultry unit in Essex 17,000 birds had been killed by the storm or would have to be destroyed. Recently harvested crops stored in damaged and unroofed buildings, silos and grainstores would be ruined. Fruit still waiting to be picked had been destroyed. Power cuts had disrupted milking at many dairy farms and threatened crops and other farm produce in cold stores. An NFU spokesman had "spoken to many farmers and growers and without exception they have said that they have never known anything like it."

The New Forest in Hampshire had been devastated and live electricity cables tangled in fallen trees were hampering forestry workers as they began the melancholy task of clearing the storm damage. Thetford Forest in Norfolk and the forests around Woodbridge in Sussex had also suffered huge losses of trees and the inhabitants of some forested areas in Northamptonshire had had to be evacuated because of the danger to life and limb from falling and damaged trees. Tree surgeons up and down the country were revving their chainsaws and contemplating the not entirely distressing prospect of at least six months' continuous work just to clear the fallen timber.

Barry Still, a tree surgeon from Kingston in south-west London, had already received "400 emergency calls, including one from a woman whose baby escaped death by a few feet when a tree crashed into her bedroom." Having dealt with that, he spent the afternoon clearing thirty fallen oak trees, all 200 years old, from a garden near the Coombe Hill golf club. Like other tree surgeons, Mr Still laid part of the blame for the massive loss of trees at the door of local councils and conservationists, "Time and again the lady who owns these oaks has asked permission to have them pruned and the council has refused because it is a conservation area. If they had not been top-heavy, they would still be there." That woman and other house owners also faced the possibility of large bills to have their fallen trees removed. Although local authorities had to clear trees blocking roads, landowners were responsible for the cost of removing trees and branches on their land that might constitute a hazard to the public.

Others had lost property, homes and businesses. A large caravan site at Hayling Island in Hampshire was completely destroyed and at Brighton a three-ton stone minaret from the roof of the Royal Pavilion smashed down into the ornate Music Room causing massive damage. Shoreham airfield was a jumble of overturned light aircraft: in all twenty-seven were damaged beyond repair. Six of the famous seven trees that gave the town of Sevenoaks its name were felled by the storm. Even away from its main track there was substantial damage and one death, as a 25 year-old died when his motorbike was blown into the central reservation of the M62 near Liverpool.

However, it was in the South that the storm's worst effects were felt and as one sardonic observer noted, "Londoners were so shaken by the terrors of the dark night that many even spoke to strangers." The streets of the capital were strewn with rubble from collapsed walls and shattered roofs, spiders' webs of toppled scaffolding blocked several streets and the

The gales ripped through protective awnings put up during restoration work on Brighton's Royal Pavilion and toppled a three-ton minaret, causing massive damage. (PA Archive/PA Photos)

crunch of broken glass was heard everywhere underfoot. The City of London had almost come to a standstill, with "an eerie calm" during the morning rush hour. Blue and white police "incident" tapes closed off huge areas of street and pavement as police patrolled with anxious looks at unstable cornices, guttering, scaffolding, gables and chimney stacks overhead. The Stock Exchange suspended share trading as computers crashed and telephone lines were cut; precious few traders made it into work that morning anyway. Those who did either stared in vain at their blank screens, only installed in the "Big Bang" the previous October, or wandered down to the trading floors and began "doing deals with other men, using that crude tool of their fathers, the human voice."

Fears for the safety of the dome brought the Central Criminal Court at the Old Bailey to a halt and very few cases went ahead at the High Court, since judges, barristers, juries and witnesses all failed to turn up. The Tower of London was also closed for the first time since the war because of the danger from damaged trees. Cheapside was "paved now with the black stuff from a roof: inch-thick slabs which had skimmed like frisbees to demolish windows a wide street away." In the Strand a huge stone eagle had been dislodged from its perch and made "its first landing on earth for perhaps 200 years."

The Mall was a wilderness of flattened and broken trees, and though the Guards in their bearskins still stood sentry by the gates of Buckingham Palace, the security fences topping the high walls around the palace gardens were garlanded with uprooted bushes, plants and flowers, and St Paul's was crowned with a "wreath of intertwined branches." Many of the beautiful plane trees that had surrounded the cathedral had been felled or badly damaged. A verger at St Paul's described the cathedral as "like a boulder in a stream. Even in moderate winds it suffers. We have a safety staff who climb up on nights like that. Last night they had to cling to each other like men on a mountain to reach the roof." Once there, they could only watch helplessly as several windows in the dome of the cathedral cracked and broke, showering torrents of broken glass 280 feet to the floor beneath.

There were bizarre sights too on every side: pot plants stranded on a traffic island, a traffic cone embedded in a window, a deckchair, fully assembled, sitting on top of a bush in St James's Park, racing skiffs used on the Serpentine now hanging like pine cones from the branches of the surrounding trees.

Electricity cables and telephone lines had been brought down right across the South, and hundreds of thousands of homes remained without power 24 hours later. Even with the help of hundreds of power crews brought in from the rest of the country, it took many days for electricity supplies to be restored to all areas. When all the insurance claims for storm damage were in, the total had reached an all time record amount – an estimated £1.9 billion – prompting an immediate steep increase in premiums.

Europe also took a pounding from the storms with severe damage in north-west Spain and France, including the derailment of a train at Santiago de Compostela. And an aircraft en route from Milan to Cologne vanished from radar screens over mountainous terrain.

Although not as severe, fresh storms the following day caused further disruption and damage, bringing down several buildings already severely damaged in the Great Storm; three people in Lancing, Sussex, were hospitalised after their house collapsed. In Southend, where 25,000 houses had been damaged in the storms, suppliers ran out of materials just hours after builders began work on repairing the damage. Southern Region services were once more disrupted and twenty rivers in Kent, Sussex and Hampshire burst their banks, adding to the woes of local residents.

Traders in the City of London soon had another reason for misery. After heavy falls the previous week, "Black Monday" – 19 October 1987 – saw the Dow Jones index plummet by over 500 points, its worst ever fall, and the crash also wiped £63 billion from the value of shares on the London Stock Exchange. The era of red braces, "Loadsamoney", "Greed is good" and "The Big Bang" had ended in a "Big Bust" – a spectacular collapse.

The official Met Office record book bears only a three word entry for that October day in 1987: "The Great Storm", and not even the oldest inhabitants of southern England could remember a storm to compare with it. The strength of the winds and the scale of destruction made it the worst storm of the twentieth century and indeed the worst in almost three centuries, since Defoe's Great Storm of 1703.

Home Secretary Douglas Hurd called it "the most widespread night of disaster since 1945" and as government and people took stock, the weather forecasters came in for heavy criticism for failing to predict the ferocity of the storm, with my alleged comment "there will not be a hurricane" drawing particular ire. The media laid siege to the home of Michael Fish and that of the Director-General of the Met Office, Dr John Houghton, and there were calls for his resignation. Explanations of the true course of events did not chime with the media's witch-hunt and so were ignored. John Houghton was perfectly correct to say that the storm was not a hurricane, because a true hurricane is defined by winds of 120 to 160 mph. But the storm did produce sustained winds of Force 12 – confusingly described as Hurricane Force on the Beaufort Scale (defined as a wind of 64 knots [74 mph] or more, sustained over a period of at least 10 minutes) – and to everyone except meteorologists it was a hurricane; it certainly felt like one to those watching their roofs disintegrate and their much-loved trees come crashing down.

The London Weather Centre claimed that the storms "came from nowhere" yet the Met Office had received four days advance warning of the risk of very high winds from "the most sophisticated weather forecasting computer in the world" at the Reading-based European Centre for Medium Range Weather Forecasting. Despite this, the Met Office "failed to anticipate the winds' severity and the route they would take".

Critics also pointed out that Dutch television viewers had been told on the Wednesday night that freak storms were expected to hit the English coast two days later. The French Meteorological Department had acted on the same warning, carried out its own checks and then issued an alert broadcast on French radio and television on the Thursday afternoon. Perhaps as a result of that forewarning, although northern France suffered even higher winds than Britain, the gales there claimed only four lives, against the eighteen deaths recorded in Britain. Georges Dhonneur, a scientist at the French Meteorological Department, said that he was convinced the warnings had helped France to escape with so few fatalities.

Dr John Houghton loyally defended his staff or, as The Times put it, "adopted the defensive posture of Gypsy Rose Lee being asked by a customer for their money back." "There was no hurricane," John Houghton said. "A hurricane can last for hours. These were intense gusts of short duration. Given the same equipment and the same data, we would make the same forecast again." Met Office spokesmen also pointed out that "we had forecast the risk of very high winds at the end of the week as long ago as Sunday, but we did fail to fill in the detail. The storm blew up over the Bay of Biscay and came in over the sea from an area where we have very, very sparse weather information. We failed to realise the rapid way the depression was deepening." The Met Office also argued that the sudden drop in pressure and changing track of the depression that brought the storm could not have been predicted far in advance and it was unfortunate that the revised forecasts went out too late in the day to be heard by most people.

Nonetheless, Mr Hurd insisted that "clearly the Met Office will want to look at their experience and our experience in the last 24 hours, to see if anything can be done to improve their predictions." In fact, if the Great Storm had been predicted in the infamous 9.30pm forecast on the BBC, it would have probably led to an even greater loss of life because more people would have been up and about, checking on their houses, relatives and neighbours and would therefore have been more vulnerable to falling slates, chimneys, masonry and trees. Although some people were killed while they slept, bed was probably the safest place to be on that wild and windy night.

In answer to the public criticism – justified or not – the Met Office extended and enhanced its coverage of the atmosphere over the ocean to the south and west of the UK by increasing the quality and quantity of observations from ships, aircraft, buoys and satellites. It also enhanced the computer models used in forecasting and revised the system of severe weather warnings. They are now much more readily and frequently issued, some would say to excess, as it now often seems as if almost any strong wind, or rainfall or snowfall will trigger yet another "Severe Weather Warning." Whether among all this crying of "Wolf", the advent of the next Great Storm will be identified as an exceptional weather event, triggering the necessary alarms, only time will tell. For that, if nothing else, we must blame the Great Storm of 1987. Michael Fish adds: "If you must, even though I wasn't there at the time, blame me. My notoriety was deeply frustrating and unpleasant at the time but, once more proving the truth of the old adage about ill winds, the Great Storm of 1987 did make its perceived arch-villain famous, and I've dined out on it ever since!"

Arguments still rage about whether the October 1987 storm could truly be said to be a hurricane – it wasn't, and nor was the far more powerful Great Storm of 1703. However, they were both savage and very damaging storms, feeding in their long track over the ocean on the latent heat generated in the Atlantic by an unusually wet and warm year and arriving with terrible force upon the shores of Britain. They have happened before and can and will happen again, particularly as global warming is raising ocean temperatures across the globe. If and when the next so-called hurricane strikes Britain – and if 1703 was a "once in 300 year" weather event, the next is already overdue – it will dwarf even the Great Storm of October 1987 and may leave hundreds or even thousands of dead in its wake.

* * * *

On May Day 1988 several walkers were killed after being struck by lightning during violent, spectacular thunderstorms over the hills. In Shropshire, a man died on Caradoc Hill, and in the Lake District four people were killed. A climber was killed on the summit of Helvellyn, another on Great Rigg, where three others were injured, and two were fatally struck at Red Tarn. At Ambleside, a bungalow burst into flames after being struck by a lightning bolt.

Just over a week later, on Monday 9 May, a devastating series of thunderstorms dumped three and a half inches of rain on the Home Counties in 24 hours, including an inch of rain

Devastating thunderstorms, such as those of May 1988, can have dramatic results. The photograph above was taken high up on the Aran Fawddwy Mountain in mid Wales and shows one of the trenches created when a lightning strike instantly vapourised moisture in the cracks of the rocks and caused them to break apart. The damage is thought to have resulted from a positive cloud to ground strike, one of the rarest and yet most powerful forms of lighting. The photograph opposite shows two trees near Ludlow that display classic signs of lightning damage. Missing bark down one side of the trunk indicates the path of the current to the ground. (Howard Kirby, IJMet/TORRO)

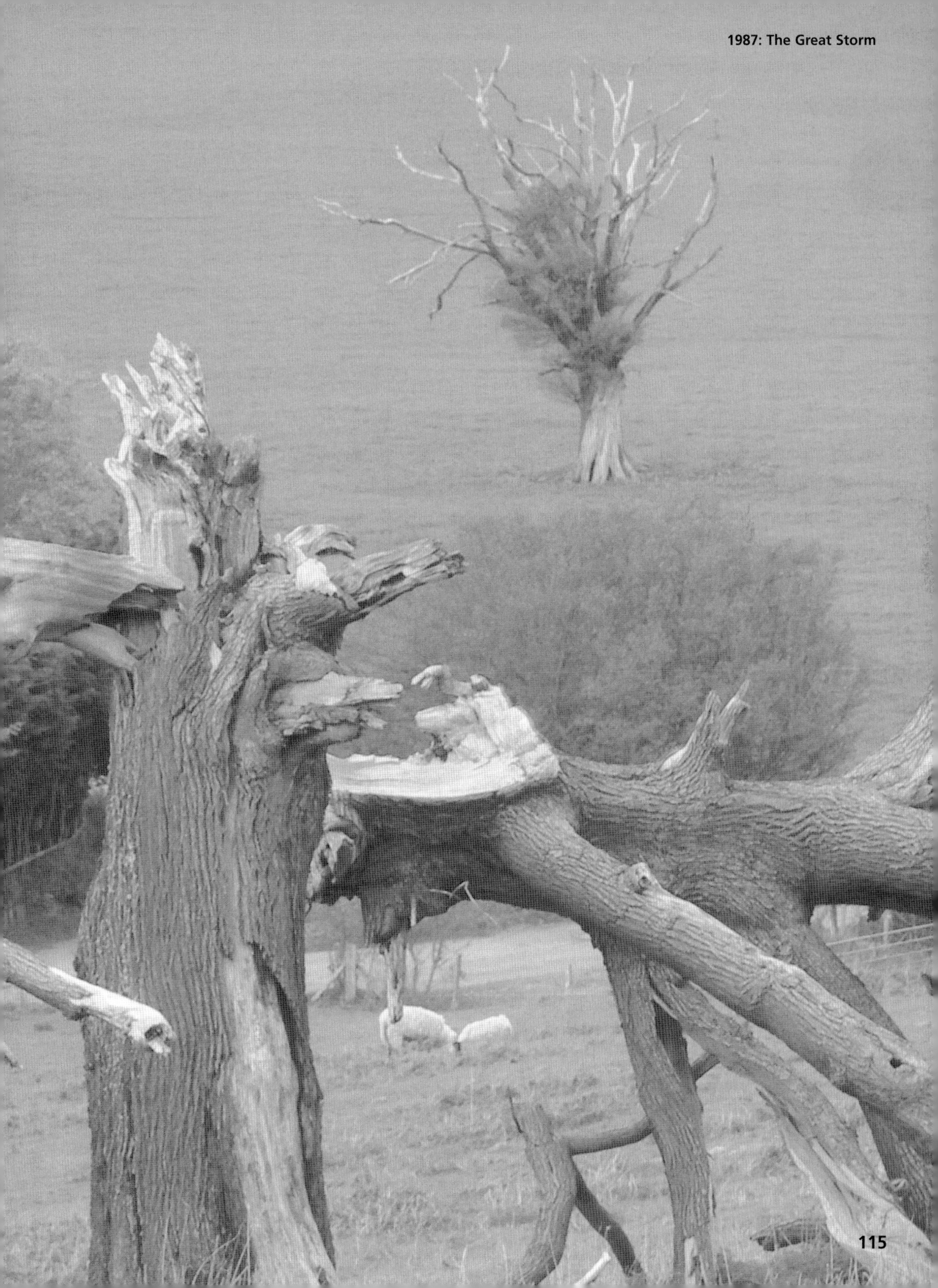

in just 25 minutes. As commuters headed into London that Monday morning, they found every western approach to the capital was under water. The resulting traffic jams stretched back more than twenty miles.

In June 1988, dust devils sent dirt and hay whirling up into the air at Marple in Cheshire and caused a fifteen minute "haystorm" on a neighbouring farm. Elsewhere there were unsubstantiated claims that "pink frogs" had fallen from the skies during a rainstorm near Worcester and that worms had fallen with rain in Sheffield.

On 3 July 1988, a tornado at Carn How in Cleveland threw a pigeon loft and a rabbit hutch over a 6 foot fence; the consequences for the occupants were not reported. A local man was also hurled 100 feet through a hedge but emerged relatively unscathed. Three days later a tornado in Wiltshire picked up a seven foot corrugated roofing sheet and carried it for 20 minutes and 2 miles before dumping it back to earth at the feet of a startled farmer. Later that month, on 24 July 1988, 2 acres of greenhouses at Spalding in Lincolnshire were destroyed by yet another tornado.

On 8 September 1988 a sandstorm dumped thousands of tons of brown sand over Britain. A violent storm in the Moroccan Sahara had whirled the sand thousands of feet into the air and high winds had then carried it to Britain.

1990 –
Burns' Night Storm

The Great Storm of October 1987 had universally been described as "a once in 200 or 300 year event" – yet just 27 months later, on 25 January 1990, another storm of equal savagery ripped across Britain. If the South had been the major sufferer in 1987, the "Burns' Night Storm" of January 1990 created havoc throughout the country.

Wild storm winds, with gusts peaking at 108 miles an hour, were created by a vicious depression with a central low pressure zone dropping like a stone to reach just 949 millibars – a phenomenon that one climatologist described as "a bomb". The ensuing storm hit the south-western tip of England at six that morning and ripped right across the country over the rest of that day. It killed forty-seven British people, blew down four million trees, damaged thousands of homes, factories and offices, left one and a half million houses without electric power and caused massive damage and disruption right across Europe.

Penzance was the first place to feel the the full force of the storm. Peter Horder, Head Gardner at Trengwainton Gardens, a National Trust property near Penzance, was woken before dawn by a window of his cottage banging in the wind. As day broke he could only watch the destruction of the gardens he had nurtured for twenty years. "I was close to tears. The wind was howling and screaming and wreaking havoc. It was heartbreaking to see the damage and be powerless to help." The gales tore up trees that had stood for 150 years and even survived the ravages of 1987. One falling tree battered down one side of a walled garden, exposing a collection of rare South American plants to the full destructive force of the storm.

The storm raged for three hours, then blasted on over Mounts Bay but long before that the cutting edge of the storm was slicing its way right across the West Country. At Launceston, a poultry farmer, Stephen Blake, had just completed the morning feed of his hens when a savage gust picked up a 120 foot fibreglass chicken house, hurled it 300 yards and smashed it to the ground in fragments. "I heard the shed start to go, then the wind picked it up like a sponge and threw it to the ground. There are pieces from one end of the village to the other." The 7,500 chickens inside were left exposed to the savage, icy blasts of the wind but worse was to follow as his other three chicken houses then collapsed where they stood, crushing 10,000 hens. The helpless Stephen Blake could only watch "ten years of my life go down the drain".

The storm hit the Lizard peninsula at 8.15 that morning. Eight people were trapped in the Polurrin Hotel at Mullion, when the roof of the hotel blew off. Emergency services were unable to reach the trapped people for several hours because fallen trees had blocked the roads. Keith Willey, a long-term Mullion inhabitant, said the wind was "the fiercest I have known in 38 years here. Half the bakery roof went and many other homes and buildings lost their roofs. Garden sheds flew everywhere and cables and telegraph poles went down."

At Cullompton in Devon, market gardener Ian Cummings had to run for his life as six acres of greenhouses holding thousands of tomato plants and chrysanthemums were shattered by the storm. "The glasshouses looked like polythene bags blowing up into the air and then exploding. When we walked back to pick over the wreckage the radios that had been left in the greenhouses were still playing. It all happened so fast."

The roof of the grandstand at Torquay United's football ground also blew off, and later that morning, at Trowbridge in Wiltshire, the lunchtime communion at the 500 year-old St James' Church was brought to an abrupt terrifying end. "We heard a creaking and a cracking in the spire," one of the congregation said. "I looked up and could not believe my eyes; the spire was just falling in. We ran out without even picking up our bags and slammed the door."

The storm roared on across the country leaving destruction on an epic scale. Hundreds of churches, town halls and other public buildings were damaged, and in London "a sizeable piece of masonry was blown from the central tower" of the Central Lobby of Parliament, connecting the House of Commons with the House of Lords. It crashed down into the Engineers' Court with "a resounding roar," heard by MPs in the nearby Members' Lobby. Dr John Cunningham, Shadow Leader of the Commons, remarked that "while no doubt the news that lumps of masonry are falling off the House may be a cause for celebration in some quarters of the country, and no doubt with the added hope that some of us don't get out before the whole thing comes down, it is a serious matter nevertheless." Meanwhile British Telecom had to issue an appeal to Londoners not to use the telephone, because a 100 per cent increase in the volume of calls, as anxious people checked on the safety of their relatives and friends, blocked lines to the emergency services.

In the North Sea a Soviet cargo ship, the Briz, lost all power and began taking in water as it drifted, helpless, towards the Dutch coast. A crewman was washed overboard and feared drowned and four others were lifted off by helicopter after the captain radioed an urgent request and prepared to abandon ship. A Dutch frigate and a lifeboat went alongside to take off the rest of the 56-man crew. Another man was washed overboard from the bulk carrier Serica, labouring in mountainous seas 200 miles off Land's End.

The Dover coastguard described the conditions in the Channel as "terrible, with hurricane-force winds". The ferry Chartres, en route from Newhaven to Dieppe with one hundred and thirty people on board, lost all power in mid-Channel and drifted helpless for an hour before jury-rigged repairs were carried out and it was able to limp on towards Dieppe. Other cross-Channel sailings were suspended and ships already underway had to ride out the storm at sea rather than risk trying to make harbour in such conditions. "It was virtually impossible to manoeuvre out of the harbour," the Dover coastguard said. The Harwich to Hook of Holland ferry waited out the storm tied to a buoy in Harwich estuary. Out in the Channel, shipping trying to steer a south-westerly course came to a standstill, unable to make any headway against the winds, while those heading north-east were bowling along at "high speed," driven on by 85 knot winds.

On land, the winds were also causing chaos to transport. The AA said that "most motorways in the South and West are littered with overturned lorries," blown onto their sides by the gales. The Severn Bridge was closed to all traffic, and the M5 closed in both directions between Taunton and Portishead. The M4 was closed at Junction 4, the Heathrow turn-off, because a van was dangling over a bridge parapet, and was also shut in the Thames Valley, where at least sixteen high-sided vehicles had blown over. The M25 was closed near Heathrow and at several other points in Surrey and Essex. The northbound M1 in Bedfordshire and Buckinghamshire had to be cleared repeatedly after a succession of lorries blew over, and on one short stretch of the M27, between Junctions 11 and 12 near Gosport in Hampshire, four lorries were left on their sides by ferocious gusts. Police banned lorries from the M40 and the M25 in Surrey, and police and motoring organisations lambasted the drivers of high-sided vehicles who ignored the conditions and police warnings and continued their journeys at speeds of up to 90 miles an hour. Chief Inspector Laurie Fray said, "I do not know how anyone who considers himself a professional driver can drive HGVs in this weather. We have had reports of some going along on two wheels. There is only one word for it: lunatic."

On the railways, Bristol Temple Meads station was closed because the roof was unsafe and tens of thousands of London commuters were left stranded as all the mainline railway stations except Victoria were closed, amid safety fears about glass falling from damaged station roofs. Even when they reopened, there were virtually no trains because of power cuts and lines blocked by fallen trees. A spokesman for Southern Region said, "We are managing to run services when and how. We will endeavour to get people home but we can not say when. We are getting there eventually." Ten Underground stations were also closed due to storm damage and seven of the nine Underground lines – the Metropolitan, Central, Piccadilly, Northern, Bakerloo, Hammersmith and City, and District lines – were partly closed because of falling masonry and trees. The West Coast main line from London to Manchester was paralysed and the Midland main line and routes to the West Country and Wales were little better. No sooner had railway workers removed a fallen power cable that was blocking the main line between Reading and Didcot than a tree was blown down on the same section of track, causing further huge delays and cancellations. A British Rail spokesman said they would be working "through the night" to repair the damage, but warned that while fallen trees could be removed relatively quickly, "re-erecting overhead power lines takes considerably longer."

Drivers making for home fared no better. At the start of the evening rush hour in London traffic was "at a virtual standstill – gridlocked" according to BBC radio's traffic reporter at Scotland Yard. Waterloo Bridge was blocked after a double-decker bus and a high-sided lorry were blown over, the Edgware Road was partially closed because of the dangerous condition of a building facing the road, and hundreds of other roads were closed because of falling masonry or simply choked with cars. Three hours later, at 7 pm, Scotland Yard's traffic cameras were still showing all the London bridges jammed; at 9 pm, the huge jams in the West End were still there, the bridges were still blocked and even the streets in the City, normally long deserted by that time of night, were filled with slow moving traffic. One commuter remarked that "usually I drive home to the Archers. Last night, as I turned into my drive, they were just starting Book At Bedtime."

No sooner had railway workers removed one fallen tree than they had to tackle another, causing huge delays and cancellations throughout the network. The catastrophic Burns' Night storm was the worst in a series in January and February 1990 that killed over one hundred people in Britain alone.

Many motorists gave up the attempt to get home; hotels throughout London were soon reporting "No Vacancies" as frustrated commuters booked every available room. Others slept on the floor of their offices, though one company had "laid in a stock of camp beds after the rail strikes last year" and put its workers up in relative comfort.

The next day soldiers and Royal Marines were drafted in to work alongside the emergency services, local authority workmen and tree surgeons to help clear the thousands of trees that still blocked A-roads and minor roads. Even when most of the main roads had been cleared, London suffered a fresh bout of traffic chaos when a building damaged in the gales and now dangerously unstable, forced the closure of the Embankment for the entire day.

Airline schedules had also suffered massive disruption as winds blew parked aircraft into each other and made loading others almost impossible. At Bristol, a Shorts 360, was blown over just after its passengers had disembarked. Dozens of pilots aboard aircraft inbound to London airports found the gusts too dangerous and diverted to airports in the North or on the Continent. However, many pilots delighted in the chance to "do some real flying" and 447 flights touched down at Heathrow that day, even though many passengers suffered airsickness from the turbulence. Said one pilot, "We would not have attempted a landing if there had been any hint of danger... but I think I would have diverted to another airfield if it had been very much stronger because as professional commercial pilots our interest is in not only the safety but the comfort of the passengers."

At the height of the storm British Airways ordered its pilots to stop take-offs until conditions eased and at 3 o'clock in the afternoon the whole airport was shut for fifteen minutes because of danger to aircraft and passengers from flying debris. At Gatwick, perimeter fences were blown down, office roofs were badly damaged and two Air Europe aircraft were seriously damaged when a ferocious gust blew both of them over onto one wing. They then fell back when the gust had passed, smashing their undercarriages. Stansted, which had taken a number of diversions from Heathrow earlier in the day, was itself forced to close down and divert all of its flights later in the afternoon.

There were hundreds of injuries and forty-seven people died in Britain and Northern Ireland as a result of the storms, with deaths recorded from Cornwall to Cleveland, and Ulster to Kent. Most ambulance crews suspended the industrial action they had been taking to go to the aid of police and army services that had been overwhelmed by the spate of emergency calls. A union spokesman said that it was "such a diabolical situation at the moment, with dozens of accidents, that it was the least we could do." However, the union's chief negotiator, Roger Poole, had to appeal to his members in the north-west of the capital to reconsider their decision to take unofficial strike action. "We strongly oppose strike action and I would say to our members that they must maintain accident and emergency service."

On land in France, the Paluel nuclear power centre near Rouen was damaged when the gales blew down a tall chimney at the site, and the collapse of pylons carrying high-tension power cables caused the shutdown of five of the six generating plants at the Gravelines nuclear power station near Calais. In the Netherlands, Schiphol airport, one of the world's busiest, suspended all flights for several hours and the Dutch railway system halted. Hundreds of commercial greenhouses were wrecked and "trees snapped like matchboxes," crushing several car drivers. A girl of thirteen from Hilversum was one of five people killed by falling trees. One man was killed when he was blown into the gears of his windmill while trying to fasten its sails against the storm.

In West Germany the storms caused massive damage in the Schleswig-Holstein and Lower Saxony regions, tore the roof from the Felde station near Lubeck, toppled billboards, scaffolding and fencing in Frankfurt and Hamburg and paralysed the Frankfurt traffic as police closed off streets because of falling debris and masonry. In Hanover, a wolf escaped from the city zoo after the winds damaged its enclosure. It was later recaptured without harm to the city's inhabitants.

Like its continental counterparts, the Met Office had given accurate advance warning of the gales on this occasion, predicting "severe gales likely to cause structural damage" fourteen hours before they struck the Cornish coast. The depression that caused the storm first appeared on Met Office computers at 8.05 on Wednesday morning, far out in the mid-Atlantic. Meteorologists monitored its progress as it moved eastwards, deepening all the time, and early that evening they put out a storm warning. "We got it right this time," a Met Office spokesman said with understandable pride. "At 6.45 pm on Wednesday we issued a warning to all television, radio stations and other media that a spell of severe weather would be experienced as a deep low pressure system moved east. We said there would be heavy snow overnight and on Thursday in Scotland, and in England and Wales, driving rain that would make road conditions difficult. We went on to say that such conditions would be enhanced by gales or severe gales and structural damage was expected in south-west and southern England." Michael Fish, the perceived villain of the piece two years earlier, issued a clear warning to viewers of the evening weather forecast this time that winds of "even storm force" were imminent. The Met Office attributed the high death toll despite their advance warnings to the fact that, unlike the Great Storm of 1987, this one had struck in daylight hours when many more people were out in the streets and more vulnerable to falling trees and flying debris. Opposition MPs, while acknowledging the accuracy of the Met Office forecast, complained that the Government had failed to learn the lessons of 1987 and did not provide adequate warnings to the public. They called for "American-style storm warnings and advice" to be given in future.

Financial losses from the storm were massive. The Forestry Commission reported losses of three million trees on its plantations alone. Uprooted and badly damaged trees in private woodlands would take the total to between four and five million. After the massive losses three years earlier this was another devastating blow to the forestry industry and to the environment. The vast majority of the flattened trees were mature ones that provided the maximum food and shelter for birds, insects and other wildlife. It would be many years before their replacements would grow to maturity. Kew Gardens, denuded of many of its specimen trees in 1987, suffered a further grievous loss of another one hundred trees including a large arbutus and a cork oak more than 80 years old, while the damage to the three nurseries housing some of Kew's rarest species was even worse than in the 1987 storm. Staff worked throughout the night by the light of car headlamps, making temporary weatherproof protection for the damaged nurseries with tarpaulins stretched over plywood or wooden frames to protect the conservation collections from the low temperatures.

Three hundred trees were lost in London's Royal Parks and the National Trust estimated the losses on its properties at 40,000 trees, including 1,500 rare mature specimens. Cedars proved particularly vulnerable, and in a reversal of 1987, ten foresters travelled from Kent and Sussex to Cornwall – one of the worst affected areas in this storm – to help with the clear-up there. At Stonor Park, near Henley in Oxfordshire, Lord Camoys went out at dawn to inspect the damage to his property. When he opened the east gate of the rear garden, he discovered "all fifteen ash trees were laid out neatly in front of him like stalks of asparagus on a plate." When faced with this calamity, the only recorded comment from Lord Camoys, clearly a man of few words, was "Crikey." At one end of the house a cypress had come down, missing the 14th century chapel by "perhaps a foot." At the other a vast cedar more than 200 years old had come crashing down; "the upended circle of its roots framed Lord Camoys like a man in the mouth of a tunnel." "I had bonfires under this tree when I was a boy," he said wistfully – he would be having another, rather larger one now. Beeches "the size of factory chimneys" lay everywhere, along with great oaks, larches, walnuts and yews.

Half of the beeches on the ridgeline behind the house had been uprooted. They pulled up great slabs of Chilterns chalk with them, which now stood vertical, still clasped by the roots, like tombstones for the trees that now lay dead.

National Trust properties like Cliveden, the former home of the Astors, and Hughenden, once Disraeli's residence, not only lost trees but also suffered structural damage both to the main house and several outbuildings. In all some five hundred National Trust properties suffered storm damage ranging from "a few lost slates to severe damage."

Many deaths and injuries were caused by falling trees. A woman from Cardiff was killed when a tree fell on her car, but her 2 month-old baby daughter was lifted unhurt from the wreckage. Another woman, aged 25, died when a tree fell on her soft-top car as she waited at a traffic light in Cheltenham, Gloucestershire, and another woman died in a similar incident at Basingstoke. Another dozen motorists died, including a lorry driver and his mate, crushed in their cab in Hemel Hempstead in Hertfordshire, and police Chief Inspector John Smith, a 51 year-old father of five, who was hit by a falling tree at Morestead near Winchester.

Three schoolgirls were also killed. One girl, 11 year-old Emily McDonald, was fatally struck by falling debris when the roof caved in at Grange Junior School in Swindon, injuring five of her schoolmates, two of them seriously; another, aged sixteen, died when a conservatory collapsed at St Brandon's School, Clevedon, near Bristol, injuring four of her classmates, one seriously. A third schoolgirl, a 15 year-old from Ware in Hertfordshire, was hit by a falling tree as she was walking home.

The young daughter of a serviceman was killed by flying debris at an Army barracks in Colerne, Wiltshire, and a woman died at Horton Cross, near Yeovil in Somerset, when a house chimney crashed through the roof of her house. Collapsing walls crushed three women to death and falling scaffolding killed three men. In Skipton in Yorkshire, a woman was blown off her moped and killed under the wheels of a car and in Oldham, Greater Manchester, a youth was drowned. One of the most horrific incidents involved Gorden Kaye, the star of BBC Television's hit comedy 'Allo 'Allo. He was in a critical condition in Charing Cross Hospital having undergone brain surgery, after scaffolding collapsed on his car as he parked near his house in Hounslow, west London. As the scaffolding fell, a plank of wood smashed through his windscreen and into his head. There were non-human casualties too: a 30-foot whale found dead on Greatstone Beach at Romney Marsh in Kent was thought to have been driven ashore by 100 mile per hour winds.

The catastrophic storm was the worst of a series in January and February of that year that killed 200 people across Europe, including 109 in Britain, 81 in France, 29 in Germany and 11 in the Netherlands. Despite a chilling warning to property owners from Geoff Salmon, managing director of Salmon Adams Hilton, a leading claims negotiator, that insurance companies "would try to avoid paying claims by declaring that buildings were badly maintained," the Burns' Night Storm was estimated to have cost insurers in Britain alone about £2.0 billion. That total was even more than the 1987 "Great Storm" but perhaps because it had been predicted by the forecasters, it has failed to have anything like the same long term impact on the national consciousness.

****

Storms continued into the spring of 1990. This dramatic picture of the North Sea hurling its might against Scarborough promenade was taken on April 4 that year.
(Scarborough Evening News)

Tornadoes form near the boundary between the up-currents and downdraughts in a thunderstorm cloud, creating violent winds that can damage all in their wake. Caravans were one of the casualties at Ponsanooth, near Falmouth, in November 1997, as they were at Hoghton, Lancashire, on 2 May 2005. The main picture shows the amazing damage that resulted when giant stakes, each weighing about a quarter of a ton, were picked up from a barn and driven into the side of a caravan. Each one was at a different angle, indicating the rotation of the tornado, and had anyone been inside they would most likely have been killed.

  The top photograph overleaf is a detail of the same caravan, showing how grit had been slammed into the metalwork as if it had been peppered by a shotgun. The lower picture depicts a buckled wall and a garage destroyed by the same tornado, which was rated at T3 on the internationally recognised T-scale (winds of between 93 and 114mph). (Samantha Hall, IJMet/TORRO)

On 14 March 1994 a rainbow was claimed to have been visible in the skies over Sheffield from nine in the morning until three in the afternoon. If true, it was the longest lasting rainbow on record.

On July 24 1994 picnickers in Berkshire were treated to a light show, as violent thunderstorms with spectacular lightning displays rolled past without shedding a drop of rain. However, the clouds then burst over Oxford and nearly three inches of rain left water flowing through the streets under the replica of the Venetian Bridge of Sighs at Hertford College. Meanwhile a single bolt of lightning killed an entire herd of twenty-two cows and their unborn calves at Thetford in Norfolk. On the same day no less than 22,000 lightning strikes were recorded over the hills of East Staffordshire and North Derbyshire

During the night of 23-24 January 1996, a prolonged ice-storm – freezing rain, glaze and ice pellets – affected much of England and Wales, causing severe damage to trees and power cables. The next morning saw a rash of broken bones as pedestrians slipped and fell on the ice-covered pavements, and traffic chaos as commuters found themselves driving on roads like ice rinks.

Later that day, 25 January, a whirlwind destroyed six caravans at Tywyn in North Wales and lifted a 35 foot mobile home right over the roof of a house, before depositing it – in fragments – on the other side.

On 7 June 1996 a tornado struck Basingstoke in Hampshire, throwing a garage right over the roof of a house.

A heatwave ended just in time to wreck the Queen's Buckingham Palace Garden Party on 23 July 1996. Ensuring a memorable day for them in more ways than one, two of Her Majesty's guests were struck by lightning. Another fourteen people were hit in various other parts of London, and a man helping to get in the hay crop on a farm in County Durham was struck by lightning and killed.

Five days later, on 28 July, light refracting from ice crystals in cirrus clouds produced the extraordinary phenomenon known as "parhelia" or "mock suns" in which three suns appeared to be shining simultaneously in the western sky.

On 7 August 1996 residents of Sheffield woke up to discover six inches of snow outside their homes. Children piled out of doors to begin making the first August snowballs they had ever constructed, though Met office pedants later insisted that that it wasn't snow at all but merely "soft hail."

On 17 May 1997 storms generated enormous hailstones, flash floods and tornados that ripped through Oxfordshire, Buckinghamshire and Bedfordshire. Further north there were flash floods in the north-east, one of which washed a farmer's entire crop of potatoes out of a field and onto the A166, blocking the road.

On 27 August 1997 the first camel ever to be killed by a lightning strike in the United Kingdom – and what were the odds against that sentence ever being uttered? – was hit

Their lightweight construction makes caravans particularly vulnerable to storms. This one was left draped around a telegraph pole like a supermarket carrier bag.

during a thunderstorm at Knowsley Safari Park near Liverpool. The park was crowded with visitors at the time but none were hurt.

Four days later, on 31 August, a security kiosk in Staffordshire exploded when it was hit simultaneously by twin bolts of lightning. Miraculously a guard, who was seated inside at the time, although showered with glass, escaped unhurt.

Those who had offered to settle their debts or complete unpleasant tasks "when pigs fly" were given an unpleasant shock on 2 September 1997 when a tornado at Newark in Nottinghamshire lifted a number of pigs from the ground and sent them hurtling 100 feet into the air. Those who believed that pigs don't bounce were also given a chance to confirm their theories when the intrepid porcine aviators came down again.

On 8 November 1997 a tornado flattened parts of Ponsanooth near Falmouth in Cornwall, ripping off roofs and destroying outbuildings, cars and caravans. It also sucked a pond dry as it passed over it.

People celebrating the New Year at Land's End on 31 December 1997 were treated to the sight of a waterspout on the sea.

On 4 January 1998, Portland in Dorset was hit by 94 knot (108 mph) winds and thunderstorms that dumped hailstones half an inch in diameter. Just along the coast, at the Isle of Wight, a yacht on its trailer was picked up by the winds and dumped 10 yards away, and three days later, on 7 January, a tornado ripped through the Sussex seaside town of Selsey, damaging hundreds of houses at an estimated cost of over £1 million.

14 February 1998 was the hottest Valentine's Day ever recorded with the temperature in Bristol reaching 18.8°C, but in County Mayo, in place of the normal winter rainfall, several tons of red sand dropped from the skies, brought all the way from Morocco by the same winds that had produced those record temperatures.

On 8 April 1998 a woman in Burton-upon-Trent had a narrow escape when ball lightning burst through her front door, but was then earthed through a radiator in her hall. On the same day residents of Swansea were treated to the sight of a spectacular waterspout. Meanwhile floods in central England were causing billions of pounds of damage to homes and businesses.

One of the largest waterspouts ever seen in Britain formed over the sea near Felixstowe, Suffolk, on 11 June 1998. It was estimated to be 2,500 feet high and 50 feet wide at its base.

On Boxing Day, 26 December 1998, fierce gales swept the country. They were particularly severe across Northern Ireland and Scotland, gusting to over 100 mph and ripping off roofs and battering down chimney stacks. Trees and telegraph poles were toppled and power lines cut. Five people were killed in the storms; thousands more remained without electricity for three days, making it a miserable Christmas.

A man from Bedwas in South Wales was struck by lightning on 29 May 1999 as he walked to his local fish and chip shop. The metal buckle on his flat cap melted and the cap itself burst into flames, but the man escaped with nothing worse than singed hair.

In December 1999 two exceptionally violent storms hit northern Europe. The worst wind-storms in generations, they were far more severe than the "Great Storm" that hit southern England in 1987 but, perhaps because a storm that has little effect on England is dismissed in Fleet Street as an irrelevance, the Great Storm of 1999 generated virtually no headlines in Britain.

A fierce gale at the start of the month had caused the strongest winds for over a century in Denmark, then just after Christmas, two depressions of unprecedented intensity developed within 36 hours, and struck with savage force. The first ravaged France and Germany, the second, hard on its heels, spread destruction along the north coast of Spain, France's west coast and across many Mediterranean countries. Gusts of 161 mph were recorded – the relatively sheltered site of Orly Airport in Paris recorded one of 107 mph – and 120 people died, 92 of them in France, many of whom were killed after the storm winds triggered avalanches. In an ironic reversal of the perception about the British "Great Storm" in 1987, Met Office and BBC forecasters correctly predicted the French storms, but they were not forecast by our French equivalents.

On 10 August 2003, the hottest day ever recorded in the UK, saw temperatures of 38.1°C at Gravesend and 38.5°C at Brogdale, near Faversham in Kent. On the same day parts of the north-east were hit by ferocious thunderstorms. Hailstones of up to 20 millimetres in diameter were found and 47 millimetres of rain fell in just twelve minutes at Carlton-in-Cleveland, 30 millimetres of it in only five minutes, the highest rainfall total in such a short period ever recorded in the UK.

Mammatus, one of the most terrifying of clouds for any airline pilot, is generally linked with huge rotating currents of air and massive turbulence. It is described in the International Weather Atlas as 'seldom seen low to middle clouds associated with severe wind squalls, hail, heavy precipitation, tornadoes and thunderstorms'. This particular example formed over London Road, Reading, during a cool April morning. (Matt Clark, IJMet/TORRO)

A police dog team searches for bodies in the aftermath of the Boscastle flood disaster on 16 August 2004. Incredibly, not a single life was lost though over a hundred occupants had to cling to roofs and trees as the 9 foot high wall of water ripped through the village, hurling cars in all directions. (PA Archive/PA Photos)

2004 – Boscastle Flood Disaster

On 16 August 2004, just over half a century after the Lynmouth flood disaster, a startlingly similar combination of weather and topography brought catastrophe to another small town on that coastline. Starting at midday, torrential rain swept along the North Cornwall coast from Tintagel to Bude. In less than two hours the ferocious thunderstorms dumped more than 2 inches of rain on Boscastle, a picturesque harbour village dating from medieval times that was packed with summer visitors. Three rivers converge on the little village at the foot of a narrow, precipitous valley, and all burst their banks almost simultaneously around 4 o'clock that afternoon.

Trees were uprooted, and cars, vans, caravans, tents, huts and outbuildings swept away in a raging wall of water, mud and debris, 9 feet high and moving at 30 to 40 miles per hour. It ripped through the town and swamped the harbour, wrecking the boats tied up at the quay. The devastation of the fabric of the town was swift and near-total. The many historic old buildings lining the river banks and clinging to the rocky walls of the valley were battered and broken by the flood and the trees and cars it hurled against them.

The most celebrated building in Boscastle, the wave-roofed, three hundred year old "Pixie House," was completely destroyed. Many of the small shops and cafes along the harbour walk also disappeared and even those that survived suffered major structural damage. At the same time, the raging waters lifted fifty cars and vans bodily out of the car park alongside the River Valency and dumped them into the harbour. Seventy year-old Mary Sharp watched "the cars come down like a duck race, bobbing along. The roar and the smell was horrible."

It was little short of miraculous that amongst this appalling devastation, not a single life was lost as more than one hundred and twenty terrified inhabitants and tourists clinging to roofs, chimneys, gutters and trees, were rescued by emergency services. Six RAF Sea King Rescue Helicopters and a Coastguard Helicopter were called up to airlift the stranded people to safety.

As the floodwaters subsided, they revealed a village whose inhabitants, buildings and even its topography had been altered forever in a few catastrophic moments.

The unstable weather of August 2004, which precipitated the Boscastle flood disaster, was not confined to Cornwall. A utility company is seen taking on a large repair job at Bramshall, near Uttoxeter, after a tornado struck on 23 August. Estimated at scale T2/3, it had a track extending over three miles and caused damage to several homes, cars, trees and farmland.
(Steve Warren, IJMet/TORRO)

* * * *

On 8 January 2005, weekend gales felled one of English cricket's most famous totems – the ancient lime tree that had famously stood within the boundary of the Kent County Cricket ground in Canterbury since it was opened in 1847. Only four cricketers – Arthur "Jacko" Watson of Sussex in 1925, the West Indies' Learie Constantine in 1928, Middlesex's Jim Smith in 1939 and the West Indies and Kent batsman Carl Hooper in 1992 – had ever managed to hit a six right over the top of the tree.

The lime tree, believed to be more than 200 years old, had been long beloved by collectors of cricketing trivia – if the ball hit the tree it counted as four runs and batsmen could not be given out caught if the ball rebounded from the tree into their hands. As a BBC correspondent, perhaps unkindly, pointed out, the tree was also "long treasured by followers

Simon Williamson, Assistant Head Groundsman, disconsolately walks away from all that remains of the famous lime tree at Kent County Cricket Club's ground at Canterbury. Felled by a gale on 8 January 2005, it was replaced by a new six-foot high tree two months later. (PA Archive/PA Photos)

of Kent cricket, who could console themselves that even when their team was unremarkable, there was at least one way they were unique."

However, when ground staff arrived for work on the Monday morning, they found only a seven foot stump with the rest of the tree scattered around it. Kent CCC chief executive Paul Millman said: "It's been in intensive care for several years and we planted a substitute about four years ago in anticipation of this sad day." Even though it was still less than six feet high, the new lime tree was eventually replanted on the playing area on 8 March 2005.

The York Minster fire in July 1984, illustrated on page 14, represented lightning damage at its most spectacular – and expensive. Yet in terms of a pyrotechnic display, this storm was undoubtedly eclipsed by that of 12 June 2006 illustrated in these two photographs and also the frontispiece of this book. Forming in the vicinity of Torbay, it created localised flooding as it moved north-east over much of Devon and Somerset. The storm produced almost continuous lightning for two hours as it passed over Exeter in the early hours, the picture opposite being taken when it was about four miles away. In the view on this page it has moved away slightly and is some six miles distant. (Matt Clark, IJMet/TORRO)

High-sided vehicles on motorways, like this lorry on the M1, are always vulnerable to being blown over during gales. Such incidents led to eleven motorways being closed in one half-hour period during the Great Storm of 18 January 2007.

The ferocious storm of 18 January 2007 proved to be merely the prelude to a year of dismal and disturbed weather with massive rainfall. An exceptionally high water table could have caused this long and deep flood on the A4110 between Lawton Cross and Bainstree Cross in Herefordshire, photographed on 6 March after a very wet spell. (Howard Kirby, IJMet/TORRO)

The Great Storms of 2007

On 18 January 2007 yet another ferocious storm carved a trail of destruction throughout Britain and northern Europe. Hurricane force winds uprooted trees, tore down power lines, damaged thousands of buildings, brought air, sea, road and rail travel grinding to a halt and killed at least forty-seven people, including fourteen in Britain, twelve in Germany, three in France, six in the Netherlands, two in Belgium, six in Poland and at least four in the Czech Republic.

Gusts of 99 miles per hour were recorded at the Old Needles Battery on the Isle of Wight, but the storms created havoc all the way to the North and Scotland, where heavy snow added to the misery. The 26-man crew of a British-registered container ship with a dangerous cargo was rescued after the ship was holed and started to founder as it tried to make its way through the Channel in the storm seas. As the 62,000 ton Napoli, carrying a cargo of explosives, began shipping water and listing dangerously, the crew abandoned ship and took to the lifeboat. They spent 90 minutes in the raft, facing waves up to 60 feet high, before two helicopters from RNAS Culdrose in Cornwall airlifted them to safety. Among the crew were two Britons, Forbes Duthie and Nicholas Coulburn, as well as seamen from Bulgaria, India and Turkey. Mr Duthie, from Inverness, said that the crew had been "desperately sea-sick and dehydrated," and it had been "like the end of the world" when the order was given to abandon ship. A coast guard tug eventually took the abandoned vessel in tow 24 hours later as it drifted towards the French coast.

In the air, Heathrow restricted the number of take-offs and landings because of safety concerns, forcing BA to cancel four hundred flights and there were wholesale cancellations and diversions at Cardiff, Manchester, Glasgow and a dozen other airports. On the roads, because of accidents and high-sided vehicles blowing over in the gales, eleven motorways – the M1, M5, M6, M18, M20, M25, M40, M42, M60, M62 and M80 – were all closed in rapid succession within one half-hour period during the afternoon, paralysing virtually the entire motorway network. South Yorkshire police estimated that twenty lorries had blown over on the MI and M18 in the county and in Lancashire, after a lorry blew over and crushed a saloon car, police ordered lorry drivers to park until the gales eased. In Kent the coast-bound M20 was closed between Junctions 11 and 12 simply to allow lorries to queue to get into the port of Dover, which had ceased operating and closed its gates because of the weather. All cross-

Channel ferry services were also suspended until the storms abated.

Hundreds of major and minor roads were blocked by falling trees or masonry, and Trafficmaster, the traffic monitoring network, described it as its "busiest single day for traffic incidents in ten years. At its peak, at 3 pm, we were monitoring 226 simultaneous incidents, the highest we have ever recorded."

Train companies cancelled hundreds of departures. London commuters were left stranded in the evening rush hour, the East Coast main line was closed between London and York and no trains ran on the West Coast line between London and Scotland. Floods blocked the line between Llandudno and Blaenau Ffestiniog, First Great Western imposed speed restrictions, leading to long delays, and London Underground services were thrown into chaos because of the number of "objects on the line": branches, whole trees, slates, masonry and other debris from the storm. That chaos was further fuelled by the closure of several main line and Underground stations. London Bridge station was closed in mid-afternoon after part of the forecourt roof collapsed, blocking a section of the main concourse. Liverpool Street station also closed after suffering similar storm damage and King's Cross Underground station had to shut after a power cut blacked out all the lights.

The storms reaped their customary harvest of uprooted trees and downed power lines, blacking out tens of thousands of homes from the South Coast to the Scottish Borders. A spokesman for one company, Scottish Power, said that they had "brought in all our linemen and they have been put on standby until it is safe for them to work. Our engineers will work through the night." Brighton Pier and Kew Gardens were both closed for safety reasons and the hallowed turf of Lord's cricket ground was left strewn with debris after the gales damaged the roof of the Tavern Stand.

Television viewers around the world saw film of the storms including the arresting sight of a woman in London blown off her feet and driven thirty yards down the street by the gales, before she came to a halt. She suffered only minor injuries, cuts and bruises as a result of her ordeal. Most of the people killed were motorists crushed by falling trees, but in North London, a toddler, two year-old Saurav Ghai, was killed yards from his home in Belsize Park, when a six-foot, brick garden wall collapsed on top of him as he was walking past with his child minder. An eyewitness said "two sections of the wall just fell. The bricks completely covered the little boy and woman had her leg trapped. Workmen across the road dropped what they were doing and ran over to pull the bricks off the boy." A passing off-duty fireman tried to resuscitate the little boy, but he had been crushed to death.

The managing director of Birmingham International Airport, 49 year-old Richard Heard, was also killed when a tree branch smashed through the windscreen of his four-wheel drive car as he travelled to work at 5.45 am. Ambulance crews were called to the crash-site on the B4373, near Bridgnorth in Shropshire, but were unable to save him.

A woman lorry driver, Christine Doran from Moston in Manchester, died when her articulated lorry was blown off the A629 Skipton by pass in North Yorkshire. The lorry careered across the road and ended upside down in a canal. The driver was dead by the time rescue crews could reach her. Several other drivers lost their lives; a German lorry driver was killed when his lorry was blown onto another vehicle on the A55 in Chester, another died when his truck hit a car on the A47 south of Ludlow in Shropshire; a passenger in a car was killed by a falling tree in Streatley, Berkshire; and one man was killed and another seriously injured when their car was hit by a fire engine on an emergency call-out to John Lennon

Airport in Liverpool, where an aircraft was about to make an emergency landing.

At Bamber Bridge, near Preston in Lancashire, an Essex man was killed as he was refuelling his car when the canopy of the garage forecourt blew off and landed on top of him. A 62 year-old Manchester man died from head injuries after the winds blew him head first into a steel shutter; a man in his eighties died of a suspected heart attack outside his home in Prenton on the Wirral as he tried to stop his fencing blowing away in the wind; and a 61 year-old man died after being hit by a falling tree in Middlewich in Cheshire. Two schoolboys were taken to hospital, one of them with serious back injuries, after a tree fell on them at St Augustine of Canterbury's Catholic High School in St Helens.

As the storm moved eastwards, gusts of up to 126 mph were recorded in the German state of Bavaria and Berlin's new main train station was shut down after a two-ton girder fell 40 metres (130 feet) from the glass facade – only erected eight months previously – onto an outdoor staircase. Mercifully there were no injuries, but elsewhere an 18-month old child died in Munich after being hit by a terrace door torn from its hinges by the wind and two fire-fighters died – one of a heart attack, the other hit by a falling tree.

The entire German national railway network was also shut down, with overhead power lines broken and fallen trees blocking many lines. A railway spokesman said, "We've never had such a situation in Germany before." Thousands of Dutch commuters were also stranded overnight when the railroad service suspended all trains because of power failures and blocked tracks.

A million households, in both Germany and the Czech Republic, suffered power failures and tens of thousands of homes in Poland and Austria were also blacked out. Peter Werner, of the Potsdam Institute for Climate Research, warned that Europe should brace itself for more frequent and more intense storms in the future, "In times of rapid climactic change, extreme events arise more frequently."

People in Britain, and particularly in South Yorkshire, did not have to wait long to see the truth of those words. In June 2007, a time of year when Britons are normally looking out the sun loungers and the Factor 30, a series of intense storms hit Britain.

Early on Friday 15 June storms swept England and Wales, dumping the equivalent of a month's rainfall in twenty-four hours and causing widespread flooding. A severe flood warning – denoting "extreme danger to life and property" – was in place on the River Don in South Yorkshire and there were twenty-seven other flood warnings.

Rescue teams using boats and specialist heat-seeking equipment searched in vain for a 17-year-old soldier, who fell into Risedale Beck and was swept away while on an exercise near Catterick Garrison, North Yorkshire. Two other soldiers also fell into the swollen river but were rescued by RAF helicopters. The emergency services were deluged by calls from people trapped inside cars and homes by flash floods.

In the Aston area of Birmingham, two hundred homes were inundated when the River Tame burst its banks and one hundred workers at a factory in Sutton Coldfield in the West Midlands were trapped by floods. Sixty were rescued by fire crews as the water level rose to six feet deep in places, but forty opted to stay put on an upper floor until the water receded.

In Wales, homes were flooded in Borth, Lampeter and Aberteifi, and a 57 year-old man from Gloucester broke his leg after his car hit surface water and overturned on the M42 in Warwickshire.

Fourteen schools were closed by flooding in Sheffield and Barnsley, a nursing home was evacuated, and in Chapeltown a 14 year-old boy was rescued from a swollen river.

The weather was a little quieter over that weekend but on the Sunday night and Monday morning, intense and very prolonged thunderstorms swept the country, causing renewed flooding. River levels were still well above normal the following weekend when an even more prolonged and torrential downpour caused some of the worst flooding in years across a swathe of England running from North-East, Yorkshire and Lincolnshire through to the West Midlands. Beginning on Sunday 24 June and continuing throughout the next day, the monsoon-like rain made it the wettest June day on record, in what was already the wettest June ever recorded. 4 inches of rain fell in 12 hours in some areas, with Yorkshire the worst affected.

A state of emergency was declared in Hull as cemeteries and the city crematorium were flooded and over seventy schools were closed. Sewage was said to be "flowing into hundreds of homes." Several lives were lost across the country but perhaps the most horrific incident occurred in the Hessle district of Hull, where a 28 year-old worker on a fish farm, Mike Barnett, trapped his foot in grating as he tried to clear a blocked manhole. As water levels continued to rise, he became submerged. A nearby householder and amateur diver, Sandra Green, took him her oxygen and diving mask, and another neighbour dived in three times to try and save him, but the water pressure was so great that they could not move him. As the flood waters rose to the man's chest and then shoulders, emergency crews began frantic attempts to rescue him, and took turns diving into the turbulent water despite the risks from boulders and branches swept down by the flood. As the man became completely submerged, he was given a breathing tube and the emergency crews began contemplating the amputation of his leg to free him, but, weak and exhausted after four hours in the water, he died from hypothermia before it could be done.

The body of a 68 year-old Worcestershire county court judge, Eric Dickinson, who went missing after phoning his wife to say that his car was being swept away by floods, was recovered by police divers from his submerged car at Bow Brook in Pershore. There was another drowning in the River Lean at Nottingham, and though a 9 year-old boy was rescued from the River Lud in Louth, Lincolnshire, after passers-by heard his screams, another boy died in Sheffield. The body of teenager, Ryan Parry, who had been swept away by the floods, was recovered from the River Sheaf, and a 68 year-old man, Peter Harding, also died in Sheffield, carried off by the raging waters as he tried to cross a flooded road. In the Brightside district of the city hundreds of workers were trapped in their offices after the River Don burst its banks and completely isolated the building. Helicopters were used to rescue the trapped workers. A rising tide of water, contaminated by raw sewage, trapped another two hundred people in the first floor canteen of the Royal Mail's distribution centre in Sheffield. A worker stuck in another office building in Brightside Lane said the road outside had become "a tributary of the River Don. Retaining walls are collapsing, several have come down and some of these walls are 100 years old. There are car bonnets submerged. There's no way of getting out, it's like a flood plain."

The Catcliffe area of Sheffield was almost completely submerged and hundreds of residents spent the next three days and nights in council emergency reception centres, but the villages of Bentley and Toll Bar near Doncaster were even more badly affected. Inundated by several feet of water, the flood waters did not drain away and, seven days after the flood, the whole of Toll Bar was still under water and fire crews using equipment borrowed from the Army and from other fire services all over the country were pumping out 50,000 litres a minute back into the River Don. Furious locals also complained that their calls for emergency help went unanswered for twenty-four hours. Relief sluice gates usually open to allow floodwater in the Ea Beck running through Bentley and Toll Bar to escape into the River Don, but the Don was already so swollen with floodwaters from Sheffield upstream that the sluice gates were held shut and the Ea Beck burst its banks, flooding the two villages. "We were sacrificed to save Doncaster town centre," shopworker Sharon Sanderson complained, "and now we have no homes. We are scared it will be like New Orleans, where it took months to re-house people."

The transport system was massively disrupted. The M1 was closed in both directions near Sheffield for thirty-six hours after fissures appeared in the wall of the Ulley dam, threatening to send a wall of millions of gallons of water crashing onto the motorway and three nearby villages, which were all hastily evacuated as engineers and fire crews began frantic attempts to pump out some of the excess water in the dam and shore up the walls. Convoys bringing thousands of tonnes of stone blocks to shore up the dam walls had to be given police escorts to get them through the traffic gridlocks caused by the closing of the motorway and scores of other roads. Almost every road in the triangle between Sheffield, Barnsley and Doncaster was closed. All trains on the East Coast mainline between Leeds and London were cancelled, with damage to embankments put at £1 million, and services from Leeds and Sheffield were still being cancelled or disrupted five days later. Rotherham station would "remain closed for at least a fortnight" according to a Network Rail spokesman. Power engineers cut off supplies to 67,000 homes as waters engulfed the Arksey power station. Thousands of other homes also lost their electricity as the floods threatened scores of substations.

Hundreds of householders in Lincolnshire and Nottinghamshire were evacuated as the flash floods added their burden to the already swollen rivers. Residents of one hundred and twenty flats in Lincoln were evacuated by dinghy as the River Witham burst its banks, seven hundred and nine houses in Worksop were also evacuated and the whole of the town centre was sealed off. Householders in Ludlow also had to leave home in a hurry after a bridge collapsed into the swollen River Corve, severing a gas main. The River Teme in Worcestershire burst its banks and the Severn was lapping at buildings in Worcester.

The sports centre in Cheltenham was flooded when a nearby lake burst its banks, and torrents of flood water coursed through the middle of the town. Fifty children on a school bus at Lydney in the Forest of Dean had to be rescued after it got stranded in flood water, and a group of disabled children had to be treated for shock after their minivan was hit by a falling tree in Manchester. Less seriously, spectators at the opening day of Wimbledon had to sit unprotected in the rain during a ninety minute weather delay after a new ban on using umbrellas on the show courts came into effect.

Fire crews in Gloucester rescued fifty dogs and twenty cats after the floods inundated the kennels where they were housed. Fire crews also had to rescue forty stranded sheep in Tewkesbury, as rising waters threatened to drown them. Three thousand space blankets were handed out to the last bedraggled revellers from the Glastonbury Festival as they waited for coaches and buses to take them home. Tractors had to pull dozens of stranded cars out of the mud and there were eight hour queues of people trying to leave the site. The helicopter carrying Shirley Bassey, one of the headliners at the festival, was forced down by bad weather as it flew her away from the festival site and had to make an emergency landing on a school

The wettest June day on record, in what was already the wettest June ever recorded, occurred on the 25th of the month when prolonged thunderstorms created monsoon-like rain that was at its most extreme over much of South and East Yorkshire. Among the worst affected communities were Bentley and Toll Bar on the Ea Beck, north of Doncaster, where floodwaters were unable to escape back into a River Don already swollen beyond its limits. The southern part of Bentley presented a grim sight on the day after the storm, with a dinghy being brought into service on Yarborough Terrace and the ambulance station behind the blue doors totally out of action. (Tim Prosser, IJMet/Torro)

Two days later on 28 June, what is normally the busy A19 through Bentley was still impassable. The large pump was brought over from Ulley Reservoir, near Rotherham, which at the height of the storm had threatened collapse and caused the M1 to be closed for thirty-six hours. (Tim Prosser, IJMet/TORRO)

By 28 June many residents of Bentley were being forced to leave their homes. Francesca Granger is carrying her neighbour's daughter Amy Jolly, aged three, along the main road through the village. (PA Wire/PA Photos)

Thorpe Marsh Power Station at Barnby Dun, north-east of Doncaster, on 27 June, with the Don resembling a giant lake. The Army was brought in to ferry sandbags across the river to protect the switch house. (Tim Prosser, IJMet/TORRO)

playing field in Camberley, Surrey.

As ever, there were a few jackals waiting to prey on the misery of others. Police had to crack down on cases of theft and looting from homes and shops that had been temporarily abandoned to the floods. In Doncaster, a group of teenagers tried to sell stolen sandbags to anxious householders, and in Hull police arrested two men on the Bransholme Estate who were allegedly impersonating council officers as a cover for stealing property from evacuated houses. Conmen and cowboy builderrs were also reported to be operating in several areas, offering "quick fix" repairs to flood victims. Many of these victims were in low-income areas and Mary Dhonau, Coordinator of the National Flood Forum, estimated that "one in four" of the flood victims had ho household contents insurance. "That's an awful lot of people," she said, "who can't afford new accommodation and they don't have the money to replace all their items. It's heartbreaking. It's bad enough being flooded, but to not have insurance is just the pits."

As the flood waters began to recede, leaving tides of mud, mountains of ruined possessions, written-off cars and damaged buildings in their wake, there was fierce controversy about government cutbacks that had led to years of postponements and cuts in flood defence schemes. Among the delayed or cancelled projects was a £100 million scheme for Leeds and plans for Sheffield, Selby, Hull and Doncaster – all badly affected by the week's flooding. Arguments also raged about the policy of allowing housebuilding on flood plains, including the proposed development of 160,000 new houses on the Thames estuary.

By the end of the week the death toll had risen to seven, with the body of an elderly woman recovered from the swollen River Severn at Ironbridge. Six hundred had also been injured and thousands of people had to be rescued from flooded houses in the worst summertime floods ever recorded in Britain. In Sheffield, 286mm (11.2 inches) of rain fell during June 2007, making it the wettest ever month since local records began in 1882 and beating the previous record by an incredible 61mm. The National Fire Service reported that their operations during the floods across the country were the biggest since the Second World War.

As this book goes to print, the extreme weather has claimed yet more swathes of England as more record breaking weather swept across the south of Britain on 20th July. Pershore in Worcestershire recorded 143mm (nearly 5 inches), which is between two and three months' rainfall in a day; 121mm of rain fell at RAF Brize Norton in the 17 hours to 5pm, with 43mm (almost 2 inches) falling in just one hour at Weir Wood in East Sussex – a month's rain in an hour. Almost one hundred and fifty flights were cancelled at Heathrow, trains were stranded, and thousands of commuters were marooned when the M5 was submerged. Many communities close to the River Severn in Gloucestershire experienced devastating floods. Ironically, tens of thousands of people were left without water in their taps as a result of flooding of one the county's main treatment stations.

(Opposite) Hull declared a state of emergency during the floods. The main A63 road into the city was barely passable near Brough on 25 June. (Simon Mills)
Flooding in the suburbs of Hull. The council estimated that no fewer than 6,000 dwellings had been damaged, easily giving the city the doubtful distinction of having the highest number of flooded properties anywhere in Britain. (BBC Look North)

(Above) In Hull and large areas of South Yorkshire the days ticked by, with streets filled with sewage and sludge and homeless families furious about what they saw as official indifference to their plight. The Prince of Wales made a well-publicised visit to Toll Bar, north of Doncaster, on 4 July, although a pair of waders might have been a useful addition to the royal wardrobe for the day. (PA Wire/PA Photos)

(Opposite. top) Inevitably, there was muttering in the 'neglected' North when massive media coverage was given to the floods that followed unprecedented rainfall in much of southern England on 20th July. One of the worst affected communities was Tewkesbury in Gloucestershire, which seen from the air was more akin to an island with the twelfth-century abbey only just clear of the surrounding sea of floodwater. (PA Wire/PA Photos)

(Lower) Tewkesbury was effectively cut off from the outside world with canoes ousting cars as a means of transport. (PA Wire/PA Photos)

The Beaufort Wind Scale

Francis Beaufort devised his scale of wind force in 1805, when serving aboard HMS Woolwich, and first mentioned it in his private log on 13 January 1806, stating that he would "hereafter estimate the force of the wind according to the following scale:"

Category	Description
0	Calm
1	Faint air just not calm
2	Light airs
3	Light breeze
4	Gentle breeze
5	Moderate breeze
6	Fresh breeze
7	Gentle steady gale
8	Moderate gale
9	Brisk gale
10	Fresh gale
11	Hard gale
12	Hard gale with heavy gusts
13	Storm

Checkpoint Controls and Cancer

Introduction

MICHAEL B KASTAN

Johns Hopkins Oncology Center, Baltimore, MD 21205

The progression of a eukaryotic cell through the cell cycle requires the integration of a large number of extracellular and intracellular signals. If the appropriate signals are not present, the cell will fail to make the transition from one phase of the cycle to the next. Cell cycle transitions depend on an ordered series of molecular events in which the initiation of one event is dependent on the successful completion of an earlier event. For example, it would be detrimental for a cell to begin replicative DNA synthesis before it has completed segregation of chromosomes in mitosis. The arrest of a cell at a particular phase of the cycle due to a lack of appropriate signals for progression is called a "cell cycle checkpoint" (Hartwell and Weinert, 1989). Such checkpoint controls presumably enhance the organism by minimizing somatic genetic alterations and/or affecting cellular survival.

These cell cycle arrest signals can be initiated either because of changes in the extracellular environment (eg alterations in nutrient status, mitogen signalling or cell-cell contact) or because of signals intrinsic to the cell (eg damage to the DNA). Many of our concepts of these dependent control mechanisms arose initially from studies in yeast, where the major control point of cell growth dependent on extracellular conditions was shown to occur in the G_1 phase of the cycle (called "start") (Hartwell *et al*, 1974), and a major control point after DNA damage appeared to be in the G_2 phase of the cycle and is dependent on the *rad9* gene (Weinert and Hartwell, 1988, 1990; Hartwell and Weinert, 1989). The basic molecular mechanics of cell cycle controls, including the sequences and/or functions of the cyclins, the cyclin dependent kinases and the cyclin dependent kinase inhibitors, appear to be highly conserved from yeast through humans. For example, the mammalian equivalent of the yeast "start" site is called the "restriction point" and similarly occurs at a discrete time in G_1 (Pardee, 1989). The DNA damage induced checkpoint controls appear to be similarly conserved in terms of sequence and function, with homologies among the yeast cell checkpoint genes *rad3* and *mec1*, the *Drosophila* gene *mei-41* and the human DNA damage response genes, *DNA-PK* and *ATM* (Hari *et al*, 1995; Hartley *et al*, 1995; Hunter, 1995; Morrow *et al*, 1995; Savitsky *et al*, 1995; Zakian, 1995). This degree of evolutionary conservation attests to the important roles of these control mechanisms in eukaryotic life and suggests that studies of these genetic and biochemical pathways in lower eukaryotes will be relevant to the biology of mammalian cells.

Cancer Surveys Volume 29: *Checkpoint Controls and Cancer*
© 1997 Imperial Cancer Research Fund. 0-87969-518-8/97. $5.00 + .00

Abnormalities in cell cycle regulation are relevant to cancer biology on at least two levels. Firstly, mutations in cell cycle control genes are selected for during tumorigenesis because they contribute to the increase in tissue cell number that is pathognomonic of tumours. Enhanced activity of gene products that drive the cell through the cycle or decreased activity of gene products that inhibit the cell cycle can contribute to the dysregulated cell growth characteristic of malignant cells. Secondly, mutations in cell cycle control genes can enhance the frequency of somatic mutations and thus contribute to the panoply of phenotypic abnormalities in tumours, such as loss of differentiation, invasiveness, angiogenesis and metastasis. Thus, cell cycle control genes appear to be both the targets of mutations during tumorigenesis and part of the engine that drives these somatic mutations.

The *TP53* gene is one of the most commonly mutated genes in human tumours (Hollstein *et al*, 1991, 1996). Elucidation of TP53 protein function has supported this concept that mutation of a gene product involved in control of cell proliferation can contribute to tumorigenesis both by disrupting normal growth controls and by enhancing the occurrence of somatic mutations. TP53 protein levels rapidly increase by a posttranscriptional mechanism following exposure of cells to DNA damaging agents (Kastan *et al*, 1991, 1992; Fritsche *et al*, 1993; Tishler *et al*, 1993; Nelson and Kastan, 1994; Huang *et al*, 1996). Appropriate cellular responses to other stresses, such as hypoxia and imbalance of nucleotide pools, also appear to rely on normal TP53 function (Graeber *et al*, 1994, 1996; Linke *et al*, 1996). The rise in TP53 protein levels after DNA damage is required for an arrest of mammalian cells in the G_1 phase of the cell cycle, prior to the restriction point, and thus prevents entry into S phase (Kastan *et al*, 1991, 1992; Kuerbitz *et al*, 1992; Kessis *et al*, 1993; Slebos *et al*, 1994); this cell cycle arrest thus prevents replication of a potentially damaged genome. TP53 also appears to influence the induction of apoptotic cell death following such cytotoxic insults (Clarke *et al*, 1993; Debbas and White, 1993; Lowe *et al*, 1993a,b; Yonish-Rouach *et al*, 1993; Canman *et al*, 1995). Thus, dysfunction of TP53 can potentially lead to both untimely replication of a damaged DNA template and inappropriate survival of cells after a cytotoxic insult (Fig. 1). Either or both of these events in a premalignant cell could contribute significantly to the progression of a cell to a fully transformed phenotype. Recent data from model systems have supported the notion that loss of TP53 mediated apoptosis is strongly selected for during tumorigenesis because of inappropriate survival of premalignant cells following imposition of certain cellular stresses such as DNA strand breaks or hypoxia (Symonds *et al*, 1994; Graeber *et al*, 1996). It has also been clearly demonstrated that loss of TP53 function contributes to increased somatic changes, including enhanced gene amplification and chromosomal aneuploidy (Livingstone *et al*, 1992; Yin *et al*, 1992; Meyn *et al*, 1994; Cross *et al*, 1995). Recent insights linking TP53 to the control of centrosome synthesis have provided another novel mechanistic insight into how TP53 dysfunction leads to genetic instability (Fukasawa *et al*, 1996).

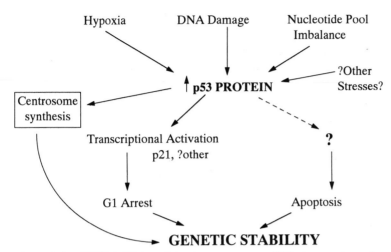

Fig. 1. Role of TP53 (p53) in cellular stress responses. DNA damage, hypoxia and nucleotide pool imbalance have all been shown to lead to increases in TP53 levels. It is conceivable that other cellular stresses may have similar consequences. Increases in TP53 levels appear to lead to either growth arrest or programmed cell death. Which of these physiological endpoints occurs can be influenced by either intracellular or extracellular factors. The TP53 dependent arrest of cells in the G_1 phase of the cell cycle appears to depend largely on the transcriptional activation of the *p21WAF1/CIP1* gene (Brugarolas *et al*, 1995; Deng *et al*, 1995; Waldman *et al*, 1995). The mechanism by which p53 induction enhances apoptosis is not known and whether transcriptional activation by TP53 is required is not clear. TP53 dysfunction can theoretically contribute to genetic instability and tumorigenesis because of either a defective cell cycle arrest or inappropriate cell survival or a combination of the two events. Loss of TP53 function also contributes to aneuploidy, a common alteration in human tumours, apparently by loss of its effects on centrosome synthesis (Fukasawa *et al*, 1996)

Less is known about the molecular mediators of other mammalian cell cycle checkpoints, but it is likely that there will be other examples of checkpoint genes that are mutated in human cancers. In addition to the immense task of replicating three billion nucleotides with fidelity in each cell cycle, the mitotic spindle and spindle poles must also function normally to maintain genetic integrity in each daughter cell. Whereas defects in surveillance of DNA could contribute to chromosomal rearrangements such as deletions, amplifications and translocations, defects in spindle surveillance could lead to mitotic non-dysjunction, leading to whole chromosome gain or loss, and defects in surveillance of the spindle poles could lead to changes in ploidy. All three of these categories of genetic abnormalities—chromosomal rearrangements, aneuploidy and polyploidy—are commonly found in tumour cells.

One of the major aims of this volume is to provide a broad view of how cell cycle control mechanisms normally function in eukaryotic cells and how their dysfunction contributes to genetic instability and malignant transformation. Detailed discussions of the mechanics of cell cycle progression are presented. Specific attention is paid to the G_1 to S transition, the mechanics and control of replicative DNA synthesis, and the transition of cells through G_2 and mitosis after DNA replication. Informative data on these issues come from a variety of experimental model systems spanning the eukaryotic evolutionary spectrum. Subsequent discussions then address how the cell uses this machinery to halt the cell cycle in response to various metabolic signals. The implications of the failure of mammalian cells to implement these protective mechanisms are then discussed, with a particular focus on malignant transformation. The contribution of cell survival in inappropriate physiological situations falls into this category. Model systems from in vitro mechanisms to cell culture systems to whole animal tumour models are utilized to provide these insights. One topic that has clear relevance to genetic instability but is not covered in this volume (because it does not fall under the rubric of cell cycle control abnormalities) is the role of mismatch repair defects in human cancers. The reader should be aware, however, of the critical role of DNA repair abnormalities in genetic instability and human tumorigenesis and can refer to other reviews for information on this topic. We hope the combination of discussions presented here will provide the reader with a framework to understand our current concepts of the mechanics of cell cycle control and cell cycle checkpoints and how dysfunction of these mechanisms contribute to tumorigenesis.

References

Brugarolas J, Chandrasekaran C, Gordon JI et al (1995) Radiation-induced cell cycle arrest compromised by p21 deficiency. *Nature* **377** 552–557

Canman CE, Gilmer T, Coutts S and Kastan MB (1995) Growth factor modulation of p53-mediated growth arrest vs. apoptosis. *Genes and Development* **9** 600–611

Clarke AR, Purdie CA, Harrison DJ et al (1993) Thymocyte apoptosis induced by p53-dependent and independent pathways. *Nature* **362** 849–852

Cross SM, Sanchez CA, Morgan CA et al (1995) A p53-dependent mouse spindle checkpoint. *Science* **267** 1353–1356

Debbas M and White E (1993) Wild-type p53 mediates apoptosis by E1A, which is inhibited by E1B. *Genes and Development* **7** 546–554

Deng C, Zhang P, Harper JW, Elledge SJ and Leder P (1995) Mice lacking p21CIP1/WAF1 undergo normal development, but are defective in G1 checkpoint control. *Cell* **82** 675–684

Fritsche M, Haessler C and Brandner G (1993) Induction of nuclear accumulation of the tumor-suppressor protein p53 by DNA-damaging agents. *Oncogene* **8** 307–318

Fukasawa K, Choi T, Kuriyama R, Rulong S and Vande Woude GF (1996) Abnormal centrosome amplification in the absence of p53. *Science* **271** 1744–1747

Graeber TG, Peterson JF, Tsai M et al (1994) Hypoxia induces accumulation of p53 protein, but activation of a G1-phase checkpoint by low-oxygen conditions is independent of p53 status. *Molecular and Cellular Biology* **14** 6264–6277

Graeber TG, Osmanian C, Jacks T et al (1996) Hypoxia-mediated selection of cells with diminished apoptotic potential in solid tumours. *Nature* **379** 88–91

Hari KL, Santerre A, Sekelsky JJ et al (1995) The mei-41 gene of *D. melanogaster* is a structural and functional homolog of the human ataxia telangiectasia gene. *Cell* **82** 815–821

Hartley KO, Gell D, Smith GCM et al (1995) DNA-dependent protein kinase catalytic subunit: a relative of phosphatidylinositol 3-kinase and the ataxia telangiectasia gene product. *Cell* **82** 849–856

Hartwell LH and Weinert TA (1989) Checkpoints: controls that ensure the order of cell cycle events. *Science* **246** 629–634

Hartwell LH, Culotti J, Pringle JR and Reid BJ (1974) Genetic control of the cell division cycle in yeast. *Science* **183** 46–51

Hollstein M, Sidransky D, Vogelstein B and Harris CC (1991) p53 mutations in human cancers. *Science* **253** 49–53

Hollstein M, Shomer B, Greenblatt M *et al* (1996) Somatic point mutations in the p53 gene of human tumors and cell lines: updated compilation. *Nucleic Acids Research* **24** 141–146

Huang L-C, Clarkin KC and Wahl GM (1996) Sensitivity and selectivity of the DNA damage sensor responsible for activating p53-dependent G1 arrest. *Proceedings of the National Academy of Sciences of the USA* **93** 4827–4832

Hunter T (1995) When is a lipid kinase not a lipid kinase? When it is a protein kinase. *Cell* **83** 1–4

Kastan MB, Onyekwere O, Sidransky D, Vogelstein B and Craig RW (1991) Participation of p53 protein in the cellular response to DNA damage. *Cancer Research* **51** 6304–6311

Kastan MB, Zhan Q, El-Deiry WS *et al* (1992) A mammalian cell cycle checkpoint pathway utilizing p53 and GADD45 is defective in ataxia-telangiectasia. *Cell* **71** 587–597

Kessis TD, Slebos RJ, Nelson WG *et al* (1993) Human papillomavirus 16 E6 expression disrupts the p53-mediated cellular response to DNA damage. *Proceedings of the National Academy of Sciences of the USA* **90** 3988–3992

Kuerbitz SJ, Plunkett BS, Walsh WV and Kastan MB (1992) Wild-type p53 is a cell cycle checkpoint determinant following irradiation. *Proceedings of the National Academy of Sciences of the USA* **89** 7491–7495

Linke SP, Clarkin KC, DiLeonardo A, Tsou A and Wahl GM (1996) A reversible, p53-dependent G0/G1 cell cycle arrest induced by ribonucleotide depletion in the absence of detectable DNA damage. *Genes and Development* **10** 934–947

Livingstone LR, White A, Sprouse J *et al* (1992) Altered cell cycle arrest and gene amplification potential accompany loss of wild-type p53. *Cell* **70** 923–935

Lowe SW, Ruley HE, Jacks T and Housman DE (1993a) p53-dependent apoptosis modulates the cytotoxicity of anticancer agents. *Cell* **74** 957–967

Lowe SW, Schmitt SW, Smith SW, Osborne BA and Jacks T (1993b) p53 is required for radiation-induced apoptosis in mouse thymocytes. *Nature* **362** 847–849

Meyn MS, Strasfeld L and Allen C (1994) Testing the role of p53 in the expression of genetic instability and apoptosis in ataxia-telangiectasia. *International Journal of Radiation Biology* **66** S141–S149

Morrow DM, Tagle DA, Shiloh Y, Collins FS and Hieter P (1995) TEL1, an *S. cerevisiae* homolog of the human gene mutated in ataxia telangiectasia, is functionally related to the yeast checkpoint gene MEC1. *Cell* **82** 831–840

Nelson WG and Kastan MB (1994) DNA strand breaks: the DNA template alterations that trigger p53-dependent DNA damage response pathways. *Molecular and Cellular Biology* **14** 1815–1823

Pardee AB (1989) G1 events and regulation of cell proliferation. Science 246 603–608

Savitsky K, Bar-Shira A, Gilad S *et al* (1995) A single ataxia telangiectasia gene with a product similar to PI-3 kinase. *Science* **268** 1749–1753

Slebos RJC, Lee MH, Plunkett BS *et al* (1994) p53-dependent G1 arrest involves pRB-related proteins and is disrupted by the human papillomavirus 16 E7 oncoprotein. *Proceedings of the National Academy of Sciences of the USA* **91** 5320–5324

Symonds H, Krall L, Remington L *et al* (1994) p53-dependent apoptosis suppresses tumor growth and progression in vivo. *Cell* **78** 703–711

Tishler RB, Calderwood SK, Coleman CN and Price BD (1993) Increases in sequence specific DNA binding by p53 following treatment with chemotherapeutic and DNA damaging agents. *Cancer Research* **53** 2212–2216

Waldman T, Kinzler KW and Vogelstein B (1995) p21 is necessary for the p53-mediated G1 arrest in human cancer cells. *Cancer Research* **55** 5187–5190

Weinert TA and Hartwell LH (1988) The RAD9 gene controls the cell cycle response to DNA damage in *Saccharomyces cerevisiae*. *Science* **241** 317–322

Weinert TA and Hartwell LH (1990) Characterization of RAD9 of *Saccharomyces cerevisiae* and evidence that its function acts post-tranlationally in cell cycle arrest after DNA damage. *Molecular and Cellular Biology* **10** 6554–6564

Yin Y, Tainsky MA, Bischoff FZ, Strong LC and Wahl GM (1992) Wild-type p53 restores cell cycle control and inhibits gene amplification in cells with mutant p53 alleles. *Cell* **70** 937–948

Yonish-Rouach E, Grunwald D, Wilder S *et al* (1993) p53-mediated cell death: relationship to cell cycle control. *Molecular and Cellular Biology* **13** 1415–1423

Zakian VA (1995) ATM-related genes: what do they tell us about functions of the human gene? *Cell* **82** 685–687

The author is responsible for the accuracy of the references.

Control of the G_1/S Transition

S I REED

Department of Molecular Biology, MB-7, Scripps Research Institute, 10550 North Torrey Pines Road, La Jolla, California 92037

INTRODUCTION

Arguably, the modern conceptualization of G_1 control in mammalian cells began with the discovery and characterization of *cdc* mutations in budding yeast (Hartwell *et al*, 1974). One mutation in particular, *cdc28-1*, conferred arrest in the G_1 phase of the cell cycle but did not impair growth, leading to the conclusion that the corresponding gene product had an essential role in the execution of a G_1 regulatory event. This G_1 restriction event has been called Start in yeast (Hartwell *et al*, 1974). After many years, and as a result of the contributions of many investigators, we now know that the *CDC28* gene encodes the prototype of a conserved family of protein kinases, known as cyclin dependent kinases (CDKs), that control most if not all of the major cell cycle transitions of eukaryotes, ranging from yeast to humans (reviewed in Pines, 1994; Lees, 1995; Morgan, 1995; Nigg, 1995). This includes the G_1/S phase transition, where it has become clear that cell cycle entry and commitment to a round of cell division is mediated by an interplay between internally and externally generated signals and a core machinery of cell cycle progression based on specialized CDKs (reviewed in Sherr, 1993; Pines, 1994; Lees, 1995; Morgan, 1995; Nigg, 1995). This short article reviews the evolution of current models of G_1 control in mammalian cells in the context of the most recent research developments.

G_1 CYCLIN DEPENDENT KINASES

As stated above, the first phenotype attributable to a loss of function mutation in a CDK was G_1 arrest conferred by a *cdc28* mutation in budding yeast

(Hartwell *et al*, 1974; Reed, 1980). Subsequently, mutations in the homologous gene of fission yeast, *cdc2*, were shown to confer primarily a G_2 arrest (Nurse and Bisset, 1981). Similarly, the first temperature sensitive mammalian CDK mutation conferred G_2 arrest (Th'ng *et al*, 1990), and CDK activity was shown to be essential for mitosis in frog eggs (Dunphy *et al*, 1988; Gautier *et al*, 1988). After a period of confusion, the apparent paradox was resolved by the discovery that CDKs form the core regulatory machinery for both the G_1/S phase and the G_2/M phase transitions. In budding yeast, this is accomplished with a single CDK and at least two distinct sets of regulatory components (reviewed in Nasmyth, 1993; Reed, 1995). Built into the CDK paradigm is the requirement for a positive regulatory subunit known as a cyclin. In budding yeast, the G_1/S phase transition is promoted by a class of G_1 cyclins known as Clns in conjunction with the Cdc28 kinase, whereas the G_2/M phase transition is promoted by a complementary class of G_2 cyclins known as Clbs, also with Cdc28 (reviewed in Nasmyth, 1993; Cross, 1995; Reed, 1995). The revelation of such a duality of CDK mediated cell cycle control in yeast led to the search for a comparable system in mammalian cells.

Mammalian G_1 Cyclins and CDKs

Although regulation of passage through G_1 is conceptually similar in yeast and mammalian cells, G_1 control in mammalian cells is mediated with an added level of complexity. Whereas both higher and lower eukaryotes utilize and regulate specialized G_1 cyclins, vertebrates also utilize and regulate several different CDK subunits for G_1 control (reviewed in Pines, 1994; Lees, 1995; Morgan, 1995; Nigg, 1995). Cyclins associated with the G_1/S phase transition in mammalian cells are D (of which there are three partially redundant isoforms), E and possibly A. The CDKs associated with G_1 control are CDK2, CDK4 and CDK6. Since the historical aspects of the discovery of G_1 CDKs are covered in a number of review articles, including those referenced above, they will not be discussed in the current article. Suffice it to say that many research avenues, both intentional and serendipitous, led to the discovery of mammalian G_1 cyclins and CDKs. Perhaps the most salient concept to emerge from this body of research is the close connection among human G_1 cyclins, CDKs and cancer (discussed below).

D Type Cyclins

Mammalian species express three cyclin D isoforms (D1, D2 and D3) in a cell specific and tissue specific pattern (reviewed in Sherr, 1993). It is assumed, but not yet proven, that all normal proliferating cells in an animal express at least one D type cyclin. The fact that the D type cyclin system is somewhat redundant is underscored by the analysis of cyclin D nullizygous mice. Whereas cyclin D1 nullizygous mice are virtually normal save for a few specific developmental defects and small size (Fantl *et al*, 1995; Sicinski *et al*, 1995), the cyclin D1/D2 doubly nullizygous mouse, although viable, exhibits a much more severe phenotype (Weinberg RA, personal communication). It is assumed that the triply nullizygous mouse will be inviable. Despite these relatively straightforward results using a mouse model, the issue of cyclin D essentially remains somewhat confused because of an apparent conflict with results obtained using cells in culture in conjunction with microinjection or electroporation approaches. Within the context of microinjecting a number of different cell lines with cyclin D1 specific antibodies or antisense constructs, cyclin D1 function was shown to be essential for the G$_1$/S phase transition (Baldin *et al*, 1993; Quelle *et al*, 1993). It is not yet clear whether the apparent conflict between the whole animal and tissue culture results resides in relatively trivial differences in experimental design or whether embryos deprived of an essential cyclin embark upon a modified programme of cyclin expression and usage. The availability of cell cultures from nullizygous animals has brought the resolution of these questions within experimental reach.

The cell cycle promoting effects of D type cyclins are thought to be mediated via activation of two structurally related CDKs designated CDK4 and CDK6 (Matsushime *et al*, 1992; Meyerson and Harlow, 1994). CDK4 and CDK6 share the interesting characteristic that they are structurally quite different from the prototypical CDKs discovered in budding and fission yeast. Furthermore, they are not histone H1 kinases, unlike the more structurally conserved members of the family. Although both kinases appear to be equally activated by D type cyclins in vitro, their relationship to each other is unclear. Many cell types apparently express both CDK4 and CDK6, although not necessarily at equal levels.

Cyclin E

The other class of cyclin associated with G$_1$/S control is that defined by cyclin E. Although encoded by a single gene, alternative splicing apparently yields at least two distinct cyclin E polypeptides that differ by the addition of a few N-terminal aminoacids (Ohtsubo *et al*, 1995). So far, no functional or regulatory difference has emerged to explain the existence of these two isoforms. Cyclin E has been reported to associate with a single catalytic partner, CDK2 (Dulic *et al*, 1992; Koff *et al*, 1992). Unlike CDK4 and CDK6, CDK2 is one of the class of histone H1 kinases that are highly conserved with respect to the prototype yeast CDKs, cdc28 and cdc2. Cyclin E appears to be expressed in all proliferating tissues and cell types, and although a nullizygous mouse has not been reported, an antibody microinjection experiment suggests that it is essential for progression into S phase in mammalian cells (Ohtsubo *et al*, 1995). This is consistent with observations made in replication competent *Xenopus* egg extracts, where immunodepletion of cyclin E blocks entry into S phase (Jackson *et al*, 1995; Rempel *et al*, 1995). These results, however, must be extrapolated to mammalian somatic cells with caution because CDK activities are regulated and utilized somewhat differently during amphibian early embryonic development. Furthermore, although antibody microinjection experiments suggest that cyclin E is essential for S phase entry, transfection with antisense oligonucleotides reveals that cyclin E is not rate limiting for the G_1/S phase transition, since the level of cyclin E could be reduced dramatically without any impact on the rate of entry into S phase (Sato K and Reed S, unpublished).

Cyclin A

Although cyclin A is not normally catagorized as a G_1 cyclin, a G_1/S role cannot be rigorously excluded. Firstly, it has been demonstrated that ectopic expression of cyclin A in G_1 can advance entry into S phase (Resnitzky et al, 1995). Secondly, although the level of cyclin A protein and associated kinase activity peaks late in G_2, there is residual cyclin A throughout G_1 that might be regulated posttranslationally, and cyclin A protein begins to accumulate before S phase (Dulic et al, 1992, 1994). Phenotypically, loss of cyclin A function via antibody microinjection causes a failure to replicate DNA (Girard et al, 1991; Pagano et al, 1992), but this has been interpreted as implicating cyclin A in progression through S phase rather than in S phase entry, largely because of the kinetics of cyclin A accumulation.

Like cyclin E, cyclin A associates with and activates CDK2 (Pines and Hunter, 1990; Elledge and Spottswood, 1991; Ninomiya-Tsuji et al, 1991; Tsai et al, 1991). This is assumed to be the principal catalytic partner of cyclin A during S phase. However, unlike cyclin E, cyclin A also can be isolated in complex with CDK1 (Pines and Hunter, 1991). It has been proposed that cyclin A/CDK2 might account for S phase specificities, whereas cyclin A/CDK1 activity might be directed towards G_2 or M phase functions. However, no definitive experiment resolving this issue has been reported.

REGULATION AND FUNCTION OF G_1 CDKs

Consistent with the central role of cell cycle regulation in the physiology of cells and organisms, there appear to be multiple layers of regulation accessible to both internal and external signals. These include modulation of both positive and negative regulatory proteins, as well as of protein phosphorylation.

Regulation of Cyclin Availability

As stated above, cyclins are requisite positive regulatory subunits of all CDKs. The determination of the cyclin A/CDK2 three dimensional structure illustrates this graphically in that cyclin A binding remodels the active site of a CDK into a catalytically competent conformation (Jeffrey et al, 1995). It is assumed that other cyclins act similarly. From the point of view of the G_1/S phase transition, therefore, the accumulation of G_1 cyclins allowing the activation of G_1 CDKs would be expected to be an important component of regulation. This view is bolstered by the observation that ectopic overexpression of D type cyclins, cyclin E or cyclin A during G$_1$ can advance entry into S phase (Ohtsubo and Roberts, 1993; Quelle *et al*, 1993; Resnitzky *et al*, 1994, 1995; Resnitsky and Reed, 1995). However, it is not clear whether modulation of cyclin availability always has an important role in the normal course of cell cycle progression. The best case for regulation of this type is for D type cyclins in the context of cell cycle exit and entry. In fact, cyclin D1 was discovered as the product of a growth factor responsive transcript in macrophages (Matsushime *et al*, 1991). Subsequently, in many cell types, removal of mitogenic stimuli and concomitant entry into a quiescent state has been associated with failure to express or maintain expression of cyclin D1 and/or other D type cyclins (reviewed in Sherr, 1993). However, not all antiproliferative signals lead to downregulation of cyclin D levels, and therefore other regulatory mechanisms must be operative (discussed below). In unrestrained cycling cells, cyclin D levels do not appear to be highly regulated, according to the results of biochemical analysis of proteins from synchronized populations (Matsushime *et al*, 1991; Baldin *et al*, 1993). However, examination of individual cells with immunofluorescence microscopy reveals that cyclin D1 undergoes dramatic, but poorly understood, changes of state through the cell cycle (Baldin *et al*, 1993; Dulic *et al*, in press). Early G$_1$ cells appear to be devoid of detectable staining, whereas mid to late G$_1$ cells stain positively in their nuclei. However, before S phase, cyclin D1 again becomes undetectable with immunofluorescence techniques and remains so through S phase. Whether these fluxes result from changes in cellular compartmentalization or from changes in association with other proteins affecting fixation remains to be determined, but they raise the possibility that cyclin availability based on parameters other than steady state level may have an important role in cyclin D regulation.

Cyclin E levels appear to be more periodic than cyclin D levels in cycling cells. This is confirmed both biochemically and cytologically (Dulic *et al*, 1992, in press; Koff *et al*, 1992; Ohtsubo *et al*, 1995). In most cellular systems, cyclin E appears in the nucleus and disappears from it later than does cyclin D1. However, in light of the observation that cyclin E levels are not rate limiting for S phase entry in normal circumstances, it is not clear how important this regulation is. Another distinction between the response of cyclin E and D1 levels to external signals is in the context of cell cycle exit. Whereas, as stated above, cyclin D1 downregulation appears to be a hallmark of quiescence, cyclin E levels persist in many cell types (Dulic *et al*, 1994; Slingerland *et al*, 1994; Hengst and Reed, 1996). In many systems, only modest cyclin E accumulation accompanies progression from G$_0$ to S phase. Mechanistically, this is likely to occur because cyclin E/CDK2 complexes are maintained in an inactive state in quiescent cells and cyclin E half life is tightly coupled to the activity of cyclin E/CDK2 complexes, thus allowing accumulation without a high biosynthetic rate (Won and Reed, 1996). However, the implications of this apparent buffering of cyclin E levels are that in the context of cell cycle entry and exit, cyclin E levels per se are not likely to be of regulatory importance. As stated above, although cyclin A levels are low in G$_0$ and G$_1$, they are detectable, and therefore in the absence of a clear definition of the role of cyclin A and the levels of cyclin A required to perform such a role, the same issues pertinent to cyclin E regulation are also pertinent to cyclin A. As for cyclin E, posttranslational modulation of cyclin A associated kinase activity may be the predominant regulatory mode in the context of G_1/S control.

Negative Regulation of CDK Activity

As suggested above, modes of CDK regulation other than cyclin availability appear to be predominant in cell cycle control. Mechanistically, there are two general strategies of negative regulation, CDK inhibitory proteins and negative phosphorylation of CDKs.

CDK Inhibitors

Two structurally related families of CDK inhibitors have been identified (reviewed in Lees, 1995; Sherr and Roberts, 1995). One of these, the Cip/Kip family, consists of three members in mammalian species that are active against virtually all G_1 CDK activities, including cyclin D/CDK4(CDK6), cyclin E/CDK2 and cyclin A/CDK2. The other family, known collectively as INK4, consists of at least four members and appears to be specific for CDK4 and CDK6. There is no detectable homology between the two classes of inhibitors.

Experiments with cells in culture indicate that Cip/Kip family members accumulate in response to negative regulatory signals or the withdrawal of positive mitogenic signals. For example, DNA damage in fibroblasts in response to ionizing or ultraviolet radiation causes a TP53 dependent increase in the inhibitor p21^{Cip1} (Dulic *et al*, 1994; El-Deiry *et al*, 1994). By mechanisms that have not yet been elucidated, the presence of DNA damage is conveyed to TP53, a specialized transcription factor responsible for p21^{Cip1} transcriptional activation (El-Deiry *et al*, 1993). Resultant elevation in p21^{Cip1} levels then leads to inhibition of all essential G_1 CDK activities, conferring G_1 arrest (Harper *et al*, 1993, 1995). This model has been partially confirmed using p21^{Cip1} nullizygous mice, in that fibroblasts from such animals have an impaired but not completely abrogated G_1 DNA damage checkpoint (Brugarolas *et al*, 1995; Deng *et al*, 1995). Therefore, not all of the TP53 dependent G_1 arrest in response to DNA damage is mediated via p21^{Cip1}. Furthermore, an essential role for p21^{Cip1} in development, implied from cell culture experiments and in situ hybridization analysis of mouse embryos (Parker *et al*, 1995), is ruled out by the apparent normal development of the nullizygous mice (Brugarolas *et al*, 1995; Deng *et al*, 1995). For example, it has been inferred from the increase in p21^{Cip1} levels observed as myocytes leave the cell cycle and fuse to form myotubes that p21^{Cip1} is important for the process (Halevy *et al*, 1995). However, p21^{Cip1} nullizygous mice show no defect in muscle development.

Tissue culture cell studies on the structurally related inhibitor p27^{Kip1} indicate a link between its regulation and the withdrawal of mitogenic signals.

For example, p27^{Kip1} levels increase in response to withdrawal of serum or growth factors in various cell types (Hengst *et al*, 1994; Nourse *et al*, 1994; Pagano *et al*, 1995; Hengst and Reed, 1996). Unlike the regulation of p21^{Cip1}, the regulation is translational and posttranslational but not at the level of transcription (Pagano *et al*, 1995; Hengst and Reed, 1996). Antisense experiments with established cell lines confirm a role for p27^{Kip1} in cell cycle exit in that targeting of p27^{Kip1} maintains cells in cycle even in low serum (Coats *et al*, 1996). Also consistent with this model is the observation that p27^{Kip1} nullizygous mice are larger than normal and suffer from multiple tissue hyperplasias (Fero *et al*, 1996; Kiyokawa *et al*, 1996; Nakayama *et al*, 1996;). On the other hand, development is largely normal, and excess cell division is limited. Furthermore, embryonic fibroblasts isolated from p27^{Kip1} nullizygous mice appear to have a normal response to serum withdrawal, in contrast to the established cell lines described above (Nakayama *et al*, 1996). Therefore, the developmental lesion associated with loss of p27^{Kip1} function is much more subtle than a catastrophic derangement of cell cycle control. The relative "normalities" of cells derived from both p21^{Cip1} and p27^{Kip1} nullizygous mice indicate that cell cycle regulation is highly redundant in vivo, that some redundancy is probably lost as cells are adapted to tissue culture and that at least some of the mechanisms of cell cycle control remain to be elucidated. Therefore, although there is now a wealth of literature describing changes in p21^{Cip1} and/or p27^{Kip1} levels in response to cell cycle regulatory signals (too extensive to describe here), the biological importance of these changes remains to be validated in vivo and using material from nullizygous mice.

Finally, one member of the Cip/Kip family that has not been imbued with important regulatory functions as a result of tissue culture studies is p57^{Kip2} (Lee *et al*, 1995; Matsuoka *et al*, 1995). Although p57^{Kip2} mRNA undergoes significant tissue specific fluctuations during mouse development (Matsuoka *et al*, 1995), no dramatic changes in p57^{Kip2} levels have been reported for tissue culture cells being subjected to proliferative or antiproliferative signals. Furthermore, observation of changes in expression pattern during development may not be informative with regard to essential functions, since p21^{Cip1} undergoes equally dramatic changes in expression during embryogenesis, but elimination of p21^{Cip1} has no apparent effect on the process.

With regard to the mechanism of action of Cip/Kip inhibitors, all share homology in an N-terminal domain of approximately 60 aminoacids shown to be essential and sufficient for CDK inhibition. The function(s) of the C-terminal portions of each is not clear. A site capable of binding the essential replication factor PCNA (proliferating cell nuclear antigen) is present near the C-terminal end of p21^{Cip1} (Waga *et al*, 1994; Chen *et al*, 1995; Luo *et al*, 1995; Nakanishi *et al*, 1995), although the function of this interaction in vivo remains to be determined. There has been some support for the idea that a stoichiometry of 2:1 of inhibitor molecules to CDK complexes is required for inhibition (Zhang *et al*, 1994; Harper *et al*, 1995). However, the determination of the three dimensional structure of the inhibitory domain of p27^{Kip1} bound to

CDK2/cyclin A by crystallographic methods argues strongly against this idea (Russo *et al*, 1996). The structure indicates that one inhibitor molecule binds to the cyclin moiety of the complex and invades the active site of the CDK in a manner that would preclude catalytic activity. Specifically, the C-terminus of the inhibitory domain displaces a strand of the β sheet that constitutes the N-terminal lobe of the kinase and extends to occupy the ATP binding site of the catalytic cleft (de Bondt *et al*, 1993; Jeffrey *et al*, 1995). Therefore, assuming that $p21^{Cip1}$ binds in the same fashion, a 1:1 stoichiometry should be fully inhibitory.

p16

Less is known about the role of INK4 family inhibitors in cell cycle control. The first member of the family to be identified, p16, accumulates specifically in cells expressing viral oncoproteins and in some tumour cells devoid of the tumour suppressor protein RB (discussed below) (Serrano *et al*, 1993; Li *et al*, 1994; Parry *et al*, 1995). However, the functional significance of these relationships is not evident, and p16 does not accumulate in cells from RB negative mouse embryos (Peters G, personal communication). The most interesting correlation is between loss of p16 and cancer (discussed below). Another related inhibitor, p15, does appear to have a definitive role in cell cycle regulation. Studies on the response of epithelial cells to the cytokine transforming growth factor β indicate that p15 is strongly induced at the transcriptional level (Hannon and Beach, 1994; Reynisdottir *et al*, 1995). This in turn leads to saturation and inhibition of CDK4 and CDK6. A secondary effect of p15 induction is displacement of $p27^{Kip1}$ from CDK4 and CDK6 complexes so that it can now stoichiometrically inhibit cyclin E/CDK2 complexes and, possibly, cyclin A/CDK2 complexes (Reynisdottir *et al*, 1996). Thus, induction of one inhibitor, p15, has the combined effect of inhibiting CDK4 and CDK6, as well as mobilizing $p27^{Kip1}$ to inhibit CDK2, even though levels of p27 are not apparently increased. The roles of two other INK4 family inhibitors, p18 and p19, are not known (Guan *et al*, 1994; Chan *et al*, 1995; Hirai *et al*, 1995). They appear to be expressed in many cell types at the G_1/S phase boundary, suggesting that they may be involved in downregulating CDK4 and CDK6 as cells progress from G_1 to S phase.

p15

p18

p19

Regulation of CDK Phosphorylation

A second mode of potential regulation of CDKs is by phosphorylation of inhibitory sites. Although all CDKs require a phosphorylation in their active site for activity, there is no evidence to date that this is of regulatory importance. On the other hand, CDK2 can be inhibited by phosphorylation on Tyr-15 and, possibly, Thr-14 (Gu *et al*, 1992; Sebastian *et al*, 1993). It has been proposed that such phosphorylation is constitutive, since antibody microinjection mediated loss of activity of the CDK T14Y15 phosphatase Cdc25A leads to G_1 arrest (Hoffmann *et al*, 1994; Jinno *et al*, 1994), although the precise kinase complex of importance in this experiment has not been identified. However, the regulatory significance of Cdc25A dependence for activation of G_1 CDKs comes from work with a human Burkitt's lymphoma derived cell line known as

Daudi. When Daudi cells are treated with the cytokine interferon α, which causes them to enter a non-proliferative state, there is a rapid transcriptional downregulation of Cdc25A, which leads to an accumulation of inactive, tyrosine phosphorylated CDK2/cyclin E and CDK2/cyclin A complexes (Tiefenbrun *et al*, 1996). Presumably, this will be a paradigm for other cytokine mediated antiproliferative responses.

Function of G$_1$ CDKs

Whereas much is known about the regulation of G$_1$ CDK activities, less has been learned about their targets. The notable exception is RB, the product of the retinoblastoma susceptibility gene. RB is a negative regulator of the G$_1$/S phase transition and is inactivated by phosphorylation on sites that match the CDK consensus (reviewed in Hinds, 1995; Weinberg, 1995). It is clear that phosphorylation of RB late in G$_1$ constitutes an important regulatory component of the G$_1$/S phase transition (reviewed in Hinds, 1995; Weinberg, 1995). Although still a matter of controversy, an increasing body of evidence suggests that the primary function of D type cyclins in conjunction with CDK4 and CDK6 is to phosphorylate RB late in G$_1$. In support of this conclusion, elimination of CDK4/CDK6 activity, either by neutralizing cyclin D or by ectopic expression of one of the INK4 inhibitors, p16, confers G$_1$ arrest only in cells that contain functional RB (Lukas *et al*, 1994, 1995a,b; Koh, *et al*, 1995). In cells rendered negative for RB function, either by mutation or (in the mouse) as a result of targeted disruption, CDK4 and CDK6 are non-essential for proliferation. Furthermore, RB is the only known efficient substrate of CDK4 and CDK6 in vitro (Matsushime *et al*, 1994; Meyerson and Harlow, 1994). The most prevalent model for RB function is that hypophosphorylated RB sequesters transcription factors that promote the G$_1$/S phase transition, particularly members of the E2F family that accumulate during G$_1$ (reviewed in Hinds, 1995; Weinberg, 1995; Sanchez and Dynlacht, 1996). Phosphorylation of RB by CDK4 and CDK6 causes release of these factors, so that genes essential for progression to S phase can be *trans*-activated. The fact that ectopic overexpression of E2F isoforms can promote forced entry of quiescent cells into S phase is consistent with the idea that release of sequestered E2F is a critical regulatory event (Johnson *et al*, 1993; De Gregori *et al*, 1995). In fact, the phosphorylation of RB is the molecular event that best corresponds to the G$_1$ restriction point (R point) proposed, based on physiological studies, to be the point where signals pertinent to growth and differentiation are integrated into the cell cycle control machinery (reviewed in Zetterberg *et al*, 1995).

There are several oversimplifications in the model for RB function and regulation outlined above. Firstly, it is clear that RB has the capacity to interact with and regulate a number of proteins, including transcription factors other than E2F. These include the proto-oncogene product ABL (Welch and Wang, 1993, 1995) and the transcription factor Elf-1 (Wang *et al*, 1993). However, the consequences of regulation of these proteins by RB remain to be clarified. Secondly, E2F consists of at least five different isoforms that form obligate heterodimers with a second group of proteins known as DP-1 (reviewed in La Thangue, 1994; Muller, 1995; Sanchez and Dynlacht, 1996). Whereas it has been reported that RB is found in complex with E2F1, E2F2 and E2F3, two other E2F isoforms, E2F4 and E2F5, are found in complex with two RB related proteins known as p107 and p130 (reviewed in Sanchez and Dynlacht, 1996). Although the functions of p107 and p130 combined with E2F4 and E2F5 appear to be repressive, the roles of these proteins in cell cycle control in vivo are not yet clear. Although cells derived from RB nullizygous embryos are clearly defective in cell cycle regulation (Jacks *et al*, 1992; Lee *et al*, 1992), as would be expected for loss of a key regulator, cells derived from p107 and p130 nullizygous mice appear to be normal (Jacks T, personal communication). Analysis of development in multiply nullizygous embryos suggests a degree of functional redundancy between the three members of the RB family, but RB remains the dominant member in terms of both development and cell cycle regulation (Jacks T, personal communication). Furthermore, although p107 and p130 appear to become phosphorylated at or near the G_1/S phase transition, the regulatory roles of phosphorylation and the kinases involved are not clear. Finally, the interrelationship between the different E2F isoforms and their respective roles in cell cycle regulation remain to be determined by loss of function experiments.

Whereas the relationship between D type cyclins and their partners, CDK4 and CDK6, and RB phosphorylation appears to be universally accepted, there is little consensus on the targets of cyclin E/CDK2 kinase. Support for the idea that cyclin E/CDK2 contributes to RB phosphorylation derives principally from experiments where plasmids expressing cyclin E and RB were transiently co-transfected into cells sensitive to inhibition by ectopic expression of RB (Hinds *et al*, 1992). The fact that RB mediated inhibition was neutralized and RB became phosphorylated was interpreted as indicative of RB phosphorylation by cyclin E/CDK2. However, the excessive overexpression and lack of synchrony in the populations analysed leave these experiments open to alternative interpretations. In fact, under conditions that are less perturbed, where D type cyclins can be shown to be effective in driving phosphorylation of endogenous RB, cyclin E activation of CDK2 was incapable of driving RB phosphorylation (Resnitzky and Reed, 1995). Yet under these conditions, cyclin E clearly advanced cells from G_1 to S phase, consistent with phosphorylation of alternative targets. Unfortunately, such alternative targets have not yet been identified, and the true function of cyclin E remains a mystery. Perhaps the most informative suggestions derive from work using frog eggs. There, in the absence of any level of G_1 growth control, cyclin E/CDK2 is essential and involved in events close to the initiation of DNA replication (Jackson *et al*, 1995; Rempel *et al*, 1995). In fact, cyclin E has been shown to co-localize with centres of DNA replication in S phase nuclei (Jackson *et al*, 1995), consistent with a direct S phase function. However, these results need to interpreted cautiously with regard to extrapolation to somatic cells. The most notable distinction between the two systems is that the embryonic isoform of cyclin A found in amphibian eggs and early embryos does not complex to CDK2, whereas cyclin A in somatic cells is primarily complexed to CDK2 (Fang and Newport, 1991; Howe *et al*, 1995). Therefore, cyclin E/CDK2 in eggs and early embryos may perform a function delegated to cyclin A/CDK2 in somatic cells. The role of cyclin E in the G$_1$/S phase transition thus remains to be elucidated.

Similarly, the role of cyclin A/CDK2 has not been established. Although cyclin A is capable of driving G$_1$ cells into S phase when expressed ectopically (Resnitzky *et al*, 1995), it is not known whether this corresponds to a normal function of cyclin A or a gain of function artefact. Microinjection experiments indicate that cyclin A is essential for some aspect of S phase (Girard *et al*, 1991; Pagano *et al*, 1992), but it is not clear whether this represents an essential initiation function or an S phase maintenance function. One potential essential function of cyclin A during S phase is the phosphorylation and concomitant inactivation of E2F/DP-1 heterodimers (Dynlacht *et al*, 1994; Krek *et al*, 1994; Xu *et al*, 1994). Impairment of this event leads to a mid S phase arrest for reasons that remain elusive. However, negative regulation of E2F by phosphorylation is unlikely to be a prerequisite for the G$_1$/S phase transition.

G$_1$ CONTROL AND CANCER

Since cancer is in part a disease of uncontrolled cellular proliferation, it seems logical at face value to presume that mutations that deregulate cell cycle control might contribute to cancer. However, a more serious consideration of what constitutes malignancy argues against such a simplistic hypothesis. In fact, it still remains to be determined whether simple deregulation of the cell cycle is a major factor in cancer. The most intriguing observation relevant to this discussion is the clear involvement of many of the genes discussed above in the context of cell cycle control in malignancy. Cyclin D1 has been recovered as an oncogene in numerous different contexts (Lammie *et al*, 1991; Motokura *et al*, 1991; Withers *et al*, 1991). p16INK4, the inhibitor of cyclin D/CDK4, is an important tumour suppressor in both familial and sporadic cancer (Kamb *et al*, 1994; Nobori *et al*, 1994; Okamoto *et al*, 1994; Koh *et al*, 1995), and mutant alleles of CDK4 that cannot be inhibited by p16 have also been implicated. In addition, p16 nullizygous mice are extremely cancer prone (Serrano *et al*, 1996). RB, the apparent target of the cyclin D/CDK4 regulatory system, is also an important tumour suppressor (reviewed in Hinds, 1995; Weinberg, 1995). To a lesser extent, cyclin E has been suggested as an oncogene on the basis of amplification of the cyclin E gene and overexpression of cyclin E in cells derived from breast carcinomas (Keyomarsi and Pardi, 1994). In the latter case, however, the contribution of cyclin E to the aetiology of the disease has not been clearly established. Equally intriguing, however, is the fact that p21[Cip1] and p27[Kip1] have not been implicated as tumour suppressors in human malignancy. Nor have the other INK4 inhibitors p15, p18 and p19. The fact that many types of mutations that would be expected to deregulate the cell cycle are not recovered as oncogenes or tumour suppressors suggests that cancer is not simply a disease of proliferation. In fact, the data available are more compatible with the idea that loss of cell cycle control can lead to cancer only in specialized circumstances where proliferation leads to override of checkpoint controls that guard genetic stability and therefore to genetic instability. This idea is supported by the observations that p16 accumulates in fibroblasts from mouse embryos only as they approach senescence and that fibroblasts from p16 nullizygous mice are inherently immortal and do not go through crisis (Beach D, personal communication). Although the contribution of genetic instability to malignancy is well documented, this hypothesis will be rigorously testable only when the molecular mechanisms underlying cell cycle checkpoint control have been elucidated in detail.

SUMMARY

On the basis of current knowledge, control of the G_1/S phase transition is largely a matter of regulating a set of specific cyclin dependent kinase (CDK) activities. In mammalian cells, the G_1/S specific CDK activities are composed of complexes between D type cyclins and either CDK4 or CDK6 and between cyclin E (and possibly cyclin A) and CDK2. A variety of internal and external signals regulate G_1/S specific CDKs by modulating cyclin availability, the levels of CDK inhibitory proteins and the phosphorylation status of CDKs. Although much is now known about the regulation of G_1/S specific CDKs, the only well characterized substrate to date is the retinoblastoma gene product, RB. Phosphorylation of RB by CDKs neutralizes its cell cycle inhibitory properties, allowing progression of G_1 to S phase. Not surprisingly, many components of the cell cycle regulatory machinery, including CDKs, CDK inhibitors and CDK substrates, are important targets of mutations that lead to human malignancy.

References

Baldin V, Lukas J, Marcote MJ, Pagano M and Draetta G (1993) Cyclin D1 is a nuclear protein required for cell cycle progression in G1. *Genes and Development* **7** 812–821

Brugarolas J, Chandrasekaran C, Gordon JI, Beach D, Jacks T and Hannon G (1995) Radiation-induced cell cycle arrest compromised by p21 deficiency. *Nature* **377** 552–557

Chan FKM, Zhang J, Cheng L, Shapiro DN and Winoto A (1995) Identification of human and mouse p19, a novel CDK4 and CDK6 inhibitor with homology to p16[ink4]. *Molecular and Cellular Biology* **15** 2682–2688

Chen J, Jackson PK, Kirschner MW and Dutta A (1995) Separate domains of p21 involved in the inhibition of Cdk kinase and PCNA. *Nature* **374** 386–388

Coats S, Flanagan WM, Nourse J and Roberts JM (1996) Requirement for p27[Kip1] for restriction point control in the fibroblast cell cycle. *Science* **272** 877–879

Cross FR (1995) Starting the cell cycle: what's the point? *Current Opionion in Cell Biology* **7**

790–797

De Bondt HL, Rosenblatt J, Jancarik J, Jones HD, Morgan DO and Kim S-H (1993) Crystal structure of cyclin-dependent kinase 2. *Nature* **363** 595–602

De Gregori J, Kowalik T and Nevins JR (1995) Cellular targets for activation of the E2F1 transcription factor include DNA synthesis- and G1/S-regulatory genes. *Molecular and Cellular Biology* **15** 4215–4224

Deng C, Zhang P, Harper JW, Elledge SJ and Leder P (1995) Mice lacking p21$^{CIP1/WAF1}$ undergo normal development, but are defective in checkpoint control. *Cell* **82** 675–684

Dulic V, Lees E and Reed SI (1992) Association of human cyclin E with a periodic G1-S phase protein kinase. *Science* **257** 1958–1961

Dulic V, Kaufmann WK, Wilson SJ *et al* (1994) p53-dependent inhibition of cyclin-dependent kinase activities in human fibroblasts during radiation-induced G1 arrest. *Cell* **76** 1013–1023

Dulic V, Stein GH, Far DF and Reed SI Biphasic nuclear localization of p21^{Cip1} and cyclin D1: evidence for a role in the G2/M transition. *Oncogene* (in press)

Dunphpy WG, Brizuela, L, Beach D and Newport J (1988) The *Xenopus cdc2* protein is a component of MPF, a cytoplasmic regulator of mitosis. *Cell* **54** 423–431

Dynlacht BD, Flores O, Lees JA and Harlow E (1994) Differential regulation of E2F transactivation by cyclin/cdk2 complexes. *Genes and Development* **8** 1772–1786

El-Deiry WS, Tokino T, Velculescu VE *et al* (1993) WAF1, a potential mediator of p53 tumor suppression. *Cell* **75** 817–825

El-Deiry WS, Harper JW, O'Connor PM *et al* (1994) WAF1/CIP1 is induced in p53-mediated G1 arrest and apoptosis. *Cancer Research* **54** 1169–1174

Elledge SJ and Spottswood MR (1991) The new human p34 protein kinase, CDK2, identified by complementation of a *cdc28* mutation in *Saccharomyces cerevisiae*, is a homolog of *Xenopus* Eg1. *EMBO Journal* **10** 2653–2659

Fang F and Newport J (1991) Evidence that that G1-S and the G2-M transitions are controlled by difference cdc2 proteins in higher eukaryotes. *Cell* **66** 731–742

Fantl V, Stamp G, Andrews A, Rosewell I and Dickson C (1995) Mice lacking cyclin D1 are small and show defects in eye and mammary gland development. *Genes and Development* **9** 2364–2372.

Fero ML, Rivkin M, Tasch M *et al* (1996) A syndrome of multiple organ hyperplasia with features of gigantism, tumorogenesis and female sterility in p27^{Kip1}-deficient mice. *Cell* **85** 733–744

Gautier J, Norbury C, Lohka M, Nurse P and Maller J (1988) Purified maturation-promoting factor contains the product of a *Xenopus* homolog of the fission yeast cell cycle control gene *cdc2$^+$*. *Cell* **54** 433–439.

Girard F, Strausfeld U, Fernandez A and Lamb NJ (1991) Cyclin A is required for the onset of DNA replication in mammalian fibroblasts. *Cell* **67** 1169–1179

Gu Y, Rosenblatt J and Morgan D (1992) Cell cycle regulation of Cdk2 activity by phorphosrylation of Thr160 and Tyr15. *EMBO Journal* **11** 3995–4005

Guan K-L, Jenkins CW, Li Y *et al* (1994) Growth suppression by p18, a p16$^{INK4/MTS1}$ and p14$^{INK4/MTS2}$-related CDK6 inhibitor, correlates with wild-type pRb function. *Genes and Development* **8** 2939–2952

Halevy O, Novitch BG, Spicer DB *et al* (1995) Correlation of terminal cell cycle arrest of skeletal muscle with induction of p21 by myoD. *Science* **267** 1018–1021

Hannon GJ and Beach D (1994) p15^{INK4B} is a potential effector of TGF-β-induced cell cycle arrest. *Nature* **371** 257–261

Harper JW, Adami GR, Wei N, Keyomarsi K and Elledge SJ (1993) The p21 Cdk-interacting protein Cip1 is a potent inhibitor of G1 cyclin-dependent kinases. *Cell* **75** 805–816

Harper JW, Elledge SJ, Keyomarsi K *et al* (1995) Inhibition of cyclin-dependent kinases by p21. *Molecular Biology of the Cell* **6** 387–400

Hartwell LH, Culotti J, Pringle JR and Reid BJ (1974) Genetic control of the cell division cycle in yeast. *Science* **183** 46–51

Hengst L and Reed SI (1996) Translation control of p27[Kip1] accumulation during the cell cycle. *Science* **271** 1861–1864

Hengst L, Dulic V, Slingerland JM, Lees E and Reed SI (1994) A cell cycle regulated inhibitor of cyclin-dependent kinases. *Proceedings of the National Academy of Sciences of the USA* **91** 5291–5295

Hinds PW (1995) The retinoblastoma tumor suppressor protein. *Current Opinion in Genetics and Development* **5** 79–83

Hinds PW, Mittnacht S, Dulic V *et al* (1992) Regulation of retinoblastoma functions by ectopic expression of human cyclins. *Cell* **70** 993–1006

Hirai H, Roussel MF, Kato J-Y, Ashmun RA and Sherr CJ (1995) Novel Ink4 proteins, p19 and p18, are specific inhibitors of cyclin D-dependent kinases CDK4 and CDK6. *Molecular and Cellular Biology* **15** 2672–2681

Hoffmann I, Draetta G and Karsenti E (1994) Activation of the phosphatase activity of human cdc25A by a cdk2-cyclin E dependent phosphorylation at the G1/S transition. *EMBO Journal* **13** 4302–4310

Howe JA, Howell M, Hunt T and Newport JW (1995) Identification of a developmental timer regulating the stability of embryonic cyclin A and a new somatic A-type cyclin at gastrulation. *Genes and Development* **9** 1164–1176

Jacks T, Fazeli A, Schmitt EM, Bronson RT, Goodell MA and Weinberg RA (1992) Effects of an *Rb* mutation in the mouse. *Nature* **359** 295–300

Jackson PK, Chevalier S, Philippe M and Kirschner MW (1995) Early events in DNA replication require cyclin E and are blocked by p21[CIP1]. *Journal of Cell Biology* **130** 755–769.

Jeffrey PD, Russo AA, Polyak K *et al* (1995) Crystal structure of a cyclin A-cdk2 complex at 2.3 Å: mechanism of CDK activation by cyclins. *Nature* **376** 313–320

Jinno S, Suto K, Nagata A *et al* (1994) Cdc25A is a novel phosphatase functioning early in the cell cycle. *EMBO Journal* **13** 1549–1556

Johnson DG, Schwarz JK, Cress WD and Nevins JR (1993) Expression of transcription factor E2F1 induces quiescent cells to enter S phase. *Nature* **365** 349–352

Kamb A, Gruis NA, Weaver FJ *et al* (1994) A cell cycle regulator potentially involved in genesis of many tumor types. *Science* **264** 436–440

Keyomarsi K and Pardee AB (1993) Redundant cyclin overexpression and gene amplification in breast cancer cells. *Proceedings of the National Academy of Sciences of the USA* **90** 1112–1116

Kiyokawa H, Kineman RD, Manova-Todorova KO *et al* (1996) Enhanced growth of mice lacking the cyclin-dependent kinase inhibitor function of p27[Kip1]. *Cell* **85** 721–732

Koff A, Giordano A, Desai D *et al* (1992) Formation and activation of a cyclin E-cdk2 complex during the G1 phase of the human cell cycle. *Science* **257** 1789–1694

Koh J, Enders GH, Dynlacht BD and Harlow E (1995) Tumour-derived p16 alleles encoding proteins defective in cell-cycle inhibition. *Nature* **375** 506–510

Krek W, Ewen ME, Shirodkar S, Arany Z, Kaelin WG and Livingston DM (1994) Negative regulation of the growth-promoting transcription factor E2F-1 by a stably bound cyclin A-dependent protein kinase. *Cell* **78** 161–172.

Lammie GA, Fantl V, Smith R *et al* (1991) D11S128, a putative oncogene on chromosome 11q13 is amplified and expressed in squamous cell and mammary carcinomas and linked to BCL-1. *Oncogene* **6** 439–444

La Thangue NB (1994) DRTF1/E2F: an expanding family of heterodimeric transcription factors implicated in cell-cycle control. *Trends in Biochemical Sciences* **19** 108–114

Lee EY-HP, Chang CW, Hu N *et al* (1992) Mice deficient for Rb are nonviable and show defects in neurogenesis and haematopoiesis. *Nature* **359** 288–294

Lee M-H, Reynisdottir I and Massague J (1995) Cloning of p57[Kip2], a cyclin-dependent kinase inhibitor with unique domain structure and tissue distribution. *Genes and Development* **9** 639–649

Lees E (1995) Cyclin dependent kinase regulation. *Current Opinion in Cell biology* **7** 773–780

Li Y, Nichols MA, Shay JW and Xiong Y (1994) Transcriptional repression of the D-type cyclin-dependent kinase inhibitor p16 by the retinoblastoma susceptibility gene product pRb. *Cancer Research* **54** 6078–6082

Lukas J, Muller H, Bartkova J *et al* (1994) DNA tumor virus oncoproteins and retinoblastoma gene mutations share the ability to relieve the cells requirement for cyclin D1 function in G1. *Journal of Cell Biology* **125** 625–638

Lukas J, Barkova J, Rohde M, Strauss M and Bartek J (1995a) Cyclin D1 is dispensable for G1 control in retinoblastoma gene-deficient cells independently of cdk4 activity. *Molecular and Cellular Biology* **15** 2600–2611

Lukas J, Parry D, Aagard L, Mann DJ, Bartkova J, Strauss M, Peters G and Bartek J (1995b) Retinoblastoma-protein-dependent cell-cycle inhibition by the tumor suppressor p16. *Nature* **375** 503–506

Luo Y, Hurwitz J and Massague J (1995) Cell-cyclin inhibition by independent CDK and PCNA binding domains in p21^{CIP1}. *Nature* **375** 159–161

Matsuoka S, Edwards M, Bai C *et al* (1995) p57^{KIP2}, a structurally distinct member of the p21^{CIP1} Cdk-inhibitor family, is a candidate tumor suppressor gene. *Genes and Development* **9** 650–662

Matsushime H, Roussel MF, Ashmun RA and Sherr CJ (1991) Colony-stimulating factor 1 regulates novel cyclins during the G1 phase of the cell cycle. *Cell* **65** 701–713

Matsushime H, Ewen ME, Strom DK *et al* (1992) Identification and properties of an atypical catalytic subunit (p34^{PSK-J3}/cdk4) for mammalian D type G1 cyclins. *Cell* **71** 323–334

Matsushime H, Quelle DE, Shurtleff SA, Shibuya M, Sherr CJ and Kato J-Y (1994) D-type cyclin-dependent kinase activity in mammalian cells. *Molecular and Cellular Biology* **14** 2066–2076

Meyerson M and Harlow E (1994) Identification of G1 kinase activity for cdk6, a novel cyclin D partner. *Molecular and Cellular Biology* **14** 2077–2086

Morgan DO (1995) Principles of CDK regulation. *Nature* **374** 131–134

Motokura T, Bloom T, Kim HG *et al* (1991) A novel cyclin encoded by a bcl-linked candidate oncogene. *Nature* **350** 512–515

Muller R (1995) Transcriptional regulation during the mammalian cell cycle. *Trends in Genetics* **11** 173–178

Nakanishi M, Robetorye RS, Adami GR, Pereira-Smith OM and Smith JR (1995) Identification of the active region of the DNA synthesis inhibitory gene p21$^{Sdi1/CIP1/WAF1}$. *EMBO Journal* **14** 555–563

Nakayama K, Ishida N, Shirane M *et al* (1996) Mice lacking p27^{Kip1} display increased body size, multiple organ hyperplasia, retinal dysplasia and pituitary tumors. *Cell* **85** 707–720

Nasmyth K (1993) Control of the yeast cell cycle by the Cdc28 protein kinase. *Current Opinion in Cell Biology* **5** 166–179

Nigg EA (1995) Cyclin-dependent protein kinases: key regulators of the eukaryotic cell cycle. *BioEssays* **17** 471–480

Ninomiya-Tsuji J, Nomoto S, Yasuda H, Reed SI and Matsumoto K (1991) Cloning of a human cDNA encoding a CDC2-related kinase by complementation of a budding yeast *cdc28* mutation. *Proceedings of the National Academy of Sciences of the USA* **88** 9006–9010

Nobori T, Miura KI, Wu DJ, Lois A, Takabayashi K and Carson DA (1994) Delection of the cyclin-dependent kinase-4 inhibitor gene in multiple human cancers. *Nature* **368** 753–756.

Nourse J, Firpo E, Flanagan WM *et al* (1994) Interleukin-2-mediated elimination of the p27^{Kip1} cyclin-dependent kinase inhibitor prevented by rapamycin. *Nature* **372** 570–573

Nurse P and Bissett Y (1981) Gene required in G1 for commitment to cell cycle and in G2 for control of mitosis in fission yeast. *Nature* **292** 558–560

Ohtsubo M and Roberts JM (1993) Cyclin-dependent regulation of G1 in mammalian fibroblasts. *Science* **259** 1908–1912

Ohtsubo M, Theodoras AM, Schumacher J, Roberts JM and Pagano M (1995) Human cyclin E, a nuclear protein essential for the G1 to S phase transition. *Molecular and Cellular Biology* **15** 2612–2624

Okamoto A, Demetrick DJ, Spillare EA *et al* (1994) Mutations and altered expression of p16[INK4] in human cancer. *Proceedings of the National Academy of Science USA* **91** 11045–11049

Pagano M, Pepperkok R, Verde F, Ansorge W and Draetta G (1992) Cyclin A is required at two points in the human cell cycle. *EMBO Journal* **11** 961–971

Pagano M, Tam SW, Theodoras AM *et al* (1995) Role of the ubiquitin-proteasome pathway in regulating abundance of the cyclin-dependent kinase inhibitor p27. *Science* **269** 682–685

Parker SB, Eichele G, Zhang P *et al* (1995) p53-independent expression of p21[Cip1] in muscle and other terminally differentiating cells. *Science* **267** 1024–1027.

Parry D, Bates S, Mann DJ and Peters G (1995) Lack of cyclin D-Cdk complexes in Rb-negative cells correlates with high levels of p16[INK4/MTS1]. *EMBO Journal* **14** 503–511

Pines J (1994) Protein kinases and cell cycle control. *Seminars in Cell Biology* **5** 399–408

Pines J and Hunter T (1990) Human cyclin A is adenovirus E1A-associated protein p60 and behaves differently from cyclin B. *Nature* **346** 760–763

Quelle DE, Ashmun RA, Shurtleff SA *et al* (1993) Overexpression of mouse D-type cyclins accelerates G1 phase in rodent fibroblasts. *Genes and Development* **7** 1559–1571

Reed SI (1980) The selection of *S. cerevisiae* mutants defective in the start event of cell division. *Genetics* **95** 561–577

Reed SI (1995) START and the G1-S phase transition in budding yeast. In: Hutchison C and Glover DM (eds). *Cell Cycle Control*, pp 40–55, IRL Press, Oxford

Rempel RE, Sleight SB and Maller JM (1995) Maternal *Xenopus* cdk2-cyclin E complexes function during meiotic and early embryonic cell cycles that lack a G1 phase. *Journal of Biological Chemistry* **270** 6843–6855

Resnitzky D and Reed SI (1995) Different roles for cyclins D1 and E in regulation of the G1-to-S transition. *Molecular and Cellular Biology* **15** 3463–3469

Resnitzky D, Gossen M, Bujard H and Reed SI (1994) Acceleration of the G1/S phase transition by expression of cyclins D1 and E with an inducible system. *Molecular and Cellular Biology* **14** 1669–1679

Resnitzky D, Hengst L and Reed SI (1995) Cyclin A-associated kinase activity is rate limiting for entrance into S phase and is negatively regulated in G1 by p27[Kip1]. *Molecular and Cellular Biology* **15** 4347–4352

Reynisdottir I, Polyak K, Iavarone A and Massague J (1995) Kip/Cip and Ink4 Cdk inhibitors cooperate to induce cell cycle arrest in response to TGF-β. *Genes and Development* **9** 1831–1845

Russo AA, Jeffrey PD, Patten AK, Massague J and Pavletich NP (1996) Crystal structure of the p27[Kip1] cyclin-dependent kinase inhibitor bound to the cyclin A-Cdk2 complex. *Nature* **382** 325–331

Sanchez I and Dynlacht BD (1996) Transcriptional control of the cell cycle. *Current Opinion in Cell Biology* **8** 318–324

Sebastian B, Kakazuka A and Hunter T (1993) Cdc25 activation of cyclin-dependent kinase by dephosphorylation of threonine-14 and tyrosine-15. *Proceedings of the National Academy of Sciences of the USA* **90** 3521–3524

Serrano M, Hannon GJ and Beach D (1993) A new regulatory motif in cell-cycle control causing specific inhibition of cyclin D/CDK4. *Nature* **366** 704–707

Serrano M, Lee H-W, Chin L, Cordon-Cardo C, Beach D and DePinho RA (1996) Role of the INK4a locus in tumor suppression and cell mortality. *Cell* **85** 26–38.

Sherr, CJ (1993) Mammalian G1 cyclins. *Cell* **73** 1059–1065

Sherr CJ and Roberts JM (1995) Inhibitors of mammalian G1 cyclin-dependent kinases. *Genes and Development* **9** 1149–1163

Sicinski P, Donaher JL, Parker SB *et al* (1995) Cyclin D1 provides a link between development and oncogenesis in the retina and the breast. *Cell* **82** 621–630.

Slingerland JM, Hengst L, Pan CH, Alexander D, Stampfer MR and Reed SI (1994) A novel inhibitor of cyclin-Cdk activity detected in transforming growth facter beta-arrested epithelial cells. *Molecular and Cellular Biology* **14** 3683–3694

Th'ng JPH, Wright PS, Hamaguchi J *et al* (1990) The FT210 cell line is a mouse G2 phase mutant with a temperature sensitive *cdc2* gene product. *Cell* **63** 313–324

Tiefenbrun N, Melamed D, Levy N *et al* (1996) Alpha interferon suppresses the cyclin D3 and *cdc25A* genes, leading to a reversible G0-like arrest. *Molecular and Cellular Biology* **16** 3934–3944

Tsai LH, Harlow E and Meyerson M (1991) Isolation of the human cdk2 gene that encodes the cyclin A- and adenovirus E1A-associated p33 kinase. *Nature* **353** 174–177

Waga S, Hannon GJ, Beach D and Stillman B (1994) The p21 inhibitor of cyclin-dependent kinases controls DNA replication by interaction with PCNA. *Nature* **369** 574–578

Wang C-Y, Petryniak B, Thompson CB, Kaelin WG and Leiden JM (1993) Regulation of the Ets-related transcription factor Elf-1 by binding to the retinoblastoma protein. *Science* **260** 1330–1335

Weinberg RA (1995) The retinoblastoma protein and cell cycle control. *Cell* **81** 323–330

Welch PJ and Wang JYJ (1993) A C-terminal protein binding domain in RB regulates the nuclear c-Abl tyrosine kinase in the cell cycle. *Cell* **75** 779–790

Welch PJ and Wang JYJ (1995) Disruption of retinoblastoma protein function by coexpression of its C pocket fragment. *Genes and Development* **9** 31–46

Withers DA, Harvey RC, Faust JB, Melnyk O, Carey K and Meeker TC (1991) Characterization of a candidate *bcl1-1* gene. *Molecular and Cellular Biology* **11** 4846–4853

Won K-A and Reed SI (1996) Activation of cyclin E/CDK2 is coupled to site-specific autophosphorylation and ubiquitin-dependent degradation of cyclin E. *EMBO Journal* **15** 4182–4193

Xu M, Sheppart CA, Peng C-Y, Yee AS and Piwnica-Worms H (1994) Cyclin A/CDK2 binds directly to E2F-1 and inhibits the DNA-binding activity of E2F-1/DP-1 by phosphorylation. *Molecular and Cellular Biology* **14** 8420–8431

Zetterberg A, Larsson O and Wiman KG (1995) What is the restriction point? *Current Opionion in Cell Biology* **7** 835–842

Zhang H, Hannon GJ and Beach D (1994) p21-containing cyclin kinases exist in both active and inactive states. *Genes and Development* **8** 1750–1758

The author is responsible for the accuracy of the references.

S Phase Damage Sensing Checkpoints in Mammalian Cells

J M LARNER[1] • **H LEE**[1] • **J L HAMLIN**[2]

[1]*Departments of Radiation Oncology and* [2]*Biochemistry, University of Virginia School of Medicine, Charlottesville, VA 22908*

INTRODUCTION

If a eukaryotic chromosome suffers a double strand break that is not repaired properly, two potentially devastating outcomes are predicted based on first principles: either (a) cell death due to the loss of an acentric fragment bearing critical genes or (b) genetic instability resulting from broken chromosome ends that participate in illegitimate recombination events (translocations, deletions, insertions, inversions or fusion of sister chromatids leading to bridge–breakage–fusion cycles). Although it is difficult to know how often cell death occurs in the context of the organism, gross chromosomal rearrangements are exceedingly rare in normal cells, suggesting either that double strand breaks are also very rare or that they are repaired efficiently.

In fact, although a relatively small radiation insult delivered to a typical cultured mammalian cell generates a significant number of single strand breaks (which the cell repairs quite efficiently; Kemp *et al*, 1984), an enormous dose is required to generate a substantial number of double strand breaks (Zaider, 1993). What, then, is the origin of the large number of double strand breaks required to explain the genetic instability (manifested as gross chromosomal rearrangements) that characterizes most tumour cells?

We believe that the answer lies in the nature of several DNA damage sensing pathways that in normal cells prevent single strand breaks from being converted to potentially lethal double strand lesions by DNA replication. The origins of single strand breaks in normal, unirradiated cells are unknown. However, errors in the rejoining reactions of topoisomerases are prime candidates, since they break and reseal the DNA backbone many thousands of times in a cell cycle during the processes of transcription, replication and deconcatenation of sister chromatids (Cozzarelli, 1977). One DNA damage sensing pathway is mediated by TP53 and, in response to ionizing radiation, either transiently arrests cells in late G_1 before entry into the S period or shunts them towards an apoptotic death (see Wahl et al, Pan et al, and O'Connor, this volume). A second, S phase damage sensing (SDS) pathway acutely downregulates DNA synthesis at individual origins of replication in response to radiation (Watanabe, 1974; Painter and Young, 1975; Walters and Hildebrand, 1975; Larner et al, 1994). Recent evidence also suggests that cells beyond the late G_1 checkpoint that have not yet entered the S period are prevented from doing so after a radiation insult (Lee et al, in press).

Since each of these checkpoints or pathways is sensitive to single strand breaks, it would appear that the cell expends considerable effort to prevent their conversion to double strand breaks by replication. We have only recently come to this field as a consequence of a longstanding interest in mammalian DNA replication. Our purpose here is to discuss the nature of the mechanisms that normally regulate DNA replication, to review what is known about the SDS pathway and the putative G_1/S checkpoint and to attempt to relate the two fields.

BACKGROUND

When mammalian cells are exposed to ionizing radiation, DNA synthesis (as measured by incorporation of [³H]thymidine into DNA) is inhibited within a matter of minutes. The dose-response curve for the inhibition of replication is biphasic, with a steep initial component at low doses and a more shallow component at higher doses (Watanabe, 1974; Painter and Young, 1975; Walters and Hildebrand, 1975; Larner et al, 1994). Makino and Okada (1975) proposed that the sensitive (steep) portion of the curve results from inhibition of replicons that were about to initiate, whereas the insensitive (shallow) component corresponds to effects on chain elongation in replicons that had already initiated.

Walters and Hildebrand (1975) confirmed this suggestion by measuring the size distribution of labelled DNA in irradiated versus non-irradiated CHO cells on alkaline sucrose gradients: 20 minutes after irradiation, there was a 55% reduction in the amount of label in DNA lighter than 4×10^7 daltons (ie nascent chains initiated after irradiation) and only a 5% reduction in DNA heavier than 4×10^7 daltons (the component initiated before radiation). These observations were confirmed and extended to HeLa and L cells by Painter and Young (1975). Assuming that there is a direct correspondence between the amount of DNA damage and the degree of inhibition, Painter calculated a radiation target size of approximately 10^9 daltons based on a D_0 of 5 Gy. He therefore proposed that a single hit might inactivate several replicons in a cluster simultaneously. This proposal was based on earlier fibre autoradiographic studies in which it appeared that 5–10 adjacent replicons can initiate replication synchronously (Huberman and Riggs, 1968).

Watanabe (1974) directly compared the effects of radiation on initiation versus chain elongation in DNA fibre autoradiographic studies on dividing mouse leukaemia cells. He noted that the rate of chain growth (eg the grain track length for a given pulse time) was only slightly affected with a D_0 of 150 Gy when measured 30 minutes after irradiation. However, in the same samples, the percentage of labelled replicons that had grain tracks less than 5 mm (those that probably initiated after irradiation) was greatly decreased and only increased to normal 110 minutes after irradiation.

In combination with many other studies, these observations have led to the conclusion that relatively low dose irradiation preferentially inhibits initiation at origins of replication, but higher doses can also affect chain elongation per se. At least two general mechanisms could explain these effects. In the first, radiation could damage the template itself, either by inducing strand breaks, crosslinking DNA to proteins or to complementary DNA or by introducing adducts that impede the movement of enzymes or enzyme complexes on the template. To accommodate the different susceptibilities of initiation and chain elongation, however, the initial melting of the helix during initiation would have to be much more sensitive than chain elongation, and the inactivation of one origin of replication would have to inhibit many other origins simultaneously. Alternatively, radiation damage to DNA could elicit a global cellular response in which a factor (presumably protein) would sense the damage and repress initiation (directly or indirectly) at all origins until the damage is repaired.

In effect, these two general models suggest that initiation of DNA synthesis is inhibited either in *cis* (model 1) or in *trans* (model 2). Until recently, it has not been possible to distinguish between these two models owing to the lack of a specific target replicon in which initiation and elongation reactions could be studied at the molecular level. However, some support for the *trans*-regulatory model has been obtained by Lamb *et al* (1989), who examined the effects of ionizing radiation on a human cell line harbouring an autonomously replicating plasmid. This plasmid contains the Epstein-Barr virus (EBV) origin of replication and a gene encoding the EBV initiation protein, EBNA1. It was shown that at a dose of 30 Gy, which is sufficient to damage the cellular genome but did not detectably damage the replicating plasmid, both cellular and viral replication were significantly inhibited. Although there are several more complicated explanations for this effect, the most straightforward interpretation is that the damage inflicted on the host DNA signals a cellular response mechanism that represses the origin in the virus and, by inference, in the host chromosomes as well (eg the response mechanism is mediated in *trans*). Cleaver *et al* (1990) have reported similar results with an autonomously replicating plasmid containing the SV40 origin of replication.

Wang (1995) showed that extracts from irradiated HeLa cells support much less DNA synthesis than mock irradiated control extracts when used as the source of replication proteins in an SV40 origin and T antigen dependent in vitro replication system. Since the template plasmid DNA was never exposed to radiation, Wang's results strongly suggest the operation of a *trans*-acting regulatory process. However, although much has been learned about the enzymology of replication from eukaryotic viral models, substantial differences exist between initiation of viral and chromosomal replicons. In particular, the raison d'être of viruses such as SV40 is to replicate as many times as possible in a single cell cycle, whereas eukaryotic chromosomal origins are designed to initiate only once per cycle.

MAMMALIAN ORIGINS OF REPLICATION

Since the ultimate target of DNA damage sensing pathways that function in S phase is thought to be the origins themselves, a brief review of eukaryotic chromosomal origins is warranted. We begin with a background discussion of origins in the budding yeast, *Saccharomyces cerevisiae,* since a considerable amount is known about them, and since genetic studies in yeast have identified several potentially conserved pathways involved in the cellular response to DNA damage. We then review the state of the mammalian chromosomal origin field.

The Yeast Paradigm

Just as in simpler prokaryotic and eukaryotic viral replicons, the synthesis of yeast chromosomes is controlled by the interaction of an initiator protein complex with genetic replicators. These replicators were identified as autonomously replicating sequence (ARS) elements by their ability to confer high frequency transformation ability on a co-linear selectable marker (Chan and Tye, 1980; Stinchcomb *et al*, 1980). Yeast replicators are spaced at intervals of about 30 kb along the chromosomal DNA fibre (Newlon and Theis, 1993) and can be activated in early, mid or late S phase (Ferguson *et al*, 1991). They are characterized by an 11 bp A/T rich consensus sequence that is necessary, but not sufficient, for proper origin function (Newlon and Theis, 1993). Because nascent strand start sites usually lie very close to the replicators themselves in these simple systems, the general term "origin" is frequently used to refer simultaneously to both entities. As we will see, however, this term may be misleading when applied to the more complicated origins of higher eukaryotic cells. In the remainder of this chapter, we use the word "origin" to refer to the position(s) where nascent strands initiate on the template in a given replicon regardless of whether or not they may contain a true genetic replicator.

The yeast *trans*-acting initiator is composed of a six subunit origin recognition complex (ORC; Bell and Stillman, 1992) that binds to replicators (ie the active chromosomal ARS elements) throughout the cell cycle (Diffley *et al*, 1994; Liang *et al*, 1995). The Cdc6/Cdc18 protein is transiently associated with ORC from the end of mitosis to the G_1/S transition (Diffley *et al*, 1994) and apparently has a critical role in initiation, since overexpression of this protein results in endoreduplication of the entire genome without intervening mitoses (Nishitani and Nurse, 1995; Piatti *et al*, 1995). There is also evidence that the Cdc7 kinase, whose activity fluctuates throughout the cell cycle, interacts with Orc2p and may thereby regulate passage through the G_1/S transition (Jackson *et al*, 1993; Hardy, 1996). This interaction, in turn, may be regulated by the Dbf4 protein kinase, which is transiently induced at the G_1/S boundary (Kitada *et al*, 1992; Jackson *et al*, 1993; Dowell *et al*, 1994).

Examples of Eukaryotic Origins

The fact that homologues of several of the subunits of ORC have been identified in higher eukaryotic cells suggests that initiation of replication may be regulated similarly in mammalian chromosomes (Gavin *et al*, 1995; Gossen *et al*, 1995; Carpenter *et al*, 1996; Takahara, 1996). Thus, the yeast paradigm will be very useful in guiding experiments on mammalian systems in the future. To our knowledge, however, no cognate DNA sequences that could correspond to genetic replicators have been isolated from higher eukaryotic chromosomes on the basis of their ability to bind to mammalian ORC proteins. However, there are probably thousands of copies of the yeast 11 bp A/T rich ARS consensus sequence sprinkled throughout mammalian genomes, since we have detected more than 20 of these sequences in a 5 kb fragment in the Chinese hamster dihydrofolate reductase locus (Leu T-H and Hamlin JL, unpublished). Furthermore, specific higher eukaryotic genomic sequences that are capable of reproducible autonomous replication have not been forthcoming (but see Frappier and Zannis-Hadjopoulos, 1987; McWhinney and Leffak, 1990; Sudo *et al*, 1990). In fact, virtually any mammalian DNA fragment greater than a few kilobases in length can replicate to some degree when cloned into a selectable vector containing a nuclear partition function (Krysan *et al*, 1989; Heinzel *et al*, 1991). Thus, the argument has been made that specific mammalian replicators either are very close together or do not exist at all (Heinzel *et al*, 1991).

Nevertheless, many aspects of replication control in higher eukaryotic cells are similar to those in yeast. The classic DNA fibre autoradiographic studies of Huberman and Riggs (1968) showed that mammalian origins are bidirectional, although active origins are spaced somewhat further apart along the chromosomal DNA fibre than in yeast (every 100 kb or so). Thus, there could be as many as 50 000 individual origins in the genome, but the autoradiographic technique sheds no light on whether these origins are fixed genetic replicators and/or whether the same loci are used from one cell cycle to the next. As mentioned above, many of the autoradiographic images obtained in these studies could be interpreted to mean that several adjacent origins fire simultaneously (Huberman and Riggs, 1968), which suggests additional levels of regulatory complexity.

In the absence of genetic evidence for the presence of replicators in higher eukaryotic chromosomes, investigations have concentrated on localizing the positions along the chromosome at which nascent strands initiate, which should be close to genetic replicators by analogy to simpler systems. Efforts to identify individual origins in a chromosomal domain of interest fall into several different categories: (a) identifying the earliest fragments in a domain of interest to be labelled with radioactive thymidine in the early S phase in synchronized cells (Heintz and Hamlin, 1982; Burhans *et al*, 1986; Leu and Hamlin, 1989; Carroll *et al*, 1993); (b) identifying the position along a template at which either leading (Handeli *et al*, 1989) or lagging (Burhans *et al*, 1990) nascent strands switch from one template to the other (which occurs at an origin); (c) determining nascent strand size and/or abundance (nascent strands from the same origin that fired at different times in different cells form a pyramid with the smallest fragments centred over the origin; Vassilev *et al*, 1990); (d) determining the copy number of various fragments along a region of interest in interphase nuclei by fluorescence in situ hybridization (those loci with the highest number of double signals must be the earliest replicating; Kitsberg *et al*, 1993); (e) mapping the direction of replication fork movement, which should diverge from an origin (Nawotka and Huberman, 1988); and (f) detecting restriction fragments containing the small replication bubbles that are centered over origins by their unique migration behaviour in two dimensional gels (Brewer and Fangman, 1987).

Candidate origins have now been identified by one or more of these techniques in the following mammalian chromosomal loci: human rDNA (Little *et al*, 1993), *MYC* (Iguchi *et al*, 1988; Leffak and James, 1989), globin (Kitsberg *et al*, 1993) and lamin B2 (Giacca *et al*, 1994); Chinese hamster rhodopsin (Gale *et al*, 1992), RPS14 (Tasheva and Roufa, 1994), dihydrofolate reductase (*DHFR*; reviewed in Hamlin *et al*, 1994), and α locus (Ma *et al*, 1990); murine adenosine deaminase (Carroll *et al*, 1993) and Syrian hamster CAD (Kelly *et al*, 1995). Potential origins have also been identified near the *Drosophila* histone (Shinomiya and Ina, 1993) and polymerase α genes (Shinomiya and Ina, 1994), in *Xenopus* rDNA genes (Hyrien and Mechali, 1993) and in the profilin loci of *Physarum* (Benard and Pierron, 1992).

The Model CHO Dihydrofolate Reductase Origin

There is presently some disagreement about the nature of the initiation reaction in higher eukaryotic origins, probably owing to the different views af-

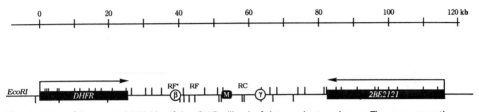

Fig. 1. Map of the central 120 kb of the CHO dihydrofolate reductase locus. The convergently transcribing *DHFR* and *2BE2121* genes are indicated as boxes flanking the 55 kb intergenic region in the *DHFR* domain. The positions of ori-β, ori-γ, and a matrix attachment region (M) are shown. Vertical lines above the map indicate the *Eco*RI sites in this locus

forded by different methods of analysis. The model replicon that was identified and characterized in our laboratory (the amplified *DHFR* locus in the CHOC 400 cell line; Milbrandt *et al*, 1981) has been studied by more replicon mapping techniques than any other locus and will serve to illustrate the point (Fig. 1).

The 240 kb *DHFR* amplicon in CHOC 400 cells is characterized by two convergently transcribed genes (*DHFR* and *2BE2121*) that are separated by a 55 kb intergenic spacer with a prominent matrix attachment region in its centre (MAR; Dijkwel and Hamlin, 1988). Early labelling studies (Heintz and Hamlin, 1982; Burhans *et al*, 1986), measurements of leading strand template bias (Handeli *et al*, 1989) and replication fork direction (Vaughn *et al*, 1990; Dijkwel *et al*, 1994) and localization of bubble containing fragments on two dimensional gels (Vaughn *et al*, 1990; Dijkwel and Hamlin, 1992; Dijkwel *et al*, 1994) all suggest a model in which replication can initiate at virtually any position within the spacer region, but more often at two positions (termed *ori-β* and *ori-γ*; Leu and Hamlin, 1989) that are separated by approximately 22 kb and which straddle the MAR (Fig. 1). When this locus was analysed by measuring lagging strand template bias (Burhans *et al*, 1990) or nascent strand sizes (Vassilev *et al*, 1990), the results were interpreted to mean that the vast majority of initiations occur in the immediate vicinity of *ori-β* (*ori-γ* was not analysed in the latter two studies but presumably would also have registered as highly preferred over other sites in the approximately 23 kb region between the MAR and the *2BE2121* gene).

The disagreement among different data sets is thus largely a quantitative matter, and all parties agree that the *ori-β* and *ori-γ* loci represent bona fide origins. The question is whether most nascent strand start sites lie within a few hundred base pairs of these sites or can be chosen from potential sites lying as much as 10–12 kb away. Several other higher eukaryotic loci appear to undergo a delocalized mode of initiation when assayed by two dimensional gel replicon mapping techniques (Delidakis and Kafatos, 1987; Heck and Spradling, 1990; Shinomiya and Ina, 1991, 1993, 1994; Hyrien and Mechali, 1993; Liang *et al*, 1993; Little *et al*, 1993).

Our current view of initiation in the *DHFR* locus is that *ori-β* and *ori-γ* each correspond to genetic replicators and constitute recognition sites for an initiator complex, which then signals entry of the multienzyme replication complex itself; the replication complex could then migrate along the template in either direction and lay down the primers for nascent strands, stochastically more often near the replicators themselves but sometimes at considerable distances. It is further suggested that transcription of the *DHFR* and *2BE2121* genes in the late G_1 and early S period (Foreman and Hamlin, 1989) may propagate a disturbance along the chromatin template that is locked in by attachment to the matrix at the MAR, thereby activating the *ori-β* and *ori-γ* loci for initiation.

Alternatively, replicators per se may not exist, as suggested by the failure to detect them in ARS assays; rather, particular regions of the chromosome (in this case, between two genes) may be recognized because of altered chromatin structure, which itself could be modified by local transcriptional activity and/or proximity to nuclear matrix attachment sites. For example, in early *Xenopus* embryos when transcription of rDNA genes is not occurring, replication in the rDNA locus initiates at random sites throughout the locus, including the body of the rDNA genes themselves; however, later in development, when active transcription on rDNA genes commences, initiation is confined to the intergenic region (Hyrien *et al*, 1995).

Chromatin context and attachment to the nuclear matrix figure importantly in both of these models, because we have recently obtained evidence that attachment to the nuclear matrix alters chromatin architecture in a way that could be critical for origin function. In quantitative studies, we have shown that only 10–15% of the *DHFR* amplicons in CHOC 400 cells are attached to the nuclear matrix at the intergenic MAR (Fig. 1) (Dijkwel and Hamlin, 1988). In addition, only 10–15% of amplicons appear to sustain active initiation events in any one S period, with the remainder being replicated passively by replication forks from adjacent active amplicons (Dijkwel and Hamlin, 1992; 1995; Dijkwel *et al*, 1994). Thus, at any given moment a fragment in the intergenic zone in the CHO *DHFR* locus will contain replication bubbles resulting from internal initiations, as well as single forks that arise from initiation sites in adjacent (active) amplicons.

These observations suggested that inefficient origin usage and failure to attach to the matrix could be related. This notion was strengthened considerably by the recent observation that *ori-β* and *ori-γ* each display prominent micrococcal nuclease hypersensitive sites, but only in those copies of the amplicon that are attached to the nuclear matrix (Pemov S and Hamlin JL, unpublished). These sites are detected in cells that are trapped at the G_1/S boundary with the replication inhibitor, mimosine, but the sites completely disappear within 90 minutes of entry into S when initiation at these loci is maximum (Mosca *et al*, 1992). Therefore, the micrococcal nuclease hypersensitive sites appear to correspond to some prereplicative state that is dissipated either just before or at the time that origins fire.

Whether the cell cycle dependent nuclease hypersensitive sites at *ori-β* and *ori-γ* represent bound regulatory proteins (eg ORC analogues) or easily melted (presumably A/T rich) DNA sequences remains to be seen. Nevertheless, these data suggest that the regulation of mammalian origins may turn out to be multilayered and quite complex; they also offer a possible explanation for why it has been difficult to recreate functional episomes by searching for mammalian ARS elements in phenotypic assays.

RADIATION EFFECTS ON DEFINED MAMMALIAN REPLICONS

With the identification of several different origins in mammalian chromosomes, as well as the advent of sophisticated replicon mapping methods, it became possible to assess directly the effect of radiation damage on defined replicons. In our laboratory, we have studied three different replicons: the early firing CHO *ori-β/ori-γ* loci (reviewed in Hamlin *et al*, 1994); the mid firing, amplified, Chinese hamster *ori-α* locus, which lies about 200 kb upstream from *ori-β/ori-γ* in the much larger amplicon of a methotrexate resistant Chinese hamster lung fibroblast (Ma *et al*, 1990); and the highly repeated rDNA locus in human cells, which consists of both early and mid firing variants (Little *et al*, 1993).

In preliminary experiments, it was shown that CHO cells, Chinese hamster lung fibroblasts and human cells each display a typical biphasic dose-response curve when subjected to increasing doses of ionizing radiation (Larner *et al*, 1994; Larner JL, unpublished). The relative immediacy of the damage arrest response was demonstrated in time course studies in which the rates of incorporation of [^{3}H]thymidine were inhibited maximally 1.5–2 hours after radiation and began to recover shortly thereafter. This result is consistent with that seen with numerous other mammalian cell lines. Each of these cell lines lacks functional TP53, as evidenced by the failure to arrest in G_1 after a radiation challenge (Larner *et al*, 1994; Lee *et al*, in press) and, in the case of CHO cells, by the presence of a missense mutation in the conserved DNA binding domain of the protein (Lee *et al*, in press). Thus, the SDS pathway is clearly TP53 independent.

We first analysed the effects of radiation on the amplified *DHFR* domain in CHO cells. Asynchronous cultures were irradiated with 9 Gy, which is the approximate dose at the inflection point between the steep and shallow components of the dose-response curve in this cell line and which results in approximately 50% inhibition of [^{3}H]thymidine incorporation within 30 minutes (Larner *et al*, 1994). After irradiation, replication intermediates were prepared and analysed by the two dimensional gel replicon mapping method (Fig. 2).

In the sample taken just before or immediately after irradiation, a clear bubble arc indicative of initiation was detected with a radioactive hybridization probe specific for a fragment containing *ori-β* (Larner *et al*, 1994); as explained above, the bubble arc is accompanied by a single fork arc containing replication forks that emanated from initiation sites elsewhere in the intergenic region or from distant amplicons. Therefore, radiation treatment per se

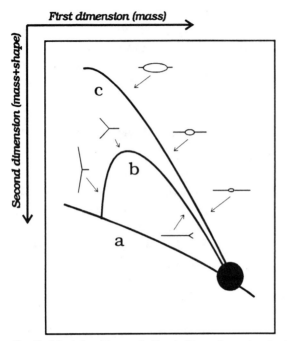

Fig. 2. Principle of the neutral/neutral two dimensional gel replicon mapping method. A restriction digest is separated on an agarose gel in the first dimension according to molecular mass. The resulting lane is excised, turned through 90° and run in a second dimension that separates according to both mass and shape. The digest is transferred to a membrane and hybridized with appropriate probes to detect the fragment of interest. Curve a represents the arc of linear, non-replicating fragments, and the large 1*n* spot corresponds to the non-replicating component of any given probed fragment. Curve b represents a fragment that is replicated passively by forks emanating from an outside origin. Curve c corresponds to a fragment with a centred origin

does not appear to fragment replication intermediates. However, within 30 minutes of radiation treatment, the bubble arc had completely disappeared and did not begin to reappear until 2.5 hours later. By contrast, the pattern of the single fork arc did not change significantly in response to radiation (Larner *et al*, 1994).

These data on the early firing *DHFR* origin are therefore in complete agreement with the proposal that radiation doses less than approximately 10 Gy primarily affect initiation while affecting chain elongation only slightly. Furthermore, the response is detectable within 15–30 minutes of a radiation challenge.

Since initiation in the *DHFR* locus in CHOC 400 cells can be detected throughout the first 2–2.5 hours of the S phase in highly synchronized cell populations (Dijkwel and Hamlin, 1992, 1995), and since the asynchronous cell population used in the studies above would contain cells at all stages of the S phase, complete inhibition of initiation at all of the *DHFR* origins in the experiment above suggests that the SDS pathway may be operative at all times in S phase. However, when the effects of radiation on CHOC 400 cells were examined in synchronized cells, the results suggested a somewhat more complicated situation.

In these experiments, cells were released from a G_0 block and allowed to progress towards the S phase in the presence of either aphidicolin, which is a chain elongation inhibitor (Huberman, 1981), or mimosine, which appears to inhibit initiation per se (probably by preventing entry of the replication machinery into the replication fork; Dijkwel and Hamlin, 1992; Mosca *et al*, 1992). The population was collected at the beginning of the S phase, and the cultures were then irradiated with 9 Gy and sampled at various times thereafter.

Surprisingly, cells irradiated before or immediately after drug removal were refractory to radiation, whether assessed by effects on [^{3}H]thymidine incorporation into DNA in the population as a whole or at the level of the *DHFR* origin on two dimensional gels (Larner *et al*, 1994). This finding indicated that cells collected near the G_1/S boundary with these drugs had passed the radiation sensitive detection point. Thus, although mimosine acts on replication at a step before the aphidicolin sensitive step (ie chain elongation; Mosca *et al*, 1992) and is able to prevent replication forks from being established (Dijkwel and Hamlin, 1992), it must function after the radiation detection step that prevents initiation.

When cells were irradiated at hourly intervals after removal of mimosine, [^{3}H]thymidine incorporation was progressively reduced, suggesting that the SDS pathway begins to respond at later times (Lee *et al*, in press). At all time points, the cell population demonstrated typical biphasic dose-response curves, except that the steep component of the curve became progressively more pronounced at later times in the S phase. However, at no time during the S phase in synchronized cells was [^{3}H]thymidine incorporation reduced to 50% of preirradiation values, as it is when an asynchronous population is irradiated with 9–10 Gy (Larner *et al*, 1994). Two dimensional gel analysis of the *DHFR* origin essentially mimicked these results in the early time points when this origin normally fires.

To determine whether radiation induced downregulation at individual origins of replication is a generalized phenomenon and not limited to early firing loci, we have examined the well characterized multicopy rDNA replicons in the lymphoblastoid Wilson cell line (Little *et al*, 1993). Like the *DHFR* origin in CHO cells, the rDNA origin is a delocalized one and is situated in the spacer region between tandem repeated genes (Little *et al*, 1993). Conveniently, a restriction fragment length polymorphism distinguishes between early and mid firing variants of this locus (Little *et al*, 1993). Using the two dimensional gel replicon mapping technique to analyse the rDNA variants in an asynchronous culture, we were able to show that initiation is immediately downregulated in response to radiation in both the early and mid firing variants (Larner JM and Hamlin JL, unpublished). Therefore, the SDS pathway functions on extranucleolar early firing origins as well as on early and mid firing rDNA origins sequestered in nucleolar organizers.

In an asynchronous population, therefore, the majority of origins appear to be prevented from firing by a radiation challenge. However, those that have passed a particular point at the beginning of the S phase are refractory to radiation and do not become maximally sensitive again for several hours. This result suggests that a significant proportion of the radiation sensitive origins that are downregulated in an asynchronous population must be in late G_1 before the mimosine sensitive step. Furthermore, since the rate of [^{3}H]thymidine incorporation is depressed for at least 2 hours after a radiation challenge, it follows that this damage sensing pathway must be able to remember the insult for about 2 hours.

Indeed, by examining an asynchronous CHOC 400 cell population at hourly intervals after radiation by fluorescence activated cell sorter analysis, it was possible to detect a substantial component of the population that was transiently arrested at the G_1/S boundary in response to radiation treatment (Lee *et al*, in press). Since this cell line does not contain functional TP53, these data provide direct evidence for a G_1/S checkpoint that functions to prevent entry into the S phase in the face of DNA damage. It is not yet clear whether this checkpoint is distinct from the damage arrest mechanism that prevents initiation at origins during the S phase itself or whether it represents a bona fide cell cycle checkpoint that regulates entry into the S phase via some signal transduction cascade that eventually impinges on origins of replication. However, since the entire S phase can be repositioned (delayed) along the time axis by this mechanism when cells are irradiated in late G_1, it fits the description of a legitimate cell cycle checkpoint (Hartwell and Weinert, 1989; Weinert, 1992; Paulovich and Hartwell, 1995).

POTENTIAL PLAYERS IN S PHASE DAMAGE SENSING PATHWAYS

Although the acute effects of radiation on DNA synthesis are now reasonably well defined, very little is known about the signal sensing mechanism itself or how signal recognition is ultimately transduced into inhibition at individual origins. However, there are several provocative findings in both yeast and mammalian cells that may prove to be relevant.

Yeast cells that carry a defective allele of *MEC1*, the homologue of the human ataxia-telangiectasia (AT) gene, *ATM,* appear to be defective in both S phase delay and in the G_2/M checkpoint, and they may be defective in the G_1 checkpoint as well (Weinert *et al*, 1994; Paulovich and Hartwell, 1995). When *S cerevisiae* is subjected to the alkylating agent, methylmethanesulphonate (MMS), the rate of replication is slowed, much as it is when mammalian cells are irradiated. However, when either *MEC1* or *MEC2/RAD53/SAD1* mutants are subjected to MMS, replication proceeds unabated, indicating that inhibition of replication is a *trans*-regulatory process and that *MEC1* and *MEC2/RAD53/SAD1* encode regulatory proteins (Allen *et al,* 1994; Paulovich and Hartwell, 1995; Sanchez *et al,* 1996). This phenotype is much like that of human fibroblasts that are defective in the *ATM* gene, since these cells fail to arrest in G_1 (Kastan *et al*, 1992), in the S phase (Painter, 1981) and probably also at the G_2 checkpoint (Hong *et al*, 1994). Since multiple checkpoints are defective in AT, the ATM gene product in mammals is thought to function early in a branch that is common to the damage sensing pathways it controls (Kastan *et al*, 1992; Leonard *et al*, 1995). Therefore, at least some aspects of the S phase damage sensing pathway may have been conserved in evolution up to mammals, and studies in yeast may help to identify some of the critical players in mammalian damage sensing signal transduction pathways.

Recent studies on mammalian cells have suggested other possible intermediaries in S phase damage sensing mechanisms. The transcription factor complex E2F is required to transcribe several genes involved in sustaining S phase events (DeGregori, 1995). E2F is held in an inactive complex by RB and is activated in late G_1 when it is released from RB by a cyclin/cyclin dependent kinase phosphorylation event (Shirodkar *et al*, 1992; Dowdy *et al*, 1993; Ewen *et al*, 1993). Sometime later in the S phase, E2F is inactivated by cyclin A kinase phosphorylation (Krek *et al*, 1994). Conceivably, if cyclin A kinase were to be lowered in response to a radiation challenge, the resulting inappropriate occupancy of active E2F on cognate promoters in the middle of the S phase could result in S phase arrest (Krek *et al*, 1994).

Although this is an attractive model, it does not suggest molecular mechanisms by which the unscheduled occupancy of certain promoters by E2F would activate the S phase checkpoint. Furthermore, cyclin A levels do not appear to be lowered in response to radiation doses less than 10 Gy (Datta *et al*, 1992); therefore, it is difficult to imagine the catastrophic insult that could mimic such a large amount of damage in vivo.

Several studies in both yeast and in mammalian cells have raised the possibility that the replication machinery itself can sense DNA damage and subsequently downregulate DNA replication. For example, radiation affects the activity of replication protein A (RPA) (Liu and Weaver, 1993). RPA is a trimeric single stranded DNA binding protein required for initiation, chain elongation and DNA repair (Blackwell and Borowiec, 1994; Waga *et al*, 1994; He *et al*, 1995; Lee and Kim, 1995; Singh and Samson, 1995). The p34 subunit is thought to be phosphorylated during the S and G_2 phases of the cell cycle, but not during G_1 (Dutta and Stillman, 1992). However, ionizing radiation may lead to unscheduled phosphorylation of RPA on p34 (Liu and Weaver, 1993). Furthermore, the radiation induced phosphorylation of p34 appears to be delayed in AT cells, which are defective in the SDS pathway (Liu and Weaver, 1993). Thus, although these observations suggest that DNA damage can lead to modification of a protein (RPA) involved in DNA repair, they do not provide an explanation for the downregulation of initiation at origins in response to radiation. Furthermore, it is presently unclear whether the change in phosphorylation status of p34 is the result or the cause of S phase arrest. Finally, it is difficult to understand how modulation of p34 could specifically affect initiation of DNA synthesis without affecting chain elongation, since RPA is involved in both.

Several other seemingly unrelated observations must ultimately be fitted into any model for an S phase damage sensing pathway. Although the SDS pathway probably exerts its regulatory role on all chromosomal origins, be they nucleolar (as in rDNA replicons) or extranucleolar, it does not appear to affect mitochondrial DNA synthesis (Cleaver, 1992). Therefore, the relevant signals must be excluded from the mitochondrion, or mitochondrial replication origins must be immune to the same signals that inhibit chromosomal origins.

Finally, it has been shown that caffeine lowers the degree of inhibition of DNA synthesis following radiation treatment (Fairchild and Cowan, 1991), whereas inhibitors of calmodulin and calmodulin dependent protein kinases have been shown to elicit an AT like radiation resistant DNA synthesis phenotype in human fibroblasts (Lum *et al*, 1993; Mirzayans *et al*, 1995). Together, these observations suggest that the SDS pathway(s) is effected by a typical signal transduction cascade involving many protein kinases.

QUESTIONS FOR THE FUTURE

Despite more than four decades of study on the effects of DNA damage on DNA synthesis, several fundamental questions remain unanswered. Among them are the following.

What Is the Nature of the Initiating Signal?

It is still not known what initial signal leads to an acute decrease in the rate of DNA replication or whether it the same signal that activates other cell cycle checkpoints. When a given DNA damaging agent is administered to cells, one or more of four (and possibly five) different checkpoints can be activated depending on cell cycle position. These include the TP53 mediated G_1 checkpoint, a proposed G_1/S transition checkpoint, the S phase damage sensing checkpoint or pathway and the G_2/M checkpoint. Recently a fifth damage responsive mechanism (termed a G_1/M checkpoint) has been suggested to operate in S *cerevisiae* in which the cell commits itself to mitosis in late G_1 in an S phase independent manner (Toyn *et al*, 1995).

The signal that activates any one or all of these pathways is not known. We assume that it is the broken DNA ends themselves, since different types of insults that lead to single strand breaks (eg ionizing radiation and hydrogen peroxide) appear to activate the same spectrum of pathways (Nelson and Kastan, 1994). However, it is conceivable that some other intermediate that is coincidentally generated (eg free radicals) could serve as a soluble indicator that DNA damage is likely. It is possible that the identity of the initial signal that is sensed may have to wait until the sensory molecules themselves are identified.

Why Preferentially Inhibit Initiation Rather Than Active Forks?

We assume that the purpose of the SDS pathway and, potentially, the G_1/S checkpoint is to prevent cells from converting single strand breaks to double strand breaks by replication. Therefore, it would seem to make more sense to immediately prevent any further fork movement to allow repair of damaged templates rather than preventing further initiations but allowing active forks to proceed to their natural stopping places (eg the two termini of a given replicon). However, several agents that reduce the rate of replication fork travel by lowering nucleotide pool levels (eg aphidicolin, mimosine and hydroxyurea) also destabilize replication intermediates (Snapka *et al*, 1991; Levenson and Hamlin, 1993; Kalejta RF and Hamlin JL, unpublished), which by itself would presumably signal DNA damage. Therefore, slowing replication forks may not be a viable option.

Does G_1/S Delay and/or the SDS Pathway Improve Cell Survival?

If G_1/S delay and/or the SDS pathway functions to prevent the conversion of single strand breaks to double strand breaks by replication, then one would predict that the exquisite sensitivity of AT cells to ionizing radiation results from their inability to downregulate replication in response to DNA damage (Painter and Young, 1980). There is evidence both for and against this proposal (Meyn *et al*, 1993; Verhaegh *et al*, 1995). In yeast it was shown that the operation of the G_1/S checkpoint dramatically increases cell survival after radiation treatment (Siede *et al*, 1994). However, it was shown more recently that human chromosome 4 can restore radiation sensitive DNA replication to an AT like CHO cell line, but the resulting transformants retained their original sensitivity to radiation (Verhaegh *et al*, 1995). Although this finding suggests that radiation resistant DNA synthesis and sensitivity to radiation are independent of one another, other explanations are possible. For example, the introduction of supernumerary copies of certain unrelated genes that are passengers on chromosome 4 could lead to increased apoptosis or other detrimental consequences that offset any gain in survival derived from restoration of the SDS pathway.

Does the SDS Pathway Affect Genomic Stability?

As with the TP53 mediated G_1 checkpoint, it would be predicted that the SDS pathway would be important in protecting genomic integrity. In fact, AT cells do undergo many more chromosomal rearrangements than normal cells, and AT patients are much more prone to develop cancers than are normal individuals (see Rotman and Shiloh, this volume:). However, as outlined above, AT cells have lost multiple damage sensing pathways, making it difficult to determine whether the loss of the S phase damage sensing pathway(s) alone is sufficient to allow genetic instability.

SUMMARY

Mammalian cells have evolved multiple responses for dealing with DNA damage. One response is to acutely downregulate DNA synthesis at the initiation step. Essentially nothing is known about the initial signal that activates this SDS pathway or the macromolecules involved in transducing the signal into the final inhibitory step at origins. Determining whether any radiation induced changes in known proteins involved in cell cycle regulation or in other signal transduction pathways are primary or secondary responses to DNA damage constitutes a major challenge to identifying members of the pathway. It may turn out to be easier to identify the final mediator in the pathway, namely the protein(s) whose interaction with origins is ultimately affected by radiation. Hopefully, mutations in SDS genes in genetically tractable systems such as S cerevisiae or Schizosaccharomyces pombe will allow the identification of homologous genes in mammals.

Most tumour cells are TP53 negative, and yet it is not clear that TP53 status influences radiation sensitivity. The SDS pathway may therefore represent an important protective mechanism that stands in the way of effective tumour cell killing by radiation therapy. It is hoped that an understanding of this pathway will provide opportunities for developing novel antineoplastic targets and/or radiation sensitizers.

Acknowledgements

We thank the other members of our laboratory for valuable discussions and encouragement. We particularly thank Carlton White and Kevin Cox for expert technical assistance. The work in our laboratory was supported by a grant to JLH from the National Institutes of Health (RO1CA52559). HL was supported in part by an American Cancer Society institutional research grant (IRG149L). JML was supported in part by a Radiological Society of North America scholar's award.

References

Allen JB, Zhou Z, Siede W, Friedberg EC and Elledge SJ (1994) The SAD1/RAD53 protein kinase controls multiple checkpoints and DNA damage-induced transcription in yeast. *Genes and Development* **8** 2401–2415

Bell SP and Stillman B (1992) ATP-dependent recognition of eukaryotic origins of replication by a multiprotein complex. *Nature* **357** 128–134

Benard M and Pierron G (1992) Mapping of a *Physarum* chromosomal origin of replication tightly linked to a developmentally-regulated profilin gene. *Nucleic Acids Research* **20** 3309–3315

Blackwell LJ and Borowiec JA (1994) Human replication protein A binds single-stranded DNA in two distinct complexes. *Molecular and Cellular Biology* **14** 3993–4001

Brewer BJ and Fangman WL (1987) The localization of replication origins on ARS plasmids in *S. cerevisiae*. *Cell* **51** 463–471

Burhans WC, Selegue JE and Heintz NH (1986) Isolation of the origin of replication associated with the amplified Chinese hamster dihydrofolate reductase domain. *Proceedings of the National Academy of Sciences of the USA* **83** 7790–7794

Burhans WC, Vassilev LT, Caddle MS, Heintz NH and DePamphilis ML (1990) Identification of an origin of bidirectional DNA replication in mammalian chromosomes. *Cell* **62** 955–965

Carpenter PB, Mueller PR and Dunphy WG (1996) Role for a *Xenopus* Orc2-related protein in controlling DNA replication. *Nature* **379** 357–360

Carroll SM, DeRose ML, Kolman JL, Nonet GH, Kelly RE and Wahl GM (1993) Localization of a bidirectional DNA replication origin in the native locus and in episomally amplified murine adenosine deaminase loci. *Molecular and Cellular Biology* **13** 2971–2981

Chan CS and Tye BK (1980) Autonomously replicating sequences in *Saccharomyces cerevisiae*. *Proceedings of the National Academy of Sciences of the USA* **77** 6329–6333

Cleaver JE (1992) Replication of nuclear and mitochondrial DNA in X-ray-damaged cells: evidence for a nuclear-specific mechanism that down-regulates replication. *Radiation Research* **131** 338–344

Cleaver JE, Rose R and Mitchell DL (1990) Replication of chromosomal and episomal DNA in X-ray-damaged human cells: a cis- or trans-acting mechanism? *Radiation Research* **124** 294–299

Cozzarelli NR (1977) The mechanism of action of inhibitors of DNA synthesis. *Annual Review of Biochemistry* **46** 641–668

Datta R, Hass R, Gunji H, Weichselbaum R and Kufe D (1992) Down-regulation of cell cycle control genes by ionizing radiation. *Cell Growth and Differentiation* **3** 637–644

DeGregori J, Kowalik T and Nevins JR (1995) Cellular targets for activation by the E2F1 transcription factor include DNA synthesis- and G1/S-regulatory genes. *Molecular and Cellular Biology* **15** 4215–4224 [erratum **15** 5846-5847]

Delidakis C and Kafatos FC (1987) Amplification of a chorion gene cluster in *Drosophila* is subject to multiple cis-regulatory elements and to long-range position effects. *Journal of Molecular Biology* **197** 11–26

Diffley JF, Cocker JH, Dowell SJ and Rowley A (1994) Two steps in the assembly of complexes at yeast replication origins in vivo. *Cell* **78** 303–316

Dijkwel PA and Hamlin JL (1988) Matrix attachment regions are positioned near replication initiation sites, genes, and an interamplicon junction in the amplified dihydrofolate reductase domain of Chinese hamster ovary cells. *Molecular and Cellular Biology* **8** 5398–5409

Dijkwel PA and Hamlin JL (1992) Initiation of DNA replication in the dihydrofolate reductase locus is confined to the early S period in CHO cells synchronized with the plant amino acid mimosine. *Molecular and Cellular Biology* **12** 3715–3722

Dijkwel PA and Hamlin JL (1995) The Chinese hamster dihydrofolate reductase origin consists of multiple potential nascent-strand start sites. *Molecular and Cellular Biology* **15** 3023–3031

Dijkwel PA, Vaughn JP and Hamlin JL (1994) Replication initiation sites are distributed widely in the amplified CHO dihydrofolate reductase domain. *Nucleic Acids Research* **22** 4989–4996

Dowdy SF, Hinds PW, Louie K, Reed SI, Arnold A and Weinberg RA (1993) Physical interaction of the retinoblastoma protein with human D cyclins. *Cell* **73** 499–511

Dowell SJ, Romanowski P and Diffley JF (1994) Interaction of Dbf4, the Cdc7 protein kinase regulatory subunit, with yeast replication origins in vivo. *Science* **265** 1243–1246

Dutta A and Stillman B (1992) cdc2 family kinases phosphorylate a human cell DNA replication factor, RPA, and activate DNA replication. *EMBO Journal* **11** 2189–2199

Ewen ME, Sluss HK, Sherr CJ, Matsushime H, Kato J and Livingston DM (1993) Functional interactions of the retinoblastoma protein with mammalian D-type cyclins. *Cell* **73** 487–497

Fairchild CR and Cowan KH (1991) Multidrug resistance: a pleiotropic response to cytotoxic drugs. *International Journal of Radiation Oncology, Biology, Physics* **20** 361–367

Ferguson BM, Brewer BJ and Fangman WL (1991) Temporal control of DNA replication in yeast. *Cold Spring Harbor Symposia on Quantitative Biology* **56** 293–302

Foreman PK and Hamlin JL (1989) Identification and characterization of a gene that is coamplified with dihydrofolate reductase in a methotrexate-resistant CHO cell line. *Molecular and Cellular Biology* **9** 1137–1147

Frappier L and Zannis-Hadjopoulos M (1987) Autonomous replication of plasmids bearing monkey DNA origin-enriched sequences. *Proceedings of the National Academy of Sciences of the USA* **84** 6668–6672

Gale JM, Tobey RA and D'Anna JA (1992) Localization and DNA sequence of a replication origin in the rhodopsin gene locus of Chinese hamster cells. *Journal of Molecular Biology* **224** 343–358

Gavin KA, Hidaka M and Stillman B (1995) Conserved initiator proteins in eukaryotes. *Science* **270** 1667–1671

Giacca M, Zentilin L, Norio P *et al* (1994) Fine mapping of a replication origin of human DNA. *Proceedings of the National Academy of Sciences of the USA* **91** 7119–7123

Gossen M, Pak DT, Hansen SK, Acharya JK and Botchan MR (1995) A *Drosophila* homolog of the yeast origin recognition complex. *Science* **270** 1674–1677

Hamlin JL, Mosca PJ and Levenson VV (1994) Defining origins of replication in mammalian cells. *Biochimica et Biophysica Acta* **1198** 85–11

Handeli S, Klar A, Meuth M and Cedar H (1989) Mapping replication units in animal cells. *Cell* **57** 909–920

Hardy CF (1996) Characterization of an essential Orc2p-associated factor that plays a role in DNA replication. *Molecular and Cellular Biology* **16** 1832–1841

Hartwell LH and Weinert TA (1989) Checkpoints: controls that ensure the order of cell cycle events. *Science* **246** 629–634

He Z, Henricksen LA, Wold MS and Ingles CJ (1995) RPA involvement in the damage-recognition and incision steps of nucleotide excision repair. *Nature* **374** 566–569

Heck MM and Spradling AC (1990) Multiple replication origins are used during *Drosophila* chorion gene amplification. *Journal of Cell Biology* **110** 903–914

Heintz NH and Hamlin JL (1982) An amplified chromosomal sequence that includes the gene for dihydrofolate reductase initiates replication within specific restriction fragments. *Proceedings of the National Academy of Sciences of the USA* **79** 4083–4087

Heinzel SS, Krysan PJ, Tran CT and Calos MP (1991) Autonomous DNA replication in human cells is affected by the size and the source of the DNA. *Molecular and Cellular Biology* **11** 2263–2272

Hong JH, Gatti RA, Huo YK, Chiang CS and McBride WH (1994) G2/M-phase arrest and release in ataxia telangiectasia and normal cells after exposure to ionizing radiation. *Radiation Research* **140** 17–23

Huberman JA (1981) New views of the biochemistry of eucaryotic DNA replication revealed by aphidicolin, an unusual inhibitor of DNA polymerase alpha. *Cell* **23** 647–648

Huberman JA and Riggs AD (1968) On the mechanism of DNA replication in mammalian chromosomes. *Journal of Molecular Biology* **32** 327–341

Hyrien O and Mechali M (1993) Chromosomal replication initiates and terminates at random sequences but at regular intervals in the ribosomal DNA of *Xenopus* early embryos. *EMBO Journal* **12** 4511–4520

Hyrien O, Maric C and Mechali M (1995) Transition in specification of embryonic metazoan DNA replication origins. *Science* **270** 994–997

Iguchi AS, Okazaki T, Itani T, Ogata M, Sato Y and Ariga H (1988) An initiation site of DNA replication with transcriptional enhancer activity present upstream of the c-myc gene. *EMBO Journal* **7** 3135–3142

Jackson AL, Pahl PM, Harrison K, Rosamond J and Sclafani RA (1993) Cell cycle regulation of the yeast Cdc7 protein kinase by association with the Dbf4 protein. *Molecular and Cellular Biology* **13** 2899–2908

Kastan MB, Zhan Q, el-Deiry WS *et al* (1992) A mammalian cell cycle checkpoint pathway utilizing p53 and GADD45 is defective in ataxia-telangiectasia. *Cell* **71** 587–597

Kelly RE, DeRose ML, Draper BW and Wahl GM (1995) Identification of an origin of bidirectional DNA replication in the ubiquitously expressed mammalian CAD gene. *Molecular and Cellular Biology* **15** 4136–4148

Kemp LM, Sedgwick SG and Jeggo PA (1984) X-ray sensitive mutants of Chinese hamster ovary cells defective in double-strand break rejoining. *Mutation Research* **132** 189–196

Kitada K, Johnston LH, Sugino T and Sugino A (1992) Temperature-sensitive cdc7 mutations of *Saccharomyces cerevisiae* are suppressed by the DBF4 gene, which is required for the G1/S cell cycle transition. *Genetics* **131** 21–29

Kitsberg D, Selig S, Keshet I and Cedar H (1993) Replication structure of the human beta-globin gene domain. *Nature* **366** 588–590

Krek W, Ewen ME, Shirodkar S, Arany Z, Kaelin Jr WG and Livingston DM (1994) Negative regulation of the growth-promoting transcription factor E2F-1 by a stably bound cyclin A-dependent protein kinase. *Cell* **78** 161–172

Krysan PJ, Haase SB and Calos MP (1989) Isolation of human sequences that replicate autonomously in human cells. *Molecular and Cellular Biology* **9** 1026–1033

Lamb JR, Petit-Frere C, Broughton BC, Lehmann AR and Green MH (1989) Inhibition of DNA replication by ionizing radiation is mediated by a trans-acting factor. *International Journal of Radiation Biology* **56** 125–130

Larner JM, Lee H and Hamlin JL (1994) Radiation effects on DNA synthesis in a defined chromosomal replicon. *Molecular and Cellular Biology* **14** 1901–1908

Lee SH and Kim DK (1995) The role of the 34-kDa subunit of human replication protein A in simian virus 40 DNA replication in vitro. *Journal of Biological Chemistry* **270** 12801–12807

Lee H, Larner JM and Hamlin JL Cloning and characterization of Chinese hamster p53 cDNA. *Gene* (in press)

Leffak M and James CD (1989) Opposite replication polarity of the germ line c-myc gene in HeLa cells compared with that of two Burkitt lymphoma cell lines. *Molecular and Cellular Biology* **9** 586–593

Leonard CJ, Canman CE and Kastan MB (1995) The role of p53 in cell-cycle control and apoptosis: implications for cancer. *Important Advances in Oncology* 33-42

Leu TH and Hamlin JL (1989) High-resolution mapping of replication fork movement through the amplified dihydrofolate reductase domain in CHO cells by in-gel renaturation analysis. *Molecular and Cellular Biology* **9** 523–531

Levenson V and Hamlin JL (1993) A general protocol for evaluating the specific effects of DNA replication inhibitors. *Nucleic Acids Research* **21** 3997–4004

Liang C, Spitzer JD, Smith HS and Gerbi SA (1993) Replication initiates at a confined region during DNA amplification in Sciara DNA puff II/9A. *Genes and Development* **7** 1072–1084

Liang C, Weinreich M and Stillman B (1995) ORC and Cdc6p interact and determine the frequency of initiation of DNA replication in the genome. *Cell* **81** 667–676

Little RD, Platt TH and Schildkraut CL (1993) Initiation and termination of DNA replication in human rRNA genes. *Molecular and Cellular Biology* **13** 6600–6613

Liu VF and Weaver DT (1993) The ionizing radiation-induced replication protein A phosphorylation response differs between ataxia telangiectasia and normal human cells. *Molecular and Cellular Biology* **13** 7222–7231

Lum BL, Gosland MP, Kaubisch S and Sikic BI (1993) Molecular targets in oncology: implications of the multidrug resistance gene. *Pharmacotherapy* **13** 88–109

Ma C, Leu TH and Hamlin JL (1990) Multiple origins of replication in the dihydrofolate reductase amplicons of a methotrexate-resistant chinese hamster cell line. *Molecular and Cellular Biology* **10** 1338–1346

McWhinney C and Leffak M (1990) Autonomous replication of a DNA fragment containing the chromosomal replication origin of the human c-myc gene. *Nucleic Acids Research* **18** 1233–1242

Makino F and Okada S (1975) Effects of ionizing radiation on DNA replication in cultured mammalian cells. *Radiation Research* **62** 37–51

Meyn MS, Lu-Kuo JM and Herzing LB (1993) Expression cloning of multiple human cDNAs that complement the phenotypic defects of ataxia-telangiectasia group D fibroblasts. *American Journal of Human Genetics* **53** 1206–1216

Milbrandt JD, Heintz NH, White WC, Rothman SM and Hamlin JL (1981) Methotrexate-resistant Chinese hamster ovary cells have amplified a 135-kilobase-pair region that includes the dihydrofolate reductase gene. *Proceedings of the National Academy of Sciences of the USA* **78** 6043–6047

Mirzayans R, Famulski KS, Enns L, Fraser M and Pateron MC (1995) Characterization of the signal transduction pathway mediating gamma ray-induced inhibition of DNA synthesis in human cells: indirect evidence for involvement of calmodulin but not protein kinase C nor p53. *Oncogene* **11** 1597–1605

Mosca PJ, Dijkwel PA and Hamlin JL (1992) The plant amino acid mimosine may inhibit initiation at origins of replication in Chinese hamster cells. *Molecular and Cellular Biology* **12** 4375–4383 [erratum **13** 1981]

Nawotka KA and Huberman JA (1988) Two-dimensional gel electrophoretic method for mapping DNA replicons. *Molecular and Cellular Biology* **8** 1408–1413

Nelson WG and Kastan MB (1994) DNA strand breaks: the DNA template alterations that trigger p53-dependent DNA damage response pathways. *Molecular and Cellular Biology* **14** 1815–1823

Newlon CS and Theis JF (1993) The structure and function of yeast autonomously replicating sequences. *Current Opinion in Genetics and Development* **3** 752–758

Nishitani H and Nurse P (1995) p65cdc18 plays a major role controlling the initiation of DNA replication in fission yeast. *Cell* **83** 397–405

Painter RB (1981) Radioresistant DNA synthesis: an intrinsic feature of ataxia telangiectasia. *Mutation Research* **84** 183–190

Painter RB and Young BR (1975) X-ray-induced inhibition of DNA synthesis in Chinese hamster ovary, human HeLa, and mouse L cells. *Radiation Research* **64** 648–656

Painter RB and Young BR (1980) Radiosensitivity in ataxia-telangiectasia: a new explanation. *Proceedings of the National Academy of Sciences of the USA* **77** 7315–7317

Paulovich AG and Hartwell LH (1995) A checkpoint regulates the rate of progression through S phase in *S. cerevisiae* in response to DNA damage. *Cell* **82** 841–847

Piatti S, Lengauer C and Nasmyth K (1995) Cdc6 is an unstable protein whose de novo synthesis in G1 is important for the onset of S phase and for preventing a 'reductional' anaphase in the budding yeast *Saccharomyces cerevisiae*. *EMBO Journal* **14** 3788–3799

Sanchez Y, Desany BA, Jones WJ, Liu Q, Wang B and Elledge SJ (1996) Regulation of TAD53 by the ATM-like kinase MEC1 and TEL1 in yeast cell cycle checkpoint pathways. *Science* **271** 357–360

Shinomiya T and Ina S (1991) Analysis of chromosomal replicons in early embryos of *Drosophila melanogaster* by two-dimensional gel electrophoresis. *Nucleic Acids Research* **19** 3935–3941

Shinomiya T and Ina S (1993) DNA replication of histone gene repeats in *Drosophila melanogaster* tissue culture cells: multiple initiation sites and replication pause sites. *Molecular and Cellular Biology* **13** 4098–4106

Shinomiya T and Ina S (1994) Mapping an initiation region of DNA replication at a single-copy chromosomal locus in *Drosophila melanogaster* cells by two-dimensional gel methods and PCR- mediated nascent-strand analysis: multiple replication origins in a broad zone. *Molecular and Cellular Biology* **14** 7394–7403

Shirodkar S, Ewen M, DeCaprio JA, Morgan J, Livingston DM and Chittenden T (1992) The transcription factor E2F interacts with the retinoblastoma product and a p107-cyclin A complex in a cell cycle-regulated manner. *Cell* **68** 157–166

Siede W, Friedberg AS, Dianova I and Friedberg EC (1994) Characterization of G1 checkpoint control in the yeast *Saccharomyces cerevisiae* following exposure to DNA-damaging agents. *Genetics* **138** 271–281

Singh KK and Samson L (1995) Replication protein A binds to regulatory elements in yeast DNA repair and DNA metabolism genes. *Proceedings of the National Academy of Sciences of the USA* **92** 4907–4911

Snapka RM, Shin CG, Permana PA and Strayer J (1991) Aphidicolin-induced topological and recombinational events in simian virus 40. *Nucleic Acids Research* **19** 5065–5072

Stinchcomb DT, Thomas M, Kelly J, Selker E and Davis RW (1980) Eukaryotic DNA segments capable of autonomous replication in yeast. *Proceedings of the National Academy of Sciences of the USA* **77** 4559–4563

Sudo K, Ogata M, Sato Y, Iguchi AS and Ariga H (1990) Cloned origin of DNA replication in c-myc gene can function and be transmitted in transgenic mice in an episomal state. *Nucleic Acids Research* **18** 5425–5432

Takahara K (1996) Mouse and human homologues of the yeast origin of replication recognition complex subunit ORC2 and chromosomal localization of the cognate human gene ORC2L. *Genomics* **31** 119–122

Tasheva ES and Roufa DJ (1994) A mammalian origin of bidirectional DNA replication within the Chinese hamster RPS14 locus. *Molecular and Cellular Biology* **14** 5628–5635

Toyn JH, Johnson AL and Johnston LH (1995) Segregation of unreplicated chromosomes in *Saccharomyces cerevisiae* reveals a novel G1/M-phase checkpoint. *Molecular and Cellular Biology* **15** 5312–5321

Vassilev LT, Burhans WC and DePamphilis ML (1990) Mapping an origin of DNA replication at a single-copy locus in exponentially proliferating mammalian cells. *Molecular and Cellular Biology* **10** 4685–4689

Vaughn JP, Dijkwel PA and Hamlin JL (1990) Replication initiates in a broad zone in the amplified CHO dihydrofolate reductase domain. *Cell* **61** 1075–1087

Verhaegh GW, Jongmans W, Jaspers NG *et al* (1995) A gene that regulates DNA replication in response to DNA damage is located on human chromosome 4q. *American Journal of Human Genetics* **57** 1095–1103

Waga S, Bauer G and Stillman B (1994) Reconstitution of complete SV40 DNA replication with purified replication factors. *Journal of Biological Chemistry* **269** 10923–10934

Walters RA and Hildebrand CE (1975) Evidence that x-irradiation inhibits DNA replicon initiation in Chinese hamster cells. *Biochemical and Biophysical Research Communications* **65** 265–271

Wang Y (1995) Regulation of DNA replication in irradiated cells by trans-acting factors. *Radiation Research* **142** 169–175

Watanabe I (1974) Radiation effects on DNA chain growth in mammalian cells. *Radiation Research* **58** 541–556

Weinert TA (1992) Dual cell cycle checkpoints sensitive to chromosome replication and DNA damage in the budding yeast *Saccharomyces cerevisiae*. *Radiation Research* **132** 141–143

Weinert TA, Kiser GL and Hartwell LH (1994) Mitotic checkpoint genes in budding yeast and the dependence of mitosis on DNA replication and repair. *Genes and Development* **8** 652–665

Zaider M (1993) A mathematical formalism describing the yield of radiation-induced single- and double-strand DNA breaks, and its dependence on radiation quality. *Radiation Research* **134** 1–8

The authors are responsible for the accuracy of the references.

Cyclins and the G₂/M Transition

M R JACKMAN • J N PINES

Wellcome/CRC Institute of Developmental Biology, Tennis Court Road, Cambridge CB2 5QX

Introduction
Mitotic cyclins and Cdks
Regulation of Cdk activity
  Thr-160 phosphorylation of Cdk1: stabilizing the primed cyclin/Cdk1 complex
  Thr-160 dephosphorylation: deactivating Cdk1
  Tyr-15/Thr-14 phosphorylation
  Tyr-15/Thr-14 dephosphorylation
Co-ordinating Wee1 and Cdc25 activity
  Wee1 downregulation
  Cdc25 upregulation
Inhibitory proteins and mitotic cyclin/Cdks
Cks family of Cdk binding proteins
Cell cycle dependent subcellular localization of mitotic cyclin/Cdk complexes
  Regulating mitotic cyclin/Cdk activity by subcellular localization
  Targeting mitotic cyclin/Cdk activity by localizing mitotic cyclins
Physiological functions of different mitotic cyclin/Cdks: substrate specificities
  Cytoplasmic events at G₂/M
  Nuclear changes at G₂/M
Summary

INTRODUCTION

At mitosis, there is a dramatic change in the entire architecture of the cell. The nuclear envelope and the Golgi apparatus disassemble, chromosomes condense, the actin and intermediate filament networks are reorganized and the microtubule network is reshaped to form the spindle. These dramatic morphological rearrangements are accompanied by a large increase in the amount of phosphoprotein within the cell, caused by a pivotal shift in the balance of kinase and phosphatase activities in the cell. In this chapter, we discuss how the cyclin dependent kinases (Cdks) essential to the initiation of mitosis are activated at the G₂/M transition and the effects they have on defined substrates during this transition.

Cancer Surveys Volume 29: *Checkpoint Controls and Cancer*
© 1997 Imperial Cancer Research Fund. 0-87969-518-8/97. $5.00 + .00

MITOTIC CYCLINS AND CDKS

In all eukaryotes, mitosis is induced when the cyclin B–p34^{cdc2} Ser/Thr protein kinase complex is activated (for reviews, see Nurse, 1990; Maller, 1991). p34^{cdc2} is the prototypical member of the cyclin dependent kinase family, and hence it is also referred to as Cdk1. Monomeric Cdk1 is inactive. To be active, Cdk1 must first bind a cyclin.

Cyclins are defined by an approximately 100 aminoacid region called the "cyclin box", through which they bind and activate Cdks (Kobayashi *et al*, 1992; Lees and Harlow, 1993). Mitotic cyclins are those that are degraded at mitosis. They have a partially conserved "destruction box" sequence at their N-terminus, which is necessary for their sudden and rapid destruction specifically in mitosis (Glotzer *et al*, 1991). This destruction is mediated by ubiquitin dependent proteolysis (Glotzer *et al*, 1991; Stewart *et al*, 1994; Sudakin *et al*, 1995). Studies in cell free systems have identified large multiprotein complexes, named the anaphase promoting complex (APC) in frogs (King *et al*, 1995) and cyclosome in clams (Hershko *et al*, 1994), that specifically ubiquitinate mitotic cyclins in a cell cycle dependent manner. The levels of mitotic cyclins increase as the cell progresses through S and G$_2$ phases and peak at mitosis. The primary mitotic cyclins are the B type cyclins. Fission yeast have one mitotic B type cyclin (cdc13) (Booher and Beach, 1987; Hagan *et al*, 1988), whereas budding yeast have up to four mitotic B type cyclins, Clbs 1–4 (Fitch *et al*, 1992; Richardson *et al*, 1992). Multicellular organisms have between one (clams and sea urchins) and four (frogs) identified B type cyclins (Minshull *et al*, 1989; Pines and Hunter, 1989; Gallant and Nigg, 1992). In somatic cells, the A type cyclins are also required for mitosis. They differ from the B types in both primary structure and in their behaviour during the cell cycle. The A type cyclins bind to a second Cdk, Cdk2, and this complex is active throughout S and G$_2$ phases and the initial stages of mitosis. Its primary function is thought to be regulation of S phase events, but cyclin A is also essential for cells to enter mitosis (Lehner and O'Farrell, 1990). In some cells, cyclin A also associates with Cdk1 in late G$_2$ phase. Both cyclin A/Cdk complexes are active at the beginning of mitosis but are turned off when cyclin A is destroyed during metaphase, at a slightly earlier stage in mitosis than cyclin B1, which is degraded when cells enter anaphase (Hunt *et al*, 1992; Lorca *et al*, 1992).

REGULATION OF CDK ACTIVITY

Cdk regulation is quite complex, but the resolution of the Cdk2 crystal structure, both as an inactive monomer and as an active cyclin bound complex, has illuminated the manner in which cyclin binding turns on a Cdk. Like other kinases, Cdk2 has a bilobar structure (Knighton *et al*, 1991a,b; De Bondt *et al*, 1993), and it binds ATP and substrate in the cleft between the two lobes. In its inactive, monomeric form, Cdk2 binds ATP in a conformation that makes impossible a nucleophilic attack by the substrate hydroxyl on the β-γ phosphate bond of ATP. In addition, a conserved part of the protein, the "T loop", obscures the catalytic cleft and prevents substrates from binding.

Genetic and biochemical analyses of Cdk activation (reviewed in Hanks and Hunter, 1995; Morgan, 1995) predicted that cyclin binding and phosphorylation/dephosphorylation of conserved residues would change the conformation of Cdk2 to match that of cAMP dependent protein kinase, PK-A, whose crystal structure had been solved as an active monomer (Knighton et al, 1991a,b; De Bondt et al, 1993). The predictions were confirmed when the atomic structure of the partially active Cdk2/cyclin A complex was solved (Jeffrey et al, 1995; reviewed in Pines, 1993; Morgan and De Bondt, 1994; Johnson et al, 1996). By binding Cdk2, cyclin A alters the way in which ATP is orientated in the catalytic cleft and causes the T loop to move away from the catalytic cleft, thus neutralizing the two major factors that keep monomeric Cdk2 inactive.

The crystal structure of the cyclin A component of the crystallized cyclin A/Cdk2 complex has also been solved on its own (Brown et al, 1995). A comparison between the two structures shows that there is effectively no change in the cyclin box when it binds a Cdk. The regions of cyclin A and Cdk2 that interact are conserved among some of the members of the cyclin and Cdk families, so it is reasonable to assume that structural changes in Cdk2 when it binds cyclin A will also apply to Cdk1 when it binds a B type cyclin. Cyclin A binds mainly to the N-terminal lobe of Cdk2 (Jeffrey et al, 1995). This lobe contains the conserved PSTAIRE motif, known to be important for Cdk2 binding to cyclin (Ducommun et al, 1991; Endicott et al, 1994).

Thr-160 Phosphorylation of Cdk1: Stabilizing the Primed Cyclin/Cdk1 Complex

When cyclin binds to its partner Cdk, it primes the Cdk by making a threonine residue in the T loop of the Cdk more accessible to a kinase, and phosphorylation at this site fully activates the cyclin/Cdk complex. (This residue is Thr-161 in human Cdk1 and Thr-160 in Cdk2.) The analogous threonine or tyrosine residue in the T loop is phosphorylated in most active kinases (Marshall, 1994). Biochemical studies indicate that Thr-161 phosphorylation occurs after Cdk1 has bound a B type cyclin (Solomon et al, 1992). From X ray crystal structures of cyclin A/Cdk2 compared with PK-A, it has been predicted that phosphorylation of Thr-160 will secure the newly positioned T loop in a cationic pocket of the C-terminal lobe away from the catalytic cleft (Jeffrey et al, 1995), and this has been confirmed in the crystal structure of Thr-160 phosphorylated Cdk2/cyclin A. Thr-160 phosphorylation could also stabilize the cyclin A/Cdk2 complex, because phosphorylated Thr-160 interacts with cyclin A. In support of this, when the equivalent threonine (Thr-167) in fission yeast cdc2 is mutated, it destabilizes binding to the fission yeast B type cyclin, cdc13 (Gould *et al*, 1991).

The kinase that phosphorylates the T loop threonine in Cdks is called Cdk activating kinase (CAK). The first CAK to be purified was another cyclin/Cdk complex—cyclin H/Cdk7 (Fesquet *et al*, 1993; Poon *et al*, 1993; Solomon *et al*, 1993; Fisher and Morgan, 1994; Makela *et al*, 1994; Tassan *et al*, 1994). However, a very different role for cyclin H/Cdk7 became apparent when cyclin H/Cdk7 was found to associate with human transcription factor IIH, a multiprotein complex needed for RNA polymerase II transcription and DNA repair (Feaver *et al*, 1994; Roy *et al*, 1994). There is still the possibility that cyclin H/Cdk7 has a dual role as both the Cdk activating kinase and as a regulator of transcription and DNA repair. However, in budding yeast, these two roles are clearly separate. The budding yeast proteins most closely related to the cyclin H/Cdk7 complex, the Ccl1 cyclin in a complex with the Kin28 protein kinase, are restricted to a role in transcription and repair, where they phosphorylate the C-terminal domain of RNA polymerase II (Cismowski *et al*, 1995). The budding yeast CAK has been identified as a 43 kDa protein kinase that is active as a monomer (Kaldis *et al*, 1996; Thuret *et al*, 1996; Espinoza *et al*, 1996). This protein kinase, Civ1-Cak1, is encoded by an essential gene and specifically phosphorylates the T loop threonine of Cdc28. The aminoacid sequence of Civ1-Cak1 shows that its distant but closest relatives are the Cdk family of kinases. However, Civ1-Cak1 is clearly not a Cdk. It is very unusual among protein kinases in not having the sequence GXGXXG in its ATP binding motif.

Thr-160 Dephosphorylation: Deactivating Cdk1

The phosphatase that reverses the CAK dependent phosphorylation of the T loop threonine is thought to be the dual specificity phosphatase, KAP (Cdk associated phosphatase) (Poon and Hunter, 1995). KAP was originally isolated in two independent yeast two hybrid screens as a protein that associated with Cdk1 and Cdk2 (Gyuris *et al*, 1993; Hannon *et al*, 1994). KAP can bind to the cyclin/Cdk complex but can dephosphorylate only a monomeric Cdk. Therefore, Cdk dephosphorylation necessarily follows cyclin degradation (Poon and Hunter, 1995). It is not clear whether KAP activity is regulated during the cell cycle, or whether it has an essential role at a particular point in the cell cycle. For example, it may be that Thr-161 dephosphorylation is important for the complete inactivation of the mitotic kinase, Cdk1, and thus for the transition from mitosis back into G_1 phase. Overexpressing KAP in human cells slows down passage through G_1 phase (Poon and Hunter, 1995), which might be through preventing the full activation of Cdk2 or Cdk4/Cdk6 by CAK. However, this requires either that CAK can phosphorylate monomeric Cdk2/Cdk4/Cdk6 or that KAP can dephosphorylate Cdk2/Cdk4/Cdk6 when they are bound to G_1 cyclins. The effect of overexpressing KAP upon progression through G_2 phase has not been described.

Tyr-15/Thr-14 Phosphorylation

Mitotic cyclin/Cdk1 complexes in vertebrates are kept inactive during late S and G$_2$ phases by phosphorylation on Cdk1 at two conserved residues, Tyr-15 and Thr-14, which are located in its active site. The Thr-14 phosphorylation inhibits Cdk1 kinase activity by interfering with ATP binding (Endicott *et al*, 1994), whereas phosphorylating Tyr-15 keeps Cdk1 inactive by interfering with the transfer of the γ phosphate from ATP to substrate (Atherton *et al*, 1993). To deregulate Cdk1 completely in vertebrates, both Tyr-15 and Thr-14 need to be mutated to non-phosphorylatable residues. Mutation of only Tyr-15 or Thr-14 causes cells to progress into a less extensive mitotic state than that observed when both the Thr-14 and Tyr-15 of Cdk1 are substituted (Krek and Nigg, 1991; Norbury *et al*, 1991).

Like T loop phosphorylation, Cdk1 is phosphorylated at Thr-14 and Tyr-15 residues only when it is bound to a cyclin (Solomon *et al*, 1990; Meijer *et al*, 1991a; Meuller *et al*, 1995a,b; Wanatabe *et al*, 1995). Fission yeast cdc2 is phosphorylated on Thr-14 and Tyr-15 by two partially redundant kinases, Wee1 and Mik1 (Russell and Nurse, 1987a,b; Lundgren *et al*, 1991; Parker and Piwnica, 1992; Den-Haese *et al*, 1995). Similarly, the *Xenopus* and human homologues of Wee1/Mik1 phosphorylate only the Tyr-15 residue of Cdk1/Cdk2 (Parker and Piwnica, 1992; McGowan and Russell, 1993; Meuller *et al*, 1995a; Watanabe *et al*, 1995). The Thr-14 of vertebrate Cdk1 is phosphorylated by a membrane associated kinase activity in frog egg extracts (Kornbluth *et al*, 1994) and in human cells (Atherton-Fessler *et al*, 1994). By contrast, *Xenopus* Wee1 is cytosolic (Kornbluth *et al*, 1994; Meuller *et al*, 1995a). A cDNA encoding the kinase responsible for Thr-14 phosphorylation was identified in a screen for *Xenopus* Wee1 homologues. This kinase was named Myt1 (membrane associated tyrosine and threonine specific cdc2 inhibitory kinase) (Meuller, 1995b), and its sequence had a putative transmembrane domain. A human Myt1 has recently been cloned and appears to localize mainly to the endoplasmic reticulum (Piwnica-Worms H, personal communication)

Tyr-15/Thr-14 Dephosphorylation

The activation of Cdk1 kinase by dephosphorylation of phospho-Tyr-15/Thr-14 is the pivotal event in triggering mitosis. Phospho-Tyr-15 and -Thr-14 are dephosphorylated by the dual specificity phosphatase, Cdc25 (Dunphy and Kumagai, 1991; Gautier *et al*, 1991; Sebastian *et al*, 1993). In fission yeast, only one cdc25 has been identified, although a second phosphatase, pyp3, also contributes towards activating cdc2 (Millar *et al*, 1992). Three isoforms of Cdc25 have been identified in human cells; Cdc25A, B and C (Galaktionov and Beach, 1991). Cdc25A protein levels and activity are highest at the end of G$_1$ phase; therefore, Cdc25A may be involved in regulating the G$_1$/S transition, perhaps by dephosphorylating phospho-Thr-14/Tyr-15 on Cdk2 associated with cyclins E and A (Hoffman *et al*, 1994; Jinno *et al*, 1994). However, whether Cdk2 is significantly phosphorylated on Thr-14/Tyr-15 at the end of G_1 phase remains a matter of debate. The function of Cdc25B is unclear, but recent studies indicate that it could be involved in activating cyclin B/Cdk1 and consequently spindle formation at prophase (Gabrielli *et al*, 1996). Cdc25C also has been shown to be important for activating cyclin/Cdk1 activity at prophase (Dunphy and Kumagai, 1991; Gautier *et al*, 1991; Millar and Russell, 1992; Clarke *et al*, 1993). Experiments using *Xenopus* extracts and *Schizosaccharomyces pombe* have shown that some checkpoint controls (including DNA damage and unreplicated DNA) (Clarke *et al*, 1995) may regulate the dephosphorylation of Tyr-15 and Thr-14 (Norbury *et al*, 1991; Smythe and Newport, 1992; Sorger and Murray, 1992). Checkpoint mechanisms are described by O'Connor in this volume and therefore will not be further discussed here.

CO-ORDINATING WEE1 AND CDC25 ACTIVITY

During interphase, Wee1 kinase activity must be greater than Cdc25 phosphatase activity in order to keep the mitotic cyclin/Cdks inactive. A PP2A like phosphatase has been implicated in keeping both Wee1 active and Cdc25C inactive during S and G_2 phases (Kinoshita *et al*, 1990; Lee *et al*, 1991; Clarke *et al*, 1993; Tang *et al*, 1993). Some studies also report that a type 1 protein phosphatase maintains Cdc25C in its inactive state (Izumi *et al*, 1992; Walker *et al*, 1992). To activate Cdk1 rapidly at prophase, the Wee1 kinase activity must be downregulated and Cdc25B/C phosphatase activity must be upregulated. This is achieved by extensive phosphorylation of both Wee1 and Cdc25 (reviewed in Dunphy, 1994).

Wee1 Downregulation

Wee1 activity can be downregulated by the Nim1/Cdr1 kinase, first identified in *S pombe* (Russell and Nurse, 1987a,b; Feilotter *et al*, 1991). Nim1/Cdr1 phosphorylates Wee1 directly on the C-terminal region that forms its kinase domain (Coleman *et al*, 1993; Parker *et al*, 1993; Wu and Russell, 1993). Nim1 homologues have not yet been found in vertebrates, and an alternative Wee1 inhibitory kinase activity has been observed in M phase *Xenopus* egg extracts. Moreover, this kinase activity inhibits Wee1 kinase activity to a greater extent than Nim1, by hyperphosphorylating the non-catalytic N-terminus of Wee1 (Tang *et al*, 1993). The human homologue of Wee1/Mik1, Wee1Hu, is hyperphosphorylated at M phase in transformed and non-transformed cells, and this hyperphosphorylation coincides with a reduction in Wee1Hu activity (McGowan and Russell, 1995; Watanabe *et al*, 1995). Wee1Hu protein levels also decrease during mitosis and G_1 phase, contributing to the downregulation of kinase activity (Watanabe *et al*, 1995). The N-terminus of Wee1Hu contains several Cdk1 phosphorylation sites. However, in vitro phosphorylation of Wee1Hu by cyclin B/Cdk1 does not affect its catalytic activity (Watanabe *et al*,

1995). It remains to be established whether phosphorylation of Wee1Hu by Cdks regulates its activity by some other method. Although Myt1 activity also decreases at mitosis (Meuller *et al,* 1995b), it is not known whether it is downregulated by hyperphosphorylation or, if so, whether this is by Nim1/Cdr kinase.

Cdc25 Upregulation

Cdc25C is activated by phosphorylation on its N-terminal regulatory domain (Izumi *et al,* 1992; Kumagai and Dunphy, 1992; Hoffmann *et al,* 1993). The kinase(s) that upregulates Cdc25B/C at mitosis has not been definitively identified but probably includes the mitotic Cdks themselves (Kumagai and Dunphy, 1992; Hoffmann *et al,* 1993). However, Cdc25C can be phosphorylated and activated in interphase *Xenopus* extracts devoid of Cdk1 and Cdk2 activity (Izumi and Maller, 1995). Recent studies suggest that the polo kinase (Fenton and Glover, 1993; Kitada *et al,* 1993) can bind and activate Cdc25 by phosphorylating it at sites distinct from those phosphorylated by Cdk1 (Kumagai and Dunphy, 1996). Polo is localized to the centrosomes in G$_2$ phase, and microinjecting antipolo antibodies into normal diploid fibroblasts prevents centrosomes from growing—perhaps by inhibiting the recruitment of γ tubulin—and from separating (Nigg E, personal communication). This arrests normal cells in G$_2$ phase and suggests that polo may be a link between the centrosome cycle and the rest of the cell cycle machinery. HeLa cells microinjected with antipolo antibodies go on to enter mitosis but are unable to form a proper spindle, indicating that transformed cells may lack the centrosomal checkpoint.

INHIBITORY PROTEINS AND MITOTIC CYCLIN-CDKS?

The most recently identified mechanism to regulate cyclin/Cdk activity is through binding specific inhibitor proteins (CKIs or CDIs) (Hunter, 1993; Hunter and Pines, 1994; Pines, 1994; Harper, this volume). However, all the CKIs identified in animal cells so far primarily inhibit the G$_1$ or S phase cyclin/ Cdks. These include the INK4 family of inhibitors, which are specific for the D type cyclins, and the p21, p27 and p57 inhibitors, which inhibit cyclin A, D and E/Cdk complexes. In yeast, cyclin B/Cdk1 inhibitor proteins have been identified, but again these inhibitors have their physiological effects primarily on the B type cyclin complexes involved in regulating the G$_1$ and S phases. Thus, in fission yeast, the 25 kDa product of the *rum1* (replication *u*ncoupled from *m*itosis) gene inhibits cdc2 complexed with cig2, a B type cyclin thought to be important in initiating DNA replication in fission yeast (Martin-Castellanos *et al,* 1996). Normally, rum1 is present only in G$_1$ cells, where it inhibits cig2-cdc2 and delays S phase until the cell has reached a threshold size. However, overexpressing rum1 in G$_2$ phase will inhibit the mitotic B type cyclin cdc13-cdc2 complex, and the lack of mitotic cyclin/Cdk activity "deceives" the cell into "thinking" it is in a G_1 state, causing it to undergo multiple rounds of replication (Moreno and Nurse, 1994; Correa-Bordes and Nurse, 1995; reviewed in Labib, 1996). In budding yeast, the Sic1 inhibitor (Mendenhall, 1993) has a role analogous to that of rum1; it inhibits the S phase cyclin B/Cdc2 kinases, Clb5-cdc28 and Clb6-cdc28, during G_1 phase and blocks DNA replication until the cells have passed Start (Schwob *et al*, 1994). At Start, Sic1 is phosphorylated by the Cln (G_1 cyclin)-Cdc28 kinases, which apparently target Sic1 for destruction by the ubiquitin dependent proteolysis pathway (Schwob *et al*, 1994; Schneider *et al*, 1996).

Nevertheless, there are indications that inhibitors of the mitotic cyclin/Cdks may exist. For example, budding yeast can still arrest before mitosis when the inhibitory tyrosine (Tyr-19) of Cdc28 is mutated to phenylalanine (Amon *et al*, 1992; Sorger and Murray, 1992). *Xenopus* interphase extracts can also inhibit cyclin B/Cdk1 in a Thr-14/Tyr-15 phosphorylation independent manner. *Xenopus* extracts remain arrested at the unreplicated DNA checkpoint when active baculovirally expressed cyclin B/Cdk1 is added, even when the mutant, non-phosphorylatable A14/F15 form of Cdk1 is used (Kumagai and Dunphy, 1995). Initial studies suggest that the Cdk1 inhibitory activity resides in a particulate fraction (Lee *et al*, 1994; Kumagai and Dunphy, 1995; Lee and Kirschner, 1996). Similarly, non-transformed cells transfected with the A14/F15 mutant of Cdk1 do not prematurely enter mitosis, whereas a percentage of transformed cells expressing this Cdk1 mutant do enter mitosis prematurely (Morgan D, personal communication). It is not known whether this Cdk1 inhibitory mechanism(s), supplementary to Tyr/Thr phosphorylation, is the same in budding yeast, frog eggs and vertebrate somatic cells, but it could be accounted for by mitotic/cyclin Cdc2 inhibitors present in G_2 phase cells.

CKS FAMILY OF CDK BINDING PROTEINS

The Cdks interact with a class of small proteins with a molecular mass of 9–13 kDa called Cks proteins. Although essential genes in yeast, the exact role of the Cks proteins in the cell cycle remains frustratingly elusive. The first member of this protein family was identified in *S pombe* as suc1, a *su*ppressor of particular temperature sensitive *cdc2* alleles (Hayles *et al*, 1986). The null allele of *suc1* showed that this is an essential gene required for cells to exit from mitosis (Moreno *et al*, 1989). *suc1* homologues have been found throughout the plant and animal kingdoms. In *Saccharomyces cerevisiae*, the homologue is encoded by *CKS1* and was identified as a Cdc28 kinase subunit (Hadwiger *et al*, 1989). Two human homologues (CKSHs 1 and 2) were subsequently cloned and shown to be very similar in aminoacid sequence and able to complement yeast Cks proteins (Richardson *et al*, 1990). The functional differences between human Cks1 and Cks2 have not been defined. Genetic and biochemical approaches to studying the physiological role of Cks proteins have to date not produced a consensus view of their exact physiological function(s) (Hindley *et al*, 1987; Hadwiger *et al*, 1989; Moreno *et al*, 1989; Tang *et al*, 1993). An analysis of different temperature sensitive alleles of *CKS1* in budding yeast showed that CKS1 has an essential role in exiting both G_1 and G_2 phase of the cell cycle (Tang *et al*, 1993). Interphase *Xenopus* cell free extracts immunodepleted of Cks1 are unable to enter mitosis. Extracts remain in interphase with Cdc2 in its Tyr-15 phosphorylated state (Patra and Dunphy, 1996), suggesting that Cks1 may be involved in the dephosphorylation of phospho-Tyr-15. By contrast, mitotic frog extracts are unable to exit mitosis in the absence of Cks proteins, and this is also true for fission yeast cells. Thus, the Cks proteins may perform (at least) two different functions in the cell cycle—at the end of G_2 phase in the dephosphorylation activation of cyclin B/Cdk1 and at the end of metaphase in the inactivation of cyclin B/Cdk1 (see Endicott and Nurse, 1995, for further discussion).

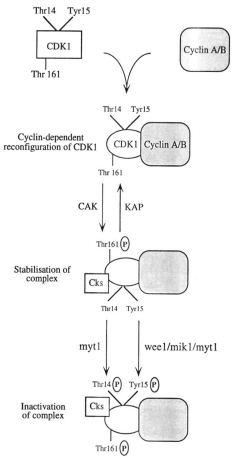

Fig. 1. Regulation of Cdk1 during the S and G₂ phases of the cell cycle. Included in this scheme is the Cks protein, although it has not yet been established at exactly what stage in the above sequence this protein associates with Cdk1

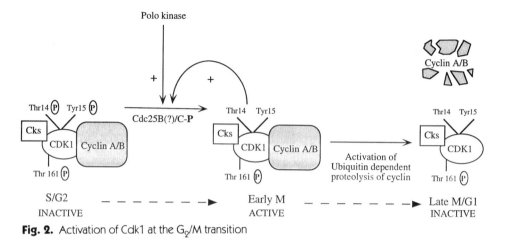

Fig. 2. Activation of Cdk1 at the G_2/M transition

That the Cks proteins may be able to perform two different functions in the cell cycle is made more likely by the finding that these proteins can potentially adopt two very different conformations. The crystal structures of fission yeast suc1 and human Cks2 have both been solved and show that the Cks proteins have a putative β hinge region that could be either extended or folded back on itself. This has implications for when the Cks protein is bound to a Cdk. The atomic structure of the CksHs1-Cdk2 complex has also recently been solved (Bourne *et al*, 1996) and shows that CksHs1 interacts solely with the C-terminal lobe of Cdk2. This means that the Cks binding site is entirely distinct from that of cyclin (Jeffrey *et al*, 1995) and agrees with experimental data showing that both proteins are able to bind the Cdk at the same time. However, this is only true when the β hinge region of the Cks folds back on itself. When the binding of the hinge is in its extended conformation, it would interfere with cyclin binding, perhaps precluding it altogether.

Apart from subtle changes in conformation at the binding interface, the structures of both Cks1 and Cdk2 hardly alter upon binding, suggesting that Cks proteins do not directly affect Cdk activity. The binding of Cdk to CksHs1 brings a highly conserved positively charged pocket of residues on Cks1 into close approximation to the catalytic cleft of Cdk2. This pocket could potentially act as a phosphate anion binding site, leading to the suggestion that Cks proteins may assist in the interaction of Cdk1/2 with regulatory proteins and substrates that have become phosphorylated. Of these, one obvious candidate is Cdc25 after it is activated by the polo kinase. Others could include components of the proteolysis machinery that become phosphorylated in mitosis.

CELL CYCLE DEPENDENT SUBCELLULAR LOCALIZATION OF MITOTIC CYCLIN/CDK COMPLEXES

There is accumulating evidence that different members of the multigene families of cell cycle regulators—such as the mitotic cyclins, the Cdc25 phosphatases and the Wee1/Mik1/Myt1 kinases—are localized to different compartments within the cell. This has obvious implications for substrate specificity and potentially for the mechanics of cell cycle regulation, because some cell cycle regulators also translocate between compartments in a cell cycle dependent manner.

There are some data on how the localization of the mitotic cyclins is specified. Human cyclin A remains mostly in the nucleus throughout interphase, although there is a centrosomal population in late G$_2$ phase. By contrast, the human B type cyclins are cytoplasmic. This difference is due to a region in the N-terminus of the B type cyclins that appears to retain the B type cyclins in the cytoplasm (Pines and Hunter, 1994). Without this "cytoplasmic retention signal" (CRS), cyclin B1 is transported to the nucleus at any stage of the cell cycle. Furthermore, when fused to cyclin A, the CRS is able to redirect cyclin A into the cytoplasm. Thus, the CRS appears to be both necessary and sufficient to keep the B type cyclins in the cytoplasm. However, a nuclear localization signal (NLS) fused to cyclin B1 will override the CRS, suggesting that the nuclear cyclins, such as cyclins A and E, do not have an endogenous NLS. We know relatively little about how the nuclear cyclins are properly localized. For cyclin A to be transported to the nucleus, it must first bind a Cdk (Maridor *et al*, 1993), but neither cyclin A nor Cdk1/2 possesses a recognizable NLS (Dingwall and Laskey, 1992). It is possible that these complexes may be transported to the nucleus by "piggy backing" on another protein with an NLS. For example, the requirement for cyclin A to bind a Cdk before being transported may correlate with its ability to bind nuclear proteins such as p107 (Gallant *et al*, 1995), p130 (Hannon *et al*, 1993) or the E2F transcription factor (Krek *et al*, 1994).

Regulating Mitotic Cyclin/Cdk Activity by Subcellular Localization

Just before nuclear envelope breakdown (NEBD), as judged from the integrity of the nuclear lamin network, cyclin B1 rapidly relocates from the cytoplasm to the nucleus (Pines and Hunter, 1991). The cell cycle dependent relocalization of a B type cyclin to the nucleus at prophase is conserved from starfish to humans (Pines and Hunter, 1991; Gallant and Nigg, 1992; Ookata *et al*, 1992). How this translocation is triggered is unclear, but phosphorylation of cyclin B1 in the CRS appears to be important (Pines J, unpublished). The manner in which the translocation is carried out is also unknown, but there are a number of other proteins that also exhibit regulated nuclear entry at this point in the cell cycle. One of these is the *Drosophila* protein, pendulin, a member of a large family of proteins involved in protein-protein interactions. Pendulin has a Cdk1 phosphorylation site adjacent to an NLS and is transported into the nucleus at the G$_2$/M transition (Kussel and Frasch, 1995), making pendulin and its homologues (Cortes *et al*, 1994; Gorlich *et al*, 1994; Yano *et al*, 1994) excellent candidates to assist in the nuclear translocation of cyclin B1 at G$_2$/M. However, no direct link with cyclin B1 nuclear relocalization has been shown.

The role, if any, of cyclin B1 translocation in the regulation of the G_2/M transition is unclear. Relocalization may function to bring cyclin B1/Cdk1 into contact with regulatory proteins and substrates at the correct point in the cell cycle. One candidate is the Cdc25C phosphatase that activates cyclin B/Cdk1 at mitosis. There is some debate as to whether Cdc25C is a nuclear or cytoplasmic protein (Millar *et al*, 1991; Seki *et al*, 1992; Heald *et al*, 1993). Therefore, it is difficult to predict whether cyclin B1/Cdk1 nuclear relocalization will bring it into contact with Cdc25C. Indeed, one set of data suggests that Cdc25C itself translocates into the nucleus in the same manner as cyclin B1/Cdk1 (Heald *et al*, 1993). The other member of the Cdc25 family that is active at mitosis is Cdc25B. Cdc25B has been reported to regulate prophase spindle assembly at G_2/M and localize to the cytoplasm (although when over-expressed, the majority appeared to be nuclear) (Gabrielli *et al*, 1996). Over-expression of a dominant negative Cdc25B mutant in various cell lines has been reported to cause cells to accumulate in G_2, with microtubules nucleating from centrosomes in a manner indicative of an interphase rather than a prophase state. It is not clear which isoform of Cdc25 first activates cyclin B/Cdk1, or whether Cdc25B and Cdc25C both activate cyclin B1/Cdk1 simultaneously. This will clearly be influenced by when exactly cyclin B1/Cdk1 is activated—before, after or during translocation to the nucleus. In starfish meiotic maturation, Cdk1 is activated before cyclin B/Cdk1 relocates to the nucleus (Ookata *et al*, 1992). However, it is not known (and for technical reasons difficult to determine precisely) where in the somatic cell cyclin B1/Cdk1 is first activated to levels that can precipitate mitosis. Further characterization of exactly when and where mitotic cyclin/Cdk1 is activated and by which Cdc25 isoform will increase our understanding of the G_2/M transition.

It is clear that neither human cyclin B1 nor chicken cyclin B2 induces the cell to enter mitosis when prematurely transported to the nucleus (Gallant and Nigg, 1992; Pines and Hunter, 1994), suggesting that initiation of mitosis is conditional on more than just B1/Cdk1 being relocalized to the nucleus. An additional level of control is certainly exerted by the Wee1/Mik1 kinase. Endogenous levels of the Cdk1 inhibitory kinase Wee1Hu have been difficult to detect by immunofluorescence, but the low levels detected suggest that it is nuclear (McGowan and Russell, 1995). This confirmed observations showing that exogenous, overexpressed Wee1Hu is a nuclear protein (Heald *et al*, 1993; McGowan and Russell, 1995). Thus, Wee1/Mik1 could inactivate any cyclin B1/Cdk1 kinase that translocates into the nucleus during interphase. However, we are left with the contradictory finding that the nuclear Wee1 is thought to phosphorylate Cdk1 after it has bound to a cyclin but that human cyclins B1 and B2 are cytoplasmic during interphase. It has been suggested that Wee1Hu transiently shuttles from the nucleus to the cytoplasm, as has been observed for some other nuclear proteins (Schmidt *et al*, 1993), but there is at present no experimental evidence for this.

These observations implied that there may be more than other members of the Wee1/Mik1 kinase family that regulate Cdk1 activity, a conclusion strengthened by the recent isolation of the membrane associated Myt1 kinase. The specific membrane compartments to which Myt1 localizes are not known, but there are data to suggest that they include the endoplasmic reticulum and possibly the Golgi apparatus (Piwnica-Worms H, personal communication). This suggests that Myt1 may be especially important in the inhibition of cyclin B2/Cdk1, which primarily associates with the Golgi apparatus (see below). At the time of writing, it is not known whether human cyclin B2 and Myt1 co-localize. However, the identification of Myt1 leaves open the possibility that there may be another member of the Wee1 kinase family that regulates Cdk1 bound to cyclin B1, because this mitotic kinase complex appears to be associated with the cytoskeleton and not the membrane compartment.

Targeting Mitotic Cyclin/Cdk Activity by Localizing Mitotic Cyclins

During interphase, human cyclins B1 and B2 are localized to strikingly different structures within the cytoplasm; cyclin B1 is localized to microtubules, whereas cyclin B2 is localized predominantly to the cytoplasmic face of the Golgi apparatus membrane (Jackman *et al*, 1995). Furthermore, the CRS regions of human cyclins B1 and B2 have a significant influence on where the cyclin is targeted in the cytoplasm (Pines and Hunter, 1994). However, the proteins to which the different B type cyclins bind in the cytoplasm have not yet been definitively identified. Human cyclin B1 may bind to microtubules via the microtubule associated protein 4 (Ookata *et al*, 1995), but it is not known how human cyclin B2 associates predominantly with the Golgi membrane.

The "odd man out" among the B type cyclins is cyclin B3. There are two major differences between cyclin B3 and the other B types. Its aminoacid sequence has elements identified in both the A and the B type cyclins (Gallant and Nigg, 1994), and it appears to be a nuclear protein (Gallant and Nigg, 1994). Cyclin B3 homologues have been found in *Drosophila* (Sigrist *et al*, 1995) and nematodes (Kreutzer *et al*, 1995), so it is undoubtedly an important element in cell cycle control. Cyclin B3 could have an important role in the phosphorylation of nuclear substrates at mitosis and could obviously be regulated by the nuclear Wee1/Mik1 kinase.

In addition to segregating cyclin/Cdks from regulatory proteins and substrates until the correct moment in the cell cycle, targeting mitotic cyclins to distinct subcellular locations will concentrate cyclin/Cdk1 activity to particular regions of the cell. This should influence the kinetics of interaction between cyclin/Cdks and their substrates. In fission yeast, the same B type cyclin (Cdc13) associated with p34^{cdc2} can initiate both S phase and M phase, implying that the same cyclin/Cdc2 complex can phosphorylate different substrates at different stages of the cell cycle. On the basis of this observation, Fisher and Nurse (1996) have proposed a quantitative cell cycle control model. This model suggests that Cdc13/Cdc2 can function differently by phosphorylating different substrates depending on its concentration within the yeast cell, and it supposes that M phase substrates may generally have a lower affinity for cyclin B/Cdc2 than S phase substrates. (As a caveat, it should be borne in mind that the model is based on experiments using yeast that lack other cyclins, such as Cig2, which are important in the normal cell cycle.) Thus, the concentration of cyclin B/Cdk1 to specific subcellular compartments in higher eukaryotes might be important in enabling phosphorylation of low affinity M phase substrates.

There is considerable evidence that the phosphatases that antagonize the effects of the cyclin/Cdks—primarily PP2A and PP1—are targeted to different subcellular regions. Therefore, controlling the localization of antagonistic kinases and phosphatases could allow gradients of phosphorylated and non-phosphorylated substrates to be established within the same cell. Such a mechanism has been suggested to establish a gradient of non-phosphorylated microtubule associated proteins around chromosomes, enabling microtubules to be stabilized around chromosomes during the formation of a mitotic spindle (Hyman and Karsenti, 1996).

PHYSIOLOGICAL FUNCTIONS OF DIFFERENT MITOTIC CYCLIN/CDKS: SUBSTRATE SPECIFICITIES

To understand fully the physiological roles of the different cyclin/Cdk1 complexes in bringing about mitosis, one must identify their in vivo substrates. The crystal structures of cyclin A/Cdk2 and monomeric Cdk2 indicate that cyclins bind to the smaller N-terminal lobe of Cdks, close to the presumed substrate recognition site. This suggests that cyclins may be able to influence the substrate specificity of their associated Cdk. Furthermore, the different subcellular localizations of mitotic cyclins suggest that in vivo mitotic cyclin/Cdk complexes may phosphorylate different proteins.

The Cdk family primarily recognizes the site Ser/Thr$^\circ$-Pro. However, this consensus is shared by a number of other "proline directed" kinases, such as the mitogen activated protein (MAP) kinase and glycogen synthase kinase 3 (Gsk3). This makes it difficult to identify unequivocally in vivo cyclin/Cdk1 substrates (Nigg, 1993). The following criteria must therefore be satisfied for an in vitro substrate of Cdk1 to qualify as a probable in vivo substrate. Firstly, the substrate should be phosphorylated in vivo at the same residues as are phosphorylated by Cdk1 in vitro and at a stage during the cell cycle when Cdk1 is active. Secondly, phosphorylation of a substrate by Cdk1 should change the activity of that substrate in a way that is consistent with its own and the cell's behaviour at that stage of the cell cycle.

Cytoplasmic Events at G$_2$/M

During prophase, the interphase microtubule network disassembles and begins to rearrange itself into a mitotic spindle. This is partly due to microtubules increasing the frequency of their transitions from growth to shrinkage ("catastrophes") and is precipitated by the activation of Cdk1 (Verde *et al*, 1990, 1992). This change in microtubule dynamics may be caused by inactivation of microtubule stabilizing factors and/or activation of destabilizing factors (reviewed in Shina, 1995). The localization of human and starfish cyclin B1 (Pines and Hunter, 1991; Ookata *et al*, 1993; Jackman *et al*, 1995) to microtubules before their nuclear relocalization suggests that cyclin B1/Cdk1 is the principal mitotic cyclin/Cdk1 complex involved in changing microtubule dynamics at G$_2$/M. Cyclin B1/Cdk1 may exert some of its effects via the phosphorylation of MAPs. In vitro cyclin B1/Cdk1 can associate with and phosphorylate purified MAP4 (Ookata *et al*, 1995), a ubiquitous MAP thought to have a role in polymerizing and stabilizing microtubules (Hirokawa, 1994). As with other MAPs (eg MAP2) (Faruki *et al*, 1992), Cdk1 dependent phosphorylation of MAP4 reduces its ability to stabilize microtubules in vitro. In vivo Cdk1 could co-operate with other kinases such as protein kinase C (Mori *et al*, 1991) and MAP kinase (Hoshi *et al*, 1992) to phosphorylate MAP4. However, MAP4 dependent stabilization of microtubules appears to be functionally redundant, because removal of MAP4 from microtubules in vivo does not delay mitosis or disrupt spindle formation (Wang *et al*, 1996).

A Cdk substrate that is potentially a more physiologically relevant regulator of microtubules has been identified. The Op18/stathmin protein has been found to be a phosphorylation dependent regulator of microtubule dynamics in vitro (Belmont and Mitchison, 1996). Op18/stathmin is phosphorylated in vitro by Cdk1 at Ser-38 (a residue phosphorylated in vivo during mitosis). It is not clear how Op18 affects microtubule dynamics, although its regulation by a Cdk appears to be crucial in the formation of a competent mitotic spindle. Belmont *et al* (1996; Belmont and Mitchison, 1996) have suggested that Op18 may increase the frequency of microtubule catastrophe, but there is as yet no evidence that this is the facet of Op18/stathmin regulated by phosphorylation. When the Cdk phosphorylation site of Op18 is mutated to alanine, a constitutively active form of Op18 is produced, and this destabilizes microtubules (Marklund *et al*, 1994; Larsson *et al*, 1995). Cells expressing the mutant form of Op18 arrest in mitosis with an aberrant mitotic spindle, showing that Op18 destabilizing activity must be turned off to allow a stable spindle to form (Marklund *et al*, 1996). Cdk1 dependent inactivation of Op18 may appear contrary to the destabilizing effects of Cdk1 on microtubules through phosphorylation of MAP4. However, the conversion of the interphase microtubule network to a stable mitotic spindle may require the careful co-ordination of various microtubule stabilizing and destabilizing activities. The identification of further microtubule stabilizing/destabilizing factors whose activity is regulated at mitosis will help build a more complete picture of how Cdk1 kinase activity co-ordinates cell cycle dependent changes in microtubule dynamics.

Changes in the phosphorylation state of centrosomal proteins and microtubule motors are also necessary for the formation of a functional mitotic spindle. During late G$_2$, centrosomes migrate to opposite sides of the nucleus,

and the microtubule nucleating properties of centrosomes change. Cyclin/
Cdk1 has been implicated in regulating the activity of many microtubule based
motors, but little was known until recently of how Cdk1 phosphorylation of microtubule motors affected their activity (Saunders, 1993; Auft and Reider,
1994; Lombillo et al, 1995; Saunders et al, 1995). Recent experiments have
begun to elucidate the manner in which cyclin/Cdks regulate motor proteins.

Several kinesin like motor proteins (KLPs) are required for spindle pole
assembly and function. Centrosome separation appears to require KLPs of the
bimC subfamily—Eg5/Xklp2 in *Xenopus* (LeGuellec et al, 1991; Boleti et al,
1996), bimC in the filamentous fungus *Aspergillus nidulans* (Enos and Morris,
1990) and cut7 in *S pombe* (Hagan and Yanagida, 1992)). In most of interphase, human Eg5 is diffusely cytoplasmic, but at G_2 phase, it associates with
the centrosomes. At mitosis, it binds to the spindle poles. Cdk1 phosphorylates
an evolutionarily conserved threonine (Thr-927) residue of HsEg5 (Blangy et
al, 1995) and substituting Thr-927 for a non-phosphorylatable residue stops
HsEg5 from associating with centrosomes and the spindle. Conversely,
cytoplasmic dynein is a minus end directed microtubule based motor that
transports organelles along microtubules during interphase. Activation of Cdk1
indirectly causes dynein and its activator dynactin to detach from the organelle
(Niclas et al, 1996), allowing dynein to perform distinct functions during
mitosis (Vaisberg et al, 1993).

As a cell enters mitosis, it rounds up in association with the reorganization
of the microfilament cytoskeleton. This may be due in part to cyclin/Cdk activity, because phosphorylation of non-muscle caldesmon—a cyclin/Cdk substrate in vitro—at mitosis weakens its affinity for actin, causing it to dissociate
from microfilaments (Yamashiro et al, 1990). However, in vitro, Cdk1 does not
phosphorylate all of the residues phosphorylated at mitosis in vivo, indicating
that an additional caldesmon kinase(s) is involved (Yamashiro et al, 1991).
Other cytoskeletal substrates of Cdk1 in vitro include the cytoplasmic intermediate filament subunits, vimentin and desmin. Phosphorylation causes the intermediate filaments to depolymerize (Chou et al, 1990; Dessev et al, 1991),
but the physiological relevance of this is unclear, because some cells maintain
an intermediate filament network cage around the mitotic spindle.

At mitosis, the Golgi apparatus disassembles from a single, asymmetrically
distributed entity into hundreds of vesicles that allow random distribution of
Golgi components between the two daughter cells (Warren, 1993). In vitro
systems have indicated that Cdk1 indirectly inhibits membrane traffic
(Thomas et al, 1992; Woodman et al, 1993) and causes the Golgi apparatus to
begin to disassemble (Misteli and Warren, 1994). This may be effected via the
Rab family. Membrane traffic pathways are thought to be regulated by the
Rab subfamily of ras like small GTP binding proteins, and Rab1Ap and Rab4p
are phosphorylated both in vitro and in vivo by Cdk1 (Bailly et al, 1992; van
der Sluijs et al, 1992), although the effect of Cdk1 dependent phosphorylation
is unclear. The membrane associated localization of human cyclin B2/Cdk1
makes this the most likely cyclin/Cdk complex to be involved in the disassemb-

ly of the Golgi apparatus in vivo, although no human cyclin B2/Cdk1 specific substrates have yet been identified.

Nuclear Changes at G$_2$/M

At the beginning of mitosis, just before NEBD, cyclin B1/Cdk1, cyclin B3/Cdk1 and cyclin A/Cdk1 are all in the nucleus, so any or all of these complexes could participate in nuclear events during the G$_2$/M transition. NEBD requires vesicularization of the nuclear envelope and disassembly of the nuclear lamina. The nuclear lamins are intermediate filaments that form a polymeric network that lines and stabilizes the inner nuclear membrane and provides attachment sites for interphase chromatin. Phosphorylation of nuclear lamins at the beginning of mitosis causes them to depolymerize and results in the breakdown of the nuclear lamina, which is necessary, but not sufficient, for NEBD to occur (Newport and Spann, 1987; Pfaller *et al*, 1991). Furthermore, when the mitosis specific phosphorylation sites are mutated, these mutant lamins are unable to disassemble (Heald and McKeon, 1990). The relocalization of B1/Cdk1 at G$_2$/M and the observation that cyclin B1/Cdk1 can induce nuclear lamina breakdown in vitro suggest that this cyclin/Cdk complex might be the lamin kinase (Peter *et al*, 1990b; Ward and Kirschner, 1990; Dessev *et al*, 1991; Peter *et al*, 1991). However, cyclin B1/Cdk1 is probably not the only kinase involved in nuclear lamina breakdown in vivo. Quite apart from the other nuclear cyclin/Cdk complexes, the β$_{II}$ isotype of protein kinase C is also translocated to the nucleus at the G$_2$/M transition and can very efficiently phosphorylate lamin B (Goss *et al*, 1994). During interphase, integral membrane proteins of the inner nuclear membrane interact with lamins and chromatin. Phosphorylation of at least two of these proteins, LAP2 (Foisner and Gerace, 1993) and the lamin B receptor (Courvalin *et al*, 1992), at mitosis abrogates their interactions with lamin and chromatin, thus promoting nuclear disassembly. Cdk1 is again unlikely to be the only kinase involved in phosphorylating these proteins.

RNA synthesis stops at mitosis in animal cells and nascent RNA polymerase (Pol) I (Weisenberger and Scheer, 1995) and Pol II (Shermoen and O'Farrell, 1991) transcripts are released from their DNA template. Cyclin B/Cdk1 phosphorylation of TATA binding protein (TBP) or associated factors can also directly inhibit Pol III mediated transcription, by affecting the initiation frequency of Pol III dependent genes. Given that transcription mediated by Pol I, II or III requires TBP, it has been speculated that Cdk activity may downregulate all forms of transcription (Gottesfeld *et al*, 1994). However, Cdk1 may also inhibit Pol I transcription at mitosis through effects on nucleolar disassembly (Weisenberger and Scheer, 1995). Nucleoli are membraneless nuclear organelles in which rRNA is transcribed and ribosomal subunits are assembled (Warner, 1990) and cyclin B/Cdk1 phosphorylates nucleolar proteins such as nucleolin and NO38 in vitro on sites that are phosphorylated in vivo during mitosis (Peter *et al*, 1990a). During mitosis, mRNA splicing also ceases, but a direct role for Cdk1 in downregulating splicing has not been shown (Gui *et al*, 1994).

A continuing uncertainty remains over the role of cyclin B/Cdk1 in chromosome condensation. At mitosis, histones H1, H2A and H3 are hyperphosphorylated, and the high mobility group proteins I, Y and P1 are phosphorylated by cyclin/Cdk1 in vitro on sites that are phosphorylated during mitosis (Meijer *et al*, 1991b; Nissen *et al*, 1991). These observations suggest a potential role for Cdk1 in chromosome condensation. However, chromosome condensation can occur at the restrictive temperature in mouse FT210 cells that have a temperature sensitive allele of Cdc2, after treatment with the phosphatase inhibitors fostreicin and okadaic acid. This suggests that an alternative kinase can induce chromosome condensation. One candidate is the NIMA (*never in mitosis*) kinase, a cell cycle regulated Ser/Thr kinase first identified in *A nidulans* (Osmani *et al*, 1988). When a stable form of NIMA kinase is expressed in fungi or human cells (O'Connell *et al*, 1994), it induces chromatin condensation without active Cdk1 (Lu and Hunter, 1995). Ye *et al* (1995) showed that NIMA that is allowed to accumulate in cells arrested in G_2 is only partially active and that its activation is doubled after hyperphosphorylation by Cdk1. Thus, it seems likely that NIMA is fully activated following Cdk1 activation and therefore is a downstream target of Cdk1. As yet, no NimA homologues have been found in organisms other than filamentous fungi. Indeed, a search of the complete *S cerevisiae* genome database revealed no homologues of NimA closer than the non-essential *Kin3* genes. The NEK2 kinase is a vertebrate Ser/Thr kinase related to NimA in the kinase domain, whose protein levels and kinase activity peak during G_2, but the physiological function of NEK2 is not yet known.

SUMMARY

The entry of a cell into mitosis is regulated by an elaborate network of kinases and phosphatases that control both for the timing of cell division and the complete reorganization of the cellular architecture. The mitotic cyclin/Cdks form part of large multiprotein complexes whose other components are only now beginning to be identified. The continuing identification of proteins that contribute to these complexes and changes in the composition of these complexes are likely to give a more integrated view of how mitotic cyclin/Cdk complexes are regulated and how they function—not only to induce mitosis, but also to aid further mitotic progression. Furthermore, assigning specific G_2/M functions to distinct mitotic cyclin/Cdk complexes will require the identification of differences in substrate specificities between the mitotic cyclin/Cdk complexes, perhaps in parallel with specific cyclin knockouts in mice. Such investigations will be complicated by potential functional overlap between mitotic cyclin/Cdk complexes in vitro and in vivo. Although cyclin/Cdk1 is thought to be the major kinase that initiates the onset of mitosis, a more complete understanding of how cells move from G$_2$ to a mitotic state will require further identification of kinases operating upstream, downstream and in parallel with Cdk1, their substrates and their relationship with one another during the G$_2$/M transition.

References

Amon A, Surana U, Muroff I and Nasmyth K (1992) Regulation of p34CDC28 tyrosine phosphorylation is not required for entry into mitosis in *S. cerevisiae*. *Nature* **355** 368–371

Atherton FS, Parker LL, Geahlen RL and Piwnica-Worms H (1993) Mechanisms of p34cdc2 regulation. *Molecular and Cellular Biology* **13** 1675–1685

Atherton-Fessler S, Liu F, Gabrielli B, Lee MS, Peng CY and Piwnica-Worms H (1994) Cell cycle regulation of the p34cdc2 inhibitory kinases. *Molecular and Cellular Biology* **5** 989–1001

Auft JG and Reider CL (1994) Centrosome and kinetechore movement during mitosis. *Current Opinion in Cell Biology* **6** 41–49

Bailly E, Pines J, Hunter T and Bornens M (1992) Cytoplasmic accumulation of cyclin B1 in human cells: association with a detergent-resistant compartment and with the centrosome. *Journal of Cell Science* **101** 529–545

Belmont LD and Mitchison TJ (1996) Identification of a protein that interacts with tubulin dimers and increases the catastrophe rate of microtubules. *Cell* **84** 623–631

Belmont L, Mitchison T and Deacon H (1996) Catastrophic revelations about Op18/stathmin. *Trends in Biochemical Sciences* **21** 197–198

Blangy A, Lane HA, d'Herin P, Harper M, Kress M and Nigg E (1995) Phosphorylation by p34cdc2 regulates spindle association of human Eg5, a kinesin-related motor essential for bipolar spindle formation in vivo. *Cell* **83** 1159–1169

Boleti H, Karsenti E and Vermos I (1996) Xklp2, a novel centrosomal kinesin-like protein required for centrosome separation during mitosis. *Cell* **84** 49–59

Booher R and Beach D (1987) Interaction between cdc13+ and cdc2+ in the control of mitosis in fission yeast; dissociation of the G$_1$ and G$_2$ roles of the cdc2+ protein kinase. *EMBO Journal* **6** 3441–3447

Bourne Y, Watson MH, Hickey MJ *et al* (1996) Crystal structure and mutational analysis of the human CDK2 complex with cell cycle regulatory protein CksHs1. *Cell* **84** 863–874

Brown NR, Noble MEM, Endicott JA *et al* (1995) The crystal structure of cyclin A. *Current Biology* **3** 1235–1247

Chou Y-H, Bischoff JR, Beach D and Goldman RD (1990) Intermediate filament reorganization during mitosis is mediated by p34cdc2 phosphorylation of vimentin. *Cell* **62** 1063–1071

Cismowski MJ, Laff GM, Solomon MJ and Reed SI (1995) KIN28 encodes a C-terminal domain kinase that controls mRNA transcription in *Saccharomyces cerevisiae* but lacks cyclin-dependent kinase activating kinase (CAK) activity. *Molecular and Cellular Biology* **15** 2983–2992

Clarke PR, Hoffmann I, Draetta G and Karsenti E (1993) Dephosphorylation of cdc25-C by a type-2A protein phosphatase: specific regulation during the cell cycle in *Xenopus* egg extracts. *Molecular and Cellular Biology* **4** 397–411

Clarke PR, Klebe C, Wittinghofer A and Karsenti E (1995) Regulation of Cdc2/cyclin B activation by Ran a Rs-related GTPase. *Journal of Cell Science* **108** 1217–1225

Coleman TR, Tang Z and Dunphy WG (1993) Negative regulation of the Wee1 protein kinase by direct action of the nim1/cdr1 mitotic inducer. *Cell* **73** 919–929

Correa-Bordes J and Nurse P (1995) p25rum1 orders S phase and mitosis by acting as an inhibitor of the p34cdc2 mitotic kinase. *Cell* **83** 1001–1009

Cortes P, Ye Z-S and Baltimore D (1994) RAG-1 interacts with the repeated amino acid motif of the human homologue of the yeast protein SRP1. *Proceedings of the National Academy of*

Sciences of the USA **91** 7633–7637

Courvalin J-C, Segil N, Blobel G and Worman H (1992) The lamin-B receptor of the inner nuclear membrane undergoes mitosis-specific phosphorylation and is a substrate for p34cdc2-type protein kinase. *Journal of Biological Chemistry* **267** 19035–19038

De Bondt HL, Rosenblatt J, Jancarik J, Jones HD, Morgan DO and Kim S-H (1993) Crystal structure of human CDK2: implications for the regulation of cyclin-dependent kinases by phosphorylation and cyclin binding. *Nature* **363** 595–-602

Den-Haese GJ, Walworth N, Carr AM and Gould KL (1995) The wee1 protein kinase regulates T14 phosphorylation of fission yeast cdc2. *Molecular and Cellular Biology* **6** 371–385

Dessev G, Iovcheva DC, Bischoff JR, Beach D and Goldman R (1991) A complex containing p34cdc2 and cyclin B phosphorylates the nuclear lamin and disassembles nuclei of clam oocytes in vitro. *Journal of Cell Biology* **112** 523–533

Dingwall C and Laskey R (1992) The nuclear membrane. *Science* **258** 942–947

Ducommun B, Brambilla P and Draetta G (1991) Mutations at sites involved in Suc1 binding inactivate Cdc2. *Molecular and Cellular Biology* **11** 6177–6184

Dunphy WG (1994) The decision ot enter mitosis. *Trends in Cell Biology* **4** 202–207

Dunphy WG and Kumagai A (1991) The cdc25 protein contains an intrinsic phosphatase activity. *Cell* **67** 189–196

Endicott JA and Nurse P (1995) The cell cycle and suc1: from structure to function? *Structure* **3** 321–325

Endicott JA, Nurse P and Johnson LN (1994) Mutational analysis supports a structural model for the cell cycle protein kinase p34. *Protein Engineering* **7** 243–253

Enos AP and Morris NR (1990) Mutation of a gene that encodes a kinesin-like protein blocks nuclear division in *A. nidulans. Cell* **60** 1019–1027

Espinoza FH, Farell A, Erdjument-Bromage H, Tempst P and Morgan DO (1996) A major cyclin-dependent-kinase-activating kinase (CAK) in budding yeast unrelated to vertebrate CAK. *Science* **273** 1714–1717

Faruki S, Doree M and Karsenti E (1992) cdc2 kinase-induced destabilization of MAP2-coated microtubules in *Xenopus* egg extracts. *Journal of Cell Science* **101** 69–78

Feaver WJ, Svejstrup JQ, Henry NL and Kornberg RD (1994) Relationship of cdk-activating kinase and RNA polymerase II CTD kinase TFIIH/TFIIK. *Cell* **79** 1103–1109

Feilotter H, Nurse P and Young PG (1991) Genetic and molecular analysis of cdr1/nim1 in *Schizosaccharomyces pombe. Genetics* **127** 309–318

Fenton B and Glover DM (1993) A conserved mitotic kinase active at late anaphase–telophase in syncytial *Drosophila* embryos. *Nature* **363** 637–640

Fesquet D, Labbe JC, Derancourt J et al (1993) The MO15 gene encodes the catalytic subunit of a protein kinase that activates cdc2 and other cyclin-dependent kinases (CDKs) through phosphorylation of Thr161 and its homologues. *EMBO Journal* **12** 3111–3121

Fisher DL and Nurse P (1996) A single fission yeast mitotic cyclin B p34cdc2 promotes both S-phase and mitosis in the absence of G$_1$ cyclins. *EMBO Journal* **15** 850–860

Fisher RP and Morgan DO (1994) A novel cyclin associates with MO15/CDK7 to form the CDK-activating kinase. *Cell* **78** 713–724

Fitch I, Dahmann C, Surana U et al (1992) Characterization of four B-type cyclin genes of the budding yeast *Saccharomyces cerevisiae. Molecular Biology of the Cell* **3** 805–818

Foisner R and Gerace L (1993) Integral membrane proteins of the nuclear envelope interact with lamins and chromosomes, and binding is modulated by mitotic phosphorylation. *Cell* **73** 1267–1279

Gabrielli Bg, De Souza CPC, Tonks IA, Clark JM and Hayward NK (1996) Cytoplasmic accumulation of cdc25B phosphatase in mitosis triggers centrosomal microtubule nucleation in Hela cells. *Journal of Cell Science* **109** 1081–1093

Galaktionov K and Beach D (1991) Specific activation of cdc25 tyrosine phosphatases by B-type cyclins: evidence for mutiple roles of mitotic cyclins. *Cell* **67** 1181–1194

Gallant P and Nigg EA (1992) Cyclin B2 undergoes cell cycle-dependent nuclear translocation and, when expressed as a non-destructible mutant, causes mitotic arrest in HeLa cells. *Journal of Cell Biology* **117** 213–224

Gallant P and Nigg EA (1994) Identification of a novel vertebrate cyclin: cyclin B3 shares properties with both A- and B-type cyclins. *EMBO Journal* **13** 595–605

Gallant P, Fry AM and Nigg EA (1995) Protein kinases in the control of mitosis: focus on nucleocytoplasmic trafficking. *Journal of Cell Science* **Supplement 19** 21–28

Gautier J, Solomon MJ, Booher RN, Bazan JF and Kirschner MW (1991) cdc25 is a specific tyrosine phosphatase that directly activates p34cdc2. *Cell* **67** 197–211

Glotzer M, Murray AW and Kirschner MW (1991) Cyclin is degraded by the ubiquitin pathway. *Nature* **349** 132–138

Gorlich D, Prehn S, Laskey R and Hartmann E (1994) Isolation of a protein that is essential for the first step of nuclear protein import. *Cell* **79** 767–778

Goss VL, Hocevar BA, Thompson LJ, Stratton CA, Burns DJ and Fields AP (1994) Identification of a nuclear betaII protein kinase C as a mitotic lamin kinase. *Journal of Biological Chemistry* **269** 19074–19080

Gottesfeld JM, Wolf VJ, Dang T, Forbes DJ and Hartl P (1994) Mitotic repression of RNA polymerase III transcription in vitro mediated by phosphorylation of a TFIIIB component. *Science* **263** 81–84

Gould KL, Moreno S, Owen DJ, Sazer S and Nurse P (1991) Phosphorylation at Thr167 is required for *Schizosaccharomyces pombe* p34cdc2 function. *EMBO Journal* **10** 3297–3309

Gui JF, Lane WS and Fu X-D (1994) A serine kinase regulates intracellular localisation of splicing factors in the cell cycle. *Nature* **369** 678–682

Gyuris JEG, Chertkov H and Brent R (1993) Cdi1, a human G$_1$ and S phase protein phosphatase that associates with Cdk2. *Cell* **75** 791–804

Hadwiger JA, Wittenberg C, Richardson HE, de Barros Lopes M and Reed SI (1989) A novel family of cyclin homologs that control G$_1$ in yeast. *Proceedings of the National Academy of Sciences of the USA* **86** 6255–6259

Hagan I and Yanagida M (1992) Kinesin-related cut7 protein associates with mitotic and meiotic spindles in fission yeast. *Nature* **356** 74–76

Hagan I, Hayles J and Nurse P (1988) Cloning and sequencing of the cyclin-related cdc13+ gene and a cytological study of its role in fission yeast mitosis. *Journal of Cell Science* **91** 587–595

Hanks SK and Hunter T (1995) Protein kinases 6: the eukaryotic protein kinase superfamily: kinase (catalytic) domain structure and classification. *FASEB Journal* **9** 576–596

Hannon G, Casso D and Beach D (1994) KAP: a dual specificity phosphatase that interacts with cyclin-dependent kinases. *Proceedings of the National Academy of Sciences of the USA* **91** 1731–1735

Hannon GJ, Demetrick D and Beach D (1993) Isolation of the Rb-related p130 through its interaction with CDK2 and cyclins. *Genes and Development* **7** 2378–2391

Hayles J, Beach D, Durkacz B and Nurse P (1986) The fission yeast cell cycle control gene cdc2: isolation of a sequence suc1 that suppresses cdc2 mutant function. *Molecular and General Genetics* **202** 291–293

Heald R and McKeon F (1990) Mutations of phosphorylation sites in lamin A that prevent nuclear lamina disassembly in mitosis. *Cell* **61** 579–589

Heald R, McLoughlin M and McKeon F (1993) Human Wee1 maintains mitotic timing by protecting the nucleus from cytoplasmically activated Cdc2 kinase. *Cell* **74** 463–474

Hershko A, Ganoth D, Sudakin V *et al* (1994) Components of a system that ligates cyclin to ubiquitin and their regulation by the protein kinase cdc2. *Journal of Biological Chemistry* **269** 4940–4946

Hindley J, Phear G, Stein M and Beach D (1987) Suc1+ encodes a predicted 13-kilodalton protein that is essential for cell viability and is directly involved in the division cycle of *Schizosaccharomyces pombe*. *Molecular and Cellular Biology* **7** 504–511

Hirokawa N (1994) Microtubule organisation and dynamics dependent on microtubule-associated proteins. *Current Opinion in Cell Biology* **6** 674–681

Hoffmann I, Clarke PR, Marcote MJ, Karsenti E and Draetta G (1993) Phosphorylation and activation of human cdc25-C by cdc2–cyclin B and its involvement in the self-amplification of MPF at mitosis. *EMBO Journal* **12** 53–63

Hoffmann I, Draetta G and Karsenti E (1994) Activation of the phosphatase activity of human cdc25A by a cdk2–cyclin E dependent phosphorylation at the G_1/S transition. *EMBO Journal* **13** 4302–4310

Hoshi M, Ohta K, Gotoh Y *et al* (1992) Mitogen-activated protein kinase catalysed phosphorylation of microtubule associated proteins, microtubule associated protein 2 and microtubule associated protein 4. *European Journal of Biochemical* **203** 43–52

Hunt T, Luca FC and Ruderman JV (1992) The requirements for protein synthesis and degradation, and the control of destruction of cyclins A and B in the meiotic and mitotic cell cycles of the clam embryo. *Journal of Cell Biology* **116** 707–724

Hunter T (1993) Braking the cycle. *Cell* **75** 839–841

Hunter T and Pines J (1994) Cyclins and cancer II: cyclin D and CDK inhibitors come of age. *Cell* **79** 573–582

Hyman AA and Karsenti E (1996) Morphogenetic properties of microtubules and mitotic spindle assembly. *Cell* **84** 401–410

Izumi T and Maller J (1995) Phosphorylation and activation of the *Xenopus* Cdc25 phosphatase in the absence of Cdc2 and Cdk2 activity. *Molecular Biology of the Cell* **6** 215–226

Izumi T, Walker DH and Maller JL (1992) Periodic changes in phosphorylation of the *Xenopus* cdc25 phosphatase regulate its activity. *Molecular Biology of the Cell* **3** 927–939

Jackman M, Firth M and Pines J (1995) Human cyclins B1 and B2 are localised to strikingly different structures : B1 to microtubules, B2 primarily to the Golgi apparatus. *EMBO Journal* **14** 1646–1654

Jeffrey PD, Russo AA, Polyak K *et al* (1995) Mechanism of CDK activation revealed by the structure of a cyclinA–cdk2 complex. *Nature* **376** 313–320

Jinno S, Suto K, Nagata A *et al* (1994) Cdc25A is a novel phosphatase functioning early in the cell cycle. *EMBO Journal* **13** 1549–1556

Johnson LN, Noble MEM and Owen DJ (1996) Active and inactive protein kinases: structural basis for regulation. *Cell* **85** 149–158

Kaldis P, Sutton A and Solomon MJ (1996) The Cdk-activating kinase (CAK) from budding yeast. *Cell* **86** 553–564

King RW, Peters J-M, Tugendreich S, Rolfe M and Hieter P (1995) A 20S complex containing CDC27 and CDC16 catalyzes the mitosis-specific conjugation of ubiquitin to cyclin B. *Cell* **81** 279–288

Kinoshita N, Ohkura H and Yanagida M (1990) Distinct, essential roles of type 1 and 2A protein phosphatases in the control of the fission yeast cell division cycle. *Cell* **63** 405–415

Kitada K, Johnston AL and Sugino A (1993) A multicopy supressor gene of the *Saccharomyces cerevisiae* G_1 cell cyclemutant gene DBF4encodes a protein kinase and is identified as CDC5. *Molecular and Cellular Biology* **13** 4445–4457

Knighton DR, Zheng JH, Ten-Eyck LF *et al* (1991a) Crystal structure of the catalytic subunit of cyclic adenosine monophosphate-dependent protein kinase. *Science* **253** 407–414

Knighton DR, Zheng JH, Ten-Eyck LF, Xuong NH, Taylor SS and Sowadski JM (1991b) Structure of a peptide inhibitor bound to the catalytic subunit of cyclic adenosine monophosphate-dependent protein kinase. *Science* **253** 414–420

Kobayashi H, Stewart E, Poon R, Adamczewski JP, Gannon J and Hunt T (1992) Identification of the domains in cyclin A required for binding to, and activation of, p34cdc2 and p32cdk2 protein kinase subunits. *Molecular Biology of the Cell* **3** 1279–1294

Kornbluth S, Sebastian B, Hunter T and Newport J (1994) Membrane localization of the kinase which phosphorylates p34cdc2 on threonine 14. *Molecular Biology of the Cell* **5** 273–282

Krek W and Nigg EA (1991) Mutations of p34cdc2 phosphorylation sites induce premature mitotic events in Hela cells—evidence for a double block to p34cdc2 kinase activation in vertebrates. *EMBO Journal* **10** 3321–3329

Krek W, Ewen ME, Shirodkar S, Arany Z, Kaelin WG and Livingstone DM (1994) Negative regulation of the growth-promoting transcription factor E2F-1 by a stably bound cyclin A-dependent protein kinase. *Cell* **78** 161–172

Kreutzer MA, Richards JP, De-Silva-Udawatta MN *et al* (1995) *Caenorhabditis elegans* cyclin A- and B-type genes: a cyclin A multigene family, an ancestral cyclin B3 and differential germline expression. *Journal of Cell Science* **108** 2415–2424

Kumagai A and Dunphy WG (1992) Regulation of the cdc25 protein during the cell cycle in *Xenopus* extracts. *Cell* **70** 139–151

Kumagai A and Dunphy WG (1995) Control of the Cdc2/cyclin B complex in *Xenopus* egg extracts arrested at a G₂/M checkpoint with DNA synthesis inhibitors. *Molecular Biology of the Cell* **6** 199–213

Kumagai A and Dunphy WG (1996) Purification and molecular cloning of Plx1, a Cdc25-regulatory kinase from *Xenopus* egg extracts. *Science* **273** 1377–1380

Kussel P and Frasch M (1995) Pendulin, a *Drosophila* protein with cell cycle-dependent nuclear localization, is required for normal cell proliferation. *Journal of Cell Biology* **129** 1491–1507

Labib K and Moreno S (1996) rum1: a CDK inhibitor regulating G1 progression in fission yeast. *Trends in Cell Biology* **6** 62–66

Larsson N, Melander H, Markland U, Osterman O and Gullberg M (1995) G₂/M transition requires multi-site phosphorylation of oncoprotein 18 by distinct protein kinase systems. *Journal of Biological Chemistry* **270** 14175–14183

Lee TA, Turck C and Kirschner MW (1994) Inhibition of cdc2 activation by INH/PP2A. *Molecular Biology of the Cell* **5** 323–338

Lee TH and Kirschner MW (1996) An inhibitor of p34cdc2/cyclin B that regulates the G₂/M transition in *Xenopus* extracts. *Proceedings of the National Academy of Sciences of the USA.* **93** 352–356

Lee TH, Solomon MJ, Mumby MC and Kirschner MW (1991) INH, a negative regulator of MPF, is a form of protein phosphatase 2A. *Cell* **64** 415–423

Lees EM and Harlow E (1993) Sequences within the conserved cyclin box of human cyclin A are sufficient for binding to and activation of cdc2 kinase. *Molecular and Cellular Biology* **13** 1194–1201

LeGuellec R, Paris J, Couterier A, Roghi C and Phillipe M (1991) Cloning by differential screening of a *Xenopus* cDNA that encodes a kinesin related protein. *Molecular and Cellular Biology* **11** 3395–3398

Lehner CF and O'Farrell PH (1990) The roles of *Drosophila* cyclins A and B in mitotic control. *Cell* **61** 535–547

Lombillo VA, Nislow C, Yen TJ, Gelfand VI and Mcintosh JR (1995) Antibodies to the kinesin motor domain and CENP-E inhibit microtubule depolymerisation-dependent motion of chromosomes in vitro. *Journal of Cell Biology* **128** 107–115

Lorca T, Labbe JC, Devault A *et al* (1992) Cyclin A-cdc2 kinase does not trigger but delays cyclin degradation in interphase extracts of amphibian eggs. *Journal of Cell Science* **102** 420–426

Lu KP and Hunter T (1995) Evidence for a NIMA-like mitotic pathway in vertebrate cells. *Cell* **81** 413–424

Lundgren K, Walworth N, Booher R, Dembski M M. K and Beach D (1991) mik1 and wee1 cooperate in the inhibitory tyrosine phosphorylation of cdc2. *Cell* **64** 1111–1122

McGowan CH and Russell P (1993) Human Wee1 kinase inhibits cell division by phosphorylating p34cdc2 exclusively on Tyr15. *EMBO J* **12** 75–85

McGowan CH and Russell P (1995) Cell cycle regulation of human WEE1. *EMBO Journal* **14** 2166–2175

Makela TP, Tassan J-P, Nigg EA, Frutiger S, Hughes GJ and Weinberg RA (1994) A cyclin associated with the CDK-activating kinase MO15. *Nature* **371** 254–257

Maller JL (1991) Mitotic control. *Current Oppinion in Cell Biology* **3** 269–275

Maridor G, Gallant P, Golsteyn R and Nigg E (1993) Nuclear localisation of vertebrate cyclin A correlates with its ability to form complexes with CDK catalytic subunits. *Journal of Cell Science* **106** 535–544

Marklund U, Osterman O, Melander H, Bergh A and Gullberg M (1994) The phenotype of a cdc2 kinase target site deficient mutant of oncoprotein 18 by distinct protein kinase systems. *Journal of Biological Chemistry* **269** 30626–30635

Marklund U, Larsson N, Melander H, Brattsand G and Gullberg M (1996) Oncoprotein 18 is a phosphorylation-responsive regulator of microtubule dynamics. *EMBO Journal* **15** 5290–5298

Marshall CJ (1994) Hot lips and phosphorylation of protein kinases. *Nature* **367** 686

Martin-Castellanos C, Labib K and Moreno S (1996) B-type cyclins regulate G_1 progression in fission yeast in opposition to the p25*rum1* cdk inhibitor. *EMBO Journal* **15** 839–849

Meijer L, Azzi L and Wang J (1991a) Cyclin B targets p34cdc2 for tyrosine phosphorylation. *EMBO Journal* **10** 1545–1554

Meijer L, Ostvold A-C, Walaas SI, Lund T and Laland SG (1991b) High mobility group (HMG) proteins I, Y and P1 as substrates of the M-phase-specific p34cdc2/cyclincdc13 kinase. *European Journal of Biochemistry* **196** 557–567

Mendenhall MD (1993) An inhibitor of p34CDC28 protein kinase activity from *Saccharomyces cerevisiae*. *Science* **259** 216–219

Meuller PR, Coleman TR and Dunphy WG (1995) Membrane localisation of the kinase which phosphorylates p34cdc2 on threonine 14. *Molecular Biology of the Cell* **6** 119–134

Meuller PR, Coleman TR, Kumagai A and Dunphy WG (1995) Myt 1: a membrane-associated inhibitory kinase that phosphorylates Cdc2 on both threonine-14 and tyrosine-15. *Science* **270** 86–90

Millar JB and Russell P (1992) The cdc25 M-phase inducer: an unconventional protein phosphatase. *Cell* **68** 407–410

Millar JB, Blevitt J, Gerace L, Sadhu K, Featherstone C and Russell P (1991) p55CDC25 is a nuclear protein required for the initiation of mitosis in human cells. *Proceedings of the National Academy of Sciences of the USA* **88** 10500–10504

Millar JB, Lenaers G and Russell P (1992) Pyp3 PTPase acts as a mitotic inducer in fission yeast. *EMBO Journal* **11** 4933–4941

Minshull J, Blow JJ and Hunt T (1989) Translation of cyclin mRNA is necessary for extracts of activated *Xenopus* eggs to enter mitosis. *Cell* **56** 947–956

Misteli T and Warren G (1994) COP-coated vesicles are involved in the mitotic fragmentation of Golgi stacks in a cell free sytsem. *Journal of Cell Biology* **125** 269–282

Moreno S and Nurse P (1994) Regulation of progression through the G_1 phase of the cell cycle by the rum1+ gene. *Nature* **367** 236–242

Moreno S, Hayles J and Nurse P (1989) Regulation of p34cdc2 protein kinase during mitosis. *Cell* **58** 361–372

Morgan DO (1995) Principles of CDK regulation. *Nature* **374** 131–134

Morgan DO and De Bondt HL (1994) Protein kinase regulation: insights from crystal struture analysis. *Current Biology* **6** 239–246

Mori A, Aizawa T, Saido H *et al* (1991) Site specific phosphorylation by protein kinase C inhibit assembly-promoting activity of microtubule associated protein 4. *Biochemistry* **30** 9341–9346

Newport J and Spann T (1987) Disassembly of the nucleus in mitotic extracts: membrane vesicularization, lamin disassembly, and chromosome condensation are independent processes. *Cell* **48** 219–230

Niclas J, Allan V and Vale RD (1996) Cell cycle regulation of dynein association with membranes modulates microtubule-based organell transport. *Journal of Cell Biology* **133** 585–593

Nigg EA (1993) Cellular substrates of p34cdc2 and its companion cyclin-dependent kinases.

Trends in Cell Biology **3** 296–301

Nissen MS, Langan TA and Reeves R (1991) Phosphorylation by cdc2 kinase modulates DNA binding activity of high mobility group I nonhistone chromatin protein. *Journal of Biological Chemistry* **266** 19945–19952

Norbury C, Blow J and Nurse P (1991) Regulatory phosphorylation of the p34*cdc2* protein kinase in vertebrates. *EMBO Journal* **10** 3321–3329

Nurse P (1990) Universal control mechanism regulating onset of M-phase. *Nature* **344** 503–508

O'Connell MJ, Norbury C and Nurse P (1994) Premature chromatin condensation upon accumulation of NIMA. *EMBO Journal* **13** 4926–4937

Ookata K, Hisanaga S, Okano T, Tachibana K and Kishimoto T (1992) Relocation and distinct subcellular localization of p34cdc2-cyclin B complex at meiosis reinitiation in starfish oocytes. *EMBO Journal* **11** 1763–1772

Ookata K, Hisanaga S, Okumura E and Kishimoto T (1993) Association of p34cdc2/cyclin B complex with microtubules in starfish oocytes. *Journal of Cell Science* **105** 873–881

Ookata K, Hisanga S, Bulinksi JC *et al* (1995) Cyclin B interaction with microtubule-associated protein 4 (MAP4)targets p34cdc2 kinase to microtubules and is a potential regulator of M-phase microtubule dynamics. *Journal of Cell Biology* **128** 849–862

Osmani SA, Pu RT and Morris NR (1988) Mitotic induction and maintenance by overexpression of a G_2-specific gene that encodes a potential protein kinase. *Cell* **53** 237–244

Parker LL and Piwnica WH (1992) Inactivation of the p34cdc2-cyclin B complex by the human WEE1 tyrosine kinase. *Science* **257** 1955–1957

Parker LL, Walter SA, Young PG and Piwnica-Worms H (1993) Phosphorylation and inactivation of the mitotic inhibitor Wee1 by the niml/cdr1 kinase. *Nature* **363** 736–738

Patra D and Dunphy W (1996) Xe-p9, a *Xenopus* Suc1/Cks homolog, has multiple essential roles in cell cycle control. *Genes and Development* **10** 1503–1515

Peter M, Nakagawa J, Dorée M, Labbé JC and Nigg EA (1990a) Identification of major nucleolar proteins as candidate mitotic substrates of cdc2 kinase. *Cell* **60** 791–801

Peter M, Nakagawa J, Dorée M, Labbé JC and Nigg EA (1990b) In vitro disassembly of the nuclear lamina and M-phase specific phosphorylation of lamins by cdc2 kinase. *Cell* **61** 591–602

Peter M, Heitlinger E, Haner M, Aebi U and Nigg EA (1991) Disassembly of in vitro formed lamin head-to-tail polymers by CDC2 kinase. *EMBO Journal* **10** 1535–1544

Pfaller R, Smythe C and Newport JW (1991) Assembly/disassembly of the nuclear envelope membrane: cell cycle-dependent binding of nuclear membrane vesicles to chromatin in vitro. *Cell* **65** 209–217

Pines J (1993) Cyclin-dependent kinases: clear as crystal? *Current Biology* **3** 544–547

Pines J (1994) Arresting developments in cell cycle control. *Trends in Biochemical Sciences* **19** 143–145

Pines J and Hunter T (1989) Isolation of a human cyclin cDNA: evidence for cyclin mRNA and protein regulation in the cell cycle and for interaction with p34cdc2. *Cell* **58** 833–846

Pines J and Hunter T (1991) Human cyclins A and B are differentially located in the cell and undergo cell cycle dependent nuclear transport. *Journal of Cell Biology* **115** 1–17

Pines J and Hunter T (1994) The differential localization of human cyclins A and B is due to a cytoplasmic retention signal in cyclin B. *EMBO Journal* **13** 3772–3781

Poon RYC and Hunter T (1995) Dephosphorylation of Cdk2 Thr160 by the cylin-dependent kinase interacting phosphatase KAP in the absence of cyclin. *Science* **270** 90–93

Poon RYC, Yamashita K, Adaczewski J, Hunt T and Shuttleworth J (1993) The cdc2-related protein p40MO15 is the catalytic subunit of a protein kinase that can activate p33cdk2 and p34cdc2. *EMBO Journal* **12** 3123–3132

Richardson H, Lew DJ, Henze M, Sugimoto K and Reed SI (1992) Cyclin-B homologs in *Saccharomyces cerevisiae* function in S phase and in G_2. *Genes and Development* **6** 2021–2034

Richardson HE, Stueland CS, Thomas J, Russell P and Reed SI (1990) Human cDNAs encoding homologs of the small p34Cdc28/Cdc2-associated protein of *Saccharomyces cerevisiae* and *Schizosaccharomyces pombe*. *Genes and Development* **4** 1332–1344

Roy R, Ademczewski P, Seroz T *et al* (1994) The MO15 kinase is associated with the TFIIH transcription factor-DNArepair factor. *Cell* **79** 1093–1101

Russell P and Nurse P (1987a) The mitotic inducer nim1$^+$ functions in a regulatory network of protein kinase homologs controlling the initiation of mitosis. *Cell* **49** 569–576

Russell P and Nurse P (1987b) Negative regulation of mitosis by wee1$^+$, a gene encoding a protein kinase homolog. *Cell* **49** 559–567

Saunders WS (1993) Mitotic spindle pole separation. *Trends in Cell Biology* **3** 432–436

Saunders WS, Koshland D, Eshel D, Gibbons IR and Hoyt MA (1995) *Saccharomyces cerevisiae* kinesin- and dynein-related proteins required for anaphase chromosome segregation. *Journal of Cell Biology* **128** 617–624

Schmidt-Zachmann MS, Dergemont C, Kuhn LC and Nigg EA (1993) Nuclear export of proteins: the role of nuclear retention. *Cell* **74** 493–504

Schneider BL, Yang Q-H and Futcher AB (1996) Linkage of replication to start by the Cdk inhibitor Sic1. *Science* **272** 560–562

Schwob E, Böhm T, Mendenhall MD and Nasmyth K (1994) The B-type cyclin kinase inhibitor p40$SIC1$ controls the G_1/S transition in *Saccharomyces cerevisiae*. *Cell* **79** 233–244

Sebastian B, Kakizuka A and Hunter T (1993) Cdc25M2 activation of cyclin-dependent kinases by dephosphorylation of threonine-14 and tyrosine-15. *Proceedings of the National Academy of Sciences of the USA* **90** 3521–3524

Seki T, Yamashita K, Nishitani H, Takagi T, Russell P and Nishimoto T (1992) Chromosome condensation caused by the loss of RCC1 function requires the Cdc25C protein that is located in the cytoplasm. *Molecular and Cellular Biology* **3** 1373–1388

Shermoen AW and O'Farrell PH (1991) Progression of the cell cycle through mitosis leads to abortion of nascent transcripts. *Cell* **67** 303–310

Shina N, Gotoh Y and Nishida E (1995) Microtubule severing activity in M-phase. *Trends in Cell Biology* **5** 283–286

Sigrist S, Jacobs H, Stratman R and Lehner CF (1995) Exit from mitosis is regulated by *Drosophila* fizzy and the sequential destruction of cyclins A, B and B3. *EMBO Journal* **14** 4827–4838

Smythe C and Newport JW (1992) Coupling of mitosis to the completion of S phase in *Xenopus* occurs via modulation of the tyrosine kinase that phosphorylates p34cdc2. *Cell* **68** 787–797

Solomon MJ, Glotzer M, Lee TH, Philippe M and Kirschner MW (1990) Cyclin activation of p34cdc2. *Cell* **63** 1013–1024

Solomon MJ, Harper JW and Shuttleworth J (1993) CAK, the p34cdc2 activating kinase, contains a protein identical or closely related to p40MO15. *EMBO Journal* **12** 3133–3142

Solomon MJ, Lee T and Kirschner MW (1992) Role of phosphorylation in p34cdc2 activation: identification of an activating kinase. *Molecular Biology of the Cell* **3** 13–27

Sorger PK and Murray AW (1992) S-phase feedback control in budding yeast independent of tyrosine phosphorylation of p34cdc28. *Nature* **355** 365–368

Stewart E, Kobayashi H, Harrison D and Hunt T (1994) Destruction of *Xenopus* cyclins A and B2, but not B1, requires binding to p34cdc2. *EMBO Journal* **13** 584–594

Sudakin V, Ganoth D, Dahan A *et al* (1995) The cyclosome, a large complex containing cyclin-selective ubiquitin-ligase activity, targets cyclins for destruction at the end of mitosis. *Molecular Biology of the Cell* **6** 185–198

Tang Z, Coleman TR and Dunphy WG (1993) Two distinct mechanisms for negative regulation of the Wee1 protein kinase. *EMBO Journal* **12** 3427–3436

Tassan J-P, Schultz SJ, Bartek J and Nigg EA (1994) Cell cyle analysis of the activity, subcellular localization and subunit composition of human CAK (CDK-activating kianse) *Journal of Cell Biology* **127** 467–478

Thomas L, Clarke PR, Pagano M and Gruenberg J (1992) Inhibition of membrane fusion in vitro via cyclin B but not cyclin A. *Journal of Biological Chemistry* **267** 6183–6187 [Erratum **267** 13780]

Thuret J-Y, Valay J-G, Faye G and Mann C (1996) Civ1 (CAK in vivo), a novel Cdk-activating kinase. *Cell* **86** 565–576

Vaisberg EA, Knooce MP and McIntosh JR (1993) Cytoplasmic dynein plays a role in mammalian mitotic spindle formation. *Journal of Cell Biology* **123** 849–858

van der Sluijs P, Hull M, Huber LA, Male P B. G and Mellman I (1992) Reversible phosphorylation-dephosphorylation determines the localization of rab 4 during the cell cycle. *EMBO Journal* **11** 4379–4389

Verde F, Labbe JC, Doree M and Karsenti E (1990) Regulation of microtubule dynamics by cdc2 protein kinase in cell-free extracts of *Xenopus* eggs. *Nature* **343** 233–238

Verde F, Dogterom M, Stelzer E, Karsenti E and Leibler S (1992) Control of microtubule dynamics and length by cyclin A- and cyclin B-dependent kinases in *Xenopus* egg extracts. *Journal of Cell Biology* **118** 1097–1108

Walker DH, DePaoli-Roach AA and Maller JL (1992) Multiple roles for protein phosphatase 1 in regulating the *Xenopus* early embryonic cell cycle. *Molecular and Cellular Biology* **3** 687–698

Wang XM, Peloquin JG, Zhai Y, Bulinski JC and Borisey GG (1996) Removal of MAP4 from microtubules in vivo produces no observable phenotype at the cellular level. *Journal of Cell Biology* **132** 345–357

Ward G and Kirschner M (1990) Identification of cell cycle-regulated phosphorylation sites on nuclear lamin C. *Cell* **61** 561–577

Warner JR (1990) The nucleolus and ribosome function. *Current Opinion in Cell Biology* **2** 521–527

Warren G (1993) Membrane partitioning during cell division. *Annual Reviews of Biochemistry* **62** 323–348

Watanabe N, Broome M and Hunter T (1995) Regulation of the human WEE1Hu CDK tyrosine 15-kinase during the cell cycle. *EMBO Journal* **14** 1878–1891

Weisenberger D and Scheer U (1995) A possible mechanism for the inhibition of ribosomal RNA gene transcription during mitosis. *Journal of Cell Biology* **129** 561–579

Woodman PG, Adamczewski JP, Hunt T and Warren G (1993) In vitro fusion of endocytic vesicles is inhibited by cyclin A-cdc2 kinase. *Molecular Biology of the Cell* **4** 541–553

Wu L and Russell P (1993) Nim1 kinase promotes mitosis by inactivating wee1 tyrosine kinase. *Nature* **363** 738–741

Yamashiro S, Yamakita Y, Ishikawa R and Matsumura F (1990) Mitosis-specific phosphorylation causes 83K non-muscle caldesmon to dissociate from microfilaments. *Nature* **344** 675–678

Yamashiro S, Yamakita Y, Hosoya H and Matsumura F (1991) Phosphorylation of non-muscle caldesmon by p34cdc2 kinase during mitosis. *Nature* **349** 169–172

Yano R, Oakes M, Yamaghishi M, Dodd J and Nomura M (1994) Yeast Srp1has homology to armadillo/plakoglobin/β-catenin and participates in apparently multiple nuclear functions including the maintenance of the nucleolar structure. *Proceedings of the National Academy of Sciences of the USA* **91** 6880–6884

Ye XS, Xu G, Pu RT *et al* (1995) The NIMA protein kinase is hyperphosphorylated and activated downstream of p34cdc2/cyclin B: coordination of two mitosis promoting kinases. *EMBO Journal* **14** 986–994

The authors are responsible for the accuracy of the references.

The DNA Replication Licensing System

PIA THÖMMES • J JULIAN BLOW
Imperial Cancer Research Fund, Clare Hall Laboratories, South Mimms, Herts EN6 3LD

ENSURING PRECISE DUPLICATION OF CHROMOSOMAL DNA

Duplication of eukaryotic chromosomes during S phase of the cell division cycle requires the co-ordinated activity of hundreds or thousands of replication forks. These are initiated at replication origins distributed throughout the genome. To ensure the complete replication of eukaryotic chromosomes, origins must be sufficiently closely spaced so that all the intervening DNA can be replicated before entry into mitosis occurs. Failure to replicate even small sections of DNA could lead to disastrous consequences as the sister chromatids are pulled apart during mitosis. On the other hand, to prevent any section of DNA being replicated more than once in a single S phase, each origin must fire no more than once in each cell cycle. Overreplication of DNA is potentially very harmful, because it would represent an irreversible genetic change, potentially leading to gene dosage problems and the risk of recombination and amplification occurring in the duplicated region.

USE OF CELL FREE EXTRACTS FROM *XENOPUS* TO STUDY DNA REPLICATION

Cell free extracts of eggs of the South African clawed toad *Xenopus laevis* support complete chromosome replication with the same cell cycle controls that exist in vivo (Blow and Laskey, 1986). In particular, DNA is precisely duplicated in each in vitro S phase (Blow and Laskey, 1986; Blow and Watson,

1987). Apart from a similar system derived from embryos of the fruit fly *Drosophila melanogaster* (Crevel and Cotterill, 1991), the *Xenopus* extract is currently the only eukaryotic cell free system that supports efficient chromosome replication, and it offers a unique opportunity to study the control mechanisms of this process. When DNA is added to the cell free system, it is assembled onto chromatin and then into structures resembling normal interphase nuclei (Lohka and Masui, 1983). The initiation of DNA replication in this system depends on prior assembly of the template DNA into chromatin and functional interphase nuclei with an intact nuclear envelope (Blow and Watson, 1987; Newport, 1987; Sheehan *et al*, 1988; Blow and Sleeman, 1990).

The early *Xenopus* embryo lacks a distinct G_1 phase of the cell cycle and initiation occurs virtually as soon as nuclear assembly has been completed, so that different nuclei may start to replicate at different times. Analysis of replication kinetics by flow cytometry showed that nuclei act as individual "units", each receiving a signal to replicate from the cytoplasm, which causes them to undergo a single burst of near synchronous initiation events (Blow and Watson, 1987). The feature that defines this unit of replication is likely to be the nuclear envelope, since all the DNA surrounded by an intact nuclear envelope starts to replicate at the same time, even if this DNA was originally derived from more than one nucleus (Leno and Laskey, 1991).

PREVENTING REREPLICATION OF DNA IN A SINGLE CELL CYCLE

With nuclei replicating individually in the *Xenopus* extract, some nuclei may be starting to replicate only after others have replicated fully; despite this, no replicated nuclei undergo more than one round of DNA replication (Blow and Watson, 1987). This means that replicated and unreplicated nuclei must differ, since only the unreplicated nuclei have the potential to replicate in the current cell cycle. These results are consistent with those of the classic cell fusion studies of Rao and Johnson (1970), in which HeLa cells at different stages of the cell cycle were fused and the replication timing of the different nuclei was measured. The results of these fusions are summarized in Fig. 1. In fusions of G_1 and S phase cells, the G_1 nucleus was induced to enter S phase earlier than it would normally have done, whereas the S phase nucleus continued replication as normal (Fig. 1a). However, these cytoplasmic S phase inducers can only act on G_1 nuclei and cannot induce G_2 nuclei to undergo another round of DNA replication. In S phase/G_2 and G_1/G_2 HeLa cell fusions, the G_2 nuclei did not replicate until after passage through mitosis (Fig. 1b,c) (Rao and Johnson, 1970). This suggests that the absence of DNA replication in normal G_2 cells is due not to the presence of diffusible inhibitors in the G_2 cytoplasm, but to an intrinsic difference between G_1 and G_2 nuclei.

The inability of replicated nuclei to rereplicate in response to the S phase inducers present can be demonstrated directly in *Xenopus* extracts. Replicated G_2 nuclei that were transferred to fresh extract did not undergo a further

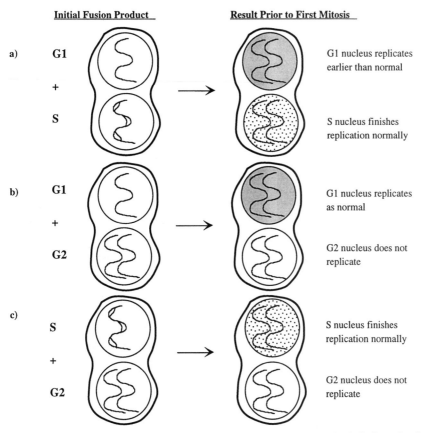

Initial Fusion Product   **Result Prior to First Mitosis**

a)   G1   G1 nucleus replicates
                                                earlier than normal

      +

      S   S nucleus finishes
                                                replication normally

b)   G1   G1 nucleus replicates
                                                as normal

      +

      G2   G2 nucleus does not
                                                replicate

c)   S   S nucleus finishes
                                                replication normally

      +

      G2   G2 nucleus does not
                                                replicate

Fig. 1. Summary of cell cycle fusion studies (Rao and Johnson, 1970). Cells from G_1, S and G_2 stages of the cell cycle were fused, giving hybrids with two nuclei. G_1 nuclei are indicated with a single chromosome, S nuclei with a chromosome containing replication bubbles and G_2 with a double chromosome. Before the first mitosis after fusion, all nuclei have fully replicated and the extent of replication performed in the hybrid is indicated by the of shading in each nucleus. Reproduced from Blow (1995) with permission

round of replication (Blow and Laskey, 1988). However, if these nuclei were allowed to pass into mitosis, the DNA efficiently rereplicated on transfer to interphase extract. Entry into mitosis therefore changed the chromosomal DNA from a refractory postreplicative state to an initiation competent prereplicative state. This change in replication competence that occurs on entry into mitosis could be mimicked by agents that caused nuclear envelope permeabilization, such as lysolecithin or phospholipase (Blow and Laskey, 1988). Similar results have been obtained in *Drosophila* extracts (Crevel and Cotterill, 1991).

THE REPLICATION LICENSING FACTOR MODEL

The observations described above can be explained by a requirement for two distinct signals for initiation at any given replication origin (Blow and Laskey,

1988; Chong *et al*, 1996). The first signal, replication licensing factor (RLF), "licenses" replication origins during late mitosis or early interphase by putting them into an initiation competent state. RLF cannot cross the nuclear envelope, so that once nuclear assembly is complete, no further origins can become licensed. Once nuclear assembly is complete, the second signal, S phase promoting factor (SPF) (Blow and Nurse, 1990; Fang and Newport, 1991; Strausfeld *et al*, 1994; Chevalier *et al*, 1995; Jackson *et al*, 1995; Strausfeld *et al*, 1996), induces licensed origins within intact nuclei to initiate and in doing so removes or inactivates the licence. Thus, in G_2, no active RLF remains in the nucleus and the nuclear envelope must be transiently permeabilized to license the DNA and allow a further round of DNA replication. As long as the licensing signal and the initiation signal act sequentially and can never act on DNA at the same time, the result will be the precise duplication of the DNA. If the two signals did act on DNA at the same time, this would lead to repetitive cycles of licensing and initiation with consequent rereplication of sections of the DNA.

Results consistent with this model were also obtained on addition of nuclei from mammalian tissue culture cells into *Xenopus* eggs (De Roeper *et al*, 1977) or egg extracts (Leno *et al*, 1992). Although intact G_1 nuclei were replicated after transfer to *Xenopus* extract, G_2 nuclei replicated only if they had been permeabilized before transfer. When permeabilized G_2 nuclei were resealed using nuclear envelope precursors without exposure to *Xenopus* extract, they remained incapable of rereplicating in *Xenopus* extract (Coverley *et al*, 1993). But replication could occur if they were exposed to *Xenopus* egg extract before resealing. Quiescent nuclei, however, may lose their licence, since quiescent (G_0) 3T3 fibroblasts behaved like G_2 nuclei and required nuclear envelope permeabilization to be competent to replicate in *Xenopus* extract (Leno and Munshi, 1994).

The licensing factor model proposes that essential initiation proteins are bound to replication origins only in M and G_1 phases and are removed from the origins once initiation of replication has occurred. DNase I footprinting of replication origins in *Saccharomyces cerevisiae* gave results consistent with this model. Two distinct footprints at yeast origins can be seen: a "prereplicative" state from the end of mitosis until the start of S phase and the "postreplicative" state from the beginning of S phase to late mitosis (Diffley *et al*, 1994). In the postreplicative state, the core region of the replication origin is bound by the origin recognition complex (Bell and Stillman, 1992; Diffley and Cocker, 1992). However, in the prereplicative state, the origin region protected from DNase I digestion is much greater, suggesting the binding of additional proteins to replication origins that are removed as the initiation of replication occurs (Diffley *et al*, 1994). In fact, the Cdc6 protein, which is necessary for origin firing in vivo, is essential for the establishment and maintenance of the prereplication complex (Cocker *et al*, 1996). The appearance of the prereplicative footprint in G_1 coincides with the time when the DNA would have become licensed.

USE OF PROTEIN KINASE INHIBITORS TO BLOCK RLF ACTIVATION

RLF activity can be inhibited by treating the *Xenopus* extract with the protein kinase inhibitors 6-dimethylaminopurine (6-DMAP) (Blow, 1993), staurosporine (Kubota and Takisawa, 1993) and olomoucine (Vesely *et al*, 1994). The activity inhibited by these treatments conforms to the four definitive features necessary and sufficient to define it as the RLF: (a) the activity modifies ("licences") DNA templates so that after a brief incubation in untreated extract, they become competent to replicate in the inhibited extracts; (b) this modification (the "licence") is required for the initiation of DNA replication; (c) the activity is unable to cross the nuclear envelope, since nuclei assembled in 6-DMAP extract are unable to replicate when transferred to untreated extract unless the nuclear envelope is transiently permeabilized; and (d) the licence is capable of supporting only a single initiation event and chromatin must be relicensed before a second round of DNA replication can be initiated.

The ability of 6-DMAP to inhibit RLF activity was used to develop an assay for RLF active in *Xenopus*. The assay involves incubation of unlicensed chromatin with fractions potentially containing RLF activity. The chromatin is then transferred to 6-DMAP treated extracts, where replication will occur only if the chromatin has been licensed in the first incubation (Blow, 1993; Chong *et al*, 1995). This assay was used to show that RLF activity is periodic in the *Xenopus* in vitro cell cycle (Blow, 1993). RLF levels are low during metaphase but rise rapidly at the metaphase-anaphase transition; during interphase, RLF levels decline again, reaching background levels by mid S phase. Since SPF activity is also periodic in the cell cycle, being detectable only in interphase (Blow and Nurse, 1990), this serves to re-enforce the spatial separation of the two activities by the nuclear envelope. In most higher eukaryotic cells, where a distinct G_1 period in the cell cycle is seen, the temporal separation of the RLF and SPF signals may also be an important mechanism for separating the two activities. This and other results suggest that the protein kinase inhibitors do not inhibit RLF activity itself but instead block the activation of RLF that normally occurs at the metaphase-anaphase transition (Blow, 1993; Blow JJ, unpublished).

To block RLF activity in *Xenopus* extracts, protein kinase inhibitors such as 6-DMAP must be added to extracts in metaphase at concentrations sufficient to inhibit mitosis promoting factor (MPF) activity (Blow, 1993; Kubota and Takisawa, 1993; Vesely *et al*, 1994). MPF induces progression from interphase to mitosis of the cell cycle and is provided by certain cyclin dependent protein kinases. The correlation between MPF and RLF inhibition is particularly significant in the case of olomoucine, which is specific for MPF like protein kinases (Vesely *et al*, 1994). This correlation implies that MPF performs two functions to allow DNA to be licensed: firstly, it causes nuclear envelope breakdown by inducing entry into mitosis and so allows RLF to gain access to chromatin, and secondly, it leads to RLF activation. Regulation of RLF activity by cyclin dependent kinases appears to be quite complex since under conditions of high MPF kinase activity, such as during metaphase, RLF is not active

(Blow, 1993). Similarly, RLF activity could be inhibited when cyclin A was added to an interphase extract. A central role for MPF in controlling RLF function is consistent with experiments that implicate corresponding cyclin dependent kinases (MPF homologues) in preventing rereplication in yeast (Broek *et al*, 1991; Hayles *et al*, 1994; Moreno and Nurse, 1994; Dahmann *et al*, 1995).

IDENTIFICATION OF RLF COMPONENTS

Two parallel approaches have been used to isolate RLF from *Xenopus* extracts. Using the 6-DMAP assay, licensing activity was followed during a chromatographic fractionation (Chong *et al*, 1995). In the first step of purification, it could be separated into two essential components, RLF-M and RLF-B. Further fractionation led to the purification of a protein complex, RLF-M, which in the presence of RLF-B is able to license the DNA and allow it to replicate in 6-DMAP treated extract. Purified RLF-M has been found to consist of several members of the MCM/P1 class of proteins including *Xl*Mcms 2, 3 and 5. During the *Xenopus* cell cycle, RLF-M associates with chromatin in G_1 and is removed during replication, consistent with its being a component of the RLF system (Chong *et al*, 1995). For replicated G_2 nuclei to rebind RLF-M and become licensed for a further round of DNA replication, the nuclear envelope must first be permeabilized and the chromatin exposed to both RLF-M and RLF-B (Chong *et al*, 1995).

In an alternative approach, Kubota et al. (1995) raised antibodies to polypeptides present on licensed, but not unlicensed, chromatin. One such protein was identified as the *Xenopus* Mcm3 protein. Proteins co-immuno-precipitating with the anti-Mcm3 antibody showed a polypeptide pattern similar to that of the RLF-M complex (Chong *et al*, 1995; Kubota *et al*, 1995). Immunodepletion of Mcm3 and associated proteins left *Xenopus* extracts unable to license DNA replication, but activity could be restored by readdition of material eluted from the antibodies. Immunofluorescence of both *Xenopus* and human cells showed Mcm3 associating with chromatin in late mitosis and being removed during S phase (Kubota *et al*, 1995).

To test directly the hypothesis that Mcm3 associated proteins are a component of licensing factor, Madine *et al* (1995a) immunodepleted *Xenopus* extracts of *Xl*Mcm3. This treatment removed a complex containing *Xl*Mcm2, 3 and 5, which resembled the polypeptide pattern of the RLF-M complex. Immunodepleted extracts were incapable of supporting the replication of *Xenopus* sperm chromatin or permeabilized nuclei prepared from G_2 HeLa cells (Madine *et al*, 1995a). By contrast, permeabilized nuclei prepared from G_1 HeLa cells still replicated efficiently in the Mcm depleted *Xenopus* extract. This suggests that active MCM/P1 proteins are bound to HeLa nuclei in G_1 but not G_2 and that these human MCM proteins can fully substitute for their *Xenopus* homologues in the cell free replication system.

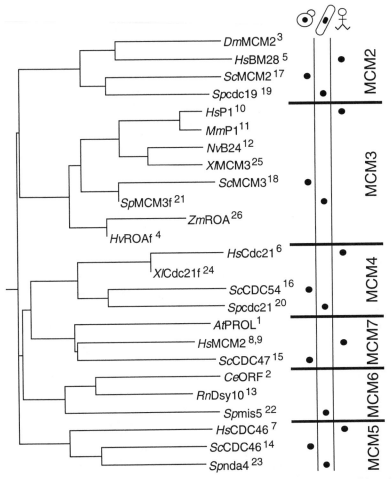

Fig. 2. Phylogenetic analysis of sequenced MCM/P1 genes and cDNAs. Full length sequences of genes in the MCM/P1 superfamily obtained using the BLAST network service at the National Center for Biotechnology Information were subjected to phylogenetic analysis using Clustal W (v1.4) (Higgins *et al*, 1991). Pairwise alignments were performed using the BLOSUM30 protein weight matrix with penalties of 10.00 for gap opening and 0.10 for gap extensions. Multiple alignments were performed using the BLOSUM protein weight series and penalties of 10.00 for gap opening and 0.05 for gap extensions. Incorporation of divergent sequences was delayed by 40%. Residue specific penalties were enabled with hydrophilic residues being defined as GPSNDQEKR. 1000 bootstrapping trials were performed with no effect on the alignment. Organisms are abbreviated: *Dm, Drosophila melanogaster; Hs, Homo sapiens; Sc, Saccharomyces cerevisiae; Sp, Schizosaccharomyces pombe; Mm, Mus musculus; Nv, Notophthalmus viridescens; Xl, Xenopus laevis; Zm, Zea mays; Hv, Hordeum vulgare; At, Arabidopsis thaliana; Ce, Caenorhabditis elegans; Rn, Rattus norvegicus.* Protein fragments are indicated in the figure by the suffix "f". Accession numbers are as follows (P=Swiss-Prot, others Genbank): [1]L39954; [2]P34647; [3]L42762; [4]Z29369; [5]X67334; [6]X74794; [7]X74795; [8]D28480; [9]X74796; [10]P25205; [11]P25206; [12]X78322; [13]U17565; [14]P29496; [15]P3813; [16]P30665; [17]P29469; [18]P24279; [19]P40377; [20]P29458; [21]P30666; [22]D31960; [23]P41389; [24]P30664; [25]U26057/D38074; [26]Z29368. (Reproduced from Chong *et al*, 1996.)

MCM/P1 PROTEINS IN OTHER ORGANISMS

Mcm (minichromosome maintenance) mutants were first isolated from the yeast *S cerevisiae* by their inability to replicate plasmids containing certain autonomously replicating sequences (yeast replication origins) (Maine *et al*, 1984). Further analysis indicated that they are defective in the initiation of DNA replication. Homologous *MCM* genes have been identified in a wide range of eukaryotes, including insects, plants, amphibians and mammals and show a high degree of sequence conservation (Fig. 2). They form a family for which we have proposed the name "*MCM/P1*" (Chong *et al*, 1996). By sequence comparison, the *MCM/P1* genes from any one organism can be grouped into six classes, *MCM2–7* (Mcm1 has been identified as a transcription factor and is distinct from the MCM/P1 class of proteins) (Chong *et al*, 1996). Although all of the *MCM/P1* genes are essential, they show a complicated pattern of suppression and synthetic lethality (Moir *et al*, 1982; Hennessy *et al*, 1991), suggesting that all the MCM/P1 proteins must interact to promote the efficient initiation of DNA replication.

Further lines of evidence suggest that MCM/P1 proteins are involved in DNA replication in metazoans. Microinjection of anti-*Mm*Mcm3 (P1) antibodies (Kimura *et al*, 1995) or anti-*Hs*Mcm2 (BM28) antibodies (Todorov *et al*, 1995) into G_1 phase mouse or human cells inhibited the onset of subsequent DNA synthesis as assessed by the incorporation of bromodeoxyuridine. Microinjection of anti-*Hs*Mcm2 (BM28) antibodies (Todorov *et al*, 1995) during S phase had no effect on DNA synthesis but inhibited cell division, suggesting that the functions of MCM/P1 proteins are also required during S phase for successful entry into mitosis.

To understand the role of MCM/P1 proteins during the initiation of replication, attention has been drawn to the cell cycle regulation of MCM/P1 proteins in higher eukaryotes. An increase in the mRNA level at the G_1/S transition has been observed for several members of the MCM/P1 family, including *Hs*Mcm3 (P1) (Thömmes *et al*, 1992; Schulte *et al*, 1995), *Mm*Mcm5 (Cdc46) (Kimura *et al*, 1995), human and mouse Mcm4 (Cdc21) and *Hs*Mcm7 (p85Mcm) (Schulte *et al*, 1995). However, this increase in mRNA does not lead to a significantly increased protein level, since this remained virtually unchanged during the cell cycle. Generally, MCM/P1 proteins are stable proteins: for example, *Hs*Mcm3(P1) showed no decrease in protein level after cells were treated with cycloheximide for 5 hours (Schulte *et al*, 1995), and the *Hs*Mcm7 (Cdc47) level was virtually unchanged after 48 hours of serum starvation (Fujita *et al*, 1996a), disappearing only after 96 hours (Fujita *et al*, 1996b). Similarly, the level of mRNA and protein for *Hs*Mcm7 (p85 Mcm) decreased during the course of differentiation in promyelocytic leukaemia cells (Schulte *et al*, 1996). MCM/P1 proteins are predominantly expressed in rapidly dividing cells such as proliferating but not quiescent *Arabidopsis thaliana* cells (Springer *et al*, 1995), oocytes of the newt *Notophthalmus viridescens* (Bucci *et al*, 1993) and cells of developing frog limbs (Buckbinder and Brown, 1992). These data suggest that MCM/P1 proteins have a major role during cell proliferation but not quiescence. The preferential expression of MCM/P1 proteins in juvenile mouse testis (Starborg *et al*, 1995) also suggests a role of MCM/P1 proteins in meiotic S phase.

SUBCELLULAR LOCALIZATION OF MCM/P1 PROTEINS

Attention had originally been drawn to MCM/P1 proteins as possible candidates for licensing factor by the observation that in the yeast *S cerevisiae*, they show a cyclical translocation between nucleus and cytoplasm. *Sc*Mcm2, *Sc*Mcm3, *Sc*Mcm5 (Cdc46) and *Sc*Mcm7 (Cdc47) proteins are nuclear during late mitosis and G_1 but then disappear from the nucleus during early S phase and can only be detected again in the nucleus in late mitosis (Hennessy *et al*, 1990; Yan *et al*, 1993; Dalton and Whitbread, 1995). This cell cycle dependent shuttling between nucleus and cytoplasm does not occur in higher eukaryotes, where the proteins are constitutively nuclear. However, MCM/P1 proteins in metazoans change their subnuclear localization throughout the cell cycle, as shown by immunofluorescence studies (Fig. 3). During nuclear extraction, a fraction of the protein population is nucleosolic and easily extractable in the presence of non-ionic detergents, whereas another part is bound to chromatin in a detergent resistant form. The detergent resistant form is observed only in cells during G_1 and S phase, with protein becoming soluble as DNA replication progresses (Kimura *et al*, 1994; Todorov *et al*, 1995; Fujita *et al*, 1996a) (Fig. 3). In synchronized mouse cells, the mouse Mcm3 protein accumulated in the heterochromatic regions of the chromosomes before the formation of replication foci (Starborg *et al*, 1995), suggesting that it is required for a preinitiation step. The granular pattern of MCM/P1 localization does not coincide with the sites of DNA synthesis, even appearing to be excluded from the sites of ongoing DNA synthesis and already replicated DNA, so that the sites of bound Mcm proteins are the ones with unreplicated DNA (Fig. 3, see arrows) (Krude *et al*, 1996). This suggests that MCM/P1 proteins associate with unreplicated chromatin in G_1 and are displaced as a result of the passage of the replication fork. These results closely parallel the cell cycle dependent association of RLF-M polypeptides with chromatin seen in *Xenopus*.

MCM/P1 PHOSPHORYLATION AND COMPLEX FORMATION

The behaviour of the Mcm proteins in binding to chromatin correlates with their phosphorylation state. *Hs*Mcm2 (BM28) (Todorov *et al*, 1995), *Hs*Mcm3 (P1) (Schulte *et al*, 1995), *Hs*Mcm4 (*Hs*Cdc21) (Musahl *et al*, 1995) and *Xl*Mcm4 (*Xl*Cdc21) (Coue *et al*, 1996) are underphosphorylated when bound to chromatin and become hyperphosphorylated when displaced. *Xl*Mcm4 (*Xl*Cdc21) exists in three different phosphorylation states, which can be clearly distinguished by their electrophoretic mobility: the mitotic form is hyperphosphorylated and does not interact with chromatin, whereas the soluble

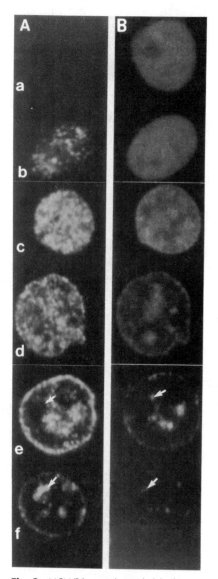

Fig. 3. MCM/P1 proteins only bind to unreplicated DNA. Nuclei at different times during S phase were analysed for DNA replication and MCM/P1 protein binding. (a) G_1/S; (b) very early S; (c) early S; (d) early-mid S; (e) mid S; (f) late S. Sites of DNA replication were visualized by bromodeoxyuridine incorporation (A), and sites of Mcm binding were immunostained with anti-HsMcm4 (Cdc21) antibodies (B). Arrows indicate spots with unreplicated DNA containing HsMcm4 bound to them. (Reproduced with modifications from Krude *et al*, 1996)

form is underphosphorylated and becomes converted to an intermediately migrating form upon binding to chromatin (Coue *et al*, 1996). Cyclin B/Cdc2 (MPF) has been suggested as the kinase that modifies *Xl*Mcm4 during mitosis, since it is present at high levels at this time, and the hyperphosphorylation can be stabilized by a stable cyclin B mutant. The phosphatase that catalyzes dephosphorylation of *Xl*Mcm4 may be phosphatase 1A or 2A (Coue *et al*, 1996).

During purification of licensing activity from *Xenopus* extracts, the proteins co-elute during several chromatographic steps and seem to form a high molecular weight complex as judged from gel filtration and glycerol gradient sedimentation (Chong *et al*, 1995). This is consistent with the observation that *Xl*Mcm2, 4, 5 and 7 can be co-immunoprecipitated with anti-*Xl*Mcm3 antibodies (Kubota *et al*, 1995; Madine *et al*, 1995a,b). A high molecular weight complex of MCM/P1 proteins has also been observed in *Drosophila* (Su *et al*, 1996). Although the individual members of the complex have been studied in detail from various organisms, little is known about the composition or tertiary structure of the complex. The most detailed studies have been performed on the human high molecular weight complex (Burkhart *et al*, 1995; Musahl *et al*, 1995; Schulte *et al*, 1995, 1996): under low salt conditions, the MCM/P1 complex sediments as a single peak of 10S. By treatment with high salt, this complex can be split into two subcomplexes, one containing Mcm 3 and 5 and the other containing the other MCM/P1 proteins.

REGULATION OF THE REPLICATION LICENSING SYSTEM

Several important questions need to be answered before we will have a comprehensive understanding of the way the replication licensing system functions. One such question concerns the role of the nuclear envelope in regulating the replication licensing system. At least in *Xenopus* (Blow and Laskey, 1988; Blow, 1993; Kubota and Takisawa, 1993; Chong *et al*, 1995) and probably in mammalian cells as well (Leno *et al*, 1992; Coverley *et al*, 1993), MCM/P1 proteins will not rebind to the replicated chromatin in G_2 without permeabilization of the nuclear envelope. However, with the exception of budding yeast, MCM/P1 proteins appear to be constitutively nuclear throughout the cell cycle. Furthermore, the MCM/P1 proteins appear to be capable of crossing the nuclear envelope, since nuclei assembled in an Mcm3 depleted *Xenopus* extract became competent to replicate following transfer to undepleted extract (Madine *et al*, 1995b).

In principle, there are two ways of explaining the apparent discrepancy that MCM/P1 proteins are part of licensing factor but are still able to cross the nuclear envelope. One possible explanation depends on the observation in *Xenopus* that in addition to the RLF-M complex of MCM/P1 proteins, another activity, RLF-B, is required for RLF-M to bind to chromatin (Chong *et al*, 1995). RLF-B, which does not contain MCM/P1 polypeptides, is in the process of fractionation (Chong JPJ, unpublished). If RLF-B were incapable of crossing the nuclear envelope in an active form, as depicted in Fig. 4, the inconsistency would be resolved. Indeed, an activity required for the assembly of MCM/P1 proteins onto chromatin ("loading factor") cannot cross the intact nuclear envelope in *Xenopus* (Madine *et al*, 1995b). RLF-B activity is also likely to show cell cycle periodicity, being active only in late mitosis and early interphase (Fig. 4) (Blow, 1993; Chong *et al*, 1995, 1996).

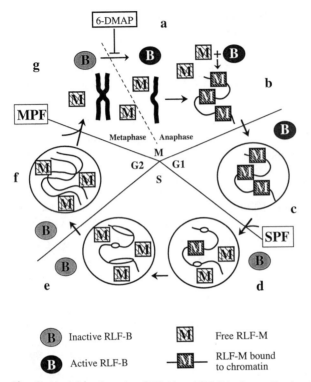

Inactive RLF-B — B
Active RLF-B — B
Free RLF-M — M
RLF-M bound to chromatin — M

Fig. 4. Model for the role of RLF-M and RLF-B in the replication licensing system. (a) On exit from metaphase, RLF-B is activated. (b) As chromosomal DNA decondenses, RLF-B allows the assembly of RLF-M onto the DNA. (c) In G_1, DNA is assembled into a nucleus, RLF-B is excluded; cytoplasmic RLF-B activity is unstable and decays. (d) S phase promoting factor triggers initiation at origins licensed by RLF-M, displacing the RLF-M. (e) Some replication origins initiate and release their RLF-M only late in S phase. (f) Fully replicated G_2 nuclei cannot rereplicate owing to the absence of RLF-B. (g) Under the action of the mitotic inducer, MPF, the cell enters mitosis, causing nuclear envelope breakdown and chromosome condensation

An alternative although not mutually exclusive explanation is that components of the replication licensing system (RLF-M or RLF-B) become inactivated within nuclei after the start of S phase. Nuclear envelope permeabilization would then serve to remove an intranuclear inhibitor of the licensing system. This inhibition might conceivably be mediated by the hyperphosphorylation of MCM/P1 proteins as they are displaced from chromatin.

As discussed above, RLF activity appears to be regulated in *Xenopus* by cyclin dependent kinases (Vesely *et al*, 1994). In addition, experiments in yeast have shown that cyclin dependent kinase activity is required in G_2 to prevent re-establishment of the prereplicative complex (Dahmann *et al*, 1995) and rereplication of DNA (Broek *et al*, 1991; Hayles *et al*, 1994; Moreno and Nurse, 1994). It therefore seems likely that the full explanation of these results will involve the cell cycle regulation of licensing factor components by cyclin dependent kinases.

The physical association between human MCM/P1 proteins and chromatin in vivo has been demonstrated by crosslinking the proteins onto DNA and separating protein-DNA complexes on caesium chloride gradients (Burkhart *et al*, 1995). In the licensing reaction in the *Xenopus* system, RLF-M and RLF-B are incubated with a chromatin template (Chong *et al*, 1995), but it is not clear onto what chromatin component the RLF-M complex is assembled. However, the MCM/P1 proteins on their own have a rather low affinity for either single or double stranded DNA (Burkhart *et al*, 1995). One simple explanation for this result is the requirement for an activity such as RLF-B, which is involved in the loading of MCM/P1 proteins onto chromatin. An additional explanation is the need for other proteins present on chromatin, such as the origin recognition complex (Bell and Stillman, 1992), onto which the MCM/P1 must be loaded (Rowles A, unpublished). Further fractionation of the *Xenopus* system should distinguish these possibilities.

SUMMARY

The *Xenopus* cell free system has proved a good model system to study in vitro DNA replication and the mechanism preventing rereplication in a single cell cycle. Studies using this system resulted in the development of a model postulating the existence of a replication licensing factor (RLF), which binds to the chromatin before the G_1-S transition of the cell cycle and is displaced during replication. The nuclear envelope prevents rebinding of RLF and hence relicensing. Nuclear envelope breakdown at mitosis is required to allow another round of replication. Protein kinase inhibitors block licensing factor activity and arrest *Xenopus* extracts in a G_2 like state. These kinase inhibitors have allowed the development of an in vitro assay leading to the biochemical purification of RLF components. RLF can be separated into RLF-B and RLF-M, the latter consisting of several members of the MCM/P1 class of replication proteins. In *Xenopus* as well as in many other eukaryotes, the binding of MCM/P1 proteins to chromatin before S phase is essential for replication to occur. The proteins are then displaced as replication proceeds. These changes in subnuclear distribution are reflected by changes in the phosphorylation status. MCM/P1 proteins do not bind to the DNA on their own but need RLF-B to be loaded onto the chromatin. Their cycling behaviour is reminiscent of the existence of a prereplicative complex at the origins of replication in yeast, suggesting that the licensing mechanism is ubiquitous in eukaryotes.

References

Bell SP and Stillman B (1992) ATP-dependent recognition of eukaryotic origins of DNA replication by a multiprotein complex. *Nature* **357** 128–134

Blow JJ (1993) Preventing re-replication of DNA in a single cell cycle: evidence for a replication licensing factor. *Journal of Cell Biology* **122** 993–1002

Blow JJ (1995) S phase and its regulation. In: Hutchison CJ and Glover DM (eds). *Cell Cycle*

Control, pp 177–205, Oxford University Press, Oxford

Blow JJ and Laskey RA (1986) Initiation of DNA replication in nuclei and purified DNA by a cell-free extract of *Xenopus* eggs. *Cell* **47** 577–587

Blow JJ and Watson JV (1987) Nuclei act as independent and integrated units of replication in a *Xenopus* cell-free DNA replication system. *EMBO Journal* **6** 1997–2002

Blow JJ and Laskey RA (1988) A role for the nuclear envelope in controlling DNA replication within the cell cycle. *Nature* **332** 546–548

Blow JJ and Nurse P (1990) A cdc2-like protein is involved in the initiation of DNA replication in *Xenopus* egg extracts. *Cell* **62** 855–862

Blow JJ and Sleeman AM (1990) Replication of purified DNA in *Xenopus* egg extract is dependent on nuclear assembly. *Journal of Cell Science* **95** 383–391

Broek D, Bartlett R, Crawford K and Nurse P (1991) Involvement of p34cdc2 in establishing the dependency of S phase on mitosis. *Nature* **349** 388–393

Bucci S, Ragghianti M, Nardi I, Bellini M, Mancino G and Lacroix JC (1993) Identification of an amphibian oocyte nuclear protein as a candidate for a role in embryonic DNA replication. *International Journal of Developmental Biology* **37** 509–517

Buckbinder L and Brown DD (1992) Thyroid hormone-induced gene expression changes in the developing frog limb. *Journal of Biological Chemistry* **267** 25786–25791

Burkhart R, Schulte D, Hu D, Musahl C, Gohring F and Knippers R (1995) Interactions of human nuclear proteins P1Mcm3 and P1Cdc46. *European Journal of Biochemistry* **228** 431–438

Chevalier S, Tassan JP, Cox R, Philippe M and Ford C (1995) Both cdc2 and cdk2 promote S phase initiation in *Xenopus* egg extracts. *Journal of Cell Science* **108** 1831–1841

Chong JPJ, Mahbubani HM, Khoo CY and Blow JJ (1995) Purification of an MCM-containing complex as a component of the DNA replication licensing system. *Nature* **375** 418–421

Chong JPJ, Thömmes P and Blow JJ (1996) The role of Mcm/P1 proteins in the licensing of DNA-replication. *Trends in Biochemical Sciences* **21** 102–106

Cocker JH, Piatti S, Santocanale C, Nasmyth K and Diffley JF (1996) An essential role for the Cdc6 protein in forming the pre-replicative complexes of budding yeast. *Nature* **379** 180–182

Coue M, Kearsey SE and Mechali M (1996) Chromatin binding, nuclear-localization and phosphorylation of *Xenopus* cdc21 are cell-cycle dependent and associated with the control of initiation of DNA-replication. *EMBO Journal* **15** 1085–1097

Coverley D, Downes CS, Romanowski P and Laskey RA (1993) Reversible effects of nuclear membrane permeabilization on DNA replication: evidence for a positive licensing factor. *Journal of Cell Biology* **122** 985–992

Crevel G and Cotterill S (1991) DNA replication in cell-free extracts from *Drosophila melanogaster*. *EMBO Journal* **10** 4361–4369

Dahmann C, Diffley JF and Nasmyth KA (1995) S-phase-promoting cyclin-dependent kinases prevent re-replication by inhibiting the transition of replication origins to a pre-replicative state. *Current Biology* **5** 1257–1269

Dalton S and Whitbread L (1995) Cell cycle-regulated nuclear import and export of Cdc47, a protein essential for initiation of DNA replication in budding yeast. *Proceedings of the National Academy of Sciences of the USA* **92** 2514–2518

De Roeper A, Smith JA, Watt RA and Barry JM (1977) Chromatin dispersal and DNA synthesis in G1 and G2 HeLa cell nuclei injected into *Xenopus* eggs. *Nature* **265** 469–470

Diffley JF and Cocker JH (1992) Protein–DNA interactions at a yeast replication origin. *Nature* **357** 169–172

Diffley JF, Cocker JH, Dowell SJ and Rowley A (1994) Two steps in the assembly of complexes at yeast replication origins in vivo. *Cell* **78** 303–316

Fang F and Newport JW (1991) Evidence that the G1-S and G2-M transitions are controlled by different cdc2 proteins in higher eukaryotes. *Cell* **66** 731–742

Fujita M, Kiyono T, Hayashi Y and Ishibashi M (1996a) Hcdc47, a human member of the Mcm family—dissociation of the nucleus-bound form during S-phase. *Journal of Biological Chemistry* **271** 4349–4354

Fujita M, Kiyono T, Hayashi Y and Ishibashi M (1996b) Inhibition of S-phase entry of human fibroblasts by an antisense oligomer against Hcdc47. *Biochemical and Biophysical Research Communications* **219** 604–607

Hayles J, Fisher D, Woollard A and Nurse P (1994) Temporal order of S phase and mitosis in fission yeast is determined by the state of the p34cdc2-mitotic B cyclin complex. *Cell* **78** 813–822

Hennessy KM, Clark CD and Botstein D (1990) Subcellular localization of yeast CDC46 varies with the cell cycle. *Genes and Development* **4** 2252–2263

Hennessy KM, Lee A, Chen E and Botstein D (1991) A group of interacting yeast DNA replication genes. *Genes and Development* **5** 958–969

Higgins DG, Bleasby AJ and Fuchs R (1991) CLUSTAL V: improved software for multiple sequence alignment. *CABIOS* **8** 189–191

Jackson PK, Chevalier S, Philippe M and Kirschner MW (1995) Early events in DNA replication require cyclin E and are blocked by p21[CIP1]. *Journal of Cell Biology* **130** 755–769

Kimura H, Nozaki N and Sugimoto K (1994) DNA polymerase alpha associated protein P1, a murine homolog of yeast MCM3, changes its intranuclear distribution during the DNA synthetic period. *EMBO Journal* **13** 4311–4320

Kimura H, Takizawa N, Nozaki N and Sugimoto K (1995) Molecular cloning of cDNA encoding mouse Cdc21 and CDC46 and characterization of the products: physical interaction between P1(MCM3) and CDC46 proteins. *Nucleic Acids Research* **12** 2097–2104

Krude T, Musahl C, Laskey RA and Knippers R (1996) Human replication proteins Hcdc21, Hcdc46 and P1mcm3 bind chromatin uniformly before S-phase and are displaced locally during DNA-replication. *Journal Of Cell Science* **109** 309–318

Kubota Y and Takisawa H (1993) Determination of initiation of DNA replication before and after nuclear formation in *Xenopus* egg cell free extracts. *Journal of Cell Biology* **123** 1321–1331

Kubota Y, Mimura S, Nishimoto S, Takisawa H and Nojima H (1995) Identification of the yeast MCM3-related protein as a component of *Xenopus* DNA replication licensing factor. *Cell* **81** 601–609

Leno GH and Laskey RA (1991) The nuclear membrane determines the timing of DNA replication in *Xenopus* egg extracts. *Journal of Cell Biology* **112** 557–566

Leno GH and Munshi R (1994) Initiation of DNA replication in nuclei from quiescent cells requires permeabilization of the nuclear membrane. *Journal of Cell Biology* **127** 5–14

Leno GH, Downes CS and Laskey RA (1992) The nuclear membrane prevents replication of human G2 nuclei but not G1 nuclei in *Xenopus* egg extract. *Cell* **69** 151–158

Lohka MJ and Masui Y (1983) Formation in vitro of sperm pronuclei and mitotic chromosomes induced by amphibian ooplasmic components. *Science* **220** 719–721

Madine MA, Khoo C-Y, Mills AD and Laskey RA (1995a) MCM3 complex required for cell cycle regulation of DNA replication in vertebrate cells. *Nature* **375** 421–424

Madine MA, Khoo CY, Mills AD, Mushal C and Laskey RA (1995b) The nuclear envelope prevents reinitiation of replication by regulating the binding of MCM3 to chromatin in *Xenopus* egg extracts. *Current Biology* **5** 1270–1279

Maine GT, Sinha P and Tye BK (1984) Mutants of *S. cerevisiae* defective in the maintenance of minichromosomes. *Genetics* **106** 365–385

Moir D, Stewart SE, Osmond BC and Botstein D (1982) Cold-sensitive cell-division-cycle mutants of yeast: isolation, properties, and pseudoreversion studies. *Genetics* **100** 547–563

Moreno S and Nurse P (1994) Regulation of progression through the G1 phase of the cell cycle by the rum1[+] gene. *Nature* **367** 236–242

Musahl C, Schulte D, Burkhart R and Knippers R (1995) A human homologue of the yeast replication protein Cdc21: interactions with other Mcm proteins. *European Journal of Biochemistry* **230** 1096–1101

Newport J (1987) Nuclear reconstitution in vitro: stages of assembly around protein-free DNA. *Cell* **48** 205–217

Rao PN and Johnson RT (1970) Mammalian cell fusion: studies on the regulation of DNA synthesis and mitosis. *Nature* **225** 159–164

Schulte D, Burkhart R, Musahl C *et al* (1995) Expression, phosphorylation and nuclear localization of the human P1 protein, a homologue of the yeast MCM3 replication protein. *Journal of Cell Science* **108** 1381–1389

Schulte D, Richter A, Burkhart R, Musahl C and Knippers R (1996) Properties of the human nuclear-protein P85mcm—expression, nuclear-localization and interaction with other Mcm proteins. *European Journal Of Biochemistry* **235** 144–151

Sheehan MA, Mills AD, Sleeman AM, Laskey RA and Blow JJ (1988) Steps in the assembly of replication-competent nuclei in a cell-free system from *Xenopus* eggs. *Journal of Cell Biology* **106** 1–12

Springer PS, McCombie WR, Sundaresan V and Martienssen RA (1995) Gene trap tagging of PROLIFERA, an essential MCM2-3-5-like gene in *Arabidopsis*. *Science* **268** 877–880

Starborg M, Brundell E, Gell K *et al* (1995) A murine replication protein accumulates temporarily in the heterochromatic regions of nuclei prior to initiation of DNA replication. *Journal of Cell Science* **108** 927–934

Strausfeld UP, Howell M, Rempel R, Maller JL, Hunt T and Blow JJ (1994) Cip1 blocks the initiation of DNA replication in *Xenopus* extracts by inhibition of cyclin-dependent kinases. *Current Biology* **4** 876–883

Strausfeld UP, Howell M, Descombes P *et al* (1996) Both cyclin A and cyclin E have S phase promoting (SPF) activity in *Xenopus* egg extracts. *Journal of Cell Science* **109** 1555–1563

Su TT, Feger G and O'Farrell PH (1996) *Drosophila* Mcm protein complexes. *Molecular Biology of the Cell* **7** 319–329

Thömmes P, Fett R, Schray B *et al* (1992) Properties of the nuclear P1 protein, a mammalian homologue of the yeast Mcm3 replication protein. *Nucleic Acids Research* **20** 1069–1074

Todorov IT, Attaran A and Kearsey SE (1995) BM28, a human member of the MCM2-3-5 family, is displaced from chromatin during DNA replication. *Journal of Cell Biology* **129** 1433–1445

Vesely J, Havlicek L, Strnad M *et al* (1994) Inhibition of cyclin-dependent kinases by purine analogues. *European Journal of Biochemistry* **224** 771–786

Yan H, Merchant AM and Tye BK (1993) Cell cycle-regulated nuclear localization of MCM2 and MCM3, which are required for the initiation of DNA synthesis at chromosomal replication origins in yeast. *Genes and Development* **7** 2149–2160

The authors are responsible for the accuracy of the references.

Cyclin Dependent Kinase Inhibitors

J WADE HARPER

Department of Biochemistry, Baylor College of Medicine, Houston, Texas 77030

Introduction
Structure and function of CKIs
    The CIP/KIP family
    The INK4 family
p21 and the multiple molecule hypothesis
Regulation of CKI abundance through antimitogenic
        pathways in vivo and in vitro
Biological roles of CKIs revealed through CKI deficient mice
    p21 and G_1 checkpoint function
    Proliferative miscues in mice lacking p27^{KIP1}
    Involvement of p16 in cancer predisposition
Questions for the future
Summary

INTRODUCTION

Progression through the eukaryotic cell cycle requires the action of positive regulatory elements that catalyze passage through particular transitions. Cyclin dependent kinases (CDKs) function in this capacity (Hunter and Pines, 1994; Sherr, 1994). These enzymes are among the most highly regulated enzymes known, reflecting their central importance in cell proliferation. Positive regulation is achieved by both cyclin association and threonine phosphorylation catalyzed by CDK activating kinase (CAK) (reviewed in Morgan, 1995). Negative regulation is achieved through phosphorylation of the catalytic subunit by Wee1/Myt1 like kinases (reviewed in Coleman and Dunphy, 1994) and through association with CDK inhibitory proteins (CKIs; reviewed in Elledge and Harper, 1994; Sherr and Roberts, 1995; Harper and Elledge, 1996).

The discovery of CKIs has provided new paradigms for understanding how extracellular and intracellular signals regulate cell cycle progression. In addition, the finding that some CKIs are tumour suppressors or are regulated by tumour suppressors has provided a direct link between cell cycle control and tumorigenesis. This chapter focuses on CKI function and regulation, with special emphasis on the roles of CKIs in development and cancer as revealed through the analysis of CKI deficient mice.

Cancer Surveys Volume 29: *Checkpoint Controls and Cancer*
© 1997 Imperial Cancer Research Fund. 0-87969-518-8/97. $5.00 + .00

STRUCTURE AND FUNCTION OF CKIs

The CIP/KIP Family

Two classes of structurally distinct CKIs have been identified in mammals: the CIP/KIP class and the INK4 class. The CIP/KIP class is typified by p21$^{\text{CIP1/WAF1}}$, the first mammalian CKI to be identified (El-Deiry et al, 1993; Harper et al, 1993; Xiong et al, 1993a). This CKI and its family members p27$^{\text{KIP1}}$ (Polyak et al, 1994; Toyoshima and Hunter, 1994) and p57$^{\text{KIP2}}$ (Lee et al, 1995; Matsuoka et al, 1995) are modular in structure and are sometimes referred to as dual specificity inhibitors (Harper and Elledge, 1996). Although this class of CKI is commonly referred to as "universal" CDK inhibitors (Xiong et al, 1993a), this description is somewhat misleading. For example, p21 is a potent inhibitor of the G_1 CDKs (CDK2, CDK3, CDK4 and CDK6), with K_i values from 0.5 to 5 nM, but is a poor inhibitor of CDC2 (K_i ~400 nM) and does not associate with CDK7/cyclin H or CDK5/p35 (Harper et al, 1995). Thus, overexpression of these inhibitors leads to arrest in G_1 (Polyak et al, 1994; Toyoshima and Hunter, 1994; Harper et al, 1995; Lee et al, 1995; Matsuoka et al, 1995). Inhibition is mediated by a conserved N-terminal approximately 60 residue domain (Harper et al, 1995; Luo et al, 1995). The interaction with CDK/cyclin complexes is likely to involve extensive regions of p21, and although results differ depending on the experimental system used, most of the available data indicate that high affinity association of p21 with CDKs requires cyclin association (Zhang et al, 1994; Goubin and Ducommun, 1995; Hall et al, 1995; Harper et al, 1995; Zhu et al, 1995). It is likely that p21 contacts both the cyclin and the kinase subunits. Alanine scanning mutagenesis of clusters of charged residues has revealed several regions of sequence that when mutated reduce, but do not abolish, association with CDK2 (Goubin and Ducommun, 1995). Only one point mutation (D52A) reduced association with CDK2 dramatically (~33-fold). This residue is in a relatively conserved region of the protein but is not conserved in mouse p57. The N-terminal domain contains a short conserved sequence motif that is also found in the CDK binding domain of an otherwise unrelated negative cell cycle regulator p107, a relative of RB (Zhu et al, 1995). When p107 is in complexes with CDK2/cyclin A, p107 becomes the exclusive substrate of the kinase and blocks phosphorylation of exogenous substrates that do not form tight complexes with the kinase (eg RB). Thus, in particular situations, p107 may behave as a CDK inhibitor. The finding that induction of p21 can lead to formation of cyclin A/CDK2/p21 complexes at the expense of p107 indicates that they share common or overlapping binding sites on the kinase (Zhu et al, 1995).

In addition, each inhibitor contains a C-terminal domain that is not required for inhibition but may serve as a binding site for other proteins. In fibroblasts, p21 participates in quaternary complexes containing a CDK, a cyclin and proliferating cell nuclear antigen (PCNA) (Xiong et al, 1993b). PCNA functions as a processivity factor for DNA polymerase δ and is required for DNA replication. Association of PCNA with these complexes is mediated through an approximately 10 aminoacid stretch at the C-terminus of p21 (Goubin and Ducommun, 1995; Warbrick *et al*, 1995). Although the biological significance of this association is not well understood at present, association of p21 with PCNA blocks polymerase δ dependent DNA replication in vitro (Flores-Rozas *et al*, 1994; Waga *et al*, 1994). Contradictory reports have appeared on the effects of p21 on excision repair. Whereas Li *et al* (1994) and Shivji *et al* (1994) have found no substantial effect of p21 in excision repair under conditions where DNA replication by polymerase δ is largely inhibited, Pan *et al* (1995) reported that p21 blocks nucleotide excision repair of DNA damaged by either ultraviolet radiation or alkylating agents and that this inhibition is reversible by PCNA. Effects on PCNA in vitro do not appear strictly to require N-terminal and central domains of p21, since a PNCA binding domain peptide is sufficient to inhibit DNA replication in vitro (Warbrick *et al*, 1995). An open question concerns the role of p21-PCNA association in the regulation of DNA replication in the normal cell cycle and during checkpoint activation. Although transient transfection of plasmids expressing the PCNA interaction domain of p21 can reduce the rate of passage through S phase (Luo *et al*, 1995), it is not clear whether the p21-PCNA interaction normally plays an essential part in checkpoint control. This question could be addressed through an analysis of cells containing inactivating mutations in the CDK binding domain of p21. In contrast to p21, p27 and p57 do not associate with PCNA. However, these two inhibitors do contain homologous C-terminal domains called the QT domain (Matsuoka *et al*, 1995). This domain may mediate interactions with other proteins.

Two p21 related CKIs have been identified in *Xenopus*, p28[Kix1] or p27[Xic1], which appear to be hybrids between CIP and KIP family members (Su *et al*, 1995; Shou and Dunphy, 1996). Although the C-terminal domains contain sequences with similarity to the QT domains of KIP1 and KIP2, these proteins can associate with PCNA, albeit much more weakly than p21. The three residues that contribute the most to PCNA binding for p21 (Warbrick *et al*, 1995) are not conserved in the *Xenopus* homologues, suggesting that there is an additional motif that can selectively recognize PCNA.

The INK4 Family

The INK4 class of CKIs is composed of p16, p15, p18 and p19 (Serrano *et al*, 1993; Guan *et al*, 1994; Hannon and Beach, 1994; Jen *et al*, 1994; Chan *et al*, 1995; Hirai *et al*, 1995). Unlike the CIP/KIP family, these inhibitors are highly selective for CDK4 and CDK6. They are composed almost entirely of ankyrin motifs. In transformed cells lacking functional RB, p16 levels are increased, and in this situation, p16 is found in complexes with CDK4 and CDK6 at the expense of D type cyclins (Serrano *et al*, 1993; Xiong *et al*, 1993b; Bates *et al*, 1994; Tam *et al*, 1994; Parry *et al*, 1995). This led to the proposal that p16 can displace cyclin D from the CDK (Serrano *et al*, 1993; Guan *et al*, 1994). However, direct displacement of cyclin D by an INK4 homologue has thus far not been demonstrated (Guan *et al*, 1994; Hall *et al*, 1995; Hirai *et al*, 1995). Addition of INK4 to CDK4/cyclin D complexes leads to rapid inhibition, and studies with p19 have demonstrated that this INK4 homologue can associate with inhibited CDK4 complexes containing cyclin D (Hirai *et al*, 1995). Together, these data indicate that INK4 homologues can associate with both cyclin associated and monomeric CDK4/6 and suggest that the molecular contacts may not be identical in the two complexes. Binding of INK4 with monomeric CDK4/6 would give a heterodimeric complex that cannot associate productively with cyclin D. Any trimeric INK4/CDK4/cyclin D complexes formed could be readily converted to the INK4/CDK complex through cyclin turnover, since it is known that D type cyclins have short half lives compared with CDK4. The development of an inducible system that allows the analysis of INK4/CDK complexes immediately after induction, before secondary events such as cyclin turnover, may help clarify the initial mechanisms of inhibition by INK4 family members.

Currently, the biochemical basis of CDK4 inhibition on p16 is not well understood. However, a p16 derived synthetic peptide corresponding to aminoacids 84–103 is sufficient to bind and inhibit CDK4 in vitro and can block cell cycle progression in tissue culture cells (Fahraeus *et al*, 1996). Although this peptide has the properties of p16, its inhibitory potency is about 15 000 times lower than full length p16, which has K_i values in the nanomolar range. This suggests that other regions of the protein may be involved in contacts that enhance the affinity. Several p16 mutants that lose their ability to inhibit CDK4 and block cell cycle progression map to the region corresponding to the peptide (see Fahraeus *et al*, 1996). One of these mutants (*G101W*), found in patients with familial melanoma, displays temperature sensitive binding to CDK4, and its ability to block cell proliferation is temperature dependent (Parry and Peters, 1996). An additional melanoma mutant (*V126D*) displays a similar phenotype (Parry and Peters, 1996). It is not yet clear whether this single p16 derived peptide is sufficient both to inhibit the activity of a preformed CDK4/cyclin D complex and to block association of CDK4 with cyclin D.

p21 AND THE MULTIPLE MOLECULE HYPOTHESIS

A curious feature of p21 is that it can be found in association with both active and inactive CDK complexes (Zhang *et al*, 1994; Harper *et al*, 1995). For example, anti-p21 immune complexes from diploid fibroblasts contain histone H1 kinase activity comparable to that found in CDK2 immune complexes (Zhang *et al*, 1994). In addition, most of the active CDK2 in these cells is associated with p21 (Harper *et al*, 1995). However, if additional p21 is added to these immune complexes, the kinase activity is completely inhibited. These and other data have led to the proposal that a single molecule of p21 can associate with the cyclin/kinase complex but that multiple inhibitor molecules (most likely two) are required to inhibit CDKs. In vitro reconstitution experiments are consistent with this idea but suggest that the p21 containing complexes may have reduced specific activity at least towards some substrates (Harper *et al*, 1995). The ability of CKI bound kinase complexes to phosphorylate substrates is potentially important, since it might allow the CKI to function in substrate targeting, perhaps through their C-terminal domains. In addition, this model is now frequently used to explain how small changes in p21 levels can dramatically alter CDK activity and cell cycle progression.

Although it is clear that p21 containing CDK complexes can phosphorylate histone H1 in vitro, it is not clear at present whether such complexes are active on their physiological substrates in vivo and if so what their specific activities are. One potential model is that p21 interacts with multiple domains of the cyclin/CDK complex, and in vitro, the CDK binding domain can "breathe" sufficiently to allow for substrate entry and catalysis, perhaps at reduced rates. To date, only histone H1 (Zhang *et al*, 1994; Harper *et al*, 1995) and a histone H1 derived peptide (Harper *et al*, 1995) have been used as substrate in these experiments, and it is conceivable that their molecular properties allow them to be phosphorylated by CDKs containing a single p21 molecule. The extent of phosphorylation may depend on the relative binding constants for histone H1 and the p21/CDK inhibitory interaction. It is conceivable that the affinity of critical cellular targets is not sufficient to compete for p21 in vivo. It is also possible that proteins such as p107 that bind directly to cyclin through a region that overlaps with the p21 binding site on cyclin (Zhu *et al*, 1995) would not be efficiently phosphorylated by CDK/cyclin complexes containing a single p21 molecule. Therefore, the ability of p21–kinase complexes to be functional is likely to be substrate dependent. In addition, it is not clear whether all CDK/cyclin complexes that associate with these inhibitors behave in the same way, since this phenomenon has been tested only with CDK2/cyclin A. Furthermore, it is not clear to what extent p27 and p57 display this property. Whereas p57 immune complexes from HeLa cells contain histone H1 kinase activity (Harper JW and Elledge SJ, unpublished), activity in p27 immune complexes has not been reported. An analysis of the activities of multiple CDK/cyclin complexes with multiple target substrates will need to be performed if we are to understand the in vivo importance of this idiosyncrasy of p21 family members. A structure of the CDK/cyclin/CKI complex will no doubt prove to be critical in understanding the complex mechanism of inhibition displayed by this class of CKIs and may suggest ways to test the relevance of this feature in vivo.

REGULATION OF CKI ABUNDANCE THROUGH ANTIMITOGENIC PATHWAYS IN VIVO AND IN VITRO

When p21 was first identified, it was envisioned that CDK inhibitors would be involved in a variety of regulatory decisions that reflect their versatility. Such decisions include checkpoint function, cell cycle timing and terminal cell cycle arrest during development. Currently, there is evidence for the involvement of one or more of the known CKIs in these processes, although it is not at all clear in many instances how the temporal and spatial expression of CKIs is controlled. One theme that is emerging is that CKI expression is highly cell type specific in vivo. Analysis of p21 and p57 expression during embryonic development and in adult tissues indicates that these two inhibitors are in general expressed in a non-overlapping pattern (El-Deiry *et al*, 1995; Matsuoka *et al*, 1995; Parker *et al*, 1995). The tissue of exception is muscle, where both of these CKIs are expressed (Matsuoka *et al*, 1995; Parker *et al*, 1995). The highest levels of expression are found in terminally differentiated cells (El-Deiry *et al*, 1995; Parker *et al*, 1995), suggesting that CKIs contribute to cell cycle arrest during development.

CKIs, in particular p21, are induced in several cell types undergoing differentiation in culture (Jiang *et al*, 1994; Steinman *et al*, 1994; Halevy *et al*, 1995; Parker *et al*, 1995; Liu *et al*, 1996). However, it is unclear whether CKI expression causes terminal differentiation or is a consequence of it. For example, p21 is induced in response to expression of MyoD (Halevy *et al*, 1995; Parker *et al*, 1995), a transcription factor that is sufficient to induce the muscle differentiation pathway and is known to cause cell cycle arrest. However, p21 is not required for cell cycle arrest or differentiation in this system, since mice lacking p21 do not have an obvious defect in muscle development (Deng *et al*, 1995). It is possible that p21 function is redundant with another cell cycle regulator in muscle. There is also evidence in other systems that CKIs can actually promote particular differentiation events. p21 is induced directly by the vitamin D receptor (Liu *et al*, 1996). Overexpression of p21 (or p27) in the absence of vitamin D in these cells leads to both cell cycle arrest and expression of differentiation markers (Liu *et al*, 1996). It is currently unclear whether induction of differentiation markers is simply dependent on cell cycle arrest in G_1 in this particular system.

Although little is known about the patterns of expression and regulation of other CKIs in vivo, tissue culture experiments, particularly with p27, are beginning to reveal other pathways that regulate CKI abundance. One theme to emerge from these studies is that p27^{KIP1} is frequently induced upon cell cycle exit in response to mitogen deprivation, antimitogenic signals or contact inhibition, and its levels are reduced when cells are stimulated to enter the cycle (Nourse *et al*, 1994; Poon *et al*, 1995; Coats *et al*, 1996; Hengst and Reed, 1996; Winston *et al*, 1996). Increased levels of p27^{KIP1} are thought to constitute a barrier to CDK activation that must be overcome during mitogen induced cell cycle progression. One mechanism by which this is achieved is the downregulation of p27^{KIP1} protein levels after growth factor stimulation. Furthermore, antisense inhibition of p27^{KIP1} expression in cycling cells prevents cell cycle withdrawal in response to serum deprivation and allows mitogen independent initiation of DNA replication (Coats *et al*, 1996). Thus, a major function of extracellular growth promoting agents is the elimination of functional p27^{KIP1}. In accord with this hypothesis, antiproliferative factors such as rapamycin antagonize the growth factor dependent decrease in p27^{KIP1}, thereby inhibiting S phase entry (Nourse *et al*, 1994). It should be noted, however, that large alterations in p27^{KIP1} abundance are not universally observed. In cycling Swiss 3T3 cells, p27 is present at low levels relative to its targets and approximately doubles in quiescent cells, still far too low to account fully for CDK inhibition (Poon *et al*, 1995). It is not clear whether these cells have lost the ability to induce p27^{KIP1} dramatically during quiescence, as observed with human diploid fibroblasts (Hengst and Reed, 1996), or whether another mechanism such as reduction in cyclin expression is the critical event in cell cycle arrest.

For some cell types, signals leading to reduced p27^{KIP1} levels can de distinguished from signals leading to increased cyclin/CDK accumulation. In both Balb/c 3T3 fibroblasts and human T lymphocytes, transition from quiescence into S phase is dependent on the sequential and synergistic action of at least two distinct mitogenic activities. CDK complexes assemble in response to the initial signal that promotes cell cycle entry but is not sufficient for passage through the restriction point. In the absence of the second signal, these complexes are inactive and are associated with p27^{KIP1} (see Firpo *et al*, 1994; Winston *et al*, 1996). Exposure to a full complement of mitogenic factors results in the reduction of p27 protein levels, a decrease in the level of Kip1 associated with the CDK holoenzyme and subsequent activation of the kinase.

What controls the levels of p27^{KIP1}? Unlike p21^{CIP1}, transcriptional induction does not appear to be a predominant mechanism for regulating p27^{KIP1} levels. *KIP1* mRNA levels are fairly constant during both cell cycle progression and cell cycle exit, indicating a posttranscriptional control of KIP1 accumulation (Hengst and Reed, 1996). Two distinct mechanisms are at play. Firstly, growth arrest leads to increased translation of KIP1 message (Hengst and Reed, 1996). Secondly, p27 half life is increased, compared with that in asynchronously proliferating fibroblasts (Pagano *et al*, 1995). In some cell types, p27^{KIP1} levels fluctuate during the cell cycle, being high in G$_1$/S and low during G$_2$/M (Hengst and Reed, 1996). Cell cycle dependent alterations in half life may contribute to the periodic decrease in KIP1 protein levels. Clues as to how p27^{KIP1} destruction is regulated have come from the finding that p27^{KIP1} is ubiquitinated in a reaction that can be performed in vitro by the Ubc2 and Ubc3 ubiquitin conjugating enzymes, called E2s (Pagano *et al*, 1995). This modification targets the protein to the proteosome for destruction. Targeted destruction of CKIs may represent a general mechanism for reducing inhibitor levels in response to mitotic signals. This is suggested by the fact that the budding yeast Cdc28p/Clbp inhibitor p40^{SIC1} is also regulated by ubiquitin mediated degradation (Schwob *et al*, 1994). In this system, mutational inactivation of Cdc4p, which is thought to function together with the ubiquitin conjugating enzyme E2 (Cdc34p), results in the accumulation of Sic1p and an inability to enter S phase. Deletion of *SIC1* suppresses the S phase arrest phenotype of the *cdc4-1*, suggesting that Sic1p is an essential target of this pathway. The identification of E3 proteins that recognize and target CKIs for destruction will help to unravel how cell cycle phase or mitogens regulate CKI destruction.

Certain cell types respond to the action of transforming growth factor beta (TGFβ) by arresting the cell cycle in G_1, but this arrest can be overcome by the action of the DNA virus oncoprotein E1A. Recent studies have provided a molecular description of the changes in CKI function that occur in response to TGFβ (Reynisdottir *et al*, 1995) as well as the mechanism by which E1A overrides this antimotitic signal (Mal *et al*, 1996). In cycling mink lung epithelial cells, $p27^{KIP1}$ levels are insufficient to completely inactivate CDK2 and CDK4 complexes required for cell cycle progression. In response to TGFβ, $p15^{INK4b}$ is transcriptionally induced (Hannon and Beach, 1994). Through either direct displacement or complex turnover, $p15^{INK4b}$ accumulates on CDK4 at the expense of $p27^{KIP1}$ and cyclin D. $p27^{KIP1}$, which is liberated from the CDK4 complexes, is then available for inhibition of CDK2/cyclin E complexes (Reynisdottir *et al*, 1995). The ability of E1A to block the action of TGFβ results from two distinct activities (Mal *et al*, 1996). Firstly, association of E1A with RB abolishes the requirement of active cyclin D/CDK4 complexes for progression through G_1, thus rendering $p15^{INK4b}$ function irrelevant (reviewed in Weinberg, 1995). Secondly, E1A associates with p27, thereby blocking its ability to inhibit cyclin E/CDK2 activity, which is required for S phase entry even in the absence of RB (Mal *et al*, 1996). Physical inactivation of CKIs may ultimately emerge as a common mechanism regulating cell cycle progression in response to mitogenic agents. Indeed, recent data suggest the existence of a heat labile inhibitor of p21 that is activated through the MYC pathway (Hermeking *et al*, 1995).

BIOLOGICAL ROLES OF CKIs REVEALED THROUGH CKI DEFICIENT MICE

Although tissue culture systems facilitate the identification of signalling pathways that regulate CKI expression, such systems do not necessarily recreate the complex biological and cellular interactions found in vivo. Thus, the true significance of a CKI to control of proliferation and development can only be realized through analysis of CKI deficient animals. In principle, this approach identifies cell types where a particular inhibitor makes a critical contribution to the balance of negative and positive growth control factors that must be maintained for homoeostasis or proper development. Furthermore, this approach will ultimately reveal the cell types that possess redundant or alternative pathways for negative growth control. The results with the three CKI knockout mice generated to date provide a wealth of information about the roles of these genes in cell cycle control, development and tumorigenesis.

p21 and G$_1$ Checkpoint Function

p21 is transcriptionally regulated by TP53 in response to DNA damage (Dulic *et al*, 1994; El-Diery *et al*, 1994), and it has been suggested in the literature that p21 may mediate some or all of the known functions of TP53 (El-Deiry *et al*, 1993, 1994). TP53 is a central factor in the DNA damage response pathway, and this activity is thought to contribute substantially to its tumour suppressor function (reviewed in Bates and Vousden, 1996). Cells lacking TP53 no longer arrest in G$_1$ in response to γ irradiation or in G$_2$ in response to activation of the mitotic spindle checkpoint. In addition, some cell types undergo apoptosis when DNA is damaged, and TP53 is required for this process as well. Analysis of p21 deficient cells indicates that p21 is required for complete G$_1$ arrest in response to DNA damage (Brugarolas *et al*, 1995; Deng *et al*, 1995) but is not required for the spindle checkpoint or for thymocytic apoptosis (Deng *et al*, 1995). One unexpected outcome of this work was the finding that the G$_1$ checkpoint with γ irradiation was only partially compromised with loss of p21 (Brugarolas *et al*, 1995; Deng *et al*, 1995), but the checkpoint in response to nucleotide pool perturbations with N-(phosphoracetyl)-L-aspartate was fully compromised (Deng *et al*, 1995). This suggests that there is a second checkpoint function that can partially compensate for p21 loss in the presence of some types of damage. In *TP53* null cells, both pathways are non-functional. Unlike *TP53* deficient mice, *p21* deficient mice do not have increased susceptibility to spontaneous tumours (Deng *et al*, 1995). This finding is in keeping with the finding that p21 is infrequently mutated in human tumours (Shiohara *et al*, 1994). However, two loss of function mutations in *p21* have been found during an analysis of prostate tumours, indicating that loss of *p21* could potentially contribute to tumorigenesis in some cell types (Gao *et al*, 1995). Together, these results indicate that the anti-apoptotic and anti-oncogenic effects of *TP53* are complex and involve much more than simply induction of p21. It is possible that *TP53's* apoptotic function or G$_2$ checkpoint function are central to its tumour suppressor function, although a contribution of the G$_1$ checkpoint function cannot be ruled out.

Proliferative Miscues in Mice Lacking p27[KIP1]

As noted above, alterations in p27 levels or utilization is a frequent consequence of changes in mitogenic signals. Mice lacking p27 display a number of phenotypes which indicate that this CKI has an important role in limiting the extent of cell proliferation but does not generally function in the process of cellular differentiation (Fero *et al*, 1996; Kiyokawa *et al*, 1996). Perhaps the most dramatic phenotype is that p27 deficient mice are substantially larger than wild type mice, indicating a general proliferative advantage in the absence of this CKI. Although all organs generally are larger than normal, organs that normally express the highest levels of p27 (spleen and thymus) are increased to the largest extent. Increased organ size reflects cell number and not cell size. Thus, cell division in these tissues is regulated largely by p27. Differentiation in these tissues appears to be normal. This has been best character-ized in the thymus, where a dramatic increase in cell number and organ size had no effect on thymic development and thymocytic differentiation. More-over, cells from all haemopoietic lineages were normally represented (Fero *et al*, 1996; Kiyokawa *et al*, 1996).

Also intriguing is the finding that *p27* deficient mice develop intermediate lobe pituitary hyperplasia or adenoma, consistent with the increased size of this organ. The absence of focal lesions surrounded by apparently normal tis-sue suggests that pituitary adenoma occurs with 100% penetrance and indi-cates that proliferation of melanotropic cells of the pars intermedia is nega-tively regulated largely through p27. However, malignant pituitary tumours have not yet been observed in these mice. This phenotype is reminiscent of $RB^{-/+}$ mice, which have pituitary tumours concomitantly with the loss of the second *RB* gene (reviewed in Williams and Jacks, 1996). However, unlike $RB^{-/+}$ mice, p27 deficient mice do not die at 9 months of age, indicating that the *RB* deficient tumours are much more aggressive. Thus, *RB* loss is not equi-valent to loss of *p27*. Although it is conceivable that pituitary pathology may contribute to altered animal size through endocrine abnormalities, alterations in the levels of growth hormone or insulin like growth factor 1 have not been observed in *p27* deficient mice (Fero *et al*, 1996; Kiyokawa *et al*, 1996).

A third phenotype observed in *p27* deficient mice is female infertility. Al-though ovarian follicles develop, they do not progress to form corpora lutea. Since p27 levels are relatively high in the corpora lutea of control mice, it is conceivable that follicle maturation by luteinizing hormone requires the action of p27 (Fero *et al*, 1996; Kiyokawa *et al*, 1996).

Involvement of p16 in Cancer Predisposition

INK4 homologues function in a pathway that appears to be a primary target for mutations that contribute to transformation. INK4 homologues specifically block the activity of CDK4/6, which are thought to regulate RB inactivation (reviewed in Sherr, 1994; Weinberg, 1995). These kinases are uniquely ac-tivated by D type cyclins. Studies carried out largely in fibroblasts indicate that RB is the sole essential substrate of D type cyclin kinases. This is based largely on the finding that INK4 homologues can block cell proliferation in $RB^{+/+}$ cells but not $RB^{-/-}$ cells (Koh *et al*, 1995; Lukas *et al*, 1995) and that anticyclin D antibodies inhibit S phase entry in $RB^{+/+}$ cells (reviewed in Sherr, 1994; Weinberg, 1995). Loss of this pathway can occur in multiple ways. Loss of p16 (reviewed in Hirama and Koeffler, 1995) or mutation of CDK4 to a form that no longer tightly associates with INK4 homologues (Wolfel *et al*, 1995; Zuo *et al*, 1996) may lead to unregulated RB phosphorylation and cell cycle entry. Alternatively, overexpression of cyclin D or CDK4 may overcome negative growth control by INK4 (reviewed in Sherr, 1994).

Given that all four INK4 homologues occupy the same position in this pathway, it is intriguing that p16[INK4a] is the only INK4 homologue strongly implicated in human cancer. INK4a and its homologue INK4b are located near each other at 9p21 (reviewed in Sherr and Roberts, 1995). This region of the genome is frequently mutated in a wide range of human cancers, and a substantial number of alleles, including the *INK4a* and *INK4b* genes, have been found to contain large deletions. However, whereas missense and non-sense mutations are frequent in *INK4a*, they are quite rare in *INK4b* (Jen *et al*, 1994). Perhaps the clearest evidence that *p16* is a true tumour suppressor comes from the finding that it is mutant in some familial melanomas (Hussus-sian *et al*, 1994; Kamb *et al*, 1994; Walker *et al*, 1995). In addition, *INK4a* mutations are frequently found in sporadic cancers of the head and neck, pancreas and oesophagus (reviewed in Hirama and Koeffler, 1995). In addition to mutations, there is also evidence that expression of p16 is blocked in some tumour types through methylation of its promoter (Herman *et al*, 1995).

A complicating feature of the *INK4a* locus is that it produces two transcripts from distinct promoters that have alternative first exons ($E1\alpha$ and $E1\beta$) but share second and third exons ($E2$ and $E3$) (Quelle *et al*, 1995). The $E1\beta$-E2-E3 transcript generates a novel protein, p19^{ARF}, from a different reading frame than that for p16 (Quelle *et al*, 1995). Although ARF does not associate with CDKs, it does block cell cycle progression in G_1 and G_2 when over-expressed and thus could formally contribute to tumour suppression at the *INK4a* locus. Although a role for p19^{ARF} is not ruled out by the mutational data, many of the data are consistent with the primary involvement of p16. Approximately 50% of all mutations affect p16 alone, with the remaining fraction affecting both p16 and p19^{ARF} (see Serrano *et al*, 1996, and references therein). In addition, mutations found in familial melanoma kindreds are frequently in $E1\alpha$. Many nonsense mutations specifically affect p16, and a number of missense mutations in p16 have been shown to affect association with CDK4/6 and cell cycle arrest. However, the fact that there are four known missense mutations in $E2$ that affect p19^{ARF} but not p16^{INK4a} leaves open the possibility that p19^{ARF} contributes to tumour suppression in some cell types (Quelle *et al*, 1995), although these alleles have not been shown to be defective in growth suppression. Because of the complexity at this locus and the paucity of data on expression of INK4a products using specific probes, we know little about the role of *INK4a* in development, although the identification of humans that are homozygous null for *INK4a* (Gruis *et al*, 1995) indicates that these genes do not have an essential role in development.

Mice have been generated that lack exons $E2$ and $E3$ and therefore delete both the p16 and p19 proteins (Serrano *et al*, 1996). As expected, these mice are viable and, consistently with data in human cancers, loss of INK4a leads to greatly enhanced tumorigenesis. Somewhat surprising is the tissue spectrum observed for spontaneous and chemically induced tumours. In contrast to humans, where melanoma, pancreatic and oesophageal tumours are frequently observed with *INK4a* mutations, mice develop primarily fibrosarcomas and lymphomas. Although *INK4a* mutations are seen in these tumour types in humans, they are relatively infrequent. This reiterates the differences in tumour type specificity frequently observed in mice and humans carrying mutations in the same genes (Harlow, 1992). An additional proliferation phenotype is observed in the spleen, where there is some proliferative expansion of the white pulp along with megakaryocytes and lymphoblasts in the red pulp (Serrano *et al*, 1996). These data suggest abnormal extramedullary haemopoiesis, and this becomes more pronounced with age.

QUESTIONS FOR THE FUTURE

Since mammalian CKIs were first identified in late 1993, there has been substantial progress on many fronts, and significant insights into mechanisms of growth control and tumour suppression have been gleaned (El-Deiry *et al*, 1993; Harper *et al*, 1993; Serrano *et al*, 1993; Xiong *et al*, 1993a). However, there are still many fundamental questions that are not understood.

One question involves the roles of INK4 proteins in development. Although some information is available on the patterns of expression of p21 and p57 during development and in adult tissues, we know little about the cell type specificity of INK4 homologues. Such information could help explain why p16, but not other INK4 proteins, functions as a tumour suppressor in many cell types. The question of functional redundancy will eventually be addressed through the generation of multiply mutant mice.

A second area where our understanding is incomplete concerns the signal transduction pathways that regulate the expression of CKIs during development. We know very little about transcriptional control mechanisms. What are the transcription factors and antimitogenic pathways that regulate induction of CKIs in terminally differentiated cells? A related question concerns the mechanisms that cause destruction of inhibitors. How is p27 destruction regulated and how is p27 recognized for destruction? Do other CKIs employ programmed destruction mechanisms? What is the role of the C-terminal domain of KIP proteins in cell cycle arrest pathways and destruction? Identification of proteins that can associate with the QT domain may help to answer these questions.

The third question concerns the function of p57^{KIP2}. Thus far, point mutations in *KIP2* have not been identified in human tumours (Kondo *et al*, 1996). However, p57 is unique among CKIs in that it is imprinted, being selectively expressed from the maternal allele in all tissues examined except the brain (Hatada and Mukai, 1995; Kondo *et al*, 1996; Matsuoka *et al*, 1996). Although the relevance of this form of regulation is not known, it is consistent with the involvement of p57 in two cancer syndromes, Beckwith-Weidemann syndrome (BWS) and Wilm's tumour II (WT2) (see Matsuoka *et al*, 1995, 1996, and references therein). BWS is characterized by organ overgrowth and cancer predisposition, WT2 by embryonic tumours, particular in the kidney. The *p57* gene is located at 11p15.5 (Matsuoka *et al*, 1995), very near lesions in these two syndromes. Both syndromes indicate that the tumour suppressor gene(s) effect is imprinted, since deletions at 11p15.5 in WT2 and balanced translocations in BWS uniquely affect the maternal allele. This, coupled with the fact that *p57* is expressed in many of the tissues most affected by these two syndromes (Matsuoka *et al*, 1995), suggests that loss of *p57* may contribute to these diseases. Generation of *p57* deficient mice should help in understanding the role of *p57* in development and tumorigenesis.

Finally, we need to solve the riddle of why there are multiple members and distinct classes of CKIs. The CIP/KIP family members have evolved to inhibit multiple G_1 CDKs, whereas the INK4 proteins inhibit only D type cyclins. Available data indicate that CDK2 and CDK3 function at steps distinct from those of CDK4 and CDK6. Why is there an inhibitor class selective for the cyclin D/RB pathway when it is clear that CIP/KIP family members are competent to arrest the cell cycle even in the absence of RB? Moreover, why does loss of *p16*, but not *p21* or *p27*, lead to widespread cancer predisposition. Does it simply reflect tissue specificity and functional redundancy or does it indicate an imbalance in kinase activity that is a reflection of the biochemical specificity of the two classes of CKIs?

SUMMARY

Progression through the eukaryotic cell cycle is regulated by the activities of a family of cyclin dependent kinases (CDKs). These kinases are negatively regulated by phosphorylation and by the action of cyclin kinase inhibitors (CKIs). In mammalian cells, two classes of CKIs have been identified, the INK4 class and the CIP/KIP class. These CKIs are versatile negative regulators of CDK function and have potential roles in development, checkpoint control and tumour suppression. Analysis of CKI knockout indicates that although these inhibitors are not generally required for survival, the phenotypes observed span the gamut of what might be expected for loss of a cell cycle inhibitor. This chapter summarizes our current understanding of the roles of CKIs in growth control.

References

Bates S and Vousden KH (1996) p53 in signaling checkpoint arrest or apoptosis. *Current Opinions in Genetics and Development* **6** 12–18

Bates S, Parry D, Bonetta K, Vousden K, Dickson C and Peters G (1994) Absence of cyclin D/cdk complexes in cells lacking functional retinoblastoma protein. *Oncogene* **9** 1633–1640

Brugarolas J, Chandrasekaran C, Gordon J *et al* (1995) Radiation-induced cell cycle arrest compromised by p21 deficiency. *Nature* **377** 552–556

Chan FKM, Zhang J, Cheng L, Shapiro D and Winoto A (1995) Identification of human and mouse p19, a novel Cdk4 and Cdk6 inhibitor with homology to p16ink4. *Molecular and Cellular Biololgy* **15** 2682–2688

Coats S, Flanagan W M, Nourse J and Roberts J M (1996) Requirement of p27Kip1 for restriction point control of the fibroblast cell cycle. *Science* **272** 877–880

Coleman TR and Dunphy WG (1994) Cdc2 regulatory factors. *Current Opinions in Cell Biology* **6** 877–882

Deng C, Zhang P, Harper J W, Elledge S J and Leder P (1995) Mice lacking p21Cip1/Waf1 undergo normal development but are defective in G_1 checkpoint control. *Cell* **82** 675–684

Dulic V, Kaufman WK, Wilson S *et al* (1994) p53-dependent inhibition of cyclin dependent kinase activitites in human fibroblasts during radiation-induced G_1 arrest. *Cell* **76** 1013–1023

El-Deiry WS, Tokino T, Velculescu VE *et al* (1993) WAF1, a potential mediator of p53 tumor suppression. *Cell* **78** 67–74

El-Deiry WS, Harper JW, O'Connor PM *et al* (1994) WAF1/CIP1 is induced in p53 mediated G_1 arrest and apoptosis. *Cancer Research* **54** 1169–1174

El-Deiry WS, Tokino T, Waldman T et al (1995) Topological control of p21WAF1/CIP1 expression in normal and neoplastic tissues. *Cancer Research* **55** 2910–2919

Elledge S J and Harper J W (1994) Cdk inhibitors on the threshold of checkpoints and development. *Current Opinions in Cell Biology* **6** 847–852

Fahraeus R, Paramio JM, Ball KL, Lain S and Lane DP (1996) Inhibition of pRb phosphorylation and cell-cycle progression by a 20-residue peptide derived from p16CDKN2/INK4A. *Current Biology* **6** 84–91

Fero ML, Rivkin M, Tasch M *et al* (1996) A syndrome of multi-organ hyperplasia with features of gigantism, tumorigenesis and female sterility in p27KIP1 deficient mice. *Cell* **85** 733–744

Firpo EJ, Koff A, Solomon M and Roberts JM (1994) Inactivation of a Cdk2 inhibitor during interleukin 2-induced proliferation of human T lymphocytes. *Molecular and Cellular Biology* **14** 4889–4901

Flores-Rozas H, Kelman Z, Dean F *et al* (1994) Cdk-interacting protein 1 directly binds with PCNA and inhibits replication catalyzed by the DNA polymerase δ holoenzyme. *Proceedings of the National Acadamy of Sciences of the USA* **91** 8655–8659

Gao X, Chen YQ, Wu N *et al* (1995) Somatic mutations of the WAF1/CIP1 gene in primary prostate cancer *Oncogene* **11** 1395–1898

Goubin F and Ducommun B (1995) Identification of binding domains on the p21Cip1 cyclin-dependent kinase inhibitor. *Oncogene* **10** 2281–2287

Gruis N, van der Velden PA, Sandkuijl A *et al* (1995) Homozygotes for CDKN2(p16) germline mutation in Dutch familial melanoma kindreds. *Nature Genetics* **10** 351–353

Guan K-L, Jenkins CW, Li Y *et al* (1994) Growth suppression by p18, a p16 and p14-related Cdk6 inhibitor, correlates with wild-type pRb function. *Genes and Development* **8** 2939–2952

Halevy O, Novitch B, Spicer D *et al* (1995) Correlation of terminal cell cycle arrest of skeletal muscle with induction of p21 by MyoD. *Science* **267** 1018–1021

Hall M, Bates S and Peters G (1995) Evidence for different modes of action of cyclin-dependent kinase inhibitors: p15 and p16 bind to kinases, p21 and p27 bind to cyclins. *Oncogene* **11** 1581–1588

Hannon G and Beach D (1994) p15 is a potential effector of TGFβ-induced cell cycle arrest. *Nature* **370** 257–261

Harlow E (1992) Retinoblastoma: for our eyes only. *Nature* **359** 270–271

Harper JW and Elledge SJ (1996) Cdk inhibitors in development and cancer. *Current Opinions in Genetics and Development* **6** 56–64

Harper JW, Adami G, Wei N *et al* (1993) The 21 Kd Cdk interacting protein Cip1 is a potent inhibitor of G_1 cyclin-dependent kinases. *Cell* **75** 805–816

Harper JW, Elledge SJ, Keyomarsi K *et al* (1995) Inhibition of cyclin dependent kinases by p21. *Molecular Biology of the Cell* **6** 387–400

Hatada I and Mukai T (1995) Genomic imprinting of p57KIP2, a cyclin-dependent kinase inhibitor, in mouse. *Nature Genetics* **11** 204–206

Hengst L and Reed SI (1996) Translational control of p27Kip1 accumulation during the cell cycle. *Science* **271** 1861–1864

Herman JG, Merlo A, Mao L *et al* (1995) Inactivation of the CDKN2/p16/MTS1 gene is frequently associated with aberrant DNA methylation in all common human cancers. *Cancer Research* **55** 4525–4530

Hermeking H, Funk JO, Reichert M, Ellwart JW and Eick D (1995) Abrogation of p53-induced cell cycle arrest by c-Myc: evidence for an inhibitor of p21WAF1/CIP1/SDI1. *Oncogene* **11** 1409–1415

Hirai H, Roussel MF, Kato JY, Ashmun RA and Sherr C (1995) Novel INK4 proteins, p19 and p18, are specific inhibitors of the cyclin dependent kinases Cdk4 and Cdk6. *Molecular and Cellular Biology* **15** 2672–2681

Hirama T and Koeffler H P (1995) Role of the cyclin-dependent kinase inhibitors in the development of cancer. *Blood* **86** 841–854

Hunter T and Pines J (1994) Cyclins and Cancer: II. *Cell* **79** 573–582

Hussussian CJ, Struewing JP, Goldstein AM *et al* (1994) Germline p16 mutations in familial melanoma. *Nature Genetics* **8** 15–21

Jen J, Harper JW, Bigner SH *et al* (1994) Deletion of p16 and p15 genes in brain tumors. *Cancer Research* **54** 6353–6358

Jiang H, Lin J, Su Z-Z *et al* (1994) Induction of differentiation in human promyelocytic HL-60 leukemia cells activates p21WAF1/CIP1 expression in the absence of p53. *Oncogene* **9** 3397–3406

Kamb A, Shattuck-Eidens D, Eeles R *et al* (1994) Analysis of the p16 gene (CDKN2) as a candidate for the chromosome 9p melanoma susceptibility locus. *Nature Genetics* **8** 22–26

Kiyokawa H, Kineman RD, Manova-Todorova KO *et al* (1996) Enhanced growth of mice lacking the cyclin-dependent kinase inhibitor function of p27Kip1. *Cell* **85** 721–732

Koh J, Enders GH, Dynlacht BD and Harlow E (1995) Tumour-derived p16 alleles encoding proteins defective in cell-cycle inhibition. *Nature* **375** 506–510

Kondo M, Matsuoka S, Uchida K *et al* (1996) Selective maternal-allele loss in human lung cancers of the maternally expressed p57KIP2 gene at 11p15.5. *Oncogene* **12** 1365–1368

Lee M-H, Reynisdottir I and Massague J (1995) Cloning of p57Kip2, a cyclin-dependent kinase inhibitor with a unique domain structure and tissue distribution. *Genes and Development* **9** 650–662

Li R, Waga S, Hannon GJ, Beach D and Stillman B (1994) Differential effects by the p21 CDK inhibitor on PCNA-dependent DNA replication and repair. *Nature* **371** 534–537

Liu M, Lee M H, Cohen M, Bommakanti M and Freedman L P (1996) Transcriptional activation of the Cdk inhibitor p21 by vitamin D3 leads to the induced differentiation of the myelomonocytic cell line U937. *Genes and Development* **10** 142–153

Lukas J, Parry D, Aagard L *et al* (1995) Retinoblastoma-protein-dependent cell-cycle inhibition by the tumour suppressor p16. *Nature* **375** 503–506

Luo Y, Hurwitz J and Massague J (1995) Cell cycle inhibition by independent CDK and PCNA binding domains in p21Cip1. *Nature* **375** 159–161

Mal A , Poon RYC, Howe PH, Toyoshime H, Hunter T and Harter ML (1996) Inactivation of p27Kip1 by the viral E1A oncoprotein in TGFβ-treated cells. *Nature* **380** 262–265

Matsuoka S, Edwards M, Bai C *et al* (1995) p57KIP2, a structurally distinct member of the p21Cip1 Cdk inhibitor family is a candidate tumor suppressor protein. *Genes and Development* **9** 650–662

Matsuoka S, Thompson JS, Edwards MC *et al* (1996) Variable imprinting of a human cyclin-dependent kinase inhibitor, p57KIP2. *Preceedings of the National Academy of Sciences of the USA* **93** 3026–3030

Morgan D (1995) Principles of Cdk regulation. *Nature* **374** 131–134

Nakayama K, Ishida N, Shrane M *et al* (1996) Mice lacking p27 (Kip1) display increased body size, multiple organ hyperplasia, retinal dysplasia, and pituitary tumors. *Cell* **85** 707–720

Nourse J, Firpo E, Flanagan WM *et al* (1994) Interleukin-2-mediated elimination of the p27Kip1 cyclin-dependent kinase inhibitor prevented by rapamycin. *Nature* **372** 570–573

Pagano M, Tan SW, Theodoras AM *et al* (1995) Role of the ubiquitin-proteosome pathway in regulating the abundance of the cyclin dependent kinase inhibitor p27. *Science* **269** 682–

685

Pan ZQ, Reardon J and Li L (1995) Inhibition of nucleotide excision repair by the cyclin-dependent kinase inhibitor p21. *Journal of Biological Chemistry* **270** 22008–22016

Parker S, Eichele G, Zhang P *et al* (1995) p53-inpedentdent expression of p21Cip1 in muscle and other terminally differentiated cells. *Science* **267** 1024–1027

Parry D and Peters G (1996) Temperature-sensitive mutants of p16CDKN2 associated with familial melanoma. *Molecular and Cellular Biology* **16** 3844–3852

Parry D, Bates S, Mann DJ and Peters G (1995) Lack of cyclin D/cdk4 complexes in Rb-negative cells correlates with high levels of p16INK4/MTS1 tumour suppressor gene product. *EMBO Journal* **14** 503–511

Polyak K, Lee MH, Erdjument-Bromage H, Tempst P and Massague J (1994) Cloning of p27^{KIP1}, a cyclin-dependent kinase inhibitor and potential mediator of extracellular antimitogenic signals. *Cell* **78** 59–66

Poon RYC, Toyoshima H and Hunter T (1995) Redistribution of the CDK inhibitor p27 between different cyclin-Cdk complexes in the mouse fibroblast cell cycle and in cells arrested with lovastatin or ultraviolet irradiation. *Molecular Biology of the Cell* **6** 1197–1213

Quelle D, Zindy F, Ashmun R and Sherr C (1995) Alternative reading frames of the INK4a tumor suppressor gene encodes two unrelated proteins capable of inducing cell cycle arrest. *Cell* **83** 993–1000

Schwob E, Bohm T, Mendenhall MD and Nasmyth K (1994) The B-type cyclin kinase inhibitor p40SIC1 controls the G_1 to S transition in *S. cerevisiae*. *Cell* **79** 233–244

Serrano M, Hannon GJ and Beach D (1993) A new regulatory motif in cell-cycle control causing specific inhibition of cyclin D/CDK4. *Nature* **366** 704–707

Serrano M, Lee H-W, Chin L, Cordon-Cardo C, Beach D and DePihno R (1996) Role of the INK4a locus in tumor suppression and cell mortality. *Cell* **85** 27–37

Sherr CJ (1994) G_1 phase progression: cycling on cue. *Cell* **79** 551–555

Sherr CJ and Roberts JM (1995) Mammalian cyclin-dependent kinase inhibitors. *Genes and Development* **9** 1149–1163

Shiohara M, El-Deiry WS, Wada M *et al* (1994) Absence of WAF1 mutations in a variety of human malignancies. *Blood* **84** 3781–3784

Shivji MKK, Grey SJ, Strausfeld UP, Wood RD and Blow J (1994) Cip1 inhibits DNA replication but not PCNA-dependent nucleotide excision-repair. *Current Biology* **4** 1062–1068

Shou W and Dunphy WG (1996) Cell cycle control by *Xenopus* p28Kix1, a developmentally regulated inhibitor of cyclin-dependent kinases. *Molecular Biology of the Cell* **7** 457–469

Steinman RA, Hoffman B, Iro A, Guillouf C, Liebermann CA and El-Houseini ME (1994) Induction of p21(WAF1/CIP1) during differentiation. *Oncogene* **9** 3389–3396

Su J-Y, Rempel RE, Erikson E and Maller JL (1995) Cloning and characterization of the *Xenopus* cyclin-dependent kinase inhibitor p27Xic1. *Proceeding of the National Academy of Sciences of the USA* **92** 10187–10191

Tam SW, Shay JW and Pagano M (1994) Differential expression and cell cycle regulation of the cyclin-dependent kinase 4 inhibitor p16Ink4. *Cancer Research* **54** 5816–5820

Toyoshima H and Hunter T (1994) p27, a novel inhibitor of G_1 cyclin-Cdk protein kinase activity, is related to p21. *Cell* **78** 67–74

Waga S, Hannon G, Beach D and Stillman B (1994) The p21 inhibitor of cyclin-dependent kinases controls DNA replication by interaction with PCNA. *Nature* **369** 574–578

Walker GJ, Hussussian CJ, Flores JF *et al* (1995) Mutations of the CDKN2/p16INK4 gene in Australian melanoma kindreds. *Human Molecular Genetics* **4** 1845–1852

Warbrick E, Lane D, Glover DM and Cox LS (1995) A small peptide inhibitor of DNA replication defines the site of interaction between the cyclin-dependent kinase inhibitor p21WAF1 and proliferating cell nuclear antigen. *Current Biology* **5** 275–282

Weinberg RA (1995) The retinoblastoma protein in cell cycle control. *Cell* **81** 323–330

Williams BO and Jacks T (1996) Mechanisms of carcinogenesis and the mutant mouse. *Current Opinions in Genetics and Development* **6** 65–70

Winston J, Dong F and Pledgar WJ (1996) Differential modulation of G_1 cyclins and the Cdk inhibitor p27Kip1 by platelet-derived growth factor and plasma factors in density arrested fibroblasts. *Journal of Biological Chemistry* **271** 11253–11260

Wolfel T, Hauer M, Schneider J *et al* (1995) A p16Ink4a-insensitive CDK4 mutant targeted by cytotoxic T lymphocutes in a human melanoma. *Science* **269** 1281–1284

Xiong Y, Hannon G, Zhang H, Casso D, Kobayashi R and Beach D (1993a) p21 is a universal inhibitor of cyclin kinases. *Nature* **366** 701–704

Xiong Y, Zhang H and Beach D (1993b) Subunit rearrangement of the cyclin-dependent kinases is associated with cellular transformation. *Genes and Development* **7** 1572–1583

Zhang H, Hannon G and Beach D (1994) p21-containing cyclin kinases exist in both active and inactive forms. *Genes and Development* **8** 1750–1758

Zhu L, Harlow E and Dynlacht B (1995) p107 uses a p21CIP1 related domain to bind cyclin/Cdk2 and regulate interacts with E2F. *Genes and Development* **9** 1740–1752

Zuo L, Weger J, Yang Q *et al* (1996) Germline mutations in the p16INK4a binding domain of Cdk4 in familial melanoma. *Nature Genetics* **12** 97–99

The author is responsible for the accuracy of the references.

Yeast Checkpoint Controls and Relevance to Cancer

T WEINERT

Department of Molecular and Cell Biology, University of Arizona, Tucson, AZ 85721

INTRODUCTION

Checkpoints are controls that regulate the progression of cell cycle events (reviewed in Hartwell and Weinert, 1989; Hartwell and Kastan, 1994). Here I discuss the checkpoints that arrest cells with damage to chromosomes. Other checkpoints recognize damage to microtubules and/or associated structures (eg kinetochores not attached to spindles) (reviewed in Murray, 1995) or other aspects of cell physiology (Lew and Reed, 1995). Normal cells with DNA damage arrest in specific stages of cell division, and arrest allows time for repair. Mutant cells with checkpoint defects fail to arrest when they have DNA damage. As a consequence, mutant cells that continue cell division suffer chromosome loss and probably other genomic rearrangements as well. This gives rise either to inviable progeny or to progeny that grow abnormally. The particular consequence to a mutant cell is complex but important, especially in cancer, and probably depends on the cell type, the extent and kind of damage and the nature of the checkpoint defect (eg G_1 defect or G_2 defect).

RELEVANCE OF CHECKPOINTS TO CANCER

The direct relevance of checkpoints to cancer is illustrated by the study of two genes, the *TP53* tumour suppressor gene and the *ATM* gene. Both *TP53* and *ATM* have roles in checkpoint controls: mutant cells fail to arrest in G_1 after DNA damage (Nagasawa *et al*, 1985; Rudolph and Latt, 1989; Kastan, 1991; Kuerbitz *et al*, 1992; Beamish and Lavin, 1994), and both mutants have other checkpoint defects as well (Painter and Young, 1980; Cross *et al*, 1995). *TP53* and *ATM* mutant cells also suffer genomic instability (Livingston *et al*, 1992; Yin *et al*, 1992; Friedberg *et al*, 1995). Finally, mutations in human *TP53* and *ATM* and mouse *p53* are associated with cancer (Malkin *et al*, 1990; Swift *et al*, 1991; Donehower *et al*, 1992; Greenblatt *et al*, 1994); the *Atm* mutant mice are still under study at the time of writing. Together, these results suggest cause and effect: mutations in checkpoint controls predispose cells to cancer by causing genomic instability (Hartwell and Kastan, 1994).

The premise of this review is that checkpoint defects are common in cancer cells and could therefore be manipulated in therapy. A checkpoint based therapeutic strategy is attractive because, in principle, checkpoint defects render the defective cells sensitive to damaging agents currently used in therapy to kill cancer cells. Using checkpoint defects in a therapeutic strategy might make cell killing more selective. Are checkpoint defective cells indeed damage sensitive? The answer is complex and underscores the need to understand the biology of checkpoint pathways. In yeast, checkpoint mutants (eg *rad9* mutants) are clearly very sensitive to DNA damaging agents, resulting in an increase of up to two logs in DNA radiation sensitivity relative to normal congenic cells (Weinert and Hartwell, 1988, 1990). In human cells, the checkpoint mutant cells tested were TP53$^-$ and were not found to be more radiation sensitive than their TP53$^+$ counterparts (see references in Powell *et al*, 1995). Differences in repair pathways between the yeast and human mutant cells tested might account for the different damage sensitivities. The yeast mutants, including *rad9*, were defective for the G_2 checkpoint (Weinert and Hartwell, 1988) and perhaps for the G_1 checkpoint as well (Siede *et al*, 1993); the TP53$^-$ human cells were defective for only the G_1 checkpoint. This suggests that in TP53$^-$ human cells, the G_2 checkpoint might compensate for defects in the G_1 checkpoint. If so, elimination of the G_2 checkpoint might increase damage sensitivity in G_1 checkpoint defective cells. Indeed, treating cells with a drug (caffeine analogues) that inhibits the G_2 checkpoint sensitized TP53$^-$ mutant cells to DNA damage to a greater extent than normal TP53$^+$ cells (Fan *et al*, 1995; Powell *et al*, 1995; Russell *et al*, 1995). In effect, the checkpoint defects sensitize cancer cells to cell killing, although the increased sensitivity requires additional defects in repair pathways (eg a G_2 checkpoint defect). Therefore, checkpoint defects may be manipulated to achieve greater damage sensitivity in cancer cells than in normal cells. What is the extent of checkpoint defects in cancer cells, and how can each be manipulated to increase its damage sensitivity? The answer to both questions requires a full understanding of checkpoint pathways (and other repair pathways) in normal and cancer cells.

YEASTS AS MODEL CELL TYPES

Yeasts are nearly optimal organisms for investigating checkpoint pathways, or any complex cellular process, in great detail (Tugendrich *et al*, 1994; Oliver, 1996). The classical and molecular genetic approaches possible in yeast allow detailed studies of gene function. Genes can be identified by classical mutant screens or by reverse genetic approaches, and the gene's activities can be manipulated in vivo to assess its role in the cell. With respect to checkpoint controls and the cell's responses to DNA damage, yeast and human cells share many of the relevant components. Conserved genes include those involved in DNA replication (Bauer and Burgers, 1990; Blanco *et al*, 1991; Cullman *et al*, 1995), in DNA repair (Friedberg *et al*, 1995) and in cell cycle control (p34CDC2/CDC28 and cyclins) (reviewed in Lees, 1995; Morgan, 1995). The conservation of genes between humans and yeasts extends to checkpoint controls as well. *MEC1* and *rad3*$^+$ are the budding and fission yeast genes, respectively, that have key roles in many checkpoint pathways. They are similar in sequence and function to the human gene *ATM* associated with checkpoint control and cancer (Savitsky *et al*, 1995). A *MEC1/rad3* homologue in *Drosophila* has also been identified, illustrating the extensive conservation of these controls through evolution (Hari *et al*, 1995).

The cell physiology of the damage responses, and a few key genes required for those responses, have now been identified in yeasts. This review describes our current understanding of the DNA damage and replication checkpoint pathways, based primarily on studies in budding yeast. Studies of fission yeast are proving equally valuable to understanding checkpoint pathways (reviewed in Sheldrick and Carr, 1993; Carr and Hoekstra, 1994; Carr, 1995). It is remarkable how similar are the pathways in the two yeasts even though they are evolutionarily as distant from each other as either is from mammals. A summary of gene homologues shared between budding and fission yeast is detailed elsewhere (Lydall and Weinert, 1996). It is worth pointing out, however, that details in the checkpoint pathways are significantly different between the two organisms. Understanding the basis for these differences may provide insights into underlying mechanisms.

Relationships between Checkpoint Pathways and Essential Cell Cycle Controls

To understand checkpoint pathways, we need to understand their role relative to the general mechanisms that regulate cell cycle events. Two types of regulators control the order of the principal events, DNA replication and mitosis (Fig. 1). One class of regulators is required for each cell division to regulate DNA replication and mitosis. This class of regulators includes the protein kinase heterodimer consisting of p34^{CDK} and associated cyclin, and other proteins that regulate p34^{CDK}-cyclin (eg in fission yeast proteins encoded by the *wee1*$^+$, *cdc25*$^+$ and *rum1*$^+$ genes and in budding yeast encoded by the *SIC1* gene; Morgan, 1995) and additional key regulators of DNA replication (in fis-

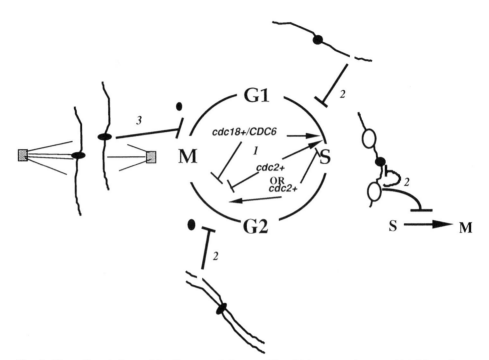

Fig. 1. The cell cycle is regulated by essential genes (1), which may activate and inhibit replication and mitosis after damage checkpoint controls delay cell cycle progression. One kind of checkpoint detects DNA damage or a delay in replication (2), and a second kind detects spindle damage (3). The exact cell cycle position of the G_2 checkpoint and the spindle assembly checkpoint (*closed circle*) is uncertain and in some cell types may be at the same stage

sion yeast cdc18⁺, in budding yeast CDC6 [the cdc18 homologue], CDC7, DBF4) (reviewed in Joachim and Deshaies, 1993; Kelley *et al*, 1993 Platti *et al*, 1995; Toyn *et al*, 1995). These genes co-ordinate the alternation between S phase and M phase, perhaps by both activation and inhibition. The details of how they do so are not yet clear (see Joachim and Deshaies, 1993; Nurse, 1994). One idea is that p34^{CDC2}-cyclin and cdc18/CDC6 initiate one event while simultaneously inhibiting the other event (Fig. 1, regulator type 1). For example, in one form, p34^{CDC2}-cyclin may activate DNA replication while inhibiting mitosis, and in another form (eg complexed to a different cyclin subunit) p34^{CDC2}-cyclin may activate mitosis while inhibiting DNA replication. The dual role of these genes is suggested by analysis of mutants with remarkable phenotypes: some mutants fail to initiate an event (eg *cdc2* mutants that arrest in G_1 and fail to initiate DNA replication; Nurse and Bissett, 1981), whereas others fail to inhibit an event (eg *cdc2* mutants, or cyclin mutants, that rereplicate DNA without mitosis; Broek *et al*, 1991; Hayles *et al*, 1994). The genes involved in DNA replication (*cdc18⁺, CDC6, CDC7, DBF4*) may also have dual roles in activation and inhibition. Together, the essential regulators may constitute the basic mechanisms regulating cell cycle progression, mechanisms that the damage responsive checkpoints may alter to arrest the cell cycle.

The second type of regulator is the damage responsive checkpoint, which arrests cells with damage. At least two types of checkpoints exist, each recognizing a different type of damage (Fig. 1, regulator 2 and 3). The DNA replication and damage checkpoint recognizes something about DNA structure. The spindle assembly checkpoint recognizes damage to spindles or to associated structures such as kinetochores or to centrosomes (reviewed in Murray, 1992, 1995). The DNA replication and damage checkpoints appear most relevant to cancer therapy because they are compromised in cancer cells and are normally not essential, and when these checkpoints are inactivated, cell death occurs only in combination with DNA damage.

NOMENCLATURE

Nomenclature in the checkpoint field can be confusing, partly because of the large number of controls that regulate the same phase of cell division and partly because of the obscure nature of the underlying mechanisms. There are distinct controls active in G_1 that prevent DNA replication and mitosis. Both can be called G_1 checkpoints; the one that prevents DNA replication is a G_1/S control, and the one that prevents mitosis is a G_1/M control. S phase checkpoints prevent mitosis and delay the progression of DNA replication; the control that delays mitosis is an S/M control, and the control that delays DNA replication is an S phase progression control. In the literature, both simpler (G_1) and more complex (G_1/S) terms are used. In most cases, the context is clear and the simpler nomenclature will suffice (eg *TP53* mutants are defective in the G_1 checkpoint), but in some cases, the more specific nomenclature is necessary to avoid ambiguity. In this review, we use the simpler nomenclature (eg G_2 checkpoint) where possible. Others have referred to the DNA replication and damage checkpoints together as DNA structure checkpoints (Carr, 1995). (For further discussion of the term "checkpoint", see Weinert and Lydall, 1996.)

DNA REPLICATION AND DAMAGE CHECKPOINTS IN YEASTS

Studies of these checkpoints in budding yeasts have led to the following general observations. Firstly, the controls arrest cell division in at least four different ways. Secondly, checkpoint genes have additional functions, including roles in DNA repair and transcriptional regulation, in addition to their roles in cell cycle arrest. The additional functions may provide clues to their biochemical functions and may also be important in considering targets for drug discovery. Thirdly, the *MEC1* gene may be a key element in checkpoint pathways and is a member of the phosphatidylinositol (PI) 3-kinase family of proteins, which have roles in DNA metabolism. Finally, some checkpoint proteins may process DNA damage, an important initial step in the checkpoint pathway.

1. Four Cell cycle Arrests: G1, S, G2 and S phase progression

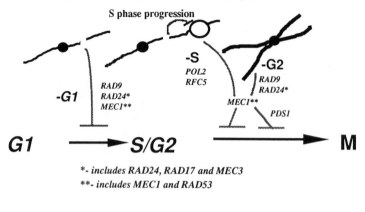

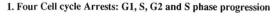

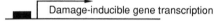

*- includes RAD24, RAD17 and MEC3
**- includes MEC1 and RAD53*

2. Role in Transcriptional Regulation

Damage-inducible gene transcription

3. Role in DNA Repair directly?

Fig. 2. Roles of checkpoint genes include four responses in cell cycle progression, transcriptional regulation and DNA repair. An additional role in DNA replication is more speculative (see text). (1) Chromosome cycle is shown with a G_1 chromosome with a DNA break, an S phase chromosome with a replication bubble and a G_2 pair of sister chromosomes, one with a DNA break. (2) Damage inducible gene with its enhancer sequence (*closed box*)

Four Distinct Cell Cycle Arrest Responses

Checkpoint controls in normal cells act in four different ways: two affect DNA replication and two affect entry into mitosis (Fig. 2). Normal cells with damage arrest in G_1, before initiation of DNA replication (Siede *et al*, 1993, 1994), and delay progression of DNA replication by slowing both initiation and elongation (Painter and Young, 1980; Larner *et al*, 1994; Paulovich and Hartwell, 1995). Furthermore, damage in S phase or arrest of DNA replication leads to arrest of mitosis (Allen *et al*, 1994; Weinert *et al*, 1994), and damage in G_2 also arrests cells before mitosis (Weinert and Hartwell, 1988; Weinert *et al*, 1994). The four responses appear to be distinct both physiologically and genetically.

Genes in Yeasts and Mammalian Cells

Nine genes have been identified in budding yeast that when mutated disrupt checkpoint controls (Fig. 2) (Weinert and Hartwell, 1988, 1993; Allen *et al*, 1994; Weinert *et al*, 1994; Navas *et al*, 1995; Sugimoto *et al*, 1996; Yamamoto *et al*, 1996b). These genes probably act in some type of signal transduction process. Clues to their roles come from studies of the four cell cycle responses. For example, *RAD9* is required for G_1 and G_2 checkpoints but not for the S phase checkpoint (Weinert and Hartwell, 1988, 1993; Siede *et al,* 1993), whereas *MEC1* is required for all three of these responses (Weinert *et al,* 1994). *POL2* is required for the S checkpoint but not for G_1 or G_2 checkpoints (Navas *et al,* 1995). (The roles of these genes in the S phase progression control have not yet been determined.) These results have several possible interpretations, and any proposed pathways must be taken as hypotheses. From what we now understand about the biochemical function of these genes, detailed below, it seems reasonable that *RAD9* and *POL2* act upstream of *MEC1* in a signal transduction pathway (Fig. 2). In sum, nine genes have been identified, some of their activities have been defined and pathways based on their order of function have been proposed. Much about the pathways remains to be defined.

The genetic complexity of the cell cycle responses extends to mammalian cells and to fission yeast as well. In mammalian cells, TP53 protein is required for the G_1 arrest but has no role in delay of S phase progression (Kastan *et al,* 1991; Kuerbitz *et al,* 1992; Larner *et al,* 1994). By contrast, the *ATM* gene is required for both G_1 arrest (Nagasawa *et al,* 1985; Rudolph and Latt, 1989; Beamish and Lavin, 1994) and delay of S phase progression (Painter and Young, 1980). Both *ATM* and *TP53* may have some limited role in the G_2 checkpoint as well (Beamish and Lavin, 1994; Fan *et al,* 1995; Powell *et al,* 1995), but whether the roles are direct or a secondary consequence of genomic instability is unknown (Paules *et al,* 1995). In fission yeast, many checkpoint genes have also been identified (reviewed in Carr, 1995; Carr and Hoekstra, 1995), and specific examples are discussed below.

Additional Roles for Checkpoint Genes

Some checkpoint genes appear to have other roles in addition to those in cell cycle arrest. *MEC1* is required for all four cell cycle arrest responses and is also essential for transcriptional induction of damage inducible genes (Kiser and Weinert, 1996). Transcriptional induction leads to increased expression of proteins involved in DNA repair, such as ribonucleotide reductase. The *RAD53* gene also has multiple roles in transcriptional induction (Allen *et al,* 1994; Kiser and Weinert, 1996). By contrast, we found that *RAD9* is not required for transcriptional induction but does contribute to it (Kiser and Weinert, 1996). The extent of the role of *RAD9* in transcriptional regulation is disputed (Abousekkra *et al,* 1996). Several checkpoint genes are also transcriptionally induced. In addition, *MEC1* is required for its own transcriptional induction, indicating the existence of a positive feedback loop in the checkpoint pathway (Kiser and Weinert, 1996).

Another role checkpoint genes may have is in DNA repair itself (Lydall and Weinert, 1995). DNA repair roles were inferred from genetic analysis: double mutants of specific genes (eg *rad9/rad24*) are more sensitive to DNA damage than is either single mutant alone. The source of the increased damage sensitivity is probably due neither to cell cycle controls nor to transcriptional regulation, because single and double mutants have the same responses. Therefore, the increased damage sensitivity of double mutants is probably due to a role in DNA repair itself. The repair pathways are probably redundant with other repair pathways because damage sensitivity of checkpoint mutants can be suppressed by imposing a delay in G_2 (Weinert T and Hartwell LH, unpublished). That checkpoint genes have roles in DNA repair is also supported by studies in fission yeast, where some alleles of fission checkpoint genes are damage sensitive but retain cell cycle arrest (Al-Khodairy et al, 1994; Griffiths et al, 1995). The identity of the repair pathway is unknown, but we speculate below that it may involve single stranded DNA and recombinational repair.

Finally, a fourth role is inferred from the observation that two checkpoint genes, MEC1 and RAD53, are essential in normal cells (Zheng et al, 1993; Kato and Ogawa, 1994). The essential function may be in some aspect of normal DNA replication, although what that role may be is unclear. Initial clues to the essential function of MEC1 come from suppressor analyses. Suppression is achieved by overexpression of three protein kinases (TEL1, RAD53 and DUN1 (Nasr et al, 1994; Morrow et al, 1995; Sanchez et al, 1996), and loss of function mutations in other genes (cdc28, cdc6, cdc7, cln1⁻cln2⁻) (Gardner R and Weinert T, unpublished; Vallen L and Cross F, personal communication). None of the suppressors tested suppress mec1's checkpoint defects (ie cell cycle progression after damage remains defective), suggesting that the checkpoint may not be the essential function (Gardner R and Weinert T, unpublished). Are there common mechanisms among these suppressors? Overexpression of the two protein kinases tested restores transcription of damage inducible genes, and several loss of function mutations slow cell division early in the cell cycle. We therefore suggest that mec1Δ cells may complete DNA replication either too slowly and/or with poor quality, generating defective progeny cells. After many rounds of cell division, mec1Δ cells accumulate damage and die. Suppressors may either provide more time for DNA replication (by slowing cell division early) or increase gene expression to minimize the effect of the damage (through transcriptional regulation). Additional evidence for some role of MEC1 in DNA replication comes from the observation that some mec1 alleles (termed esr1 by Kato and Ogawa, 1994) have elevated levels of mitotic recombination, which suggests that mec1 mutants make errors during DNA replication. MEC1 is also required to delay DNA replication progression after damage (Paulovich and Hartwell, 1995).

In sum, checkpoint genes have other roles in addition to those in cell cycle control; these include transcriptional regulation of genes involved in repair, DNA repair and an essential function, perhaps in DNA replication. The multiple functions may be important for two reasons. Firstly, they may provide clues to the biochemical mechanisms of the checkpoint proteins (eg that some proteins may be involved directly in DNA repair). Secondly, the many roles of specific proteins may influence the choice of molecular targets in drug discovery: inactivation of a specific protein may have many direct consequences in addition to inactivation of a cell cycle checkpoint.

Roles of Sensors, Signallers and Targets

The cell cycle arrest responses may be viewed as signal transduction processes in which genes act in three steps: detection of DNA damage, transduction of an inhibitory signal and finally arrest by modification of target molecule(s). Each step may involve genes termed here the "sensors", "signallers" and "targets", respectively. We first discuss putative signallers and targets, then focus most of our discussion on sensor genes and the initial events in the checkpoint pathway.

Signaller Genes

Two budding yeast genes, *MEC1* and *RAD53*, are candidate signaller genes. The evidence for their roles in signalling is indirect. Firstly, both encode putative protein kinases: *MEC1* is a member of the PI3-kinase family (Hunter, 1995), discussed further below. *RAD53* encodes a typical protein kinase (Stern *et al*, 1991). As protein kinases, they may transmit signals as in other protein kinase cascades. There is some biochemical evidence supporting the notion of a signal cascade in checkpoint pathways. Rad53p becomes phosphorylated after DNA damage, and phosphorylation requires *MEC1* (Sanchez *et al*, 1996; Sun *et al*, 1996). This suggests that *MEC1* acts upstream of *RAD53*. The second type of evidence that suggests roles as signallers is that mutants in either gene have the strongest phenotypes: *mec1* and *rad53* mutants have defects in all four checkpoint responses and in transcriptional induction of repair genes, and each shows strong damage sensitivity (Allen *et al*, 1994; Weinert *et al*, 1994; Kiser and Weinert, 1996).

That *MEC1* and *RAD53* are signaller genes that act downstream from damage should be considered a reasonable hypothesis, although other interpretations of their order of function are possible. For example, we speculate that *RAD53* may instead act on DNA damage as a sensor, perhaps by processing damage, whereas *MEC1* acts downstream from *RAD53* as a signaller. Support for this hypothesis is only indirect; *rad53* mutants have a genetic interaction with a DNA repair mutation, *rad16*, which *mec1* mutants do not have. *rad16/rad53* mutants have low viability and are more sensitive to UV irradiation than is either single mutant (Kiser and Weinert, 1996). *rad16/mec1* mutants, by contrast, have similar viability and damage sensitivity to *mec1* single mutants. The interpretations of these genetic interactions are many, but at the least they suggest *RAD53* cannot simply act downstream from *MEC1*, rather that its role is more complex. Secondly, the fission yeast putative *RAD53* homologue, called *cds1+*, does not seem to have a central role in checkpoint control (Murakami and Okayama, 1995), whereas the *MEC1* homologue, *rad3+*, does have a central role (Jimenez *et al*, 1992). The roles of

MEC1 and *RAD53* and their order of function remain key unanswered questions in the checkpoint pathways.

MEC1 *and PI3-Kinases*

The *MEC1* gene is particular intriguing. *MEC1* encodes a putative PI3-kinase, a family of kinases with specificity for phosphatidylinositol and/or protein (Hunter, 1995). A subgroup of the PI3-kinases includes the mammalian genes *DNA-PK* and *ATM*, and the yeast genes *TEL1* and *MEC1* (and their homologues in fission yeast and in *Drosophila*, *rad3+*, *MEI41*). All have roles in DNA metabolism. Study of DNA-PK, the only member studied at the biochemical level (Gottlieb and Jackson, 1993; Morozov *et al*, 1994; Hartley *et al*, 1995), may provide insight into the possible biochemical functions of Mec1p. DNA-PK is a protein kinase activated by DNA damage, an activity that requires association with two additional proteins called KU70 and KU80. KU70, KU80 and DNA-PK appear to form a heterotrimer, wherein KU subunits act as "sensors" by binding directly to DNA damage and thereby activating DNA-PK catalytic activity. Possible in vivo substrates for DNA-PK kinase activity are now being examined (for table of substrates phosphorylated in vitro, see Anderson, 1993; Boubnov and Weaver, 1995). KU70, KU80 and DNA-PK do appear to have roles in DNA metabolism, because mutants in mice and in hamster cells have defects in repair and in an immunoglobulin recombination event (Rathmell and Chu, 1994; Taccioli *et al*, 1994; Blunt *et al*, 1995). DNA-PK is not reported to have a role in control of the cell cycle in response to damage.

The *MEC1* gene is also intriguing because of its similarity in sequence and function to the *ATM* gene, a human gene associated with the disease ataxia telangiectasia and with cancer (Savitsky *et al*, 1995; reviewed in Hunter, 1995; and see Rotman and Shiloh, this volume). Both *MEC1* and *ATM* genes have roles in the G_1 checkpoint/S phase progression and perhaps the G_2 checkpoint (discussed above), as well as in transcription induction after damage (Papathanasiou *et al*, 1991; Kiser and Weinert, 1996). Another yeast gene, *TEL1*, is more similar at the sequence level to *ATM* than is *MEC1* (Morrow *et al*, 1995) (*tel1* mutants have shorter telomeres, as do AT mutant cells; Greenwell *et al*, 1995). It appears that *MEC1* and *TEL1* may mediate responses in yeast that *ATM* and some additional gene(s) mediate in human cells. In yeast, *MEC1* and *TEL1* share control of some DNA repair responses. The roles in repair are inferred from the observation that *mec1/tel1* double mutants have greater damage sensitivity than *mec1* mutants alone (*tel1* mutants have no detectable damage sensitivity; Morrow *et al*, 1995; Sanchez *et al*, 1996). Furthermore, increased gene dosage of the *TEL1* gene suppresses the essential defects in a *mec1* null mutant (Morrow *et al*, 1995; Sanchez *et al*, 1996). *TEL1* does not appear to have any role in cell cycle control; increased gene dosage of *TEL1* does not restore cell cycle arrest to *mec1* mutants (Gardner R and Weinert T, unpublished). Instead, the effects of *TEL1* on *mec1* may be by restoring transcriptional regulation (Gardner R and Weinert T, unpublished).

Together, these studies provide a common view of the role of this family of PI3-kinases in DNA metabolism. Each gene product may respond to DNA damage (possibly by interacting with sensor proteins) and mediate downstream events, possibly by phosphorylation of specific protein substrates (possible PI phosphorylation by DNA-PK has not yet been identified) (Hartley *et al*, 1995). The downstream events each PI3-kinase mediates may differ. For example, *MEC1* mediates cell cycle arrest and transcriptional induction, whereas *TEL1* mediates some aspect of telomere metabolism and transcriptional induction. *ATM* mediates G_1 arrest, S phase progression and some feature of telomere metabolism, whereas DNA-PK regulates other aspects of DNA repair. The substrates that PI3-kinase members phosphorylate, and the description of their cellular roles, will be a focus of research in the coming years.

Targets

The molecular targets of checkpoint pathways are molecules that when altered (eg phosphorylated and degraded) probably lead directly to changes in cell physiology (eg in cell cycle arrest, in transcription regulation or in DNA replication). Few targets have been identified. In transcriptional regulation, one possible target is the product of the *DUN1* gene, a protein kinase required for induction of some damage inducible genes (Zhou and Elledge, 1993). Dun1p becomes phosphorylated after damage, and phosphorylation requires *RAD53* (Allen *et al*, 1994). It is unknown whether all damage inducible transcription mediated by checkpoint genes requires *DUN1*.

There are only a few clues to molecular targets for the other checkpoint gene mediated responses. Genes with known roles in DNA replication, repair or cell cycle control have been tested. A subunit of the single stranded DNA replication protein A (RPA) becomes phosphorylated after damage, and phosphorylation requires *MEC1* (Brush G, Morrow D, Hietes P and Kelly T, unpublished). The role of RPA phosphorylation in DNA replication and repair is unknown.

Another possible target is the central cell cycle regulator p34^{CDK}-cyclin, and evidence in many cell types suggests a role in checkpoint pathways. In fission yeast, specific mutations in *cdc2*$^+$ or mutations in a regulator, *cdc25*$^+$, lead to defects in the DNA replication checkpoints. For example, *cdc2-3w* alleles enter mitosis when DNA replication is blocked—an S phase checkpoint defect (Enoch and Nurse, 1990). The role of *wee1* in the G_2 checkpoint after DNA damage, if any, is unclear: one report suggests that *wee1* mutants are partially defective for the G_2 arrest, another finds no defect (Rowley *et al*, 1992; Barbet and Carr, 1993). Another allele of *cdc2* has a genetic interaction with a fission yeast checkpoint gene, *chk1*, which has a role in the G_2 checkpoint (Walworth *et al*, 1993). In human cells, overexpression of regulators of p34^{cdc2} causes premature entry of S phase cells into mitosis (Heald *et al*, 1993). In many studies of mammalian cells with damage, there is a correlation between phosphorylation of p34^{CDC2} and damage induced arrest at the G_2 checkpoint (see O'Connor *et al*, 1993, and references therein), and a mutation in *CDC2* that prevents inhibitory phosphorylation partially inactivates the G_2 checkpoint (Jin *et al*, 1996). In frog embryos, addition of excess $p34^{CDC2}$-cyclin, or of CDC25 or a heterologous cyclin drives S phase arrested extracts (cycling extracts treated with aphidicolin) into mitosis (Dasso and Newport, 1990; Kumagai and Dunphy, 1991, 1995; Kornbluth *et al*, 1992). The response to DNA damage and the role of $p34^{CDC2}$-cyclin in arrest in frog embryos has not been reported. All of these results are consistent with a model in which $p34^{CDC2}$-cyclin is a target of inhibition by the S and G_2 checkpoints following DNA replication blocks and DNA damage. It is unclear whether $p34^{CDC2}$-cyclin is the direct target of inhibition. Mutations that activate $p34^{CDC2}$-cyclin may override arrest that occurs by a distinct mechanism.

The roles of $p34^{CDC2}$-cyclin in checkpoint controls may require a biochemical description of the mechanism(s) of its inactivation following damage or replication block. For the G_1 checkpoint, such evidence is available. Transcriptional induction of p21 and its inhibition of $p34^{CDk2}$–cyclin E can explain in part G_1 arrest after damage (Dulic *et al*, 1994; Deng *et al*, 1995). $p34^{CDK4}$–cyclin D may also have some role in a G_1 checkpoint after ultraviolet irradiation (Terada *et al*, 1995). By contrast, a biochemical mechanism for the S and G_2 checkpoints has not yet been identified, although several inhibitors of G_2/M transition have been identified and could have roles in checkpoint controls (Kumagai and Dunphy, 1995; Lee and Kirschner, 1996).

The role of $p34^{CDC2}$-cyclin in checkpoint pathways becomes confusing when one considers studies on the budding yeast checkpoint pathways (the cdc^+ homologue in budding yeast is called *CDC28*). Studies in budding yeast thus far have failed to identify any role for $p34^{CDC28}$ in cell cycle arrest (Amon *et al*, 1992; Sorger and Murray, 1992; Stueland *et al*, 1993) or in transcriptional regulation of damage inducible genes (Kiser and Weinert, 1996). For example, mutants of $p34^{CDC28}$ that fail to arrest at the S phase checkpoint in fission yeast still arrest in budding yeast. There are many plausible explanations for the contrasting results in budding yeast: the appropriate checkpoint inactivating mutations in $p34^{CDC28}$-cyclin have not been tested; arrest mechanisms are redundant in budding yeast; and most simply, $p34^{CDC28}$-cyclin is not the target of checkpoint controls in budding yeast. We have offered another, more heretical, explanation: $p34^{CDC2}$-cyclin may be involved in the processing of DNA damage, a role necessary in some but not all cell types (Lydall and Weinert, 1995; Weinert and Lydall, 1996). The role of $p34^{CDC2}$-cyclin remains a central and unsolved part of checkpoint pathways.

In budding yeast, only a single gene, *PDS1*, is to date a bona fide candidate for a downstream target gene. The *PDS1* gene was initially identified in mutants that prematurely separated their sister chromatids when cells were blocked in metaphase (Yamamoto *et al*, 1996a). *pds1* mutants have additional phenotypes, and most unexpectedly are defective for both DNA damage (G_2 checkpoint) and spindle assembly checkpoint pathways (Yamamoto *et al*, 1996b). These roles of *PDS1* in checkpoint controls have several implications.

Firstly, the spindle assembly and G_2 checkpoints converge in budding yeast. All previous spindle assembly and G_2 checkpoint mutants had defects specific for only one type of pathway. The PDS1 protein, required for both, may act downstream in a common pathway where they converge.

The relative arrest stages of DNA damage and spindle assembly checkpoint are important in considering the mechanism of arrest. In cases where cells arrest in the same stage, the mechanisms of arrest may be related. Thus, in budding yeast, the two controls may act at the same stage and perhaps by related mechanisms. One previous study in budding yeast was consistent with the two controls acting at the same stage (Weinert and Hartwell, 1988; but see Wood and Hartwell, 1982). In fission yeast, it is also possible that DNA damage and spindle assembly checkpoint controls act at the same stage in the G_2 phase (Alfa et al, 1990; Ford et al, 1994). Curiously, in mammalian cells, the *TP53* gene is implicated directly in the G_1 checkpoint (see above) and also in the G_2 DNA damage checkpoint (Fan et al, 1995; Powell et al, 1995) and in the spindle assembly checkpoint (Cross et al, 1995). The defects in the DNA damage and checkpoints could be a secondary consequence of genomic instability (see Paules et al, 1995). In mammalian cells, the DNA damage and spindle assembly checkpoints seem most clearly to act in distinctive G_2 and M phases, respectively, because the cell morphologies in arrested cells clearly differ, and when the cells are irradiated, the mitotic index decreases, indicating that M phase cells do not arrest (see Carlson, 1969). Finding an explanation for the different arrest points in yeasts and mammalian cells will be important to understanding their mechanism.

A second result from analysis of *pds1* mutants, with interesting implications, is that *pds1* mutants are not defective for the S phase checkpoint; mutant cells arrest when DNA replication is blocked. This result suggests that the mechanisms of arrest differ in the downstream targets for S and G_2 checkpoints (and see Fig. 2). Clearly, the role(s) of Pds1p will be important in understanding checkpoint pathways.

AN INITIAL STEP IN CHECKPOINT PATHWAYS: SENSORS AND PROCESSING OF DAMAGE

The initial event in the checkpoint pathway is recognition of damage, a step probably carried out by proteins that act directly on DNA. Two types of sensor genes have been identified in budding yeast—those that act in the S phase checkpoint, possibly recognizing a stalled replication fork, and those that act in the G_1 and G_2 checkpoints and recognize DNA damage.

Sensor Genes in the S/M Phase Checkpoint

At least two genes. *POL2* and *RFC5*, may recognize a delay in DNA replication to initiate the arrest response (Navas et al, 1995; Sugimoto et al, 1996). Both of these genes encode proteins that have essential roles in DNA replication; RFC5 is part of the replication complex that interacts with primer template, and DNA polymerase ε has some as yet undefined role in replication. Specific mutants in either gene are viable but fail to arrest when DNA replication is blocked. Since the corresponding proteins clearly act on DNA, these proteins may act as sensors. *POL2* and *RFC5* are not required for arrest at the G_2 checkpoint in G_2 cells with damage; therefore, they may detect some specific aspect of a block to DNA replication.

Sensor Genes in the G_1 and G_2 Checkpoints

The genes *RAD9*, *RAD24*, *RAD17* and *MEC3* have roles complementary to those of *POL2* and *RFC5*: the *RAD/MEC* genes are required for arrest in G_2 after DNA damage but are not required for arrest in S phase when DNA replication is blocked. Two of the four G_2 checkpoint genes encode proteins that may act directly on DNA: *RAD17* encodes a putative $3' \to 5'$ exonuclease that appears to degrade DNA in vivo as well (Lydall and Weinert, 1995), and *RAD24* encodes a protein that has some homology with replication accessory proteins, including RFC subunits (Griffiths *et al*, 1995; Lydall D and Weinert T, unpublished). Overexpression of the *RAD24* gene also shows genetic interactions with DNA polymerase δ and an RFC subunit encoded by *cdc44*, further indicating Rad24p interaction with DNA (Weinert TA and Hartwell LH, unpublished).

We speculate that the specificity of *POL2* and *RFC5* for the S phase checkpoint and of *RAD/MEC* genes for the G_2 checkpoint (and G_1 checkpoint) is due to their functions on DNA. A stalled replication fork and DNA damage have different structures that necessitate recognition by different proteins and thus create roles for different sensor genes. The checkpoint specificity of *POL2/RFC5* and the *RAD/MEC* genes may instead (or in addition) be due to different mechanisms of arrest between S and G_2 checkpoints (see discussion of *PDS1* above).

Sensor Genes May Process DNA Damage

Analysis of budding yeast checkpoint genes leads us to an hypothesis presented in Fig. 3. In this model, initial damage is not sufficient for arrest; rather, the damage must be processed to an intermediate structure. The intermediate serves several functions, including activation of signallers for cell cycle arrest and participation in some repair pathway. The evidence in support of this model is discussed below.

Three lines of evidence suggest that initial damage must be degraded or processed to cause cell cycle arrest. Firstly, pyrimidine dimers formed after UV irradiation do not trigger the checkpoint response in either G_1 or G_2 (Siede *et al*, 1994; Admire A, Kiser G and Weinert T, unpublished). To signal arrest, the dimers must be acted on by excision repair proteins (eg *RAD1* and *RAD2*) or by photolyase. Presumably, the dimer adduct is not recognized by

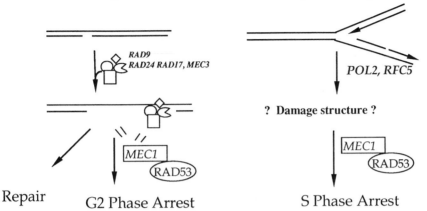

Fig. 3. Two pathways for processing DNA damage depend on initial damage. In A, a single strand nick is converted to a gap by four checkpoint proteins encoded by *RAD9*, *RAD24*, *RAD17* and *MEC3*. They degrade double stranded DNA to single stranded DNA, generating a substrate that acts in both DNA repair and cell cycle arrest. Signaller proteins encoded by *MEC1* and *RAD53* recognize some feature of the damage intermediate. In B, a stalled replication fork is processed by proteins encoded by *POL2* and *RFC5*. The damaged structure generated activates signaller proteins

checkpoint control proteins. Secondly, double strand breaks also do not appear sufficient for arrest, rather they too are processed, in this case by DNA degradation. Double strand breaks are normally degraded to a 3′ single strand that is an intermediate in recombination (Haber, 1992). Is degradation to 3′ single strands required for arrest? In one mutant, *rad50S*, double strand breaks made during meiotic recombination are not degraded, and significantly, cells do not arrest (Alani *et al*, 1990). (Degraded double strand breaks do cause arrest that requires the mitotic checkpoint genes; Lydall *et al*, 1996.) These results suggest that degradation of double stranded DNA to single stranded DNA may be linked to cell cycle arrest. Perhaps the single stranded DNA itself forms part of the signal for arrest.

A third observation also suggests that processing of initial damage is linked to arrest. In this case, damage by degradation is processed by the checkpoint proteins themselves (Lydall and Weinert, 1995). In one type of damage, an initial DNA lesion in double stranded DNA is converted to single stranded DNA; this degradation requires the checkpoint genes. (The exact nature of the initial lesion, generated by a temperature sensitive *cdc13* mutant, is unknown. It is either a nick in double stranded DNA or an exposed chromosome end; Garvik *et al*, 1995.) The cells with single stranded DNA arrest at the G_2 checkpoint; both arrest and degradation require checkpoint genes. The checkpoint gene products required for degradation included *RAD17* and *RAD24* mentioned above, genes that encode a putative 3′5′ exonuclease and a putative replication accessory protein, respectively. (*RAD9* has an enigmatic effect on degradation: degradation occurs more rapidly in *rad9* mutants than in *RAD9+* cells, as if Rad9p were acting in wild type cells as a partial inhibitor of degradation.) These results suggest that the action of checkpoint proteins on DNA damage, in this case in degrading double strand DNA to single strand, is linked to cell cycle arrest.

All three observations—on processing of UV damage, on degradation of double strand breaks and on checkpoint gene roles in degradation of some damage—suggest that the processing of initial damage may be a key initial step in the checkpoint pathways. We suggest that degradation generates single stranded DNA, which serves several functions—as an intermediate in DNA repair and as an intermediate in signalling cell cycle arrest and transcriptional activation. Figure 3A shows an example of degradation mediated by checkpoint proteins. In some cases, degradation may occur by a checkpoint gene independent mechanism, although the product of degradation is recognized by the checkpoint proteins (Lydall *et al*, 1996). In Fig. 3B, we suggest that *POL2* and *RFC5* may have analogous roles in the S phase checkpoint, processing and/or recognizing damage due to a stalled replication fork.

This model is clearly speculative. Does such an intermediate exist? We can only infer its presence from the observations that checkpoint proteins participate in degradation, that degradation correlates with arrest and that single stranded DNA is a substrate in many repair and recombination pathways (Kowalczykoski *et al*, 1994; Friedberg *et al*, 1995). If this intermediate exists, what is its structure? Checkpoint proteins are probably involved (the Rad17p exonuclease, Rad24p resembling RFC protein) and may interact with proteins involved in replication and repair, such as RFC, PCNA, RFA and DNA polymerases. The biochemical dissection of this possible intermediate has not yet begun. Are all forms of damage processed into a common intermediate, or are there multiple intermediates that differ according to the nature of the initial damage? A common intermediate would certainly simplify the biochemical analysis. How might such an intermediate signal cell cycle arrest and participate in repair? Perhaps a specific domain on a checkpoint protein, when associated with damage, serves as the immediate signal. We infer that this intermediate has a role in repair because checkpoint mutants have an element of damage sensitivity independent of their roles in cell cycle arrest. Perhaps the single stranded DNA provides a structure for recombinational repair mediated by RecA like yeast proteins.

The model and connection with DNA repair suggest that checkpoint proteins may have evolved as a repair process. As mitosis and the need for controls to co-ordinate events arose, the repair proteins may have taken on additional roles in cell cycle control.

SOS PATHWAY IN BACTERIA AS A PARADIGM

In bacteria, the responses to DNA damage are co-ordinated in a manner analogous (although probably not homologous) to the model presented in Fig. 3 for eukaryotic checkpoint controls. In bacteria, the so called "SOS" pathway controls the responses to DNA damage that result in increased DNA repair and cell survival (Little and Mount, 1982). In the SOS pathway, initial damage (eg double strand breaks and UV dimers) is not sufficient; the damage must be processed to single stranded DNA. Depending on the initial damage, different pathways of processing are required to generate single stranded DNA (Sassanfar and Roberts, 1990 and references therein). For example, the RecBC enzyme converts a double strand break to single stranded DNA but does not convert UV dimers to single stranded DNA. Single stranded DNA is the intermediate that activates the RecA protein by binding, and RecA then mediates downstream events (transcriptional derepression, DNA repair, cell cycle alterations). The general concept that initial damage is processed to a common intermediate(s), which then mediates multiple responses, is analogous in the two responses pathways. The details of the SOS and eukaryotic responses probably differ—for example, RecA per se is not required for arrest in eukaryotic cells.

CANCER CELLS

In human cells, relatively little is known about the genes controlling checkpoint pathways. Two genes that do have roles are *TP53* and *ATM*, and we speculate here on their possible roles in relation to our view of the yeast controls. Since TP53 can interact with single stranded DNA (Janyaraman and Prives, 1995; Lee *et al*, 1995; Huang *et al*, 1996), TP53 may act as a sensor to detect DNA damage much like we think the yeast G_2 checkpoint proteins detect damage. The sensing of damage by TP53 could be linked to prior processing of damage, just as UV damage and meiotic double strand break need to be processed before checkpoint proteins may signal arrest. UV damage must be excised to stabilize TP53 protein, an event linked to other TP53 mediated events (Nelson and Kastan, 1994).

The *ATM* gene has sequence and functional similarities to both *MEC1* and *TEL1* genes, as discussed above. *ATM* appears to have roles in checkpoint controls similar those of *MEC1* in yeast, with the exception that *ATM* is not required for the S/M checkpoint and has a limited role in the G_2 checkpoint. The hypomorphic nature of the G_2 checkpoint defect in AT mutant cells is curious and has two possible explanations. Firstly, *ATM* may be partially redundant with another *ATM* like gene: a putative human homologue of *MEC1* has been identified (Cimprich *et al*, 1996). Secondly, the partially defective G_2 checkpoint in AT mutant cells may be due to defects in damage processing. If AT mutant cells are partially defective for processing of initial damage, this would explain why AT mutant cells that incur damage earlier in S phase do arrest in the subsequent G_2, whereas cells that incur damage in G_2 do not arrest efficiently. This possible kinetic defect in processing could be compensated for by slowing the cells in G_2, which might then restore arrest to AT mutant cells.

Other cancer cells (Paules *et al*, 1995) and, most strikingly, mismatch repair defective mammalian cells (Hawn *et al*, 1995), appear also to be hypomorphic for the G_2 checkpoint. Mismatch repair defective strains are actually less sensitive to such damage than are normal strains. To explain these observations, we suggest the following scenario: initial mutations may compromise DNA repair by affecting the processing of damage. Such mutations may simultaneously increase resistance to some types of damage and increase genomic instability by compromising checkpoint controls.

PROSPECTS FOR CHECKPOINT BASED CLINICAL STRATEGIES

Implementation of checkpoint based strategies in cancer will require the answers to several questions. Firstly, to what extent are cancer cells checkpoint defective? Do different cancers have different spectrums of checkpoint defects, necessitating different therapeutic strategies? Secondly, once checkpoint defects in cancer cells are defined, how can they be manipulated to enhance cell killing? The studies of *TP53* mutant cells, in which a second checkpoint defect induced by drug treatment preferentially enhanced cell killing, offer one of probably several strategies. Further research on checkpoint pathways in yeast may suggest additional specific strategies.

SUMMARY

Checkpoint controls arrest cells with defects in DNA replication or DNA damage. For several reasons, checkpoint controls may be relevant to ontogeny and treatment of cancer. Firstly, mutations in two human genes, *TP53* and *ATM*, give rise to cellular defects in cell cycle checkpoints and are associated with cancer. Secondly, although checkpoint defects potentially render the cell damage sensitive, they may do so only in combination with other defects in the cell's response to damage. Therefore, manipulation of checkpoint defects, requiring a description of normal and mutant pathways, will be required for this type of therapeutic approach. Those pathways are being described in yeast cells. In budding yeast, the study of checkpoint genes has led to the view that these genes have many roles in the cellular responses to DNA damage, including roles in arrest in multiple stages of cell cycle, in transcriptional induction of repair genes, in DNA repair itself and additionally some undefined role in DNA replication. The checkpoint pathways and proteins that carry out these responses may consist of sensor proteins that detect damage, signaller proteins that transduce an inhibitory signal and target proteins that are altered to arrest cell division (or cause other changes in cell behaviour). Yeast genes that may act at each step have been identified, leading to a working model of checkpoint pathways. An initial step in the pathway may involve the processing of damage to an intermediate that signals arrest and acts in DNA repair. Human checkpoint pathways may have defects in processing damage as well.

References

Abousekkra A, Vialard JE, Morrison D *et al* A novel role for the budding yeast *RAD9* checkpoint gene in DNA damage-dependent transcription. *EMBO Journal* **15** 3912–3922

Alani E, Padmore R and Kleckner N (1990) Analysis of wild-type and rad50 mutants of yeast suggests an intimate relationship between meiotic chromosome synapsis and recombination **61** 419–436

Alfa CE, Ducommun B, Beach D and Hyams JS (1990) Distinct nuclear and spindle pole body populations of cylin-cdc2 in fission yeast. *Nature* **347** 680–682

Al-Khodairy G, Fotou E, Sheldrick KS, Griffiths DJF, Lehmann AR and Carr AM (1994) Identification and characterization of new elements involved in checkpoint and feedback controls in fission yeast. *Molecular Biology of the Cell* **5** 147–160

Allen JB, Zhou Z, Siede W, Friedberg EC and Elledge SJ (1994) The SAD1/RAD53 protein kinase controls multiple checkpoints and DNA damage-induced transcription in yeast. *Genes and Development* **8** 2416–2428

Amon A, Surana U, Muroff I and Nasmyth K (1992) Regulation of p34*cdc28* tyrosine phosphorylation is not required for entry into mitosis in *S. cerevisiae*. **355** 368–371

Anderson CW (1993) DNA damage and the DNA-activated protein kinase. *Trends in Biochemical Sciences* **18** 433–437

Barbet NC and Carr AM (1993) Fission yeast wee1 protein kinase is not required for DNA damage-dependent mitotic arrest. *Nature* **364** 824–827

Bauer GA and Burgers PMJ (1990) Molecular cloning, structure and expression of the yeast proliferating cell nuclear antigen gene. *Nucleic Acids Research* **18** 261–265

Beamish H and Lavin MF (1994) Radiosensitivity in ataxia-telangiectasia: anomalies in radiation-induced cell cycle delay. *Internation Journal of Radiation Biology* **65** 175–184

Blanco L, Bernard M, Blasco A and Salas M (1991) A general structure for DNA-dependent DNA polymerases. *Gene* **100** 27–38

Blunt T, Finnie NJ, Taccioli GE *et al* (1995) Defective DNA-dependent protein kinase activity is linked to V(D)J recombination and DNA repair defects associated with the murine SCID mutation. *Cell* **80** 813–823

Boubnov NV and Weaver DT (1995) scid cells are deficient in Ku and replication protein A phosphorylation in the DNA-dependent protein kinase. *Molecular and Cellular Biology* **15** 5700–5706

Broek D, Bartlett R, Crawford K and Nurse P (1991) Involvement of p34CDC2 in establishing the dependency of S phase on mitosis. *Nature* **349** 388–393

Carlson JG (1969) X-ray-induced prophase delay and reversion of selected cells in certain avian and mammalian tissues in culture. *Radiation Research* **37** 15–30

Carr AM (1995) DNA structure checkpoints in fission yeast. *Seminars in Cell Biology* **6** 65–72

Carr AM and Heokstra MF (1994) DNA repair and cell biology, two complementary strands of a proliferating story. *Trends in Cell Biology* (in press)

Cimprich KA, Shin TB, Keith CT and Schreiber SL (1996) cDNA cloning and gene mapping of a candidate human cell cycle checkpoint protein. *Proceedings of the National Academy of Sciences of the USA* **93** 2850–2855

Cross SM, Sanchez CA, Morgan CA *et al* (1995) A p53-dependent mouse spindle checkpoint. *Science* **267** 1353–1356

Cullman G, Fien K, Kobayashi R and Stillman B (1995) Characterization of the five replication factor C genes of *Saccharomyces cerevisiae*. **15** 4661–4671

Dasso M and Newport JW (1990) Completion of DNA replication is monitored by a feedback system that controls the initiation of mitosis in vitro: studies in *Xenopus*. *Cell* **61** 811–823

Deng C, Zhang P, Harper JW, Elledge SJ and Leder P (1995) Mice lacking p21 cip1/waf1 undergo normal development, but are defective in G_1 checkpoint control. *Cell* **82** 675–84

Donehower LA, Harvey M, Slagle BL *et al* (1992) Mice deficient for p53 are developmentally normal but susceptible to spontaneous tumours. *Nature* **356** 215–221

Dulic V, Kaufman WK, Wilson SJ *et al* (1994) p53-dependent inhibition of cyclin-dependent kinase activities in human fibroblasts during radiation-induced G_1 arrest. *Cell* **76** 1013–1023

Enoch T and Nurse P (1990) Mutation of fission yeast cell cycle control genes abolishes dependence of mitosis on DNA replication. *Cell* **60** 665–673

Fan S, Smith ML, Rivet DJ *et al* (1995) Disruption of p53 function sensitizes breast cancer MCF-7 cells to cisplatin and pentoxifylline. *Cancer Research* **55** 1649–1654

Ford JC, Al-Khodairy F, Fotou E, Sheldrick S, Griffiths JF and Carr AM (1994) *S. pombe* 14-3-3 homologues encode as essential function required for the DNA damage checkpoint. *Science* **265** 533–535

Friedberg EC, Walker GC and Siede W (1995) *DNA Repair and Mutagenesis*, ASM Press

Garvik B, Carson M and Hartwell L (1995) Single-stranded DNA arising at telomeres in cdc13 mutants may consistitue a specific signal for the *RAD9* checkpoint. *Molecular and Cellular Biology* **15** 6128–6138

Gottlieb TM and Jackson SP (1993) The DNA-dependent protein kinase: requirement for DNA ends and association with ku antigen. *Cell* **72** 131–142

Greenblatt MS, Bennett WP, Hollstein M and Harris CC (1994) Mutations in the p53 tumor suppressor gene: clues to cancer etiology and molecular pathogenesis. *Cancer Research* **54** 4855–4878

Greenwell PW, Kronmal SL, Porter SE, Gassenhuber J, Obermaier B and Petes TD (1995) *TEL1*, a gene involved in controlling telomere length in *S. cerevisiae*, is homologous to the human ataxia telangiectasia gene. *Cell* **82** 823–829

Griffiths DJG, Barbet NC, McCready S, Lehmann AR and Carr AM (1995) Fission yeast rad17: a homolog of budding yeast RAD24 that shares regions of seqeunce similarity with DNA polymerase accessory proteins. *EMBO Journal* **14** 101–112

Haber JE (1992) Exploring the pathways of homologous recombination. *Current Opinions in Cell Biology* **4** 401–412

Hari LK, Santerre A, Sekelsky JJ, McKim KS, Boyd JB and Hawley RS (1995) The mei-41 gene of *D. melanogaster* is a structural and functional homolog of the human ataxia telangiectasia gene. *Cell* **82** 815–821

Hartley KO, Gell D, Smith GCM *et al* (1995) DNA-dependent protein kinase catalytic subunit: A relative of phosphatidylinositol 3-kinase and the ataxia telangiectasia gene product. *Cell* **82** 849–856

Hartwell LH and Kastan MB (1994) Cell cycle control and cancer. *Science* **266** 1821–1828

Hartwell LH and Weinert TA (1989) Checkpoints: controls that ensure the order of cell cycle events. *Science* **246** 629–634

Hawn MT, Umar A, Arenthers JM *et al* (1995) Evidence for a connection between the mismatch repair system and the G_2 cell cycle checkpoint. *Cancer Research* **55** 3721–3725

Hayles J, Fisher D, Woollard A and Nurse P (1994) Temporal order of S phase and mitosis in fission yeast is determined by the state of th p34CDC2-mitotic B cyclin complex. *Cell* **78** 813–822

Heald R, MeLoughlin M and McKeon F (1993) Human Wee1 maintains mitotic timing by protecting the nucleus from cytoplasmically activated cdc2 kinase. *Cell* **74** 463–474

Huang LC, Clarkin KC and Wahl GM (1996) Sensitivity and selectivity of the DNA damage sensor responsible for activating p53-dependent G_1 arrest. *Proceedings of the National Academy of Sciences of the USA* **93** 4827–4832

Hunter T (1995) When is a lipid kinase not a lipid kinase? When it is a protein kinase. *Cell* **83** 1–4

Janyaraman L and Prives C (1995) Activation of p53 sequence-specific DNA binding by short single strands of DNA requires the p53 c-terminus. *Cell* **81** 1021–1029

Jimenez G, Yucel J, Rowley R and Subramani S (1992) The rad3+ gene of *Schizosaccharomyces pombe* is involved in multiple checkpoint function and in DNA repair. *Proceedings of the National Academy of Sciences of the USA* **89** 4952–4956

Jin P, Gu Y and Morgan DO (1996) Role of inhibitory CDC2 phosphorylation in radiation-induced G$_2$ arrest in human cells. *Journal of Cell Biology* **134** 963–970

Joachim JL and Deshaies RJ (1993) Exercising self-restraint: discouraging illicit acts of S and M in eukaryotes. *Cell* **74** 223–226

Kastan MB, Onyekwere O, Sidransky D, Vogelstein B and Craig RW (1991) Participation of p53 protein in the cellular response to DNA damage. *Cancer Research* **51** 6304–6311

Kato R and Ogawa H (1994) An essential gene, ESR1, is required for mitotic cell growth, DNA repair and meiotic recombination in *Saccharomyces cerevisiae*. *Nucleic Acids Research* **22** 3104–3112

Kelley TJ, Martin GS, Forsburg SL, Stephensen RJ, Russo A and Nurse P (1993) The fission yeast cdc18+ gene product couples S phase to START and mitosis. *Cell* **74** 371–382

Kiser GL and Weinert TA (1996) Distinct roles of yeast *MEC* and *RAD* checkpoint genes in transcriptional induction after DNA damage and implications for function. *Molecular Biology of the Cell* **7** 703–718

Kornbluth S, Smythe C and Newport JW (1992) In vitro cell cycle arrest induced by using artificial DNA templates. *Molecular and Cellular Biology* **12** 3216–3223

Kowalczykoski SC, Dixon DA and Rehrauer WM (1994) Biochemistry of homologous recombination in *Escherichia coli*. *Microbiological Reviews* 401–465

Kuerbitz SJ, Plunkett BS, Walsh WV and Kastan MB (1992) Wild-type p53 is a cell cycle checkpoint determinant following irradiation. *Proceedings of the National Academy of Sciences of the USA* **89** 7491–7495

Kumagai A and Dunphy WG (1991) The cdc25 protein controls tyrosine dephosphorylation of the cdc2 protein in a cell-free system. *Cell* **64** 903–914

Kumagai A and Dunphy WG (1995) Control of the Cdc2/cyclin B complex in *Xenopus* egg extracts arrested at a G$_2$/M checkpoint with DNA synthesis inhibitors. *Molecular Biology of the Cell* **6** 199–213

Larner JM, Lee H and Hamlin JL (1994) Radiation effects of DNA synthesis in a defined chromosomal replicon. *Molecular and Cellular Biology* **14** 1901–1908

Lee S, Elenbaas B, Levine A and Griffith J (1995) p53 and its 14kDa C-terminal domain recognize primary DNA damage in the form of insertion/deletion mismatches. *Cell* **81** 1013–1020

Lee TH and Kirschner MW (1996) An inhibitor of p34cdc2/cyclin B that regulates the G$_2$/M transitionin *Xenopus* extracts. *Proceedings of the National Academy of Sciences of the USA* **93** 352–356

Lees E (1995) Cyclin dependent kinase regulation. *Current Opinion in Cell Biology* **7** 773–80

Lew DJ and Reed SI (1995) A cell cycle checkpoint monitors cell morphogenesis in budding yeast. *Journal of Cell Biology* **129** 739–49

Little JW and Mount DW (1982) The SOS regulatory system of *Escherichia coli*. *Cell* **29** 11–22

Livingston LR, White A, Sprouse J, Livanos E, Jacks T and Tisty TD (1992) Altered cell cycle arrest and gene amplification potential accompany loss of wild-type p53. *Cell* **70** 923–935

Lydall D and Weinert T (1995) Yeast checkpoint genes in DNA damage processing: implications for repair and arrest. *Science* **270** 1488–1491

Lydall D and Weinert T (1996) From DNA damage to cell cycle arrest and suicide: a budding yeast perspective. *Current Opinion in Genetics and Development* **6** 4–11

Lydall D, Nikolsky Y, Bishop DK and Weinert T (1996) A meiotic recombination checkpoint controlled by mitotic checkpoint genes. *Nature* **383** 840–843

Malkin D, Li FP, Strong LC *et al* (1990) Germ-line p53 mutations in a familial syndrome of breast cancer, sarcomas and, other neoplasms. *Nature* **250** 1233–1238

Morgan DO (1995) Principles of CDK regulation. *Nature* **374** 131–134

Morozov VE, Falzon M, Anderson CW and Kuff EL (1994) DNA-dependent protein kinase is activated by nicks and larger single-stranded gaps. *Journal of Biological Chemistry* **269** 16684–16688

Morrow DM, Tagle DA, Shiloh Y, Collins FS and Hieter P (1995) TEL1, an *S. cerevisiae* homolog of the human gene mutated in ataxia telangiectasia, is functionally related to the yeast checkpoint gene MEC1. *Cell* **82** 831–840

Murakami H and Okayama H (1995) A kinase from fission yeast responsible for blocking mitosis in S phase. *Nature* **374** 817–819

Murray AW (1992) Creative blocks: cell-cycle checkpoints and feedback controls. *Nature* **359** 599–604

Murray AW (1995) The genetics of cell cycle checkpoints. *Current Opinion in Genetics and Development* **5** 5–11

Nagasawa H, Latt SA, Lalande ME and Little JB (1985) Effects of x-irradiation on cell-cycle progression, induction of chromosomal aberrations and cell killing in ataxia-telangiectasia (AT) fibroblasts. *Mutation Research* **148** 71–82

Nasr F, Becam A, Lonimski PP and Herbert CJ (1994) YBR1012 an essential gene from *S. cerevisiae*: construction of an RNA antisense conditional allele and isolation of a multicopy suppressor. *Comptes Rendus de l'Académie des Sciences, Paris* **317** 607–613

Navas TA, Zhou Z and Elledge SJ (1995) DNA polymerase epsilon links the DNA replication machinery to the S phase checkpoint. *Cell* **80** 29–39

Nelson WG and Kastan MB (1994) DNA strand breaks: the DNA template alterations that trigger p53-dependent DNA damage response pathways. *Molecular and Cellular Biology* **14**

Nurse P (1994) Ordering S phase and M phase in the cell cycle. *Cell* **79** 547–550

Nurse P and Bissett Y (1981) Gene regulated in G_1 for commitment to cell cycle and in G_2 for control of mitosis in fission yeast. *Nature* **292** 558–560

O'Connor PM, Ferris DK, Pagano M *et al* (1993) G_2 delay induced by nitrogen mustard in human cells affects cyclin A/cdk2 and cyclin B1/cdc2-kinase complexes differently. *Journal of Biological Chemistry* **268** 8298–8308

Oliver SG (1996) From DNA sequence to biological function. *Nature* **379** 597–600

Painter RB and Young RB (1980) Radiosensitivity in ataxia-telangiectasia: a new explanation. *Proceedings of the National Academy of Sciences of the USA* **77** 7315–7317

Papathansiou MA, Kerr NC, Robbins JH *et al,* (1991) Induction by ionizing radiation of the gadd45 gene in cultured human cells: lack of mediation by protein kinase. *Molecular and Cellular Biology* **11** 1009–1016

Paules RS, Levedakou EN, Wilson SJ *et al* (1995) Defective G_2 checkpoint function in cells from individuals with familial cancer syndromes. *Cancer Research* **55** 1763–1773

Paulovich AG and Hartwell LH (1995) A checkpoint regulates the rate of progression through S phase in *S. cerevisiae* in response to DNA damage. *Cell* **82** 841–847

Platti S, Lengauer C and Nasmyth K (1995) Cdc6 is an unstable protein whose de novo synthesis in G_1 is important for the onset of S phase and for preventing a 'reductional' anaphase in the budding yeast *S. cerevisiae*. *EMBO Journal* **14** 3788–3799

Powell SN, DeFrank JS, Connell P *et al* (1995) Differential sensitivity of p53⁻ and p53⁺ cells to caffeine-induced radiosensitization of G_2 delay. *Cancer Research* **55** 1643–1648

Rathmell WK and Chu G (1994) Involvement of the KU autoantigen in the cellular response to DNA double-strand breaks. *Proceedings of the National Academy of Sciences of the USA* **91** 7623–7627

Rudolph NS and Latt SA (1989) Flow cytometric analysis of x-ray sensitivy in ataxia-telangiectasia. *Mutation Research* **211** 31–41

Russell KJ, Wiens LW, Demers GW, Galloway DA, Plon SE and Groudine M (1995) Abrogation of the G_2 checkpoint results in differential radiosensitization of G_1 checkpoint-deficient and G_1 checkpoint-competent cells. *Cancer Research* **55** 1639–1642

Sanchez Y, Desany BA, Jones WJ, Jiu Q, Wang B and Elledge S (1996) Regulation of *RAD53* by the *ATM*-like kinases *MEC1* and *TEL1* in yeast cell cycle checkpoint pathways. *Science* **271** 357–360

Sassanfar M and Roberts JW (1990) Nature of the SOS-inducing signal in *Escherichia coli*. *Journal of Molecular Biology* **212** 79–96

Savitsky K, Bar-Shira A, Gilad S *et al* (1995) A single ataxia telangiectasia gene with a product similar to PI-3 kinase. *Science* **268** 1749–1752

Sheldrick KS and Carr AM (1993) Feedback controls and G_2 checkpoints: fission yeast as a model systems. *BioEssays* **15** 775–782

Siede W, Friedberg AS and Freidberg EC (1993) RAD9-dependent G_1 arrest defines a second checkpoint for damaged DNA in the cell cycle of *Saccharomyces cerevisiae*. *Proceedings of the National Academy of Sciences of the USA* **90** 7985–7989

Siede W, Friedberg AS, Dianova I and Friedberg EC (1994) Characterization of G_1 Checkpoint control in the yeast *Saccharomyces cerevisiae* following exposure to DNA-damaging agents. *Genetics* **138** 271–281

Sorger PK and Murray AW (1992) S-phase feedback control in budding yeast independent of tyrosine phophorylation of p34*cdc28*. *Nature* **355** 365–368

Stern DF, Zheng P, Beidler DR and Zerillo C (1991) Spk1, a new kinase from *Saccharomyces cerevisiae*, phosphorylates proteins on serine, threonine and, tyrosine. *Molecular and Cellular Biology* **11** 987–1001

Stueland CS, Lew DJ, Cismowski MJ and Reed SI (1993) Full activation of p34CDC28 histone H1 kinase activity is unable to promote entry into mitosis in checkpoint-arrested cells of the yeast *Saccharomyces cerevisiae*. *Molecular and Cellular Biology* **13** 3744–3755

Sugimoto K, Shimomura T, Hashimoto K, Araki H and Sugino A (1996) Rfc5, a subunit required for DNA replication, is involved in coupling DNA replication to mitosis in budding yeast. *Proceedings of the National Academy of Sciences of the USA* **93** 7048–7052

Sun Z, Fay DS, Marini F, Foiani M and Stern DF (1996) Spk1/Rad53 is regulated by Mec1-dependent protein phosphorylation in DNA replication and damage checkpoint pathways. *Genes and Development* **10** 395–406

Swift M, Morrel D, Massey RB and Chase CL (1991) Incidence of cancer in 161 families affected by Ataxia-telangiectasia. *New England Journal of Medicine* **325** 1831–1836

Taccioli GE, Gottlieb TM, Blunt T *et al* (1994) DU80: product of the XRCC5 gene and its role in DNA repair and V(D)J recombination. *Science* **265** 1442–1445

Terada Y, Tatsuka M, Jinno S and Okayama H (1995) Requirement for tyrosine phosphorylation of Cdk4 in G_1 arrest induced by untraviolet irradiation. *Nature* **376** 358–362

Toyn JH, Johnson AL and Johnston LH (1995) Segregation of unreplicated chromosomes in *Saccharomyces cerevisiae* reveals a novel G_1/M phase checkpoint. *Molecular and Cellular Biology* **15** 5312–5321

Tugendrich S, Bassett DE, McKusick V A, Boguski MS and Hieter P (1994) Genes conserved in yeast and humans. *Human Molecular Genetics* **3** 1509–1517

Walworth N, Davey S and Beach D (1993) Fission yeast chk1 protein kinase links the rad checkpoint pathway to cdc2. *Nature* **363** 368–371

Weinert TA and Hartwell LH (1988) The *RAD9* gene controls the cell cycle response to DNA damage in *Saccharomyces cerevisiae*. *Science* **241** 317–322

Weinert TA and Hartwell LH (1990) Characterization of *RAD9* of *Saccharomyces cerevisiae* and evidence that its function acts posttranslationally in cell cycle arrest after DNA damage. *Molecular and Cellular Biology* **10** 6554–6564

Weinert TA and Hartwell LH (1993) Cell cycle arrest of *cdc* mutants and specificity of the *RAD9* checkpoint. *Genetics* **134** 63–80

Weinert TA and Lydall D (1996) Pathways and puzzles in the DNA damage and replication checkpoints in yeast, In: Nickoloff JA and Hoekstra M (eds). *DNA Damage and Repair—Biochemistry, Genetics and Cell Biology,* Humana Press

Weinert TA, Kiser GL and Hartwell LH (1994) Mitotic checkpoint genes in budding yeast and the dependence of mitosis on DNA replication and repair. *Genes and Development* **8** 652–665

Wood JS and Hartwell LH (1982) A dependent pathway of gene functions leading to chromosome segregation in *Saccharomyces cerevisiae*. *Journal of Cell Biology* **94** 718–726

Yamamoto A, Guacci V and Koshland D (1996a) Pds1p is required for faithful execution of anaphase in yeast, *Saccharomyces cerevisiae*. *Journal of Cell Biology* **133** 85–97

Yamamoto A, Guacci V and Koshland D (1996b) Pds1p, an inhibitor of anaphase in budding yeast, plays a critical role in the APC and checkpoint pathways. *Journal of Cell Biology* **133** 99–110

Yin Y, Tainsky MA, Bischoff FZ, Strong LC and Wahl GM (1992) Wild-type p53 restores cell cycle control and inhibits gene amplification in cells with mutant p53 alleles. *Cell* **70** 937–948

Zheng P, Fay DS, Burton J, Xiao J, Pinkham JL and Stern DF (1993) SPK1 is an essential S-phase specific gene of *Saccharomyces cerevisiae* that encodes a nuclear serine/threonine/tryosine kinase. *Molecular and Cellular Biology* **13** 5829–5842

Zhou Z and Elledge SJ (1993) DUN1 encodes a protein kinase that controls the DNA damage response in yeast. *Cell* **75** 1119–1127

The author is responsible for the accuracy of the references.

The Anaphase Promoting Complex

A M PAGE • P HIETER

Department of Molecular Biology and Genetics, Johns Hopkins School of Medicine, 725 North Wolfe St, Baltimore, MD 21205

INTRODUCTION

Eukaryotic cell division is achieved with high fidelity by progression through an ordered series of cellular events known as the cell cycle. Each cell derived from a previous cell division undergoes a period of growth (G_1 phase), replicates its DNA (S phase), undergoes a second growth phase (G_2) and then enters mitosis (M phase), whereupon it divides to form two daughter cells. Progression through the cell cycle is controlled by a tightly regulated protein kinase activity. Maturation promoting factor (MPF) was first characterized as an activity that promoted multiple cell cycles in *Xenopus* and sea urchin eggs, and its activity was shown to require both a regulatory subunit needed for kinase activation, known as cyclin, and a catalytic subunit, a cyclin dependent kinase (CDK). Cyclin/CDK complexes phosphorylate specific substrates at appropriate stages during the cell cycle, driving the cellular events necessary to progress from one stage of the cell cycle to the next. Cyclin dependent CDK phosphorylation promotes very different events at different stages of the cell cycle. For cells preparing to undergo mitosis, mitotic cyclin/CDK complexes promote spindle pole body duplication, chromatin condensation and spindle assembly, whereas G_1/S phase cyclin/CDK complexes promote initiation and completion of DNA replication. In lower eukaryotes, a limited number of

CDKs can pair with a variety of different cyclins, each of which promotes the transition to a different phase of the cell cycle, whereas higher eukaryotes maintain a greater number of CDKs, which pair more selectively with their cyclin partners (reviewed in Murray and Hunt, 1993).

Cyclin synthesis is both necessary (Minshull *et al*, 1989) and sufficient (Murray and Kirschner, 1989) to drive the cell cycle, and consequently cyclin synthesis and degradation must occur in a periodic manner. At each stage of the cell cycle, cyclin levels must reach a certain threshold before the quiescent cyclin/CDK complex can be activated by a positive feedback mechanism. Once activated, the active cyclin/CDK complex can phosphorylate its specific substrate(s), promoting transition to the next stage of the cell cycle. After the cyclin/CDK complexes have phosphorylated their substrate(s), cyclins are rapidly degraded. Cells that are unable to degrade cyclins arrest and cannot make the transition into the next stage of the cell cycle.

UBIQUITIN MEDIATED CYCLIN DEGRADATION DRIVES THE CELL CYCLE TO COMPLETION

The inactivation of MPF is a crucial event that must occur to allow cells to exit mitosis and complete the cell cycle. The process of MPF inactivation occurs by the ubiquitin mediated proteolysis of cyclins. In both *Xenopus* extracts (Murray *et al*, 1989) and budding yeast (Ghiara *et al*, 1991), expression of a nondegradable truncated form of cyclin B arrests the cell cycle before completion of mitosis. Therefore, cyclin degradation is necessary for MPF inactivation, thus allowing cells to exit mitosis and enter interphase of the next cycle. Furthermore, both truncated (Glotzer *et al*, 1991) and full length cyclin B (Mahaffey *et al*, 1995) are polyubiquitinated. Additionally, a short N-terminal portion of cyclin B, named the destruction box, is essential for cyclin ubiquitination and is sufficient to cause an unrelated protein to become polyubiquitinated and degraded during mitosis (Glotzer *et al*, 1991). The destruction box of mitotic cyclins A and B is highly conserved between many species and contains the consensus sequence RXXLGXIXN.

Ubiquitination—a Brief Excursion

In the ubiquitination pathway (Fig. 1), the C-terminus of ubiquitin is covalently linked to a cysteine residue of a non-specific ubiquitin activating enzyme, known as an E1. This reaction forms a thioester bond and requires ATP. The E1-ubiquitin conjugate then transfers its ubiquitin moiety to another ubiquitin conjugating enzyme, known as an E2. Ubiquitin can then be transferred from the E2 directly onto the ε-amino groups of lysine residues of target proteins, but often ubiquitination of target proteins requires a specific ubiquitin protein ligase, or E3. E3 proteins are thought to confer substrate specificity for ubiquitination of target proteins in conjunction with non-specific E2 enzymes. The E3 may receive ubiquitin from the E2 and then transfer it to the target

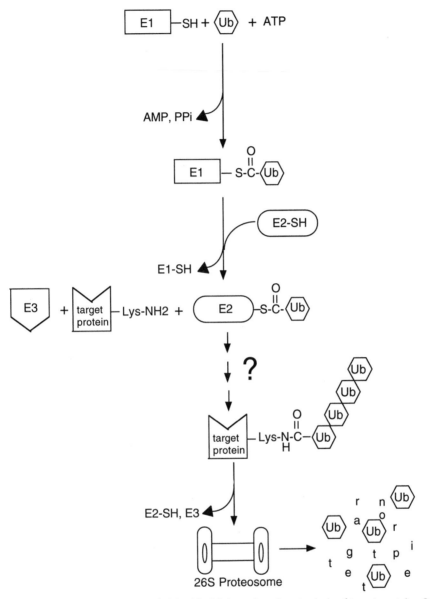

Fig. 1. Generalized pathway of ubiquitin (Ub) mediated proteolysis of target proteins. Ppi = pyrophosphate. The letters a,e,g,i,n,o,p,r,t are from the words "target protein" and represent the degradation of the target protein into its aminoacid components

protein, or it may simply facilitate the transfer of ubiquitin from the E2 directly to the target. After the initial ubiquitin molecule is attached to the target, additional ubiquitin molecules are added to lysines of the first ubiquitin moiety, forming polyubiquitin chains. The polyubiquitinated protein is then degraded by a large complex of proteins known as the proteasome, and the ubiquitin molecules are recycled and attached to new targets (for a more detailed review of ubiquitin mediated proteolysis, see Ciechanover 1994).

Cyclin Destruction

Several lines of investigation have expanded the hypothesis that cyclin degradation occurs by the ubiquitination pathway in a cell cycle regulated manner. Polyubiquitination was shown to be essential for the destruction of cyclin when the application of a methylated ubiquitin analogue incapable of forming polyubiquitin chains inhibited cyclin degradation (Hershko *et al*, 1991). In addition, Luca and Ruderman (1989) demonstrated that cyclin proteolysis required protein synthesis in a cell free extract made from cells in early interphase, but not from cells in mitosis. The destruction of two different mitotic cyclins, cyclin A and cyclin B, was initiated at slightly different times during the cell cycle, depending on the cell cycle phase of the cytoplasm and not that of the cyclins themselves (Luca and Ruderman, 1989). One very important discovery came when two groups working in very different eukaryotic systems were able to refute the previous hypothesis that inactivation of the mitotic cyclin B/CDK complex (in mitosis, cyclin B is paired with the quintessential mitotic CDK p34^{cdc2}) was necessary for sister chromatid separation. Holloway and colleagues discovered that addition of a truncated form of cyclin B lacking the destruction box to *Xenopus* extracts caused MPF to remain active but allowed chromosomes on the mitotic spindle to progress from metaphase to anaphase. Addition of either the destruction box portion of cyclin B or polyubiquitination inhibitors to extracts, however, inhibited both MPF inactivation and sister chromatid separation, suggesting that destruction of both mitotic cyclins and the proteins involved in maintaining sister chromatid adhesion occurs by the same mechanism (Holloway *et al*, 1993). Surana *et al* (1993) found that expression of Clb2p (a mitotic budding yeast homologue of cyclin B) lacking a destruction box caused budding yeast cells to arrest not in metaphase but in anaphase with separated DNA and elongated spindles.

A MITOSIS SPECIFIC PROTEIN COMPLEX WITH E3 ACTIVITY IS ESSENTIAL FOR CYCLIN UBIQUITINATION

Next, researchers began to identify the components necessary for cyclin ubiquitination in vitro. By partially fractionating clam oocytes, Hershko *et al* (1994) showed that the in vitro ubiquitination of a cyclin destruction box–protein A fusion required both E1 and E2 activities, as well as an E3 activity of high molecular weight. The E3 activity was active only when purified from M phase extracts, but neither the E1 nor the E2 activity was cell cycle specific. Furthermore, addition of cdc2 protein to purified E3 stimulated the ubiquitination reaction, even when the E3 activity was purified from interphase oocytes. This finding suggested that cdc2 kinase activity promotes mitosis specific activation of the E3 complex, either directly by cdc2 mediated phosphorylation of E3 components or indirectly by phosphorylation of other regulators. Thus, the active cyclin B/cdc2 kinase complex appears to catalyze its own inactivation at the end of mitosis. Subsequently, two laboratories were able to purify partially the E3 complex from different systems. Sudakin *et al* (1995) isolated from clam oocytes a 1500 kDa complex, named the cyclosome, that ubiquitinated cyclins A and B in an in vitro system composed of an E1, E2, ubiquitin and an ATP regenerating system. King *et al* (1995) similarly purified a 20S complex from *Xenopus* oocytes, named the anaphase promoting complex (APC), that also supported cyclin B ubiquitination in vitro. Both groups found that the E3 complex was active only when purified from mitotic extracts. The term APC is now used widely throughout the literature to describe the complex previously identified as the cyclosome.

While others were biochemically defining the E3 activity, researchers in our laboratory were studying three proteins, encoded by the *CDC16*, *CDC23* and *CDC27* genes in the budding yeast *Saccharomyces cerevisiae* (Icho and Wickner, 1987; Sikorski *et al*, 1990, 1991, 1993). *CDC16*, *CDC23* and *CDC27* were first identified in the collection of temperature sensitive cell division cycle mutants (Hartwell *et al*, 1970). At the non-permissive temperature, *cdc16*, *cdc23* and *cdc27* mutants all arrest as large budded cells with short spindles, a 2N DNA content and unsegregated chromosomes. Unlike most medial nuclear division *cdc* mutants, the terminal arrest phenotypes of *cdc16*, *cdc23* and *cdc27* mutants are independent of the *RAD9* mediated checkpoint control, suggesting a role in chromosome segregation rather than DNA metabolism (Weinert and Hartwell, 1993). Cdc16p, Cdc23p and Cdc27p are members of a large (20S) complex and physically associate with each other, as well as with themselves, as shown by co-immunoprecipitation, the yeast two hybrid system (Lamb *et al*, 1994) and co-migration on sucrose gradients (Lamb J, personal communication).

Homologues of *CDC16* and *CDC27* have been found in several other eukaryotic species. The fission yeast *Schizosaccharomyces pombe* contains a *CDC16* homologue, *cut9*[+] (Samejima *et al*, 1993; Samejima and Yanagida, 1994) and a *CDC27* homologue, *nuc2*[+] (Hirano *et al*, 1988, 1990). Like their counterparts in *S cerevisiae*, temperature sensitive mutations in either gene cause arrest in metaphase with condensed chromosomes. The *nuc2-663* mutation causes arrest in metaphase with a shortened spindle, is hypersensitive to caffeine and cannot mate. The *cut9-665* mutation arrests in mitosis with unseparated chromosomes but can complete postanaphase events such as spindle degradation, septum formation and cytokinesis. At present, there is no known homologue of *CDC23* in *S pombe*. A homologue of *CDC27*, *BimA*, exists in the filamentous fungus *Aspergillus nidulans*. BimA protein localizes to spindle pole bodies during all phases of the cell cycle (Mirabito and Morris, 1993), and like the *cdc27-1* and *nuc2-663* temperature sensitive mutants, the *bimA1* mutant arrests in mitosis, producing cells with condensed chromatin and short, intact mitotic spindles (O'Donnell *et al*, 1991).

Tugendreich and co-workers cloned human homologues of *CDC16* and *CDC27* and showed that CDC16Hs and CDC27Hs proteins both localized to the centrosome and mitotic spindle in mammalian cell lines, consistent with their putative roles as either structural or regulatory components of the spindle. However, microinjection of anti-CDC27Hs antibodies (but not anti-CDC16Hs antisera) arrested HeLa cells at metaphase with intact spindles, suggesting that the role of CDC27Hs is regulatory rather than structural and that the function of CDC27Hs takes place after spindle formation (Tugendreich *et al*, 1995). Anti-CDC16Hs and anti-CDC27Hs antisera recognize two components of the 20S *Xenopus* complex (King *et al*, 1995). Immunodepletion of the 20S complex with anti-CDC27Hs antisera ablated cyclin ubiquitination activity, and the immunoprecipitated complex was sufficient to function as an E3 in in vitro ubiquitination assays. These findings proved that Cdc16p and Cdc27p were components of the E3 complex and suggested that the Cdc16p/Cdc23p/Cdc27p complex promotes the transition from metaphase to anaphase by ubiquitinating mitotic cyclins and targeting them for destruction.

In a separate effort, Stephan Irniger in Kim Nasmyth's lab conducted an elegant genetic screen in budding yeast which showed that *CDC16* and *CDC23* is essential for cyclin B (Clb2p) degradation in vivo (Irniger *et al*, 1995). Previously, genetic screens for proteins that control the turnover of mitotic cyclins were complicated by the fact that mutations in proteins involved in cyclin destruction could not be distinguished from those that simply cause cells to arrest in mitosis for other reasons. Irniger's screen took advantage of the fact that in budding yeast, cyclin destruction activity continues after mitosis and cell division into G_1 of the next cell cycle. In this screen, mutagenized cells were arrested in G_1 by methionine induced repression of G_1 cyclins. Expression of a *CLB2-lacZ* fusion under the control of the GAL1 promoter was induced, and the mutagenized cells were shifted to the non-permissive temperature. At the non-permissive temperature, mutants that could not degrade cyclin B properly formed blue colonies, whereas colonies competent for cyclin proteolysis remained white. Irniger and co-workers isolated temperature sensitive mutants of *CDC16* and CDC23, as well as in *CSE1*, whose protein product is implicated in chromosome stability but does not seem to be part of the APC. Surprisingly, a temperature sensitive allele of *CDC27* (*cdc27-1*) failed to stabilize the Clb2p-LacZ fusion in this assay. Later, Irniger and Nasmyth (unpublished) showed that other mitotic (Clb3p) and S phase (Clb5p) cyclins are stabilized in G_1 arrested cells in APC mutants, suggesting that, in budding yeast, all B type cyclins may be substrates of the anaphase promoting complex. Zachariae and Nasmyth (1996) followed up the cyclin stabilization screen in budding yeast by demonstrating that cyclin ubiquitination is inhibited in vivo in APC mutants. Wild type extracts prepared from cells arrested in G_1 with yeast mating pheromone ubiquitinated epitope tagged Clb2p and Clb3p, but extracts prepared from yeast strains carrying mutations in APC components (*cdc23-1*, *cdc16-123* and *cdc27-1*) failed to ubiquitinate them. Furthermore, cyclin ubiquitination in wild type extracts was cell cycle stage specific and was severely reduced in extracts made from cells arrested in G_2 or mitosis. The in vivo stabilization of Clb2p-LacZ fusions and inhibition of cyclin B ubiquitination in APC mutants provide the most compelling evidence that the role of the APC is to promote the degradation of mitotic cyclins.

SEVERAL APC COMPONENTS ARE MEMBERS OF THE TPR PROTEIN FAMILY

The three well characterized APC components, Cdc16p, Cdc23p and Cdc27p, belong to a family of proteins that contain a repeated 34 aminoacid sequence known as a tetratricopeptide repeat (TPR) domain (Hirano *et al*, 1990; Sikorski *et al*, 1990; Goebl and Yanagida, 1991; Mirabito and Morris, 1993; Lamb *et al*, 1995). Several TPR proteins are involved in cell cycle control, and proteins containing the TPR motif have been linked to a variety of cellular processes including, in budding yeast, transcriptional regulation (Ski3p, Ssn6p), mitochondrial import (Mas70p), stress response (Sti1p) and peroxisomal import (Pas10p). TPR proteins usually contain multiple TPR repeats, and these repeats are often clustered in a tandem array. Although the aminoacid sequence can vary considerably between TPR repeats in different proteins, or even between different TPR domains in the same molecule, individual consensus residues (...W..LG..Y........A...F..A....P..) are highly conserved at the levels of sequence similarity and spacing (Fig. 2). These conserved residues define the TPR consensus motif, and they fall into two clusters or subdomains, A and B, based on their position in the TPR repeat. Subdomain A contains a consensus tryptophan residue at position 4 and a small aliphatic aminoacid, usually leucine at position 7, a glycine or alanine at position 8 and a consensus phenylalanine or tyrosine at position 11. Subdomain B contains an alanine at position 20, a tyrosine or phenylalanine at position 24 and an alanine at position 27. In addition, a proline is conserved at position 32. The remaining residues may be highly divergent among different TPR domains in the same protein, but many are generally conserved between corresponding TPR domains in orthologous proteins of different species. Such conservation suggests that the specific function of a given TPR domain in a specific protein has been maintained throughout evolution. Computer assisted structure prediction analysis and circular dichroism data both suggest that each subdomain of the TPR repeat forms an α helix (Hirano *et al*, 1990; Goebl and Yanagida, 1991; Sikorski *et al*, 1990). The conserved hydrophobic residues are predicted to lie on the same face of the helix, forming a surface for hydrophobic interaction between a subdomain A and a subdomain B from different repeats, possibly via a "knob and hole" mechanism. The opposite face of the helix often contains charged residues, potentially providing an additional mechanism for stabilization of tertiary structure.

TPR domains are essential for the function of the proteins in which they reside, since removal of even a single TPR domain can ablate protein function (Sikorski *et al*, 1993). In addition, many of the original temperature sensitive mutations in genes encoding the TPR proteins, including *nuc2-663*, *cut9-665*

A.

```
                      1 2 3 4 5 6 7 8 9 10 11 12 13 14 15 16 17 18 19 20 21 22 23 24 25 26 27 28 29 30 31 32 33 34
CDC23Sc    TPR6       T N A W T L M G H E  F  V  E  L  S  N  S  H  A  A  I  E  C  Y  R  R  A  V  D  I  C  P  R  D
CDC27Sc    TPR5       P E T W C C I G N L  L  S  L  Q  K  D  H  D  A  A  I  K  A  F  E  K  A  T  Q  L  D  P  N  F
CDC16Sc    TPR7       H L P K L F L G M Q  F  M  A  M  N  S  L  N  L  A  E  S  Y  F  V  L  A  Y  D  I  C  P  N  D
nuc2+      TPR7       Y N A W Y G L G M V  Y  L  K  T  G  R  N  D  Q  A  D  F  H  F  Q  R  A  A  E  I  N  P  N  N
cut9+      TPR9       A A T W A N L G H A  Y  R  K  L  K  M  Y  D  A  A  I  D  A  L  N  Q  G  L  L  L  S  T  M  D
BimA       TPR7       Y N A W Y G L G T V  Y  D  K  M  G  K  L  D  F  A  E  Q  H  F  R  N  A  A  K  I  N  P  S  N
BimA       TPR5       P E A W C A V G N S  F  S  H  Q  R  D  H  D  Q  A  L  K  C  F  K  R  A  T  Q  L  D  P  H  F
CDC27Dm    TPR6       Y V S Y T L L G H E  L  V  L  T  E  E  F  D  K  A  M  D  Y  F  R  A  A  V  V  R  F  P  R  H
CDC27Hs    TPR7       Y N A W Y G L G M I  Y  Y  K  Q  E  K  F  S  L  A  E  M  H  F  Q  K  A  L  D  I  N  P  Q  S
TPR Consensus residues X X X W X X L G X X Y  X  X  X  X  X  X  X  X  A  X  X  X  F  X  X  A  X  X  X  X  P  X  X
                           |------------ Subdomain A ------------|         |------------ Subdomain B ------------|
```

B.

```
TPR5   nuc2+     P E S W C I L A N W F S L Q R E H S Q A I K C I N R A I Q L D P T F
       BimA      P E A W C A V G N S F S H Q R D H D Q A I K C I N R A I Q L D P D F
       CDC27Dm   P V T W C V S G N C F S L Q K E H E T A I K F F K R A V Q V D P D F
       CDC27Hs   P E A W C A A G N C F S L Q R E H D I A I K F F Q R A I Q V D P N Y
       CDC27Sc   P E T W C C I G N L L S L Q K D H D A A I K A F E K A T Q L D P N F

TPR7   nuc2+     Y N A W Y G L G M V Y L K T G R N D Q A D F H F Q R A A E I N P N N
       BimA      Y N A W Y G L G T V Y D K M G K L D F A E Q H F R N A A K I N P S N
       CDC27Dm   Y N A W Y G I G T I Y S K Q E K Y E L A E I H Y V K A L K I N P Q N
       CDC27Hs   Y N A W Y G L G M I Y Y K Q E K F S L A E M H F Q K A L D I N P Q S
       CDC27Sc   Y N A W Y G L G T S A M K L G Q Y E E A L L Y F E K A R S I N P V N

TPR6   CDC16Sc   A A A W L G F A H T Y A L E G E Q D Q A L T A Y S T A S R F F P G M
       cut9+     G P A W I G F A H S F A I E G E H D Q A I S A Y T T A A R L F Q G T
       CDC16Hs   G P A W I A Y G H S F A V E S E H D Q A M A A Y F T A A Q L M K G C

TPR7   CDC16Sc   H L P K L F L G M Q F M A M N S L N L A E S Y F V L A Y D I C P N D
       cut9+     H L P Y L F L G M Q H M Q L G N I L L A N E Y L Q S S Y A L F Q Y D
       CDC16Hs   H L P M L Y I G L E Y G L T N N S K L A E R F F S Q A L S I A P E D
```

Fig. 2. TPR consensus residues are conserved in TPR domains of different proteins (A), and many non-consensus residues are conserved between species within the same TPR repeat in homologous proteins (B)

and *cdc23-1,* cause changes within consensus residues in TPR domains (Hirano *et al*, 1990; Sikorski *et al*, 1993; Samejima and Yanagida, 1994). These mutations indicate that TPR segments are important, but their precise biochemical roles remain a matter of speculation. Intramolecular interactions between different TPR domains on the same molecule may contribute to the tertiary structure of TPR containing proteins, as has been proposed for *nuc2+* (Hirano *et al*, 1990). Alternatively, TPR repeats may mediate intermolecular interactions between TPR proteins and non-TPR proteins or between different members of the TPR protein family, as exemplified by the interaction between Cdc16, Cdc23p and Cdc27p. In the case of the APC, one function of TPR domains is to mediate binding to other members of the complex, since a mutation in TPR domain 7 of *CDC27Sc* (*cdc27-663*, the same glycine to asparagine change found in the original *nuc2-663* mutant) has a deleterious effect on Cdc23p binding (Lamb *et al*, 1994).

THE COMPLETE CYCLOSOME

A joint effort between several laboratories is under way to identify and characterize the remaining components of the APC (Fig. 3). Using antisera raised against human CDC27 and CDC16 proteins, Peters and King have immuno-

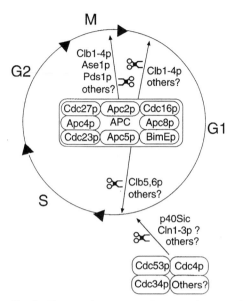

Fig. 3. The anaphase promoting complex contains Cdc27p, Cdc16p, Cdc23p and BimE protein and targets B type cyclins and other cell cycle related proteins for ubiquitin mediated proteolysis

precipitated the active complex from *Xenopus* oocytes (Peters *et al*, 1996). On a silver stained gel, the complex appears to be composed of at least eight major peptides. Microsequencing of these peptides revealed that three of the bands represent homologues of CDC16, CDC23 and CDC27, confirming the earlier co-immunoprecipitation results in budding yeast. A fourth band is homologous to Tsg24, the murine homologue of *A nidulans* BimE (both described below), a putative negative regulator of mitosis. To our knowledge, the four remaining bands (designated APC2, APC4, APC5 and APC8) do not match previously characterized proteins (Peters *et al*, 1996). Work is under way to determine whether these uncharacterized components have functional homologues in S *cerevisiae*. In addition, genetic screens in lower eukaryotes such as budding or fission yeast may identify other APC components not identified by immuno-precipitation.

The *bimE* gene of *A nidulans* was originally identified via a temperature sensitive mutation, *bimE7*, that causes cells to enter mitosis prematurely at the restrictive temperature. *bimE7* mutants aberrantly enter mitosis even if they are previously arrested in G_2 or S phase with other temperature sensitive mutations, and they arrest with condensed chromatin and mitotic spindles (Osmani *et al*, 1988). The *bimE7* mutation also causes cells arrested in S phase with hydroxyurea to form mitotic spindles. The fact that the *bimE7* mutant can bypass a G_2 or S phase arrest suggests that bimE protein acts as a negative regulator of mitosis. The *bimE* gene is essential, and its disruption led to a mitotic arrest phenotype, similar to the original mutant. The *bimE* gene has been cloned, and its open reading frame encodes a protein with three putative membrane spanning domains (Engle *et al,* 1990). A murine homologue of *bimE, Tsg24,* was isolated in a screen for cDNAs strongly expressed in murine germ cells (Starborg *et al,* 1994). Tsg24 protein is strongly expressed at interphase of 3T3 fibroblasts, and it appears to localize to the centromeres of chromosomes in a cell cycle independent manner (Jorgensen PM and Hoog C, unpublished).

The potential role of *bimE* in the APC as a putative negative regulator of mitosis is intriguing. The premature entry of *bimE7* mutants into mitosis cannot be explained easily by a failure to degrade mitotic cyclins, since the chromosome condensation and mitotic spindle formation that occur in *bimE7* mutants in *Aspergillus* are probably caused by the accumulation and activation, rather than destruction, of mitotic cyclin/CDK complexes, and these events have already occurred by the time cyclins are degraded. One possible explanation is that the premature entry into mitosis seen in *bimE7* mutants is caused by premature degradation of S and G$_2$ phase cyclins. Alternatively, BimE protein may act to prevent accumulation or activation of mitotic cyclins in response to checkpoints that monitor completion of upstream events such as DNA replication and proper spindle assembly, and so non-functional BimE protein may permit premature activation of mitotic cyclin/CDK complexes. In addition, *bimE* is also essential for the successful completion of mitosis, since *bimE7* mutants fail to enter anaphase. Recently, the *S cerevisiae* homologue of BimE/Tsg24 was identified by Kim Nasmyth's laboratory as another gene whose mutants stabilize the Clb2p-LacZ fusion protein (Zachariae *et al,* 1996). The relationship between the two putative roles of *bimE,* the negative regulation of mitosis and mitotic cyclin degradation, remains to be determined.

REGULATION OF THE APC

The APC appears to be regulated by phosphorylation during mitosis. A comparison between *Xenopus* APC components immunoprecipitated from interphase and mitotic extracts reveals that the three bands corresponding to BimE/Tsg24, CDC27 and at least one other protein all migrate considerably more slowly when in the mitotic state. Treatment with phosphatase restored their migration profile to that seen in interphase extracts (Peters *et al,* 1996). In addition, CDC16Hs has also been shown to be phosphorylated in a cell cycle dependent manner (Tugendreich S, Page A, Campbell M *et al,* unpublished).

The addition of cdc2 kinase to a cyclosome fraction from clam oocytes activates cyclin ubiquitination even when the cyclosome fraction is prepared from interphase extracts (Hershko *et al,* 1994). Furthermore, the addition of Mg^{++} ATP to a cyclosome fraction prepared from mitotic extracts stimulates cyclin ubiquitination to levels above those of untreated mitotic extracts. These findings suggest that phosphorylation events activate APC activity. The kinetics of interphase cyclosome activation by cdc2/p34 kinase reveal a lag period, suggesting that the cyclosome is not phosphorylated directly by cdc2/p34, but by other kinases that are themselves regulated by cdc2 protein (Hershko et al, 1994; Sudakin et al, 1995). One putative candidate for such a kinase in mammalian cells is a polo like kinase, Plk (Clay et al, 1993; Lake and Jelinek, 1993; Golsteyn et al, 1994,1995; Hamanaka et al, 1994, 1995). Murine Plk is predominantly active at G_2 and M phases, and it localizes to the mitotic spindle and centrosomes. Co-immunoprecipitation and yeast two hybrid data suggest that PLK interacts physically with Cdc27 and Cdc16, but so far it does not seem to correspond to any of the eight proteins immunoprecipitated from *Xenopus* (Kotani S, Tugendreich S, Hieter P et al, personal communication). We do not know whether different kinases specifically phosphorylate different APC components.

The APC must be deactivated to allow for the reaccumulation of mitotic cyclins as cells prepare to enter mitosis. There is no evidence that APC activity is regulated at the transcriptional or translational level, nor is there evidence that the APC itself is deactivated by cell cycle dependent proteolysis, since known APC components seem to be maintained at comparable levels throughout the cell cycle (Lamb J, personal communication). Since the APC seems to be activated by phosphorylation, it is likely that cells inactivate the APC complex by dephosphorylation. Indeed, phosphatase activity is essential for the successful completion of mitosis in several organisms (Doonan and Morris, 1989; Ohkura et al, 1989; Axton et al, 1990; Hisamoto et al, 1994), although no known APC components contain or are associated with phosphatase activities. An okadaic acid sensitive phosphatase activity that inactivates cyclosomes isolated from mitotic clam extracts has been described, but so far the identity of the protein responsible for this activity is unknown (Lahav-Baratz et al, 1995). The 3F3/2 antibody, an antiphosphoepitope antibody thought to recognize kinetochore components not properly attached to the mitotic spindle (Gorbsky and Ricketts, 1993), can immunoprecipitate human CDC27 and CDC16, and yet neither protein seems to display the 3F3/2 epitope (Tugendreich S, Page A, Campbell M et al, unpublished). Microinjection of anti-human CDC27 antibodies does not alter the 3F3/2 signal, suggesting that CDC27 is not required for establishing the kinetochore checkpoint. Given the available data, one potential model is that an APC component, perhaps BimE protein, is phosphorylated until all of the chromosomes are properly aligned on the metaphase plate. The phosphorylation of this component would inhibit APC mediated ubiquitination of mitotic cyclins and sister chromatid adhesion proteins until its inhibition is released by a phosphatase. Kinases activated by mitotic checkpoints in response to DNA damage, a perturbed spindle or misaligned chromosomes would maintain phosphorylation on the regulatory component of the APC until any damage was repaired, and subsequent dephosphorylation of the regulatory component would allow APC activation and subsequent entry into anaphase. The co-ordination of both phosphorylation and phosphatase events on different APC subunits may therefore be necessary to both activate and inactivate the APC.

NON-CYCLIN SUBSTRATES

Recent work in several laboratories suggests that the role of the APC is not limited to the degradation of mitotic cyclins. As mentioned earlier, the work of Holloway *et al* demonstrated that cyclin degradation was not necessary for exit from mitosis but that sister chromatid separation could be inhibited by over-expression of the cyclin destruction box. In budding yeast, Irniger and Nas-myth have found that failure to undergo anaphase in *cdc23-1* mutants can be overcome by causing cells to pass through mitosis without DNA replication (and therefore without sister chromatids) and into a "reductional anaphase", implying that a major function of Cdc23p is the separation of sister chromatids (Irniger *et al*, 1995). Thus, the APC appears to be responsible for the degradation of other substrates in addition to mitotic cyclins, perhaps including those proteins directly involved in sister chromatid adhesion. One such candidate is Pds1p, a protein shown to be necessary for sister chromatid adhesion in budding yeast. In wild type cells, Pds1p is degraded as cells exit from mitosis, and Pds1p contains a region that is weakly homologous to the destruction box of Clb2p. *cdc23-1* and *cdc16-123* mutations stabilize Pds1p in vivo, clearly implicating it as an APC substrate (Cohen-Fix *et al*, 1996). Another candidate is Ase1p, a spindle protein required for spindle pole body separation, mutations in which cause arrest in anaphase. Ase1p is stabilized in *cdc23* mutants in budding yeast, and mitotic *Xenopus* APC extracts can degrade yeast Ase1p (Juang YL, Peters JM, McLaughlin ME *et al*, unpublished).

A ROLE IN G₁ AND S PHASE?

Accumulating evidence implicates the APC in the destruction of proteins required for progression through both G_1 and S phases. A screen in budding yeast for temperature sensitive mutants that rereplicate their DNA but fail to undergo cell division identified new alleles of *CDC16* and *CDC27* (Heichman and Roberts, 1996). One untested hypothesis is that the APC may be responsible for the destruction of an unknown "licensing factor" that restricts DNA replication to one round per cell cycle. Alternatively, the failure of APC mutants to degrade B type cyclins in G_1 may promote unregulated entry into S phase and uncontrolled DNA replication (Irniger S and Nasmyth K, unpublished). The latter hypothesis is supported by the fact that the *nuc2-663* mutant fails to arrest in G_1 after nitrogen starvation and accumulates with a 2N DNA content (Kumada *et al*, 1995).

In budding yeast, a complex of proteins, including Cdc4p, Cdc34p and Cdc53p, has been implicated in the turnover of G_1 cyclins (Cln1p, Cln2p and Cln3p) (Willems *et al*, 1996; Mathias R, Steussy N, Goebl M *et al*, unpublished; Verma R, Feldman R, Correll C *et al*, unpublished). Cdc34p exhibits a ubiquitin conjugating activity (Goebl *et al*, 1988), and the complex may act as a G_1/S phase specific version of the APC. The Cdc4p/Cdc34p/Cdc53p complex is also essential for the destruction of p40^{SIC1}, a negative regulator of cyclin B/CDK complexes (Schwob E, Bohm T and Nasmyth K, unpublished; Verma R, Feldman R, Correll C *et al*, unpublished). Cln2p, Cln3p and p40^{SIC1} all contain PEST sequences (aminoacid sequences rich in proline, glutamic acid, serine and threonine), which enhance, but do not seem to be required for, degradation. Preliminary data suggest that components of the APC may also interact physically with members of the Cdc4p/Cdc34p/Cdc53p complex (Kraig KL, Patton EE, Willems AR *et al*, unpublished), but these findings need to be verified. Additional components of this "G$_1$/S cyclosome" have not yet been identified, and it will be interesting to see whether this complex and the APC do indeed share common subunits. The activities of the two complexes complement each other nicely: the Cdc4p/Cdc34p/Cdc53p complex becomes active at the stage of the cell cycle when the cyclosome is inactive, and vice versa.

THE APC AND CANCER—A POSSIBLE LINK?

As the APC is an essential machinery for cell cycle regulated cyclin degradation, it may have a potential role in the loss of control of cell proliferation in mammalian cells. At present, however, a direct link between defects in APC components and oncogenesis has not been established. None of the known APC components has been directly implicated as oncogenes. Both G$_1$/S phase cyclins (cyclins C, D and E) and, less frequently, mitotic cyclins (cyclins A and B) are overexpressed or aberrantly stabilized in a variety of tumour cell types, including breast tumours (cyclins A, B1, C, D1, D2, D3 and E) and hepatocellular carcinoma (cyclin A) (reviewed in Hunter and Pines, 1991, 1994; Buckley *et al*, 1993). One mechanism by which the normal cell cycle is altered is the increased transcription of cyclin genes, producing aberrantly high cyclin mRNA levels in tumour cells. In eukaryotic cells, failure to degrade mitotic cyclins (cyclins A and B) would be expected to arrest cells in mitosis, rather than promote uncontrolled progression through the cell cycle. The aberrant regulation of cyclins C, D and E is more frequently implicated in cancer pathogenesis; the levels of these G$_1$/S phase cyclins are controlled transcriptionally, and their degradation appears to be mediated by PEST sequences rather than by the destruction box. Perhaps cyclin C, D and E destruction is mediated by a mammalian equivalent of the Cdc4p/Cdc34p/Cdc53p complex.

Despite the lack of direct evidence implicating APC defects and a failure to degrade mitotic cyclins as a primary cause of transformation, it is easy to see how aberrant APC function might contribute to neoplasia in more subtle ways. Firstly, cells that have lost the ability to regulate APC function negatively might try to complete mitosis with damaged chromosomes or mitotic spindles, leading to the genomic instability that can promote oncogenesis. Mutations in regulatory components of the APC could result in a failure of the checkpoint machinery to halt APC mediated cyclin degradation before successful completion of events before mitotic checkpoints. Tumour cells often have more than the normal two spindle poles, suggesting that this particular checkpoint has been bypassed. Secondly, overexpression of mitotic cyclins, coupled with a failure to degrade them, could lead to aberrant progression through the cell cycle. Keyomarsi and Pardee (1993) have found that in some breast tumour cells, mitotic cyclins A and B are overexpressed in G_1, before the appearance of normal G_1/S phase cyclins. These authors propose that overexpressed mitotic cyclins can function redundantly and non-specifically in cancer cells, substituting for G_1/S cyclins and co-opting their cellular functions. Keyomarsi and Pardee suggest that such overexpression can bypass G_1/S checkpoints and promote uncontrolled passage through the cell cycle. In budding yeast, ectopic G_1 overexpression of a stable form of Clb2p lacking a destruction box can promote entry into S phase in cells lacking G_1/S cyclin (Cln1p/Cln2p/Cln3p) function (Amon *et al*, 1994). In cancer cells, the effects of high levels of mitotic cyclins in G_1 would be exacerbated by alleles of, or mutations in, APC components that reduced the ability of the APC to degrade cyclin efficiently. Such partially functional alleles might be sufficient to degrade normal levels of mitotic cyclins but could be overwhelmed by cyclin overexpression. This hypothesis is consistent with the argument that the APC is necessary in G_1 to degrade mitotic cyclins and that failure to degrade them causes spurious entry into S phase (Irniger S and Nasmyth K, unpublished).

Finally, CDC27Hs appears to interact with the retinoblastoma (*RB*) tumour suppressor gene in the yeast two hybrid system (Chen *et al*, 1995). *RB* is a negative regulator of S phase in mammalian cells, binding to and inhibiting E2-F1, a transcriptional activator important for cell proliferation. RB protein is hypophosphorylated in G_1 or G_0 cells and becomes phosphorylated before cells enter S phase, probably by the cyclin D/CDK4 complex. A glutathione-S-transferase–CDC27Hs fusion binds preferentially to the underphosphorylated form of RB, but this interaction has not been confirmed by direct co-immunoprecipitation. Substitution of an aspartic acid residue for the essential glycine of the sixth TPR domain, the same mutation found in *nuc2-663*, rendered the two hybrid interaction temperature sensitive in yeast. The biological significance of these results is unclear, and the APC has not been implicated in the ubiquitination of RB.

If mutations in APC components are implicated in transformation, it will probably be in conjunction with mutations in, or aberrant regulation of, other cell cycle proteins. We have only just begun to discover how the APC interacts with the complicated regulatory machinery of the mammalian cell cycle, and little is known about its relationship to the plethora of newly discovered cyclin-CDK inhibitors or how mutations in these inhibitors might in turn affect APC regulation. Indeed, we may find additional mammalian APC components that are not present in the APC of lower eukaryotes or have no homology with their functional homologues in *Xenopus* or *S cerevisiae*. A thorough investigation of the mammalian cyclosome is therefore crucial for the complete identification of all its components and the elucidation of their interactions with other cell cycle players.

SUMMARY

We have proposed a preliminary model of how the anaphase promoting complex functions throughout the cell cycle, but despite the flurry of recent publications characterizing the APC—its components, regulation and substrate specificity—many fundamental questions remain to be answered. Firstly, the remaining components of the APC need to be identified and characterized. We do not know if all cyclosome components are conserved in all eukaryotes, or if higher eukaryotes, having a more complicated cell cycle machinery, maintain additional subunits for more sophisticated functional and regulatory control. In addition, we need to determine the identity of the various kinases and phosphatases that regulate the APC itself. The biochemistry of individual APC components is also a mystery, and a specific biochemical function has not been assigned to any known members of the complex. It is not at all clear which subunit(s) of the complex actually recognizes the E2 enzyme and which subunit(s) recognizes the cyclin destruction box. It is likely that many cyclosome substrates remain to be identified, and it will be interesting to determine whether all cyclosome substrates require a destruction box for their degradation or whether the APC recognizes other determinants of protein instability. Finally, we assume that the APC degrades mitotic cyclins in all proliferating cells, but whether it degrades unique cell cycle related substrates in specific tissues is unclear. Furthermore, nothing is known about APC function during meiosis, or whether the APC degrades other substrates that are not related to the cell cycle. This is an exciting and rapidly developing field in the exciting world of cell cycle biology. We expect that new findings will surely reveal many interesting surprises about this essential protein complex.

Acknowledgements

We thank members of the Hieter laboratory for helpful suggestions and stimulating discussion and John Lamb and Jane Roskams Hieter for a critical reading of the manuscript. We also thank all the researchers who graciously communicated results prior to publication.

References

Amon A, Irniger S and Nasmyth K (1994) Closing the cell cycle circle in yeast: G2 cyclin proteolysis initiated at mitosis persists until the activation of G1 cyclins in the next cycle. *Cell* **77** 1037–1050

Axton JM, Dombradi V, Cohen PT and Glover DM (1990) One of the protein phosphatase 1 isoenzymes in *Drosophila* is essential for mitosis. *Cell* **63** 33–46

Buckley MF, Sweeney KJ, Hamilton A *et al* (1993) Expression and amplification of cyclin genes in human breast cancer. *Oncogene* **8** 2127–2133

Chen PL, Ueng YC, Durfee T, Chen KC, Yang-Feng T and Lee WH (1995) Identification of a human homologue of yeast nuc2 which interacts with the retinoblastoma protein in a specific manner. *Cell Growth and Differentiation* **6** 199–210

Ciechanover A (1994) The ubiquitin-proteasome proteolytic pathway. *Cell* **79** 13–21

Clay FJ, McEwen SJ, Bertoncello I, Wilks AF and Dunn AR (1993) Identification and cloning of a protein kinase-encoding mouse gene, Plk, related to the polo gene of *Drosophila*. *Proceedings of the National Academy of Sciences of the USA* **90** 4882–4886

Cohen-Fix O, Peters JM, Kirschner MW and Koshland D (1996) Anaphase initiation in *Saccharomyces cerevisiae* is controlled by the APC-dependent degradation of the anaphase inhibitor Pds1p. *Genes & Development* **10** 3081–3093

Doonan JH and Morris NR (1989) The bimG gene of *Aspergillus nidulans*, required for completion of anaphase, encodes a homolog of mammalian phosphoprotein phosphatase 1. *Cell* **57** 987–996

Engle DB, Osmani SA, Osmani AH, Rosborough S, Xin XN and Morris NR (1990) A negative regulator of mitosis in *Aspergillus* is a putative membrane-spanning protein. *Journal of Biological Chemistry* **265** 16132–16137

Ghiara JB, Richardson HE, Sugimoto K *et al* (1991) A cyclin B homolog in S. *cerevisiae*: chronic activation of the Cdc28 protein kinase by cyclin prevents exit from mitosis. *Cell* **65** 163–174

Glotzer M, Murray AW and Kirschner MW (1991) Cyclin is degraded by the ubiquitin pathway. *Nature* **349** 132–138

Goebl M and Yanagida M (1991) The TPR snap helix: a novel protein repeat motif from mitosis to transcription. *Trends in Biochemical Sciences* **16** 173–177

Goebl MG, Yochem J, Jentsch S, McGrath JP, Varshavsky A and Byers B (1988) The yeast cell cycle gene CDC34 encodes a ubiquitin-conjugating enzyme. *Science* **241** 1331–1335

Golsteyn RM, Schultz SJ, Bartek J, Ziemiecki A, Ried T and Nigg EA (1994) Cell cycle analysis and chromosomal localization of human Plk1, a putative homologue of the mitotic kinases *Drosophila* polo and *Saccharomyces cerevisiae* Cdc5. *Journal of Cell Science* **107** 1509–1517

Golsteyn RM, Mundt KE, Fry AM and Nigg EA (1995) Cell cycle regulation of the activity and subcellular localization of Plk1, a human protein kinase implicated in mitotic spindle function. *Journal of Cell Biology* **129** 1617–1628

Gorbsky GJ and Ricketts WA (1993) Differential expression of a phosphoepitope at the kinetochores of moving chromosomes. *Journal of Cell Biology* **122** 1311–1321

Hamanaka R, Maloid S, Smith MR, O'Connell CD, Longo DL and Ferris DK (1994) Cloning and characterization of human and murine homologues of the *Drosophila* polo serine-threonine kinase. *Cell Growth and Differentiation* **5** 249–257

Hamanaka R, Smith MR, O'Connor PM *et al* (1995) Polo-like kinase is a cell cycle-regulated kinase activated during mitosis. *Journal of Biological Chemistry* **270** 21086–21091

Hartwell LH, Culotti J and Reid B (1970) Genetic control of the cell-division cycle in yeast. I. Detection of mutants. *Proceedings of the National Academy of Sciences of the USA* **66** 352–359

Heichman KA and Roberts JM (1996) The yeast CDC16 and CDC27 and CDC27 genes restrict DNA replication to once per cell cycle. *Cell* **85** 39–48

Hershko A, Ganoth D, Pehrson J, Palazzo RE and Cohen LH (1991) Methylated ubiquitin inhibits cyclin degradation in clam embryo extracts. *Joural of Biological Chemistry* **266** 16376–16379

Hershko A, Ganoth D, Sudakin V *et al* (1994) Components of a system that ligates cyclin to ubiquitin and their regulation by the protein kinase cdc2. *Journal of Biological Chemistry* **269** 4940–4946

Hirano T, Hiraoka Y and Yanagida M (1988) A temperature-sensitive mutation of the *Schizosaccharomyces pombe* gene nuc2+ that encodes a nuclear scaffold-like protein blocks spindle elongation in mitotic anaphase. *Journal of Cell Biology* **106** 1171–1183

Hirano T, Kinoshita N, Morikawa K and Yanagida M (1990) Snap helix with knob and hole: essential repeats in S. *pombe* nuclear protein nuc2+. *Cell* **60** 319–328

Hisamoto N, Sugimoto K and Matsumoto K (1994) The Glc7 type 1 protein phosphatase of *Saccharomyces cerevisiae* is required for cell cycle progression in G2/M. *Molecular and Cellular Biology* **14** 3158–3165

Holloway SL, Glotzer M, King RW and Murray AW (1993) Anaphase is initiated by proteolysis rather than by the inactivation of maturation-promoting factor. *Cell* **73** 1393–1402

Hunter T and Pines J (1991) Cyclins and cancer. *Cell* **66** 1071–1074

Hunter T and Pines J (1994) Cyclins and cancer. II: Cyclin D and CDK inhibitors come of age. *Cell* **79** 573–582

Icho T and Wickner RB (1987) Metal-binding, nucleic acid-binding finger sequences in the CDC16 gene of *Saccharomyces cerevisiae*. *Nucleic Acids Research* **15** 8439–8450

Irniger S, Piatti S, Michaelis C and Nasmyth K (1995) Genes involved in sister chromatid separation are needed for B-type cyclin proteolysis in budding yeast. *Cell* **81** 269–278

Keyomarsi K and Pardee AB (1993) Redundant cyclin overexpression and gene amplification in breast cancer cells. *Proceedings of the National Academy of Sciences of the USA* **90** 1112–1116

King RW, Peters JM, Tugendreich S, Rolfe M, Hieter P and Kirschner MW (1995) A 20S complex containing CDC27 and CDC16 catalyzes the mitosis-specific conjugation of ubiquitin to cyclin B. *Cell* **81** 279–288

Kumada K, Su S, Yanagida M and Toda T (1995) Fission yeast TPR-family protein nuc2 is required for G1-arrest upon nitrogen starvation and is an inhibitor of septum formation. *Journal of Cell Science* **108** 895–905

Lahav-Baratz S, Sudakin V, Ruderman JV and Hershko A (1995) Reversible phosphorylation controls the activity of cyclosome-associated cyclin-ubiquitin ligase. *Proceedings of the National Academy of Sciences of the USA* **92** 9303–9307

Lake RJ and Jelinek WR (1993) Cell cycle- and terminal differentiation-associated regulation of the mouse mRNA encoding a conserved mitotic protein kinase. *Molecular and Cellular Biology* **13** 7793–7801

Lamb JR, Michaud WA, Sikorski RS and Hieter PA (1994) Cdc16p, Cdc23p and Cdc27p form a complex essential for mitosis. *EMBO Journal* **13** 4321–4328

Lamb JR, Tugendreich S and Hieter P (1995) Tetratrico peptide repeat interactions: to TPR or not to TPR? [Review] *Trends in Biochemical Sciences* **20** 257–259

Luca FC and Ruderman JV (1989) Control of programmed cyclin destruction in a cell-free system. *Journal of Cell Biology* **109** 1895–1909

Mahaffey DT, Yoo Y and Rechsteiner M (1995) Ubiquitination of full-length cyclin. *FEBS Letters* **370** 109–112

Minshull J, Blow JJ and Hunt T (1989) Translation of cyclin mRNA is necessary for extracts of activated xenopus eggs to enter mitosis. *Cell* **56** 947–956

Mirabito PM and Morris NR (1993) BIMA, a TPR-containing protein required for mitosis, localizes to the spindle pole body in *Aspergillus nidulans*. *Journal of Cell Biology* **120** 959–968

Murray AW and Hunt T (1993) *The Cell Cycle: An Introduction*, WH Freeman, New York

Murray AW and Kirschner MW (1989) Cyclin synthesis drives the early embryonic cell cycle. *Nature* **339** 275–280

Murray AW, Solomon MJ and Kirschner MW (1989) The role of cyclin synthesis and degradation in the control of maturation promoting factor activity. *Nature* **339** 280–286

O'Donnell KL, Osmani AH, Osmani SA and Morris NR (1991) bimA encodes a member of the tetratricopeptide repeat family of proteins and is required for the completion of mitosis in *Aspergillus nidulans*. *Journal of Cell Science* **99** 711–719

Ohkura H, Kinoshita N, Miyatani S, Toda T and Yanagida M (1989) The fission yeast dis2+ gene required for chromosome disjoining encodes one of two putative type 1 protein phosphatases. *Cell* **57** 997–1007

Osmani SA, Engle DB, Doonan JH and Morris NR (1988) Spindle formation and chromatin condensation in cells blocked at interphase by mutation of a negative cell cycle control gene. *Cell* **52** 241–251

Peters JM, King RW, Hoog C and Kirschner MW (1996) Identification of BIME as a subunit of the anaphase-promoting complex. *Science* **274** 1199–1201

Samejima I and Yanagida M (1994) Bypassing anaphase by fission yeast cut9 mutation: requirement of cut9+ to initiate anaphase. *Journal of Cell Biology* **127** 1655–1670

Samejima I, Matsumoto T, Nakaseko Y, Beach D and Yanagida M (1993) Identification of seven new cut genes involved in *Schizosaccharomyces pombe* mitosis. *Journal of Cell Science* **105** 135–143

Sikorski RS, Boguski MS, Goebl M and Hieter P (1990) A repeating amino acid motif in CDC23 defines a family of proteins and a new relationship among genes required for mitosis and RNA synthesis. *Cell* **60** 307–317

Sikorski RS, Michaud WA, Wootton JC, Boguski MS, Connelly C and Hieter P (1991) TPR proteins as essential components of the yeast cell cycle. *Cold Spring Harbor Symposia on Quantitative Biology* **56** 663–673

Sikorski RS, Michaud WA and Hieter P (1993) p62cdc23 of *Saccharomyces cerevisiae:* a nuclear tetratricopeptide repeat protein with two mutable domains. *Molecular and Cellular Biology* **13** 1212–1221

Starborg M, Brundell E, Gell K and Hoog C (1994) A novel murine gene encoding a 216-kDa protein is related to a mitotic checkpoint regulator previously identified in *Aspergillus nidulans*. *Journal of Biological Chemistry* **269** 24133–24137

Sudakin V, Ganoth D, Dahan A *et al* (1995) The cyclosome, a large complex containing cyclin-selective ubiquitin ligase activity, targets cyclins for destruction at the end of mitosis. *Molecular Biology of the Cell* **6** 185–197

Surana U, Amon A, Dowzer C, McGrew J, Byers B and Nasmyth K (1993) Destruction of the CDC28/CLB mitotic kinase is not required for the metaphase to anaphase transition in budding yeast. *EMBO Journal* **12** 1969–1978

Tugendreich S, Tomkiel J, Earnshaw W and Hieter P (1995) CDC27Hs colocalizes with CDC16Hs to the centrosome and mitotic spindle and is essential for the metaphase to anaphase transition. *Cell* **81** 261–268

Weinert TA and Hartwell LH (1993) Cell cycle arrest of cdc mutants and specificity of the RAD9 checkpoint. *Genetics* **134** 63–80

Williams A, Lanker S, Patton E *et al* (1996) Cdc53 targets phosphorylated G_1 cyclins for degradation by the ubiquitination proteolytic pathway. *Cell* **86** 453–463

Zachariae W and Nasmyth K (1996) TPR proteins required for anaphase progression mediate ubiquitination of mitotic B-type cyclins in yeast. *Molecular Biology of the Cell* **7** 791–801

Zachariae W, Shin TH, Galova M, Obermaier B and Nasmyth K (1996) Identification of sub-units of the anaphase-promoting complex of *Saccharomyces cerevisiae*. *Science* **274** 1201–1204

The authors are responsible for the accuracy of the references.

Mammalian G_1 and G_2 Phase Checkpoints

P M O'Connor

Laboratory of Molecular Pharmacology, Division of Basic Sciences, National Cancer Institute, National Institutes of Health, Bethesda, MD 20892

INTRODUCTION

During the past 5 years, we have seen significant improvements in our understanding of the mechanisms governing DNA damage induced cell cycle arrest. Much of this has come about through the convergence of information from diverse model systems including yeast, frog, mouse and human cells. In mammalian cells, cell cycle progression is halted in G_1 and/or G_2 phases following DNA damage, and for certain DNA damaging agents, a prominent S phase delay has also been reported (Table 1, Fig. 1). Delay at these points in the cell cycle appears to be governed in large part by a series of control systems commonly termed "checkpoints" (Hartwell and Weinert, 1989; Murray, 1992; Hartwell and Kastan, 1994). Activation of these DNA damage checkpoints has been proposed to extend the time for DNA repair to occur before entry into S phase or mitosis. This in turn would suppress the replication of

Cancer Surveys Volume 29: *Checkpoint Controls and Cancer*
© 1997 Imperial Cancer Research Fund. 0-87969-518-8/97. $5.00 + .00

TABLE 1. DNA damaging anticancer agents, targets and cell cycle perturbations

Anticancer agent	Target	Mode of action	Cell cycle arrest
Cisplatin Carboplatin Nitrogen mustard Melphalan Chlorambucil Mitomycin C	DNA	DNA crosslinking	G_2, S>G_1 phase
Gamma rays Bleomycin	DNA	DNA strand breaks	G_1, G_2>S phase
Camptothecin Topotecan CPT-11	topoisomerase I	stabilization of Topo I-DNA cleavable complexes	G_2, S>G_1 phase
Etoposide Adriamycin Amsacrine	topoisomerase II	stabilization of Topo II-DNA cleavable complexes	G_1, G_2>S phase

damaged DNA templates and the likelihood of their segregation into daughter cells. Cell cycle arrest can thus contribute to the fidelity with which genetic information is passed from one generation to the next, and the loss of these control systems can often have dramatic consequences on genome stability and cell survival following genotoxic stress (Hartwell, 1992; Livingstone *et al*, 1992; Yin *et al*, 1992; Hartwell and Kastan, 1994).

The importance of cell cycle delay in combating the deleterious consequences of DNA damage has been illustrated in a number of engineered model systems from yeast to human cells. However, the growing complexity of DNA damage response pathways and their roles in other cellular processes including DNA repair and apoptosis, coupled with cell type specific responses, lead us to expect a continuing evolution of our ideas about checkpoint control processes. Not least among these are that checkpoint proteins thought to be involved solely in cell cycle arrest in response to DNA damage might also participate in DNA repair.

In the current review, I focus on an exploration of the mechanisms used by mammalian cells to arrest cell cycle progression in G_1 and G_2 phases following DNA damage. Emphasis will be placed on the interaction of the DNA damage response pathways with a series of cyclin dependent kinases (CDKs), which with the fuel of phosphorylation drive a cell through all major cell cycle transitions. I also describe some common defects found in these DNA damage pathways in cancer cells and briefly comment on the impact of these alterations on radiosensitivity and chemosensitivity.

THE G_1 CELL CYCLE CHECKPOINT

Arrest of cell cycle progression in G_1 phase following DNA damage requires the function of the TP53 tumour suppressor protein (Kastan *et al*, 1991; Kuer-

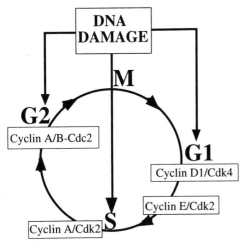

Fig. 1. Mammalian DNA damage checkpoints: braking the cell cycle engine. DNA damage induces the arrest of cell cycle progression in G$_1$ and/or G$_2$ phase. Delayed progression through S phase has also been noted for some DNA damaging agents (see Table 1). Cell cycle arrest is mediated by a series of control systems commonly termed checkpoints (Hartwell and Weinert, 1989). To induce cell cycle arrest, these checkpoints must prevent the formation and/or activation of the cyclin dependent kinases which can be thought of as the engine of the cell cycle (Murray, 1992). When activated, these cyclin dependent kinases can drive cells into S phase and mitosis. Each transition is regulated by a separate set of cyclin dependent kinases (Nurse, 1990; Murray, 1992; Morgan, 1995). The G$_1$/S phase transition is regulated in part by cyclin D1/CDK4 and cyclin E/CDK2. Cyclin D1/CDK4 represents one of several cyclin D/CDK complexes implicated in the G$_0$/G$_1$ to S phase transition in mammalian cells. Others are cyclin D2 and D3 complexed with CDK4 or CDK6 (Sherr, 1994; Weinberg, 1995). Active cyclin A/CDK2 promotes entry into S phase (Pagano *et al*, 1992). Cyclin A/CDC2 and cyclin B/CDC2 promote entry into mitosis (Pines and Hunter, 1989; Pagano *et al*, 1992). DNA damage induced G$_1$ arrest has been associated with inhibition of cyclin E/CDK2 and cyclin D1/CDK4 kinase activity (Dulic *et al*, 1994; El-Deiry *et al*, 1994). G$_2$ arrest induced by DNA damage has been associated with inhibition of cyclin A/CDC2 and cyclin B1/CDC2 complexes, whereas cyclin A/CDK2 complexes remain active in the G$_2$ arrest cells (Lock and Ross, 1990a,b; O'Connor *et al*, 1993b). The routes by which DNA damage checkpoints inhibit these kinases are discussed in the text. Redrawn with modifications from O'Connor and Fan (in press)

bitz *et al*, 1992). Whereas cells with an intact TP53 pathway exhibit γ ray induced G$_1$ arrest, cells with *TP53* gene mutations, which occur commonly in human cancer (Hollstein *et al*, 1991; Levine *et al*, 1991), or cells expressing viral oncogenes, whose products bind to and inactivate TP53, lack a G$_1$ arrest response to DNA damage (Kastan *et al*, 1991; Kuerbitz *et al*, 1992; Kessis *et al*, 1993; O'Connor *et al*, 1993b; Zambetti and Levine, 1993). The mechanism by which TP53 induces cell cycle arrest can best be understood from the transcriptional activation function of the TP53 tumour suppressor (Pietenpol *et al*, 1994). A schematic representation of the domain structure of TP53 is shown in Fig. 2. TP53 encodes a 393 aminoacid nuclear phosphoprotein that can be divided into at least three domains: an N-terminal acidic *trans*-activating domain; a central evolutionarily conserved DNA binding domain; a complex

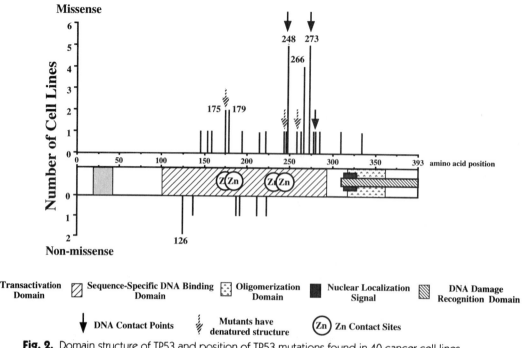

Fig. 2. Domain structure of TP53 and position of TP53 mutations found in 40 cancer cell lines. Highlighted are the N-terminal *trans*-activating domain (aminoacids 20-42), the central conserved DNA binding domain (aminoacids 100–293) and the C-terminal oligomerization domain (aminoacids 319–360) of TP53. Also indicated are potential nuclear localization sequences (aminoacids 316–325) and a putative DNA damage recognition domain (aminoacids 311–393) in the C-terminal domain, as well as aminoacid positions that contact zinc in the central DNA binding domain (aminoacids Cys-176, His-179, Cys-238 and Cyst-242). The locations of *TP53* mutations in 40 human cancer cell lines (O'Connor PM, Jackman J, Bae I *et al*, unpublished) are shown relative to these functionally important *TP53* domains. Mutations at aminoacids that make contact with DNA are shown as solid arrows. Mutation positions associated with TP53 denaturation are shown as hatched arrows. The number of cell lines showing either missense or non-missense (deletion, truncation or frameshift) mutation is shown pointing above or below the TP53 diagram, respectively. With two exceptions, *TP53* mutations were found clustered into the central conserved DNA binding domain of TP53. TP53 domain structure redrawn with modifications from O'Connor PM, Jackman J, Bae I *et al*, unpublished)

C-terminal domain that houses potential nuclear localization sequences, a homotetramerization domain and a putative DNA damage recognition domain. The majority of mutations found in human tumours are missense mutations that are clustered into the central DNA binding domain (Friend, 1994; Greenblatt *et al*, 1994). These mutations block TP53 DNA binding and *trans*-activating activity and have been termed "loss of function" mutations (Zambetti and Levine, 1993). *TP53* mutations are commonly associated with cellular accumulation of the TP53 protein, and evidence exists of a possible "gain of function" for certain *TP53* mutations (Zambetti and Levine, 1993). TP53 also has the capacity to repress gene transcription of genes containing TATA elements (Mack *et al*, 1993). Activation of TP53 following DNA damage is associ-

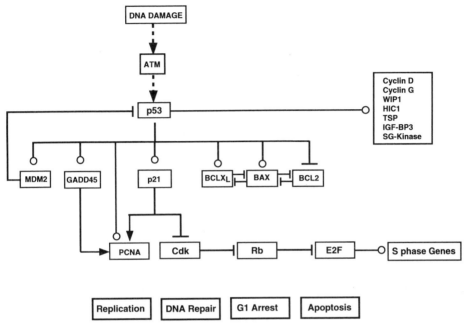

Fig. 3. Schematic view of some elements of the TP53 pathway. The TP53 (p53) pathway orchestrates a number of cellular responses to DNA damage, including G$_1$ phase arrest, apoptosis, induction of DNA repair and direct inhibition of DNA replication. DNA damage induced TP53 activation requires the ATM gene product and TP53 activation results in transcriptional upregulation of a number of downstream effector genes: *MDM2, GADD45, p21, BAX*. Also shown are a number of other recently identified *TP53* regulated genes as well as the *TP53* dependent transcriptional repression of one gene: *BCL2*. The p21 gene product of the *CIP1/WAF1* gene binds to and inhibits CDK activity, linking the TP53 pathway to the cell cycle engine. G$_1$ arrest results in part from a failure of G$_1$/S phase CDKs to phosphorylate the RB gene product, which represses genes regulated by E2F. RB independent events are also necessary for TP53 to induce G$_1$ arrest (see text). p21 also binds PCNA, implicating p21 in the control of DNA replication and/or DNA repair following DNA damage. MDM2 provides a negative feedback function to inhibit TP53 transcriptional activity. GADD45, like p21, can also bind PCNA, and this interaction has implicated GADD45 in DNA replication and DNA repair. TP53 can also transcriptionally induce the apoptosis inducing gene, *BAX*, and the apoptosis suppressing gene, *BCLX$_L$*. The BAX gene product binds to and inhibits the anti-apoptotic effect afforded by BCL2 and BCLX$_L$. Open circles represent stimulation of transcription. Closed arrows represent interactions between two components. Broken arrows represent the relay of a signal generated by DNA damage. Blunt ended lines represent inhibition

ated with accumulation of the normally short lived protein in cells (Kastan *et al*, 1991; Maltzman and Czyzyk, 1984). This accumulation appears not to be mediated at the transcriptional level. Rather, the wild type TP53 protein becomes stabilized following DNA damage. TP53 *trans*-activates a number of different genes including *GADD45, MDM2* and *CIP1/WAF1* (Fig. 3). These genes contain TP53 consensus binding sites (Kern *et al*, 1991; El-Deiry *et al*, 1992; Funk *et al*, 1992) located in either intronic (*GADD45, MDM2*) or promoter (*CIP1/WAF1*) regions of these genes. A description of the functions of these gene products is given below.

TP53 Dependent G$_1$ Arrest through p21

One gene in particular has been focused upon as the means by which TP53 induces G$_1$ arrest: *CIP1/WAF1*. El-Deiry *et al* (1993), in a search for *w*ild type TP53 *a*ctivated *f*actors, discovered *WAF1*, a gene that contained TP53 binding elements in its promoter. When *WAF1* was overexpressed in cells, it induced growth arrest. With remarkable timing, this same gene was isolated independently by Harper *et al* (1993), using a yeast two hybrid selection for CDK2 *i*nteracting *p*roteins (CIP1). In these latter studies, the *CIP1* gene was found to encode a potent inhibitor of CDK activity, and CIP1/WAF1 turned out to be the p21 protein originally observed in CDK immune complexes from untransformed cells (Xiong *et al*, 1992). p21 was also discovered through a number of other routes, including a factor that was markedly upregulated in senescent cells (Noda *et al*, 1994). For simplicity, I refer to this inhibitor as p21. The discovery that p21 was a potent inhibitor of the CDKs provided a potential mechanism by which TP53 could inhibit the cell cycle engine in G$_1$ phase. In support of this proposition, El-Deiry *et al* (1994) and Dulic *et al* (1994) found that cyclin E/CDK2 complexes were indeed inactivated in a TP53 dependent manner following DNA damage and that binding of p21 to this kinase correlated with its inactivation. CIP1/WAF1 can also inhibit a number of other CDKs important for cell cycle progression, including cyclin D1/CDK4, cyclin A/CDK2 and to some extent cyclin B1/CDC2 (Harper *et al*, 1995). The N-terminal half of p21 bears striking similarity to two other CDK inhibitor proteins: p27^{Kip1} (Polyak *et al*, 1994; Toyoshima and Hunter, 1994) and p57^{Kip2} (Lee M-H *et al*, 1995; Matsuoka *et al*, 1995), suggesting that this family of CDK inhibitor proteins might interact and inhibit CDK activity through similar mechanisms. However, neither p27^{Kip1} nor p57^{Kip2} appears to be induced by TP53 (Fan S and O'Connor PM, unpublished).

Functional Domains of p21

The Cyclin/CDK2 Interaction Domain

The structural organization and mechanism by which p21 inhibits CDK activity has been an area of intense investigation. Figure 4 illustrates the domain structure of p21 as well as similarities to the crystal structure of the p27^{Kip1} inhibitor (p27) bound to cyclin A/CDK2 (Russo *et al*, 1996). The N-terminal half of p21 is critical for binding to and inhibition of the CDKs, whereas the C-terminal half of p21 is involved in binding to at least proliferating cell nuclear antigen (PCNA), a protein required for DNA replication and DNA repair (Prelich *et al*, 1987; Shivji *et al*, 1992; Flores-Rozas *et al*, 1994; Chen J *et al*, 1995; Luo *et al*, 1995). Using a series of overlapping peptides spanning the complete p21 sequence, Chen *et al* (1996) suggested that two separate N-terminal domains in p21 were required for cyclin E/CDK2 binding and inhibition. The first of these domains (aminoacids 17–26) was later shown to be in

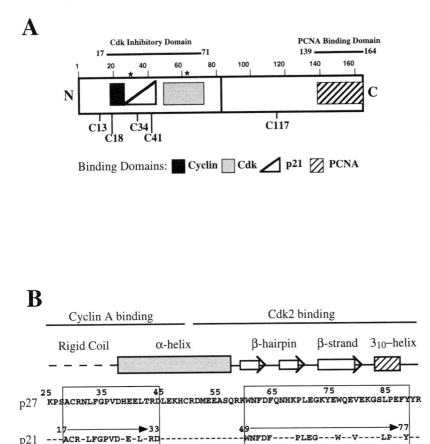

Fig. 4. Important p21 binding domains and similarities between p21 and p27. (A) Diagram of the 164 aminoacid p21 protein with regions involved in binding to cyclin/CDK complexes and PCNA. The ruler shows the number of aminoacids in p21 and stars above the ruler show the position of a codon 31 (serine/arginine) polymorphism reported by several groups and a mutation (codon 63) in a cancer cell line that has been reported to suppress p21 activity (Bhatia *et al*, 1995). Also highlighted are two regions in the N-terminus of p21 involved in cyclin and CDK interaction (aminoacids 17–26 and 49–71, respectively) and the positions of five cysteine residues, four of which are localized to the N-terminus of p21 and may indicate a potential zinc finger domain (El-Deiry *et al*, 1993). Also shown is a p21 region that as a peptide can bind recombinant p21 (aminoacids 26–45). A peptide spanning aminoacids 15–40 antagonizes p21 binding to and inhibition of cyclin E/CDK2 kinase activity (Chen *et al*, 1996). (B) Diagram of p27 showing secondary structure of the CDK-inhibitory domain and aminoacid identity with p21 (Russo *et al*, 1996). The rigid coil and part of the α helix of p27 bind to cyclin A (aminoacids 25–50); the β hairpin/strand structures of p27 along with the 3$_{10}$ helix of p27 co-operate to distort the CDK catalytic domain and displace ATP from the kinase. Aminoacid positions of identity with p21 are shown below the diagram, and noteworthy is the high aminoacid identity of the cyclin binding region between p21 and p27. Redrawn with modification from Chen *et al* (1996) and Russo *et al* (1996)

volved in binding to the cyclin (Fotedar *et al*, 1996; Lin *et al*, 1996), and the second domain (aminoacids 49–71) was found to interact with the CDK2 subunit (Fig. 4) (Goubin and Ducommun, 1995; Nakanishi *et al*, 1995; Fotedar *et al*, 1996; Lin *et al*, 1996). The actual docking sites for p21 on the cyclin and CDK2 kinase subunits have yet to be resolved. However, the recent and elegantly solved crystal structure of the p27/cyclin A/CDK2 complex (Russo *et al*, 1996) provides tantalizing clues as to the expected orientation. Co-crystallization of cyclin A/CDK2 with the 69 aminoacid CDK inhibitory domain of p27 revealed that the initial N-terminal half of p27 has an extended structure that consists of a ten aminoacid rigid coil (aminoacids 26–35) that binds to a shallow groove in the cyclin box region of cyclin A. Within this rigid coil lies the "LFG motif", which is conserved in p21 and p57^{Kip2}. The importance of this region for binding to and inhibition of CDK activity had been suggested from earlier studies of p27 mutants (Luo *et al*, 1995) and by the ability of peptides spanning this region to block p21 interaction and inhibition of cyclin E/CDK2 (Chen *et al*, 1996). The LFG motif serves as an anchor in p27 complex formation with cyclin A/CDK2 (Russo *et al*, 1996). Extending from the rigid coil of p27 lies an amphipathic helix, the hydrophobic face of which packs against the surface of the cyclin box. The helix then extends away from the cyclin and is directed to CDK2. Interactions between p27 and CDK2 are complex and involve the β hairpin, β strand and 3$_{10}$ helix of p27. Together, these structures clamp around the β sheet of CDK2, distort and widen the CDK2 catalytic cleft, allowing the 3$_{10}$ helix of p27 to enter into and occupy most of the available space within the cleft. Importantly, the 3$_{10}$ helix of p27 mimics the position of ATP in the cleft with the side chains of the conserved Tyr-88 residue making CDK2 contacts that mimic the purine base of ATP. Furthermore, the backbone carbonyl groups of Phe-87 and Arg-90 residues of p27 make hydrogen bonds with the CDK2 Lys-33, a residue critical for γ phosphate transfer. These contacts would normally be made by the ATP phosphates (Russo *et al*, 1996). Taken together, these results show how p27 and presumably other family members (p21, p57^{Kip2}) would bind to and inhibit cyclin A/CDK2.

Thr-160 phosphorylation on CDK2, which is required for kinase activity (Morgan, 1995), can also be inhibited by p27 and p21 (Polyak *et al*, 1994; Aprelikova *et al*, 1995). This inhibition most likely results from steric hindrance, since p21 does not inhibit the CDK activating kinase (CAK) (Harper *et al*, 1995). The crystal structure of p27/cyclin A/CDK2 supports the steric hindrance model by finding that the far C-terminal region of the p27 fragment used for crystallization lies close to the Thr-160 loop of CDK2 (Russo *et al*, 1996).

It has been suggested that two or more molecules of p21 per CDK complex are required to inhibit kinase activity (Zhang *et al*, 1994). However, in the crystal structure of cyclin A/CDK2/p27, only a single molecule was observed, and this appeared to be sufficient to inhibit the kinase (Russo *et al*, 1996). In addition, recent studies from Chen *et al* (1996) and Fotedar *et al* (1996) suggest that a single contiguous region of p21 spanning the complete N-terminal inhibitory domain is required for inhibition of the H1 kinase activity of cyclin E/CDK2 and cyclin A/CDK2 complexes.

The original sequencing of *WAF1* revealed a potential zinc finger consisting of four cysteine residues located at positions 13, 18, 34 and 41 of human p21 (El-Deiry *et al*, 1993). However, only one of these cysteines is conserved in p27 and p57^{Kip2}, and this cysteine lies within the ten aminoacid rigid coil of p27 that interacts with the cyclin box. Two of these cysteines lie within a peptide of p21 that Chen *et al* (1996) found could bind recombinant p21 protein in enzyme linked immunosorbent assays. The effects of mutation of these cysteine residues on p21 function has yet to be studied.

The PCNA Interaction Domain

The ability of p21 to bind also to PCNA suggests that p21 might co-ordinately regulate CDK and PCNA activity. p21 has been shown to inhibit PCNA dependent SV40 DNA replication in vitro, but transfection of cells with the C-terminal half of p21, which contains the PCNA binding domain, has little or no effect on mammalian DNA synthesis (Flores-Rozas *et al*, 1994; Waga *et al*, 1994; Luo *et al*, 1995; Nakanishi *et al*, 1995). These results suggest that the primary purpose of the p21-PCNA interaction may not be to inhibit DNA replication per se. An alternative possibility could be that p21 regulates some aspect of PCNA function in nucleotide excision repair. A potential role of the p21-PCNA interaction in DNA repair is, however, complicated by mixed findings. In some in vitro studies, p21 did not affect nucleotide excision repair activity (Flores-Rozas *et al*, 1994; Waga *et al*, 1994), but in a separate study, p21 was found to inhibit nucleotide excision (Pan *et al*, 1995). Such differences may depend on the levels of active p21 in each system, and thus it might prove useful to assess DNA repair activity in cells with normal versus no p21 expression. Indeed, recent studies in human cells lacking genes that encode p21 have revealed that such cells have reduced DNA repair activity, as evidenced in host cell reactivation experiments (MacDonald *et al*, 1996; Fan *et al*, in press). Furthermore, cells lacking p21 genes tend to be more sensitive to agents that induce DNA damage repaired through the nucleotide excision repair pathway (MacDonald *et al*, 1996; Fan *et al*, in press). Such studies suggest an involvement of p21 in DNA repair. If this is the case, then p21 possesses dual roles in both the G$_1$ checkpoint and DNA repair. It will be important to investigate this possibility further as well as to explore the nucleotide excision repair activity of extracts made from cells lacking p21 genes.

Additional Components Contributing to G$_1$ Arrest?

p21 does not appear to be the sole mediator of TP53 dependent G$_1$ arrest in murine cells, since embryonic fibroblasts from p21$^{-/-}$ mice are not completely deficient in G$_1$ arrest following γ irradiation (Brugarolas, *et al*, 1995; Deng *et al*, 1995). These findings contrasted with observations in human carcinoma HCT-116 cells disrupted for p21 function (Waldman *et al*, 1995). In these human cells, no G$_1$ arrest was observed following p21 disruption. This could mean that human and murine cells differ in their regulation of G$_1$ arrest or, alternatively, the transformed nature of the human cancer cell line used by Waldman *et al* (1995) might be defective for any additional component required for G_1 arrest. Additional candidates that might participate in TP53 dependent G_1 arrest are described below.

GADD45

GADD45 was discovered as an ultraviolet (UV) light inducible transcript in Chinese hamster ovary cells through a subtractive hybridization approach (Fornace *et al*, 1988). *GADD45* was later found to be a TP53 inducible gene that contained a TP53 binding element in its third intron (Kastan *et al*, 1992). GADD45 has been shown to block re-entry of cells into S phase following serum stimulation (Smith *et al*, 1994), suggesting that GADD45 might contribute to G_1 checkpoint regulation. A potential mechanism by which GADD45 might mediate its biological effects was suggested by the finding that GADD45, like p21, could bind to PCNA (Smith *et al*, 1994; Hall *et al*, 1995). GADD45 competes with p21 in binding to PCNA, suggesting that GADD45 and p21 share a common or overlapping binding site on the PCNA molecule (Chen IT *et al*, 1995a). GADD45 was found to stimulate the DNA resynthesis step of nucleotide excision repair in vitro (Smith *et al*, 1994), and recent studies from Smith *et al* (1996) report that antisense GADD45 expression in colon carcinoma RKO cells reduces host cell reactivation of UV damaged CAT reporter plasmids. The exact role of GADD45 in DNA repair is still under investigation (see also Kazantev and Sancar, 1995; Kearsey *et al*, 1995).

WIP1

Fiscella *et al* (1996), using a differential display procedure to isolate TP53 inducible mRNAs, discovered WIP1, a *wild* type TP53 induced *phosphatase which, like GADD45, is as growth suppressive as p21. Transfected WIP1 can delay cells in G_1 phase, opening up another potential candidate contributing to G_1 checkpoint regulation. WIP1 is a type 2C phosphatase that requires magnesium for function and is relatively insensitive to inhibition by okadaic acid. WIP1 is a relatively low abundance protein that localizes to the nucleus and causes growth arrest independent of retinoblastoma protein (RB) function. Some speculative targets for this phosphatase include the dephosphorylation of CDK substrates or possibly the Thr-160 position on CDK2, which when dephosphorylated inactivates kinase activity (Morgan, 1995).

Cyclin D1

Although cyclin D1 is most often associated with growth promotion, Pagano *et al* (1994) found that cyclin D1 overexpression induced G_1 arrest. Chen *et al* (1995c) found that TP53 activation induced the accumulation of cyclin D1. This accumulation was not due to stabilization of the normally short lived cyclin D1 protein, rather TP53 induced cyclin D1 mRNA. The cyclin D1 promoter was found to contain a potential TP53 binding site that was modestly active in TP53 dependent *trans*-activation assays. p21 was also capable of inducing cyclin D1 expression (Chen X *et al*, 1995). A possible route by which p21 might activate cyclin D1 expression is through the RB protein, which in its hypophosphorylated (active) form has been shown to induce cyclin D1 transcription (Muller *et al*, 1994) (presumably a consequence of p21 dependent inhibition of CDK activity). Thus, TP53 might induce cyclin D1 expression both directly and indirectly. Given the confounding ability of cyclin D1 both to promote and to inhibit cell cycle progression, it will be worthwhile exploring further the effects of cyclin D1 on cells.

CDK4 Tyrosine Phosphorylation and the G_1 Checkpoint

Terada *et al* (1995) have shown that the CDK4 Tyr-17 residue became phosphorylated following UV irradiation of wild type TP53 rat kidney fibroblasts. When the equivalent position in CDC2 is phosphorylated (Tyr-15) CDC2 is inactivated (Nurse, 1990; Morgan, 1995). Although the effects of CDK4 tyrosine phosphorylation on kinase activity are unknown, Terada *et al* (1995) showed that stable transfection of a CDK4 mutant that could not be tyrosine phosphorylated (Y17F) abrogated UV induced G_1 arrest. These results implicated CDK4 tyrosine phosphorylation as an important component of the G_1 checkpoint. However, how does this CDK4 mutant abrogate G_1 arrest if p21 is functional in these cells? p21 is a potent inhibitor of cyclin D/CDK4 (Harper *et al*, 1995), and one might expect that p21 would inhibit cyclin D/CDK4 independently of tyrosine phosphorylation status. One possibility is that p21 cannot bind to and/or inhibit the CDK4 mutant. Alternatively, the overexpressed CDK4 mutant might titrate p21 away from endogenous cyclin/CDK complexes, allowing such complexes to remain active and drive cells into S phase. Supporting this latter possibility, Latham *et al* (1996) have shown that stable overexpression of wild type and, importantly, kinase inactive versions of CDK4/CDK6 was able to sequester p21 and abrogate TP53 induced growth arrest.

ABL and the TP53 Dependent G_1 Checkpoint

An additional component implicated in TP53 dependent G_1 arrest is ABL. Kufe and colleagues (Yuan *et al*, 1996) reported that ABL suppresses CDK2 activity in a TP53 dependent manner and that cells lacking ABL function were deficient in G_1 arrest following irradiation. The CDK2 inhibitory effects of ABL did not require the p21 gene product. In order for ABL to exert its effects, it needed to bind to TP53, which in turn activated TP53 *trans*-activation function. However, ABL-TP53 binding was insufficient to inactivate CDK2. Additionally, ABL kinase activity was required. Further studies are needed to determine the downstream events that ABL impacts upon to cause G_1 arrest.

Is TP53 a Sensor of DNA Damage?

Nelson and Kastan (1994) convincingly showed that DNA strand breaks are sufficient and probably necessary for DNA damaging agents to activate TP53. This might explain why DNA base damaging agents such as nitrogen mustard and cisplatin induce TP53 relatively slowly compared with γ irradiation or the topoisomerase II inhibitor, etoposide (Fan *et al*, 1994). Extending the studies of Nelson and Kastan (1994), Huang *et al* (1996) showed that TP53 dependent G_1 arrest could be induced by nuclear injection of linerized plasmid DNA. This was the case whether the DNA was cut by enzymes that produce blunt ends, 5'overhangs or 3'overhangs. Similar results were also obtained using circular DNA containing a large (2.9 kb) gap or single stranded circular DNA. However, nuclear injection of supercoiled, nicked plasmid DNA or circular DNA with a small gap (25 nucleotide gap) did not induce G_1 arrest. The length of the linear DNA was also important for TP53 activation, since a 49 base pair oligonucleotide was effective at inducing G_1 arrest, but a 27 base pair fragment or single stranded oligonucleotides up to 49 nucleotides in length did not induce G_1 arrest.

Recent studies have suggested that TP53 might itself act as the direct DNA damage sensor (Jayaraman and Prives, 1995; Lee S *et al*, 1995). TP53 has the ability to recognize and stably associate, in vitro, with short single stranded DNA regions (<40 nucleotides) and insertion/deletion mismatch lesions. However, the fact that short single stranded oligonucleotides (17–49 nucleotides) did not induce G_1 arrest when microinjected into the nucleus of a wild type TP53 cell line (Huang *et al*, 1996) suggests that other components might have to assemble with TP53 on to the damaged DNA template to activate G_1 arrest. Particular emphasis has been placed on the C-terminus 83 aminoacids of TP53 for participation in DNA damage sensing (Jayaraman and Prives, 1995; Lee S *et al*, 1995). This region, apart from containing the TP53 tetramerization domain, also appears to contain a negative regulatory domain that might lock the unphosphorylated TP53 tetramer into an inactive state (see Hupp *et al*, 1995). An allosteric model of negative regulation of TP53 has been presented by Hupp *et al* (1995). These workers suggest that TP53 C-terminal modifications that reduce the net basic charge of this region could displace a TP53 C-terminal inhibitory element from its interaction with other regions of TP53, thus allowing sequence specific DNA binding by the TP53 tetramer.

A complication to the model of TP53 being a direct DNA damage sensor is where to place the *ATM* gene in the TP53 pathway (Savitsky *et al*, 1995). This gene product is required for TP53 protein accumulation and activation following γ irradiation and might act upstream of *TP53* in the G_1 checkpoint pathway (Kastan *et al*, 1992). The *ATM* gene bears homology with a number of genes found in other organisms, including *Mec1*, *Rad3*, *TEL1* and *Mei41* (see Zakian, 1995). The protein products of all these genes have sequences indicative of phosphatidylinositol-3 kinases, suggesting a possible signalling method by which these gene products might communicate with other effectors of DNA damage checkpoint control. Whether *ATM* associates with TP53 at sites of DNA damage remains to be explored.

The Retinoblastoma Protein and the G_1 Checkpoint

A number of workers have speculated on the pivotal role of the RB protein in G_1 arrest induced by TP53. RB is one of a number of pocket proteins that can transcriptionally repress genes containing E2F binding elements (see Fig. 3) (Nevins, 1992; Helin et al, 1993; Weinberg, 1995). Such genes include dihydrofolate reductase, thymidine kinase, cyclin A, B-myb and DNA polymerase alpha. These E2F regulated genes have recognizable functions for S phase, and RB mediated repression would suppress S phase entry by limiting components necessary for DNA replication. RB can be inactivated through phosphorylation mediated, at least in part, by the CDKs (Kato et al, 1993; Weinberg, 1995). p21 has been presumed to induce G_1 arrest by inhibiting the CDKs that would inactivate RB. RB does not, however, appear to be the sole downstream mediator of TP53/p21 induced G_1 arrest since overexpression of either TP53 or p21 in Saos-2 cells, which are RB defective, still induces G_1 arrest (Harper et al, 1995). In addition, DNA damage can still induce G_1 arrest in RB defective cells, albeit at a reduced level compared with normal cells (Slebos et al, 1994). Although p21 can inhibit both cyclin E/CDK2 and cyclin D1/CDK4 activity, only cyclin D1/CDK4 activity is dispensable for the G_1/S phase transition in RB defective cells (Tam et al, 1994). These results suggest that cyclin E/CDK2 has G_1/S phase promoting function(s) other than RB inactivation. Presumably, p21 targets cyclin E/CDK2 for inhibition to prevent this function from being manifested following DNA damage. Isolation of the additional target(s) of cyclin E/CDK2 action will allow further insights into G_1/S regulation and the mechanism by which TP53/p21 induce G_1 arrest independently of RB function.

G_1 Delay: Terminal Arrest or Recovery

G_1 delay following DNA damage allows an extended period of time for DNA repair to occur before entry into S phase. However, once DNA repair has been completed, the cell, if required, must recover from this stasis and re-enter the cycle. To do so, the cell must co-ordinate several events, including inactivation of TP53 and p21 and regeneration of active cyclin/CDK complexes. The transcriptional inactivation of TP53 could take place through MDM2, a TP53 induced gene product that binds to the N-terminal trans-activating domain of TP53 and blocks TP53 transcriptional activity (Fig. 3) (Wu et al, 1993; Chen, et al, 1994). This negative feedback loop provides a useful mechanism to limit the time of TP53 exposure following DNA damage. Indeed, the deregulated high level expression of MDM2, seen in some cancers, could enable such cancers to evade wild type TP53 function (Oliner et al, 1992). Once TP53 has been inactivated, the longevity of G_1 arrest is then limited to the stability of the newly synthesized TP53 regulated gene products that induce G_1 arrest. The maintenance of high p21 levels correlates with prolonged G_1 arrest following irradiation of human diploid fibroblasts (Di Leonardo et al, 1994) and lymphoma cells (Bae et al, 1995). Indeed, in studies with lymphoma cells, p21 levels were maintained long after p21 mRNA levels had returned to baseline,

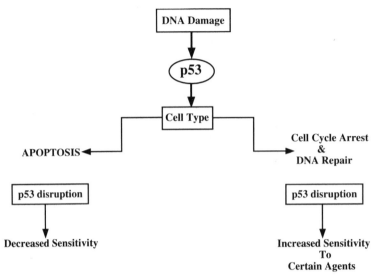

Fig. 5. Importance of cellular context in the outcome of TP53 activation. Disruption of TP53 function in cell types in which apoptosis is the dominant response of TP53 activation often leads to decreased radiosensitivity and chemosensitivity. However, disruption of TP53 function in cell types not inherently prone to TP53 mediated apoptosis has alternative effects. In most cases studied, TP53 disruption in this cell type does not alter radiosensitivity but can sensitize cells to DNA crosslinking agents such as cisplatin. Cisplatin sensitization in this cellular context is believed to result from deficiencies in DNA repair and/or cell cycle arrest (O'Connor and Fan, in press)

and as long as p21 levels remained high, cells remained arrested in G_1 phase (Di Leonardo *et al*, 1994; Bae *et al*, 1995).

TP53 Dependent Apoptosis

Although some cell types respond to wild type TP53 activation by arresting in G_1 phase, other cell types respond by undergoing apoptosis (Fig. 5) (Michalovitz *et al*, 1990; Martinez, *et al*, 1991; Yonish-Rouach *et al*, 1991; Clarke *et al*, 1993; Lowe *et al*, 1993a,b; Fan *et al*, 1994, 1995). The mechanistic basis for these cell type differences remains to be determined; however, some insights come from studies in cells overexpressing myc or E2F-1 (Qin *et al*, 1994; Wagner *et al*, 1994; Wu and Levine, 1994). These transcription factors along with the transcription factor, B-myb (Lin *et al*, 1994), can drive cells into S phase in spite of active TP53. Although apoptosis in S phase commonly results from this bypass, S phase entry may not be essential, since TP53 can induce apoptosis in myc deregulated cells arrested in G_1 by isoleucine withdrawal (Wagner *et al*, 1994). Whether "priming" a cell for S phase, by activating genes required for S phase progression, is a sufficient signal for apoptosis remains to be investigated. Cancer cell types that do not normally undergo apoptosis following wild type TP53 overexpression may not harbour deregulated E2F or c-myc or alternatively may combat apoptosis by alternative mechanisms such as overexpression of BCL2 (Walton *et al*, 1993) or suppression of BAX induction (Zhan *et al*, 1994). TP53 mediated apoptosis can be suppressed by growth factors, which in turn produce a more stable G_1 arrest (Canman *et al*, 1994, and references therein). These results suggest that growth factor signals impinge on the decision making process that determines G_1 arrest and/or apoptosis, the outcome of which is to sway judgement towards G_1 arrest and survival.

The *BAX* gene, which encodes a product that counters ability of Bcl2 to protect cells against apoptosis (Oltvai *et al*, 1993), is transcriptionally activated by wild type TP53 (Fig. 3) (Miyashita *et al*, 1994; Selvakumaran *et al*, 1994; Miyashita and Reed, 1995). *BAX* is upregulated only in those wild type TP53 lines that commit to apoptosis following TP53 activation (Zhan *et al*, 1994). In addition, in those cells that commit to apoptosis, Bcl2 mRNA levels decline following TP53 activation (Selvakumaran *et al*, 1994; Zhan *et al*, 1994). These findings suggest that TP53 dependent apoptosis might arise as the balance between BAX and BCL2 is tipped towards BAX. Enriching this model are the recent results of Zhan *et al* (1996), who have found that $BCLX_L$ mRNA levels are also induced by wild type TP53 in cells that commit to TP53 dependent apoptosis. Like BCL2, $BCLX_L$ protects against apoptosis (Boise *et al*, 1993). These results suggest that BAX induced apoptosis can be modulated independently of BCL2 and that $BCLX_L$ might raise the threshold for BAX induced apoptosis (see Fig. 3). Despite these findings, it is still unclear how the apoptotic machinery becomes activated by BAX. Studies into the requirement of TP53 dependent *trans*-activation for apoptosis have also yielded conflicting results, with some studies suggesting that the *trans*-activating activity of TP53 may not be required for apoptosis in all cell types (Caelles *et al*, 1994).

The Importance of Cellular Context to the Outcome of TP53 Activation

When Apoptosis Is the Dominant TP53 Response

TP53 disruption not only abrogates G_1 arrest following DNA damage but can also suppress apoptosis in susceptible cell types. Key experiments performed on murine embryonic fibroblasts transformed with adenovirus E1A revealed that such cells were sensitized to apoptosis induced by ionizing radiation and some chemotherapeutic agents, and wild type TP53 function was required for this sensitization (Lowe *et al*, 1993a). A tumour type that responds in a similar manner to TP53 activation is Burkitt's lymphoma. Burkitt's lymphoma is characterized by a chromosome translocation, which in most cases transfers the *MYC* gene, normally located on chromosome 8, into the immunoglobulin heavy chain locus located on chromosome 14 (Magrath, 1990). This translocation leads to deregulated MYC expression, and studies in a series of human Burkitt's lymphoma cell lines revealed that those cell lines with a functionally intact TP53 pathway were, on average, more sensitive to ionizing radiation, etoposide, nitrogen mustard and cisplatin than cell lines with mutant TP53 (O'Connor *et al*, 1993a; Fan *et al*, 1994). The decreased sensitivity of the mutant TP53 cell lines correlated with an evasion of TP53 mediated apoptosis, indicating that the mutant TP53 lines might have a survival advantage over the lymphoid lines with wild type TP53. Ionizing radiation induced G_1 arrest was clearly evident in the wild type TP53 cell lines prior to apoptosis, suggesting that G_1 arrest can occur in cells that will ultimately die of apoptosis. Whether the cells undergoing apoptosis were those that prematurely escaped G_1 arrest despite persistent TP53 activation was not determined. TP53 disruption has been associated with decreased sensitivity to DNA damage in a number of other situations (Brown *et al*, 1993; Clarke *et al*, 1993; Lee and Bernstein, 1993; Lowe *et al*, 1993a,b), and a case for TP53 mediated apoptosis has been made in each situation. Such findings have tended to promote the single impression that TP53 disruption confers decreased sensitivity to DNA damaging agents. However, as discussed below, "cellular context" and the type of DNA damaging agent now appear to be important factors in determining the outcome of TP53 activation (Fig. 5).

When Apoptosis Is Not the Dominant TP53 Response

In contrast to studies in lymphoid cells, most workers have failed to observe any impact of TP53 disruption on the radiosensitivity of some epithelial and fibroblast cells. Such studies have included experiments on murine embryonic fibroblasts, colon carcinoma cells, breast cancer cells and head and neck cancer cell lines (Brachman, *et al*, 1993; Slichenmyer *et al*, 1993; Fan *et al*, 1995; Powell *et al*, 1995; Fan *et al*, in press). Such findings have brought into focus "cellular context" as an important aspect of TP53's involvement in radiosensitivity (O'Connor and Fan, in press) (Fig. 5) and at the same time suggested that radiosensitivity in epithelial cells is not modulated through G_1 checkpoint function. In epithelial cells, the S and G_2 checkpoints might be the major DNA damage checkpoints that determine radiosensitivity (see below). Investigations into the role of TP53 function in the chemosensitivity of epithelial cells revealed quite a different response pattern. Using the breast cancer MCF-7 and colon cancer RKO cell lines, Fan *et al* (1995) showed that TP53 disruption, although not affecting radiosensitivity, sensitized these cells to the DNA crosslinking agent, cisplatin. Similar results were also obtained in colon cancer HCT-116 cells (Fan *et al*, in press), murine embryonic fibroblasts from TP53 "knockout" mouse (Hawkins *et al*, 1996), normal human diploid fibroblasts disrupted for TP53 function (Hawkins *et al*, 1996) and a cisplatin resistant ovarian cancer cell line (Brown *et al*, 1993).

Cisplatin induced DNA damage is repaired primarily through nucleotide excision repair, and consistent with a defect in this process in TP53 disrupted cells, Fan *et al* (1995) showed that TP53 defective MCF-7 cells were less able to repair cisplatin damaged CAT reporter plasmids transfected into the cells. Supporting the contention of defective DNA repair in TP53 defective cells, Smith *et al* (1995) also found that colon cancer RKO cells with disrupted TP53 function were less able than control cells to repair UV irradiated plasmids, and Ford and Hanawalt (1995) have found that TP53 defective cells are deficient in overall genome repair compared with wild type TP53 cells. Such findings join a growing list of observations pointing to an involvement of the TP53 pathway in DNA repair. Some other observations include (a) the ability of TP53 to bind to the nucleotide excision repair gene product, ERCC3 (Wang *et al*, 1994), (b) the ability of TP53 to bind transcription-replication repair factors (Wang *et al*, 1995), (c) the recently reported 3′–5′exonuclease activity of TP53 (Mummenbrauer *et al*, 1996), (d) the strand reannealing activity of TP53 (Oberosler *et al*, 1993) and (e) the ability of TP53 to *trans*-activate the *PCNA* gene, whose product is required for nucleotide excision repair (Shivakumar *et al*, 1995, Morris *et al*, 1996).

Additional investigations of downstream effectors of the TP53 pathway have also revealed evidence of a role for p21 and GADD45 in DNA repair (Smith *et al*, 1994, 1995, 1996; MacDonald *et al*, 1996; Fan *et al*, in press). Importantly, dual roles for TP53 and p21 in cell cycle arrest and DNA repair suggest that an economical sharing of components might physically link cell cycle arrest to the DNA repair machinery. Such a possibility is not unique, since similar thoughts have surfaced through yeast genetic studies of the G_2 checkpoint response to DNA damage (Lydall and Weinert, 1995). The dual role of TP53 in cell cycle arrest and DNA repair supports the "guardian of the genome" function of TP53 (Lane, 1992), which suggests that wild type TP53 could suppress genome instability through cell cycle arrest and DNA repair (see also Hartwell, 1992; Livingstone *et al*, 1992; Yin *et al*, 1992).

THE G_2 CELL CYCLE CHECKPOINT

DNA Damage Blocks CDC2 Activation

In some of the earliest biochemical studies on the mammalian G_2 checkpoint, Lock and Ross (1990a,b) showed that DNA damage induced G_2 arrest was associated with suppression of CDC2 kinase activity. A possible mechanism by which DNA damage could prevent CDC2 activation was later presented (O'Connor *et al*, 1992, 1993b; Lock *et al*, 1992). On the basis of the distinctive migration of the phosphorylated forms of CDC2 in sodium dodecyl sulphate gels, it was proposed that DNA damage suppressed the removal of inhibitory phosphorylations from the Thr-14 and Tyr-15 positions of CDC2 (Fig. 6). Cyclin A/CDC2 complexes, which are also required for mitotic entry (Pagano *et al*, 1992), are also suppressed by DNA damage; however, the activity of cyclin A/CDK2 complexes continued to rise in G_2 arrested CA46 cells (O'Connor *et al*, 1993b). A block to the removal of CDC2 inhibitory phosphorylations could occur through suppression of the CDC2 activating phosphatase CDC25C and/or by upregulation of CDC2 inhibitory kinases, such as Wee1 or Mik1 (Fig. 6). Wee1 activity, rather than being upregulated following exposure to γ rays, is maintained at interphase levels in the G_2 arrested cells (Orlandi L, Wang P and O'Connor PM, unpublished).

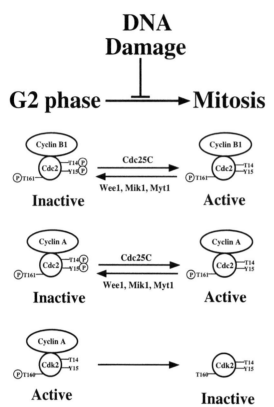

Fig. 6. DNA damage induced G$_2$ arrest is associated with inhibition of cyclin/CDC2 kinase activity. When activated, cyclin A/CDC2 and cyclin B/CDC2 complexes positively regulate progression into mitosis, which in turn is associated with loss of the cyclin A protein and associated kinase activity (Nurse, 1990; Murray, 1992; Pagano *et al*, 1992; Morgan, 1995). Cyclin A/ and cyclin B/CDC2 activation requires the removal of inhibitory phosphorylations from the catalytic domain of CDC2, and this is carried out by the CDC25C phosphatase. CDC2 inhibitory phosphorylations are imposed by Wee1, Mik1 and/or Myt1 kinases. DNA damage induced G$_2$ arrest blocks the dephosphorylation/activation of cyclin A/CDC2 and cyclin B/CDC2 complexes, and cyclin A/CDK2 complexes remain active in the DNA damaged cells (Lock and Ross, 1990a,b; O'Connor *et al*, 1993b)

DNA Damage Blocks Activation of the CDC2-CDC25C Autocatalytic Feedback Loop

CDC2 and CDC25C have been suggested to interact in an autocatalytic feedback loop to bring about rapid activation of the CDC2 kinase (Fig. 7) (Hoffmann *et al*, 1993). This positive feedback loop is initiated upon binding of CDC25C to a small fraction of the excess pool of CDC2/cyclin B1 complexes. As a result of CDC2 dephosphorylation, the now active CDC2 would focus its initial kinase activity towards CDC25C itself (CDC2's first substrate?). Hyperphosphorylation of CDC25C has been correlated with upregulation of CDC25C activity (Hoffmann *et al*, 1993; O'Connor *et al*, 1994), which could in turn allow this relatively low abundance CDC25C protein the extra activity it

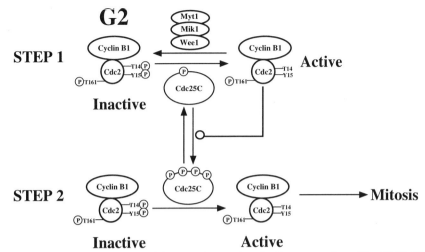

Fig. 7. DNA damage induced G$_2$ arrest is associated with suppression of the CDC2-CDC25C autocatalytic feedback loop. CDC2 has been proposed to interact with CDC25C in an autocatalytic feedback loop that would bring around the rapid activation of the master regulator of mitotic entry, the CDC2 kinase (Hoffmann *et al*, 1993). This autocatalytic feedback loop forms through two steps. In the first step, the feedback loop is initiated by interaction of CDC25C with a small proportion of hyperphosphorylated-CDC2/cyclin B complexes. The resultant CDC2 dephosphorylation at Thr-14 and Tyr-15 residues activates CDC2 kinase activity, which then phosphorylates and activates the bound CDC25C phosphatase. In the second step, maximally active CDC25C dephosphorylates/activates the remaining hyperphosphorylated-CDC2/cyclin B1 complexes in interphase cells, which promotes mitotic entry. DNA damage prevents CDC25C from reaching its most active/hyperphosphorylated state (O'Connor *et al*, 1994). This could be because the G$_2$ checkpoint prevents CDC2–CDC25C interaction or counteracts CDC25C action through upregulation of CDC2 inhibitory kinases and a CDC25C inhibitory phosphatase. Alternative possibilities include suppression of cyclin B1 levels (Muschel, 1991, 1993). Open circles indicate stimulation of an event. Arrows indicate conversion of one form into another form

might need to dephosphorylate rapidly the remaining hyperphosphorylated CDC2/cyclin B complexes in G$_2$ cells. Consistent with CDC25C activation being integrated into the G$_2$ checkpoint response, it was found that DNA damaged cells blocked CDC25C from reaching its most active/hyperphosphorylated state (O'Connor *et al*, 1994). Given the ability of CDC2 to phosphorylate CDC25C, one might expect that if CDC25C had productive access to CDC2, it would become hyperphosphorylated in G$_2$ arrested cells. Such a state could be achieved even if CDC2 was only transiently activated and then immediately downregulated by Wee1/Mik1. Since CDC25C does not become hyperphosphorylated in G$_2$ arrested cells, either CDC2-CDC25C interaction does not occur or there must be an upregulation of a CDC25C inhibitory phosphatase that rapidly converts CDC25C back into its hypophosphorylated state (Fig. 6). Consistent with the possibility that the G$_2$ checkpoint prevents CDC25C-CDC2 interaction, it was found that the interaction between CDC2 and CDC25C, which occurred transiently at the G$_2$/M border in CA46 cells, did not occur in DNA damaged G$_2$ arrested cells (O'Connor *et al*, 1994). A block to this interaction could occur in one of several ways, including checkpoint dependent anchoring of CDC25C and CDC2 at different intracellular locations until commitment to mitosis or through an inhibitor that sterically hinders CDC25C interaction with CDC2. CDC25C in interphase human lymphoma CA46 cells, like BHK cells, is located in the cytoplasm (Seki *et al*, 1992; Heald *et al*, 1993), whereas CDC25C in interphase HeLa cells is located in the nucleus (Millar *et al*, 1991; O'Connor *et al*, 1994). Given these localization differences, it is difficult to present a universal model for CDC25C location in mammalian cells. If one of the CDC2 inhibitory kinases were to associate tightly with CDC2 during interphase, this could provide a means of blocking access to CDC25C. However, investigations with Wee1 suggest that this kinase does not tightly associate with CDC2 during interphase or following DNA damage (Orlandi L and O'Connor PM, unpublished). Investigations into the binding of CDC2 to two other CDC2 inhibitory kinases, Mik1 and/or Myt1, have not been presented.

CDC2 Thr-14 and Tyr-15 Phosphorylation: Is It Necessary for G$_2$ Arrest?

The importance of CDC2 inhibitory phosphorylations in preventing premature activation of CDC2 and entry into mitosis in mammalian cells was elegantly demonstrated by Krek and Nigg (1991). In these studies, HeLa cells transfected with CDC2 mutants that lacked the ability to be phosphorylated at Thr-14 and Tyr-15 (CDC2 T14/A14, Y15/F15) underwent premature chromosome condensation (PCC) in interphase. Such studies were conducted in exponentially growing cultures, and the effect of DNA damage or inhibition of DNA synthesis on mutant CDC2 induced PCC was not investigated. It will no doubt be important to do this since studies in the budding yeast and fission yeast differ in respect to the importance of CDC2 tyrosine phosphorylation for the checkpoint response to inhibition of DNA replication and DNA damage. In fission yeast, changing CDC2 Tyr-15 to a phenylalanine abrogates the checkpoints that would normally inhibit mitotic entry following inhibition of DNA synthesis or DNA damage. However, mutation of the equivalent tyrosine position in CDC28 of budding yeast (Y19) has no effect upon viability of the yeast or upon checkpoints that monitor DNA replication or DNA damage (Enoch and Nurse, 1990; Amon *et al*, 1992; Sorger and Murray, 1992). An important experiment conducted by Sorger and Murray (1992) showed that placing CDC2-Y15F into budding yeast did not abrogate S or G$_2$ checkpoint control, whereas placing CDC28-Y19F into fission yeast was capable of disrupting these checkpoints. These results suggest that the two yeasts organize S and G$_2$ checkpoint control differently and that fission yeast relies more upon inhibition of CDC2 activity for cell cycle arrest than budding yeast. Budding yeast may have developed other means to prevent CDC28 from carrying out its function, such as blocking access of CDC28 to its substrates. Alternatively, budding yeast may conduct checkpoint control through a parallel pathway required, in connection with CDC28, to mediate mitotic entry. *Aspergillus nidulans* provides an example. In this organism, CDC2 activation by itself is insufficient for mitotic progression. The nimA kinase must also be activated, and this activation is suppressed through checkpoint control (Ye *et al*, 1996). Such findings encourage continued investigation of T14/Y15 phosphorylation in G$_2$ checkpoint control in mammalian cells since it would be helpful to understand which of the yeast checkpoint models is appropriately conserved in mammals.

Additional Mechanisms Contributing to G$_2$ Arrest: Cyclin B1 Downregulation

In HeLa cells, cyclin B1 mRNA and protein levels accumulate from very low levels at the G$_1$/S phase border to peak levels in G$_2$/M, and most of this accumulation occurs as cells enter G$_2$/M phase (Pines and Hunter, 1989). Although not seen in all cell types, it is clear that γ irradiation of S phase synchronized HeLa cells suppresses cyclin B1 mRNA and protein accumulation (Muschel *et al*, 1991, 1993). Lack of accumulation does not appear to be due to a late S phase block prior to the point that cyclin B1 accumulation would normally commence. Instead, in the G$_2$ arrested cells, both cyclin B1 mRNA stability and promoter activity are suppressed by γ irradiation (Maity *et al*, 1995). Reduced levels of cyclin B1 would limit the level of cyclin B1/CDC2 complexes in G$_2$ arrested cells, perhaps below the threshold level required for mitotic entry. Consistent with this possibility, Muschel and colleagues, using an inducible cyclin B1 expression system, have shown that overexpression of cyclin B1 in HeLa cells accelerates the rate at which γ irradiated G$_2$ arrested cells entered mitosis (Kao *et al*, in press). Induction of cyclin B1, although shortening G$_2$ delay, did not eliminate it (Kao *et al*, in press). Such findings are consistent with additional regulatory mechanisms contributing to G$_2$ checkpoint control in HeLa cells. The most likely point of control would be at the level of CDC2/cyclin B1 binding and/or activation of these complexes. Indeed, Lock and Keeling (1993) and Metting and Little (1995) have correlated G$_2$ arrest in HeLa cells with a failure to remove inhibitory phosphorylations from CDC2/cyclin B1 complexes. Assessing the level of cyclin B1/CDC2 complexes and phosphorylation status of CDC2 in these newly formed complexes could reveal whether the additional remaining G$_2$ delay seen in HeLa cells overexpressing cyclin B1 is due to the imposition of CDC2 inhibitory phosphorylations in the DNA damaged cells.

Taken together, it now appears possible that two independent mechanisms of checkpoint control might co-operate at the level of cyclin B1/CDC2 to maintain DNA damaged cells in G$_2$ phase: CDC2 inhibitory phosphorylations and suppression of cyclin B1 levels. Further studies of these mechanisms in normal and transformed cells will be necessary to establish fully their importance in G$_2$ checkpoint regulation.

The *ATM* Gene and the G_2 Cell Cycle Checkpoint

Early studies from David Scott's laboratory on cells from a patient with ataxia-telangiectasia (AT) found that such cells exhibited less G_2 delay following exposure to X rays than cells from a normal individual (Zampetti-Bosseler and Scott, 1981; Scott and Zampetti-Bosselar, 1982). These findings, in conjunction with those of Painter and Young (1980) and Houldsworth and Lavin (1980), who found that AT cells exhibited radioresistant DNA synthesis, led to the proposition that AT cells are defective in DNA damage checkpoint control. As described above, Kastan *et al* (1992) later showed that AT cells were defective in activation of the TP53 dependent G_1 checkpoint response to ionizing radiation. Consolidating defective G_1/S phase checkpoint control in AT, Barlow *et al* (1996) have shown that disruption of the murine *Atm* locus did indeed result in radioresistant DNA synthesis. The integrity of the G_2 checkpoint was not, however, assessed in *Atm* deficient mouse cells, possibly because of the poor in vitro growth of cells. Such studies could prove useful, since a number of workers have disagreed on whether the G_2 checkpoint is completely defective in AT cells. Indeed, in some cases, AT cells can exhibit prolonged G_2 arrest following DNA damage (Bates and Lavin, 1989; Hong *et al*, 1994).

A Role for TP53 in G_2 Checkpoint Control?

Using inducible systems for wild type TP53 expression, Stewart *et al* (1995) and Agarwal *et al* (1995) have shown that TP53 can induce G_2 delay. Thus, although wild type TP53 function is not required for G_2 arrest following DNA damage (Kastan *et al*, 1991; Kuerbitz *et al*, 1992; O'Connor *et al*, 1993b), TP53 might contribute an extra layer of protection to G_2 checkpoint control. In accord with this possibility, a number of workers have found that chemical inhibitors of G_2 checkpoint function are less effective at overriding G_2 arrest in cells with intact TP53 function (Fan *et al*, 1995; Powell *et al*, 1995; Russell *et al*, 1995). Although these studies point to a possible involvement of TP53 in G_2 checkpoint control, the mechanistic basis of this involvement is presently unknown. A splice variant of TP53 is preferentially expressed in G_2 phase (Kulesz-Martin *et al*, 1994), but involvement of this TP53 variant in G_2 checkpoint regulation has not been presented. A potential candidate for TP53 induced G_2 delay would be p21, which is capable of inhibiting cyclin B1/CDC2 kinase activity, albeit poorly (Harper *et al*, 1993, 1995). In opposition to a role for p21 in G_2 arrest, transfection studies have failed to reveal p21 induced G_2 arrest (Harper *et al*, 1995). TP53 also exhibits transcriptional repression activity (Zambetti and Levine, 1993), and prolonged overexpression of TP53 in the inducible systems explored (Agarwal *et al*, 1995; Stewart *et al*, 1995) could downregulate components required for the G_2/M transition, for example, cyclin B1. Whether pulsed induction of TP53 would induce G_2 arrest to the extent seen with prolonged induction of TP53 remains to be determined.

G$_2$ Checkpoint Control and Survival following DNA Damage

The length of G$_2$ delay following DNA damage has been correlated with cell survival in some yeast strains (Weinert and Hartwell, 1988; Hartwell and Weinert, 1989) and some mammalian cells (McKenna et al, 1991). In these situations, cells that failed to arrest or cells that only briefly delayed progression were found to be more sensitive to DNA damage. Such findings complied with the paradigm that G$_2$ delay ensures completion of DNA repair before cell division and that cells that enter mitosis prematurely are more likely to suffer the deleterious consequences of DNA damage (Hartwell and Kastan, 1994). In accord with this view, chemical agents that abrogate G$_2$ delay enhance the cytotoxic potency of DNA damaging agents (Lau and Pardee, 1982; Fingert et al, 1986). However, a number of studies have indicated that prolonged G$_2$ arrest does not always translate into a survival advantage, and prolonged G$_2$ arrest has in some cases been correlated with enhanced sensitivity to DNA damaging agents (Bates and Lavin, 1989; Hong et al, 1994; Fan et al, in press). This latter case could reflect the inability of cells to repair DNA damage and thus "turn off" the G$_2$ checkpoint signal that halts progression. This prolonged arrest could be thought of as the cell's attempt to protect against a potentially lethal premature mitosis.

A number of investigators have reported attenuated G$_2$ arrest in cancer cells or cells expressing viral oncogenes (O'Connor et al, 1992; Fan et al, 1995; Kaufmann et al, 1995). However, despite these observations, there do not appear to be clear examples of mammalian lines completely devoid of DNA damage induced G$_2$ arrest. This is in contrast to the generation of a number of mutants in yeast that completely fail to G$_2$ arrest following irradiation, as well as the common ablation of the TP53 dependent G$_1$ checkpoint in cancer cells (Hartwell and Kastan, 1994). Such differences may reflect the evolution of redundant pathways regulating G$_2$ arrest in mammalian cells. Alternatively, the genomic instability of cancer cells, which may in part be fuelled through defective checkpoint control (Hartwell, 1995), might eventually select against cells completely devoid of G$_2$ checkpoint function.

G$_2$ Checkpoint Abrogation and TP53 Function

The cytotoxicity of DNA damaging agents can be enhanced by agents that disrupt G$_2$ checkpoint control (Lau and Pardee, 1982). Whether such agents had any preferential activity against certain cancer cells was shown by the findings that the methylxanthines, caffeine and pentoxifylline, preferentially abrogated G$_2$ checkpoint function in cells with defective TP53 (Fan et al, 1995; Powell et al, 1995; Russell et al, 1995). This preferential action was also correlated with preferential killing of TP53 defective cancer cells exposed to one of these methylxanthines and a DNA damaging agent. Such results were among the first to point to a pharmacologically exploitable vulnerability in TP53 defective cells. However, caffeine is too toxic for clinical application, and although pentoxifylline is presently undergoing clinical trials, tolerable plasma levels are limited to 30–50 μM (Dezube *et al*, 1990), well below that needed for G_2 checkpoint abrogation in vitro (500 μM to 2 mM). Another agent that might prove useful is UCN-01. This agent is also capable of preferentially abrogating the G_2 checkpoint in cells with disrupted TP53 (Wang *et al*, 1996). UCN-01 is approximately 10 000-fold more potent than pentoxifylline, and plasma levels of UCN-01 needed to abrogate G_2 checkpoint function can be achieved in rodents without marked toxicity. UCN-01 has just begun phase I clinical trials. The molecular basis of differences in sensitivity of TP53 intact cells versus TP53 disrupted cells to these G_2 checkpoint abrogators remains to be determined but could result from differences in drug uptake or through differences in G_2 checkpoint regulation.

SUMMARY

This present review explores the mechanisms for DNA damage induced G_1 and G_2 arrest in mammalian cells. The complexity of the TP53 pathway is attested to by the variety of genes regulated by TP53, many of which require further investigation to bring their importance into focus. One gene intensely studied, *p21*, has been linked to the G_1 arrest mechanism and may, like *TP53*, be involved in some aspect of DNA repair. The outcome of TP53 activation for cell survival is equally complex and relies much upon cellular context and the type of DNA damaging agent employed. Although TP53 may participate in sensing DNA damage, additional components are likely to be required. Much of the focus on defining the mechanism of G_2 arrest in mammalian cells has concentrated on the cyclin B1/CDC2 kinase. Activation of this kinase is suppressed by DNA damage, and this may result from the imposition of inhibitory phosphorylations on the CDC2 kinase as well as downregulation of cyclin B1 levels. The logical point where the G_2 checkpoint interacts with the CDC2-CDC25C autocatalytic loop to prevent CDC2 activation remains to be defined and could involve inhibition of CDC25C-CDC2 interaction. It is hoped that moving upstream of CDC2 towards the point where DNA damage is sensed by the cell will uncover homologues of yeast components implicated in G_2 checkpoint control. The finding that certain G_2 checkpoint abrogators preferentially synergize with DNA damaging agents in cells with defective TP53 provides a potential pharmacological route through which TP53 defective cells might be targeted for destruction. Further exploration of this vulnerability might prove useful for future anti-cancer drug discovery efforts.

References

Agarwal ML, Argarwal A, Taylor WR and Stark GR (1995) p53 controls both the G2/M and the G1 cell cycle checkpoints and mediates growth arrest in human fibroblasts. *Proceedings of the National Academy of Sciences of the USA* **92** 8493–8497

Amon A, Surana U, Muroff I and Nasmyth K (1992) Regulation of p34Cdc28 tyrosine phosphorylation is not required for entry into mitosis in S. *cerevisiae*. *Nature* **355** 368–371

Aprelikova O, Xiong Y and Liu ET (1995) Both p16 and p21 families of cyclin-dependent kinase (CDK) inhibitors block the phosphorylation of cyclin-dependent kinases by the CDK-activating kinase. *Journal of Biological Chemistry* **270** 18195–18197

Bae I, Fan S, Bhatia K, Kohn KW, Fornace AJ Jr and O'Connor PM (1995) Relationships between G1 arrest and stability of the p53 and p21Cip1/Waf1 proteins following γ-irradiation of human lymphoma cells. *Cancer Research* **55** 2387–2393

Barlow C, Hirotsune S, Paylor R *et al* (1996) Atm-deficient mice: a paradigm of ataxia telangiectasia. *Cell* **86** 159–171

Bates PR and Lavin MF (1989) Comparison of gamma-radiation induced accumulation of ataxia telangiectasia and control cells in G2 phase. *Mutation Research* **218** 165–170

Bhatia K, Fan S, Spangler G *et al* (1995) A mutant p21 cyclin-dependent kinase inhibitor isolated from a Burkitt's lymphoma. *Cancer Research* **55** 1431–1435

Boise LH, Gonzalez-Garcia M, Postema CE *et al* (1993) bcl-x, a bcl2 related gene that functions as a dominant regulator of apoptotic death. *Cell* **74** 597–608

Brachman DG, Beckett M, Graves D, Haraf D, Vokes E and Weichselbaum RR (1993) p53 mutation does not correlate with radiosensitivity in 24 head and neck cancer cell lines. *Cancer Research* **53** 3667–3669

Brown R, Clugston C, Burns P *et al* (1993) Increased accumulation of p53 protein in cisplatin-resistant ovarian cell lines. *International Journal of Cancer* **55** 678–684

Brugarolas J, Chandrasekaran C, Gordon JI, Beach D, Jacks T and Hannon GJ (1995) Radiation-induced cell cycle arrest compromised by p21 deficiency. *Nature* **377** 552–557

Caelles C, Helmberg A and Karin M (1994) p53-dependent apoptosis is not mediated by transcriptional activation of p53-target genes. *Nature* **370** 220–223

Canman CE, Gilmer TM, Couts SB and Kastan MB (1995) Growth factor modulation of p53-mediated growth arrest versus apoptosis. *Genes and Development* **9** 600–611

Chen CY, Oliner JD, Zhan Q, Fornace AJ Jr, Vogelstein B and Kastan MB (1994) Interactions between p53 and MDM2 in a mammalian cell cycle checkpoint pathway. *Proceedings of the National Academy of Sciences of the USA* **91** 2684–2688

Chen IT, Smith ML, O'Connor PM and Fornace AJ Jr (1995) Direct interaction of Gadd45 with PCNA and evidence for competitive interaction of Gadd45 and p21Waf1/Cip1 with PCNA. *Oncogene* **11** 1931–1937

Chen IT, Akamatsu M, Smith ML *et al* (1996) Characterization of p21Cip1/Waf1 peptide domains required for cyclin E/Cdk2 and PCNA interaction. *Oncogene* **12** 595–607

Chen J, Jackson PK, Kirschner MW and Dutta A (1995) Separate domains of p21 involved in the inhibition of Cdk kinase and PCNA. *Nature* **374** 386–388

Chen X, Bargonetti J and Prives C (1995) p53, through p21 (WAF1/CIP1), induces cyclin D1 synthesis. *Cancer Research* **55** 4257–4263

Cho Y, Gorina S, Jeffrey PD and Pavletich NP (1994) Crystal structure of a p53 tumor suppressor-DNA complex: understanding tumorigenic mutations. *Science* **265** 346–356

Clarke AR, Purdie CA, Harrison DJ *et al* (1993) Thymocyte apoptosis induced by *p53*-dependent and independent pathways. *Nature* **362** 849–852

Deng C, Zhang P, Harper JW, Elledge SJ and Leder P (1995) Mice lacking p21CIP1/WAF1 undergo normal development, but are defective in G1 checkpoint control. *Cell* **82** 675–684

Dezube BJ, Eder JP and Pardee AB (1990) Phase I trial of escalating pentoxifylline dose with constant thiotepa. *Cancer Research* **50** 6806–6810

Di Leonardo A, Linke SP, Clarkin K and Wahl GM (1994) DNA damage triggers a prolonged p53-dependent G1 arrest and long-term induction of Cip1 in normal human fibroblasts. *Genes and Development* **8** 2540–2551

Dulic V, Kaufmann WK, Wilson SJ *et al* (1994) p53-dependent inhibition of cyclin-dependent kinase activities in human fibroblasts during radiation-induced G1 arrest. *Cell* **76** 1013–23

El-Deiry WS, Kern SE, Pietenpol JA, Kinzler KW and Vogelstein B (1992) Definition of a consensus binding site for p53. *Nature Genetics* **1** 45–49

El-Deiry WS, Tokino T, Velculescu VE *et al* (1993) WAF1, a potential mediator of p53 tumor suppression. *Cell* **75** 817–825

El-Deiry WS, Harper JW, O'Connor PM *et al* (1994) WAF1/CIP1 is induced in p53-mediated G1 arrest and apoptosis. *Cancer Research* **54** 1169–1174

Enoch T and Nurse P (1990) Mutation of fission yeast cell cycle control genes abolishes dependence of mitosis on DNA replication. *Cell* **60** 665–673

Fan S, El-Deiry WS, Bae I *et al* (1994) p53 gene mutations are associated with decreased sensitivity of human lymphoma cells to DNA damaging agents. *Cancer Research* **54** 5824–5830

Fan S, Smith ML, Rivet DJ II *et al* (1995) Disruption of p53 function sensitizes breast cancer MCF-7 cells to cisplatin and pentoxifylline. *Cancer Research* **55** 1649–54

Fan S, Chang JK, Smith ML, Duba D, Fornace AJ Jr and O'Connor PM Cells lacking CIP1/WAF1 genes exhibit preferential sensitivity to cisplatin and nitrogen mustard. *Oncogene* (in press)

Fingert HJ, Chang JD and Pardee AB (1986) Cytotoxic, cell cycle and chromosomal effects of methylxanthines in human tumor cells treated with alkylating agents. *Cancer Research* **46** 2463–2467

Flores-Rozas H, Keiman Z, Dean FB *et al* (1994) Cdk-interacting protein-1 (Cip1, Waf1) directly binds to proliferating cell nuclear antigen and inhibits DNA replication catalyzed by the DNA polymerase δ holoenzyme. *Proceedings of the National Academy of Sciences of the USA* **91** 8655–8659

Ford JM and Hanawalt PC (1995) Li-Fraumeni syndrome fibroblasts homozygous for p53 mutations are deficient in global DNA repair but exhibit normal transcription-coupled repair and enhanced UV resistance. *Proceedings of the National Academy of Sciences of the USA* **92** 8876–8880

Fornace AJ Jr, Alamo I and Hollander MC (1988) DNA damage-inducible transcripts in mammalian cells. *Proceedings of the National Academy of Sciences of the USA* **85** 8800–8804

Fotedar R, Fitzgerald P, Rousselle T *et al* (1996) p21 contains independent binding sites for cyclin and cdk2: both sites are required to inhibit cdk2 kinase activity. *Oncogene* **12** 2155–2164

Friend S (1994) p53: a glimpse at the puppet behind the shadow play. *Science* **265** 334–335

Funk WD, Pak DT, Karas RH, Wright WE and Shay JW (1992) A transcriptionally active DNA-binding site for human p53 protein complexes. *Molecular and Cellular Biology* **12** 2866–2871

Goubin F and Ducommun B (1995) Identification of binding domains on the p21Cip1 cyclin-dependent kinase inhibitor. *Oncogene* **10** 2281–2287

Greenblatt MS, Bennett WP, Hollstein M and Harris CC (1994) Mutations in the p53 tumor suppressor gene: clues to cancer etiology and molecular pathogenesis. *Cancer Research* **54** 4855–4878

Hall PA, Kearsey JM, Coates PJ, Norman DG, Warbrick E and Cox LS (1995) Characterization of the interaction between PCNA and Gadd45. *Oncogene* **10** 2427–2433

Harper JW, Adami GR, Wei N, Keyomarsi K and Elledge SJ (1993) The p21 Cdk-interacting protein Cip1 is a potent inhibitor of G1 cyclin-dependent kinases. *Cell* **75** 805–816

Harper JW, Elledge SJ, Keyomarsi K *et al* (1995) Inhibition of cyclin-dependent kinases by p21. *Molecular Biology of the Cell* **6** 387–400

Hartwell L (1992) Defects in a cell cycle checkpoint may be responsible for the genomic instability of cancer cells. *Cell* **71** 543–546

Hartwell L and Weinert T (1989) Checkpoints: controls that ensure the order of cell cycle events. *Science* **246** 629–634

Hartwell LH and Kastan MB (1994) Cell cycle control and cancer. *Science* **266** 1821–1828

Heald R, McLoughlin M and Mckeon F (1993) Human Wee1 maintains mitotic timing by protecting the nucleus from cytoplasmically activated Cdc2 kinase. *Cell* **74** 463–474

Helin K, Harlow E and Fattaey A (1993) Inhibition of E2F1 transactivation by direct binding of the retinoblastoma protein. *Molecular and Cellular Biology* **13** 6501–6508

Hoffmann I, Clarke PC, Marcote MJ, Karsenti E and Draetta G (1993) Phosphorylation and activation of human Cdc25C by Cdc2-cyclin B and its involvement in the self-amplification of MPF at mitosis. *EMBO Journal* **12** 53–63

Hollstein M, Sidransky D, Vogelstein B and Harris CC (1991) p53 mutations in human cancers. *Science* **253** 49–52

Hong J-H, Gatti RA, Huo YK, Chiang CS and McBride WH (1994) G2/M phase arrest and release in ataxia telangiectasia and normal cells after exposure to ionizing radiation. *Radiation Research* **140** 17–23

Houldsworth J and Lavin MF (1980) Effect of ionizing radiation on DNA synthesis in ataxia telangiectasia cells. *Nucleic Acids Research* **8** 3709–3720

Huang L-C, Clarkin KC and Wahl GM (1996) Sensitivity and selectivity of the DNA damage sensor responsible for activating p53-dependent G1 arrest. *Proceedings of the National Academy of Sciences of the USA* **93** 4827–4832

Hupp TR, Sparks A and Lane DP (1995) Small peptides activate the latent sequence-specific DNA binding function of p53. *Cell* **83** 237–245

Jayaraman L and Prives C (1995) Activation of p53 sequence-specific DNA binding by short single strands of DNA requires the p53 C-terminus. *Cell* **81** 1021–1029

Kao GD, McKenna WG, Maity A, Blank K and Muschel RJ Cyclin B1 availability is a rate-limiting component of the radiation-induced delay in HeLa cell. *Cancer Research* (in press)

Kastan MB, Onyekwere O, Sidransky D, Vogelstein B and Craig RW (1991) Participation of p53 protein in the cellular response to DNA damage. *Cancer Research* **51** 6304–6311

Kastan MB, Zhan Q, El-Deiry WS *et al* (1992) A mammalian cell cycle checkpoint pathway utilizing p53 and GADD45 is defective in ataxia-telangiectasia. *Cell* **71** 587–597

Kato J, Matsushime H, Hiebert SW, Ewen ME and Sherr CJ (1993) Direct binding of cyclin D to the retinoblastoma gene product (pRb) and pRb phosphorylation by the cyclin D-dependent kinase CDK4. *Genes and Development* **7** 331–342

Kaufmann WK, Levedakou EN, Grady HL, Paules RS and Stein GH (1995) Attenuation of G2 checkpoint function precedes human cell immortilization. *Cancer Research* **55** 7–11

Kazantev A and Sancar A (1995) Does p53 up-regulated Gadd45 protein have a role in excision repair? *Science* **10** 1003–1004

Kearsey JM, Shivji MK, Hall PA and Wood RD (1995) Does the p53-up-regulated Gadd45 protein have a role in excision repair? *Science* **10** 1004–1005

Kern SE, Kinzler KW, Bruskin A *et al* (1991) Identification of p53 as a sequence-specific DNA-binding protein. *Science* **252** 1708–1711

Kessis T, Slebos RJ, Nelson WG *et al* (1993) Human papillomavirus 16 E6 expression disrupts the p53 mediated cellular response to DNA damage. *Proceedings of the National Academy of Sciences of the USA* **90** 3988–3992

Krek W and Nigg EA (1991) Mutations in p34Cdc2 phosphorylation sites induce premature mitotic events in HeLa cells: evidence for a double block to p34Cdc2 kinase activation in vertebrates. *EMBO Journal* **10** 3331–3341

Kuerbitz SJ, Plunkett BS, Walsh WV and Kastan MB (1992) Wild-type *p53* is a cell cycle checkpoint determinant following irradiation. *Proceedings of the National Academy of Sciences of the USA* **89** 7491–7495

Kulesz-Martin MF, Lisafeld B, Huang H, Kisiel ND and Lee L (1994) Endogenous p53 protein generated from wild-type alternatively spliced p53 RNA in mouse epidermal cells. *Molecular and Cellular Biology* **14** 1698–1708

Lane DP (1992) p53, guardian of the genome. *Nature* **358** 15–16

Latham KM, Eastman SW, Wong A and Hinds PW (1996) Inhibition of p53-mediated growth arrest by overexpression of cyclin-dependent kinases. *Molecular and Cellular Biology* **16** 4445–4455

Lau CC and Pardee AB (1982) Mechanism by which caffeine potentiates lethality of nitrogen mustard. *Proceedings of the National Academy of Sciences of the USA* **79** 2942–2946

Lee JM and Bernstein A (1993) *p53* mutations increase resistance to ionizing radiation. *Pro-

ceedings of the National Academy of Sciences of the USA **90** 5742–5746

Lee M-H, Reynisdottir I and Massague J (1995) Cloning of p57Kip2, a cyclin-dependent kinase inhibitor with unique domain structure and tissue distribution. *Genes and Development* **9** 539–649

Lee S, Elenbaas B, Levine A and Griffith J (1995) p53 and its 14 kDa C-terminal domain recognize primary DNA damage in the form of insertion/deletion mismatches. *Cell* **81** 1013–1020

Levine A, Momand J and Finlay CA (1991) The p53 tumor suppressor gene. *Nature* **351** 453–456

Lin D, Fiscella M, O'Connor PM *et al* (1994) Constitutive expression of B-myb can bypass p53-induced Waf1/Cip1-mediated G1 arrest. *Proceedings of the National Academy of Sciences of the USA* **91** 10079–10083

Lin J, Reicher X, Wu X and Levine AJ (1996) Analysis of wild-type and mutant p21WAF1 gene activities. *Molecular and Cellular Biology* **16** 1786–1793

Livingstone LR, Whitem A, Sprousem J, Livanosm E, Jacks T and Tlsty TD (1992) Altered cell cycle arrest and gene amplification potential accompany loss of wild-type p53. *Cell* **70** 923–935

Lock RB (1992) Inhibition of p34cdc2 kinase activation, p34cdc2 tyrosine dephosphorylation and mitotic progression in Chinese hamster ovary cells exposed to etoposide. *Cancer Research* **52** 1817–1822

Lock RB and Ross WE (1990a) Inhibition of p34Cdc2 kinase induced by etoposide or irradiation as a mechanism of G2 arrest in Chinese hamster ovary cells. *Cancer Research* **50** 3761–376

Lock RB and Ross WE (1990b) Possible role for p34Cdc2 kinase in etoposide-induced cell death of Chinese hamster ovary cells. *Cancer Research* **50** 3767–3771

Lock RB and Keeling PK (1993) Responses of HeLa and Chinese hamster ovary p34Cdc2/cyclin B-kinase in relation to cell cycle perturbations induced by etoposide. *International Journal of Oncology* **3** 33–42

Lowe SW, Ruley HE, Jacks T and Houseman DE (1993a) p53-dependent apoptosis modulates the cytotoxicity of anticancer agents. *Cell* **74** 957–967

Lowe SW, Schmitt EM, Smith SW, Osborne BA and Jacks T (1993b) p53 is required for radiation-induced apoptosis in mouse thymocytes. *Nature* **362** 847–849

Luo Y, Hurwitz J and Massague J (1995) Cell cycle inhibition by independent Cdk and PCNA binding domains in p21Cip1. *Nature* **375** 159–161

Lydall D and Weinert T (1995) Yeast checkpoint genes in DNA damage processing: implications for repair and arrest. *Science* **270** 1488–1491

MacDonald RE, Wu GS, Waldman T and El-Deiry WS (1996) Repair defect in p21WAF1/CIP1 –/– human cancer cells. *Cancer Research* **56** 2250–2255

Mack DH, Vartikar J, Pipas JM and Laimins LA (1993) Specific repression of TATA-mediated but not initiator-mediated transcription by wild-type p53. *Nature* **363** 281–283

McKenna WG, Iliakis G, Weiss MC, Bernhard EJ and Muschel RJ (1991) Increased G2 delay in radiation-resistant cells obtained by transformation of primary rat embryo cells with the oncogenes H-ras and V-myc. *Radiation Research* **125** 283–287

Magrath I (1990) The pathogenesis of Burkitt's lymphoma. *Advances in Cancer Research* **5** 133–270

Maity A, McKenna WG and Muschel RJ (1995) Evidence for post-transcriptional regulation of cyclin B1 mRNA in the cell cycle and following irradiation in HeLa cell. *EMBO Journal* **14** 603–609

Maltzman W and Czyzyk L (1984) UV irradiation stimulates levels of p53 cellular antigen in non transformed mouse cells. *Molecular and Cellular Biology* **4** 1689–1694

Martinez J, Georgoff I, Martinez J and Levine AJ (1991) Cellular localization and cell cycle regulation by a temperature mutant p53 protein. *Genes and Development* **5** 151–159

Matsuoka S, Edwards M, Bai C *et al* (1995) p57Kip2, a structurally related distinct member of the p21Cip1 Cdk inhibitor family, is a candidate tumor suppressor. *Genes and Development* **9** 650–662

Metting NF and Little JB (1995) Transient failure to dephosphorylate the Cdc2-cyclin B1 complex accompanies radiation-induced G2 phase arrest in HeLa cells. *Radiation Research* **143** 286–292

Michalovitz D, Halevy O and Oren M (1990) Conditional inhibition of transformation and of cell proliferation by a temperature sensitive mutant p53. *Cell* **62** 671–680

Millar JB, Blevitt J, Gerace L, Sadhu K, Featherstone C and Russel P (1991) p55Cdc25C is a nuclear protein required for the initiation of mitosis in human cells. *Proceedings of the National Academy of Sciences of the USA* **88** 10500–10504

Miyashita T and Reed JC (1995) Tumor suppressor p53 is a direct transcriptional activator of the human bax gene. *Cell* **80** 293–299

Miyashita T, Krajewski S, Krajewski M *et al* (1994) Tumor suppressor p53 is a regulator of bcl2 and bax gene expression in vitro and in vivo. *Oncogene* **9** 1799–1805

Morgan DO (1995) Principles of CDK regulation. *Nature* **374** 131–134

Morris GF, Bischoff JR and Mathews MB (1996) Transcriptional activation of the human proliferating cell nuclear antigen promoter by p53. *Proceedings of the National Academy of Sciences of the USA* **93** 895–899

Muller H, Lukas J, Schneider A *et al* (1994) Cyclin D1 expression is regulated by the retinoblastoma protein. *Proceedings of the National Academy of Sciences of the USA* **91** 2945–2949

Mummenbrauer T, Janus F, Muller B, Wiesmuller L, Deppert W and Grosse F (1996) p53 protein exhibits 3′-to-5′ exonuclease activity. *Cell* **85** 1089–1099

Murray AW (1992) Creative blocks: cell-cycle checkpoints and feedback controls. *Nature* **359** 599–604

Muschel RJ, Zhang HB, Iliakis G and McKenna WG (1991) Cyclin B expression in HeLa cells during the G2 block induced by ionizing radiation. *Cancer Research* **51** 5113–5117

Muschel RJ, Zhang HB and McKenna WG (1993) Differential effect of ionizing radiation on the expression of cyclin A and cyclin B in HeLa cells. *Cancer Research* **53** 1128–1135

Nakanishi M, Robetorge RS, Adami GR, Pereira-Smith OM and Smith JR (1995) Identification of the active region of the DNA synthesis inhibitor gene p21(sdi1/cip1/WAF1). *EMBO Journal* **14** 555–563

Nelson WG and Kastan MB (1994) DNA strand breaks: the DNA template alterations that trigger *p53*-dependent DNA damage response pathways. *Molecular and Cellular Biology* **14** 1815–1823

Nevins JR (1992) E2F: a link between the Rb tumor suppressor protein and viral oncogenes. *Science* **258** 424–429

Noda A, Ning Y, Venable SF, Pereira-Smith OM and Smith JR (1994) Cloning of senescent cell-derived inhibitors of DNA synthesis using an expression screen. *Experimental Cell Research* **211** 90–98

Nurse P (1990) Universal control mechanism regulating onset of M-phase. *Nature* **344** 503–508

Oberosler P, Hloch P, Ramsperger U and Stahl H (1993) p53-catalyzed annealing of complementary single-stranded nucleic acids. *EMBO Journal* **12** 2389–2396

O'Connor PM and Fan S DNA damage checkpoints: implications for cancer therapy. In: Meijer L, Guidet S and Vogel L (eds). *Progress in Cell Cycle Research*, Plenum Press, New York (in press)

O'Connor PM, Ferris DK, White GA *et al* (1992) Relationships between cdc2 kinase, DNA crosslinking, and cell cycle perturbations induced by nitrogen mustard. *Cell Growth and Differentiation* **3** 43–52

O'Connor PM, Ferris DK, Pagano M *et al* (1993a) G2 delay induced by nitrogen mustard in human cells affects cyclin A/cdk2 and cyclin B1/cdc2-kinase complexes differently. *Journal of Biological Chemistry* **268** 8298–8308

O'Connor PM, Jackman J, Jondle D, Bhatia K, Magrath I and Kohn KW (1993b) Role of the p53 tumor suppressor gene in cell cycle arrest and radiosensitivity of Burkitt's lymphoma cell lines. *Cancer Research* **53** 4776–80

O'Connor PM, Ferris DK, Hoffmann I, Jackman J, Draetta G and Kohn KW (1994) Role of the cdc25C phosphatase in G2 arrest induced by nitrogen mustard. *Proceedings of the National Academy of Sciences of the USA* **91** 9480–9484

Oliner JD, Kinzler KW, Meltzer PS, George DL and Vogelstein B (1992) Amplification of a gene encoding a p53-associated protein in human sarcomas. *Nature* **358** 80–83

Oltvai ZN, Millman CL and Korsmeyer SJ (1993) Bcl2 heterodimerizes in vivo with a conserved homolog, Bax, that accelerates programmed cell death. *Cell* **74** 609–619

Pagano M, Pepperkok R, Verde F, Ansorge W and Draetta G (1992) Cyclin A is required at two points in the human cell cycle. *EMBO Journal* **11** 961–971

Pagano M, Theodoras AM, Tam SW and Draetta GF (1994) Cyclin D1-mediated inhibition of repair and replicative DNA synthesis in human fibroblasts. *Genes and Development* **8** 1627–1639

Painter RB and Young BR (1980) Radiosensitivity in ataxia telangiectasia: a new explanation. *Proceedings of the National Academy of Sciences of the USA* **77** 7315–7317

Pan ZQ, Reardon JT, Li L, Flores-Rozas H, Legerski R, Sancar A and Hurwitz J (1995) Inhibition of nucleotide excision repair by the cyclin-dependent kinase inhibitor p21. *Journal of Biological Chemistry* **270** 22008–22016

Pietenpol JA, Tokino T, Thiagalingam S, El-Deiry WS, Kinzler KW and Vogelstein B (1994) Sequence-specific transcriptional activation is essential for growth suppression by p53. *Proceedings of the National Academy of Sciences of the USA* **91** 1998–2002

Pines J and Hunter T (1989) Isolation of a human cyclin cDNA: Evidence for cyclin mRNA and protein regulation in the cell cycle and for interaction with p34Cdc2. *Cell* **58** 833–846

Polyak K, Lee M-H, Erdjument-Bromage H *et al* (1994) Cloning of p27Kip1, a cyclin-dependent kinase inhibitor and potential mediator of extracellular antimitogenic signals *Cell* **78** 59–66

Powell SN, DeFrank JS, Connell P *et al* (1995) Differential sensitivity of p53(–) and p53(+) cells to caffeine-induced radiosensitization and override of G2 delay. *Cancer Research* **55** 1643–1648

Prelich G, Kostura M, Marshak DR, Mathews MB and Stillman B (1987) Functional identity of proliferating cell nuclear antigen and a DNA polymerase delta auxiliary protein. *Nature* **326** 471–475

Qin X-Q, Livingston DM, Kaelin WG Jr and Adams PD (1994) Deregulated transcription factor E2F-1 expression leads to S phase entry and p53-mediated apoptosis. *Proceedings of the National Academy of Sciences of the USA* **91** 10918–10922

Russell KJ, Wiens LW, Galloway DA and Groudine M (1995) Abrogation of the G2 checkpoint results in differential radiosensitization of G1 checkpoint-deficient and competent cells. *Cancer Research* **55** 1639–1642

Russo AA, Jeffrey PD, Patten A, Massague J and Pavletich NP (1996) Crystal structure of the p27Kip1 cyclin-dependent kinase inhibitor bound to the cyclin A-CDK2 complex. *Nature* **382** 325–331

Savitsky K, Bar-Shira A, Gilad S *et al* (1995) A single ataxia telangiectasia gene with a product similar to PI-3 kinase. *Science* **268** 1749–1753

Scott D and Zampetti-Bosseler F (1982) Cell cycle dependence of mitotic delay in X-irradiated normal and ataxia-telangiectasia fibroblasts. *International Journal of Radiation Biology* **42** 679–683

Seki T, Yamashita K, Nishitani H, Takagi T, Russel P and Nishimoto T (1992) Chromosome condensation caused by loss of RCC1 function requires the cdc25C protein that is located in the cytoplasm. *Molecular Biology of the Cell* **3** 1373–1388

Selvakumaran M, Lin HK, Miyashita T *et al* (1994) Immediate early up-regulation of bax expression by p53 but not TGFβ1: a paradigm for distinct apoptotic pathways. *Oncogene* **9** 1791–1798

Sherr CJ (1994) G1 phase progression: cycling on cue. *Cell* **79** 551–555

Shivakumar CV, Brown DR, Deb S and Deb SP (1995) Wild-type human p53 transactivates the human proliferating cell nuclear antigen promoter. *Molecular and Cellular Biology* **15** 6785–6793

Shivji MK, Kenny MK and Wood RD (1992) Proliferating cell nuclear antigen is required for nucleotide excision repair. *Cell* **69** 367–374

Slebos RJC, Lee MH, Plunkett B *et al* (1994) p53-dependent G1 arrest involves pRb-related proteins and is disrupted by the human papillomavirus 16 E7 oncoprotein. *Proceedings of the National Academy of Sciences of the USA* **91** 5320–5324

Slichenmyer WJ, Nelson WG, Slebos RJ and Kastan MB (1993) Loss of a p53-associated G1 checkpoint does not decrease cell survival following DNA damage. *Cancer Research* **53** 4164–4168

Smith ML, Chen IT, Zhan Q, *et al* (1994) Interaction of the p53-regulated protein Gadd45 with proliferating cell nuclear antigen. *Science* **266** 1376–1380

Smith ML, Chen IT, Zhan Q, O'Connor PM and Fornace AJ Jr (1995) Involvement of the p53 tumor suppressor in repair of u.v.-type DNA damage. *Oncogene* **10** 1053–1059

Smith ML, Kotny HU, Zhan Q, Sreenath A, O'Connor PM and Fornace AJ Jr (1996) Antisense GADD45 expression results in decreased DNA repair and sensitizes cells to UV-irradiation or cisplatin. *Oncogene* **13** 2255–2263

Sorger PK and Murray AW (1992) S-phase feedback control in budding yeast independent of tyrosine phosphorylation of p34Cdc28. *Nature* **355** 365–367

Stewart N, Hicks GG, Paraskeas F and Mowat M (1995) Evidence for a second cell cycle block at G2/M by p53. *Oncogene* **10** 109–115

Tam S W, Theodoras AM, Shay JW, Draetta GF and Pagano M (1994) Differential expression and regulation of cyclin D1 protein in normal and tumor human cells: association with Cdk4 is required for cyclin D1 function in G1 progression. *Oncogene* **9** 2663–2674

Terada Y, Tatsuka M, Jinno S and Okayama H (1995) Requirement for tyrosine phosphorylation of Cdk4 in G1 arrest induced by ultraviolet light. *Nature* **376** 358–362

Toyoshima H and Hunter T (1994) p27, a novel inhibitor of G1 cyclin-Cdk protein kinase activity, is related to p21. *Cell* **78** 67–74

Waga S, Hannon GJ, Beach D and Stillman B (1994) The p21 inhibitor of cyclin-depndent kinases controls DNA replication by interacting with PCNA. *Nature* **369** 574–578

Wagner AJ, Kokontis J M and Hay N (1994) Myc-mediated apoptosis requires wild-type p53 in a manner independent of cell cycle arrest and the ability of p53 to induce p21Waf1/Cip1. *Genes and Development* **8** 2817–2830

Waldman T, Kinzler KW and Vogelstein B (1995) p21 is necessary for the p53-mediated G1 arrest in human cancer cells. *Cancer Research* **55** 5187–5190

Walton MI, Whysong D, O'Connor PM, Hockenbery D, Korsmeyer SJ and Kohn KW (1993) Constitutive expression of human Bcl2 modulates nitrogen mustard and camptothecin induced apoptosis. *Cancer Research* **53** 1853–1861

Wang Q, Fan S, Eastman A, Worland PJ, Sausville EA and O'Connor PM (1996) UCN-01: a potent abrogator of G2 checkpoint function in cancer cells with disrupted p53. *Journal of the National Cancer Institute* **88** 956–965

Wang XW, Forrester K, Yeh H, Feitelson MA, Gu JR and Harris CC (1994) Hepatitis B virus X protein inhibits p53 sequence-specific DNA binding, transcriptional activity, and association with transcription factor ERCC3. *Proceedings of the National Academy of Sciences of the USA* **91** 2230–2234

Wang XW, Yeh H, Schaeffer L *et al* (1995) p53 modulation of TFIIH associated nucleotide excision repair activity. *Nature Genetics* **10** 188–195

Weinberg RA (1995) The retinoblastoma protein and cell cycle control. *Cell* **81** 323–330

Weinert TA and Hartwell LH (1988) The RAD9 gene controls the cell cycle response to DNA damage in *Saccharomyces cerevisiae*. *Science* **241** 317–322

Wu X and Levine AJ (1994) p53 and E2F-1 cooperate to mediate apoptosis. *Proceedings of the*

National Academy of Sciences of the USA **91** 3602–3606

Wu X, Bayle JH, Olson D and Levine AJ (1993) The p53-mdm-2 autoregulatory feedback loop. *Genes and Development* **7** 1126–1132

Xiong Y, Zhang H and Beach D (1992) D type cyclins associate with multiple protein kinases and the DNA replication and repair factor PCNA. *Cell* **71** 505–514

Ye XS, Fincher RR, Tang A, O'Donnell KO and Osmani SA (1996) Two S phase checkpoint systems, one involving the function of both BIME and Tyr15 phosphorylation of p34Cdc2, inhibit NIMA and prevent premature mitosis. *EMBO Journal* **15** 3599–3610

Yin Y, Tainsky MA, Bischoff FZ, Strong LC and Wahl GM (1992) Wild-type p53 restores cell cycle control and inhibits gene amplification in cells with mutant p53 alleles. *Cell* **70** 937–948

Yonish-Rouach E, Resnitzky D, Lotem J, Sachs L, Kimchi A and Oren M (1991) Wild-type p53 induces apoptosis of myeloid leukemic cells that is inhibited by interleukin-6. *Nature* **352** 345–347

Yuan Z-M, Huang Y, Whang Y *et al* (1996) Role for c-Abl tyrosine kinase in growth arrest response to DNA damage. *Nature* **382** 272–274

Zakian RV (1995) ATM-related genes: what do they tell us about functions of the human gene? *Cell* **82** 685–687

Zambetti GP and Levine AJ (1993) A comparison of the biological activities of wild-type and mutant p53. *FASEB Journal* **7** 855–65

Zampetti-Bosseler F and Scott D (1981) Cell death, chromosome damage and mitotic delay in normal human, ataxia telangiectasia and retinoblastoma fibroblasts after X-irradiation. *International Journal of Radiation Biology* **39** 547–558

Zhan Q, Fan S, Bae I *et al* (1994) Induction of bax by genotoxic stress in human cells correlates with normal p53 status and apoptosis. *Oncogene* **9** 3743–51

Zhan Q, Alamo I, Yu K *et al* (1996) The apoptosis-associated γ-ray response of BCL-XL depends on normal p53 function. *Oncogene* **13** 2287–2293

Zhang H, Hannon GJ and Beach D (1994) p21-containing complexes exist both in active and inactive states. *Genes and Development* **8** 1750–1758

The author is responsible for the accuracy of the references.

Maintaining Genetic Stability through TP53 Mediated Checkpoint Control

G M WAHL[1] • S P LINKE[1,2] • T G PAULSON[1,2] • L-C HUANG[1]

[1]*Gene Expression Laboratory, The Salk Institute for Biological Studies, La Jolla, CA 92037;*
[2]*Department of Biology, University of California, San Diego, La Jolla, CA 92093*

INTRODUCTION

Even to untrained observers, cancer cells within a single tumour exhibit widely variable cellular and nuclear morphologies. Their heterogeneity becomes even more apparent after analysis with nucleic acid probes or markers for cell surface proteins (see eg Dulbecco, 1989; Kallioniemi *et al*, 1992, 1993; Visakorpi *et al*, 1995). This diversity arises from a multistep clonal selection process during which numerous genetic alterations accumulate within each genome. Genetic alterations that occur commonly in cancer cells, such as gene amplification, arise at rates that are often too low to be measured in normal cells (Tlsty *et al*, 1989; Tlsty, 1990; Wright *et al*, 1990b). One explanation for the apparent incompatibility between the low mutation rate in normal cells and the heterogeneity and multiple mutations in tumour cells is that one or more of the mutations incurred during cancer progression generate an underlying genomic instability that increases the rate at which variants arise during clonal expansion (Nowell, 1976; Loeb, 1991). This leads to the prediction that the controls that regulate genome duplication and segregation would be inactivated during neoplasia (Nowell, 1976; Hartwell and Weinert, 1989; Loeb, 1991).

Studies in yeast and mammalian cells show that controls in each cell cycle phase impact significantly on genetic stability. Some controls comprise "checkpoints", which ensure that all processes in one cell cycle phase are completed before the next phase begins (Hartwell and Weinert, 1989; Hartwell and Kastan, 1994). Checkpoints that respond to DNA damage and monitor the status of repair may be organized as "genome guardian complexes" composed of biosynthetic, repair and cell cycle proteins (Allen *et al*, 1994; Carr and Hoekstra, 1995; Navas *et al*, 1995). Among the checkpoints that impact on genetic stability are the following four. Firstly, a mechanism exists to prevent entry into S phase if the nucleotide precursor pool is inadequate for genome duplication (see below and Linke *et al*, 1996). Without such control, the probability of chromosome breakage, changes in DNA methylation and other alterations is increased (Windle *et al*, 1991; Baylin, 1992; Di Leonardo *et al*, 1993; Almasan *et al*, 1995a). Secondly, DNA damage induces arrest to prevent replication and/or segregation of damaged chromosomes (see below and Kastan *et al*, 1991; Cross *et al*, 1995). Thirdly, the status of DNA replication is monitored to prevent incompletely duplicated genomes from undergoing mitosis (Enoch *et al*, 1992; Weinert *et al*, 1994). Mitosis with a partially replicated genome could lead to chromosome breakage and formation of unstable dicentric chromosomes, which could fuel further genomic instability through bridge-breakage-fusion cycles (McClintock, 1984). Finally, all events required for chromosome segregation are monitored, including duplication and correct alignment of the centrosomes, assembly of the spindle and correct positioning of the chromosomes along the metaphase plate (Hoyt *et al*, 1991; Li and Murray, 1991; Cross *et al*, 1995; Fukasawa *et al*, 1996). Inappropriate chromosome segregation resulting from an aberration of any mitotic process would lead to aneuploidy and future karyotypic instability.

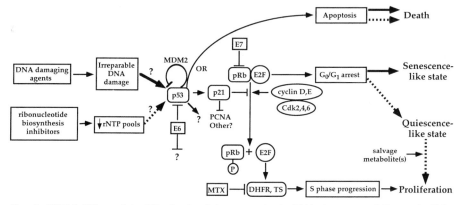

Fig. 1. TP53 (p53) mediated G_1 checkpoints responsive to DNA damage and ribonucleotide depletion. Arrows indicate induction or causation and T bars indicate inhibition. Heavy solid arrows denote DNA damage induced events and heavy dotted arrows denote ribonucleotide depletion induced events. Increases in TP53 protein leads to increases in p21 through transcriptional activation. p21 can then inhibit cyclin/CDK complexes that normally phosphorylate RB (pRb). Hypophosphorylated RB remains associated with E2F, preventing transcription of genes required for S phase progression, such as *DHFR* and *TS (thymidylate synthase)*. DNA damaging agents may cause irreparable DNA damage, leading to a prolonged or permanent arrest resembling cellular senescence (Di Leonardo *et al*, 1994). A damage independent G_1 arrest mediated by TP53 is induced by ribonucleotide depletion (Linke *et al*, 1996). However, this arrest is reversed by addition of salvage metabolites and more closely resembles quiescence. In certain cells (eg thymocytes or the myeloid leukaemia cell line ML-1) both nucleotide depletion and irradiation can induce apoptosis (Clarke *et al*, 1993; Di Leonardo *et al*, 1993; Nelson and Kastan, 1994). The E6 protein derived from oncogenic papillimaviruses causes degradation of TP53 through the ubiquitin pathway, thereby inactivating the TP53 dependent arrest/apoptosis pathway, whereas the E7 protein inactivates RB (as well as related proteins such as p107 and p130) and complexes consisting of cyclin A and E2F. Mdm2 inhibits TP533 function by binding the N-terminus of TP53 and preventing association with the basal transcription apparatus, and perhaps by other mechanisms as discussed in Fig. 2. dNTP biosynthesis inhibitors such as methotrexate (MTX) probably cause an early S phase arrest independently of TP53 status by directly inhibiting enzymes such as DHFR and TS. p21 may also affect cell cycle progression by mechanisms independent of CDK inhibition, such as through binding to proteins such as proliferating cell nuclear antigen (PCNA) (Chen *et al*, 1995)

Defects either in cell cycle checkpoints or in the machinery that detects and repairs DNA damage decrease genetic stability and consequently increase the probability of tumour formation. Germline mutations in mismatch repair (MMR) (see eg Fishel *et al*, 1993; Leach *et al*, 1993) or nucleotide excision repair (NER) (Cleaver, 1989, 1994) genes can lead to the development of colon and skin cancers, respectively. Germline mutations in the *TP53* checkpoint gene confer a predisposition to the early development of cancer in multiple organ systems in humans (Malkin *et al*, 1990; Srivastava *et al*, 1990) and mice (Donehower *et al*, 1992), and more than 50% of sporadic human cancers exhibit *TP53* mutations (Levine *et al*, 1991). Checkpoints are biochemical pathways that include a sensor to detect and transmit a signal from the process or structure being monitored. This signal is relayed to the cell cycle arrest and/or repair systems (Fig. 1). Therefore, mutations in any of the factors involved in a checkpoint might produce similar or identical phenotypes. This may explain why tumours and cell lines with wild type *TP53* genes exhibit defective TP53 checkpoint responses (Kastan *et al*, 1992; Livingstone *et al*, 1992; Nagasawa *et al*, 1995).

Diverse checkpoint functions for TP53 have been proposed, which could explain its frequent mutation in cancer. TP53 mediates a G_0/G_1 cell cycle arrest or apoptosis in response to DNA damage generated by γ radiation, ultraviolet (UV) radiation and some DNA damaging chemotherapeutic drugs (see eg below and Kastan *et al*, 1991; Clarke *et al*, 1993; Lowe *et al*, 1993b; Lu and Lane, 1993; Zhan *et al*, 1993). In addition, TP53 can mediate G_0/G_1 arrest in the absence of detectable DNA damage during ribonucleotide depletion (see below and Linke *et al*, 1996). Environmental stresses such as hypoxia, which occurs frequently at tumour sites, can also induce TP53 dependent apoptosis (Graeber *et al*, 1994, 1996), perhaps because reduced oxygen affects pyrimidine pools (Palace and Lawrence, 1995) or induces DNA breakage (Russo *et al*, 1995). Although TP53 mediated G_0/G_1 arrest cannot be induced in late G_1 cells (Di Leonardo *et al*, 1994), in vitro evidence suggests that TP53 may inhibit processive DNA synthesis in response to DNA damage and may be directly involved in DNA repair (Bakalkin *et al*, 1994; Smith *et al*, 1994, 1995b; Ford and Hanawalt, 1995; Wang *et al*, 1995). TP53 has also been reported to possess an intrinsic exonuclease function (Mummenbrauer *et al*, 1996) and to bind recombinases such as RAD51 (Sturzbecher *et al*, 1996), which may facilitate its putative repair functions. Furthermore, some studies indicate that TP53 overexpression can induce G_2 arrest (Agarwal *et al*, 1995; Stewart *et al*, 1995), although there is strong evidence that it does not participate directly in the G_2 checkpoint triggered by DNA damage (Kaufmann, 1995; Paules *et al*, 1995). Finally, TP53 appears to contribute to the control of centrosome replication (Fukasawa *et al*, 1996) and to the prevention of DNA rereplication when chromosome segregation is hindered by mitotic spindle inhibitors (Cross *et al*, 1995). Thus, TP53 may help maintain genetic stability by preventing replication of damaged DNA, replication under conditions that can cause DNA damage or mutations and rereplication of DNA that could lead to aneuploidy.

This chapter provides a perspective for understanding how cell cycle control and DNA repair contribute to genetic stability. We focus on the G_0/G_1 functions of TP53 as they are the best characterized. Given that the TP53 pathway is aberrant in such a large proportion of human cancers, a deeper understanding of the mechanisms of TP53 function and the consequences of its inactivation may lead to therapeutic strategies specifically targeted at this crucial cell cycle regulator and the pathways in which it participates.

MECHANISMS AND CONSEQUENCES OF TP53 DEPENDENT G_0/G_1 CELL CYCLE ARREST INDUCED BY IONIZING RADIATION

TP53 Mediated G_0/G_1 Arrest Maintains Genetic Stability by Eliminating Cells with Damaged DNA

Controversy has arisen concerning the mechanism by which the DNA damage inducible cell cycle arrest limits genetic variation. One theory, supported by data from cell cycle analyses in the human myeloid leukaemia cell line ML-1, is that the arrest provides additional time for repair (Kastan *et al*, 1991; Lane, 1992). This is analogous to temporary arrests induced by cell cycle checkpoint proteins in yeast, including RAD9 (Weinert and Hartwell, 1988), MEC1 (Paulovich and Hartwell, 1995) and HUS1 (Enoch *et al*, 1992).

The arrest for repair model leads to the prediction that cells with a functional TP53 pathway (TP53+) should have enhanced DNA repair, leading to decreased chromosomal aberrations and increased survival compared with cells with a non-functional TP53 pathway (TP53−). However, analyses of isogenic TP53+ and TP53− mouse or human cells show little difference in chromosome aberration frequency immediately after γ irradiation or in spontaneous or induced mutation frequencies (Di Leonardo *et al*, 1994; Bouffler *et al*, 1995; Nishino *et al*, 1995; Sands *et al*, 1995; Gupta *et al*, 1996). The model also predicts that TP53+ cells should be more radioresistant than TP53− cells. However, TP53− cells are typically significantly more radioresistant than TP53+ cells (see eg Arlett *et al*, 1988; Su and Little, 1992; Lee and Bernstein, 1993; McIlwrath *et al*, 1994; Tsang *et al*, 1995; Gupta *et al*, 1996; Yount *et al*, 1996), although a few studies show no correlation between radiosensitivity and TP53 status (see eg Slichenmyer *et al*, 1993). It is also interesting to note that the DNA damage induced G_2 delay and cellular radiosensitivity are separable phenotypes in fission yeast, showing that the capacity to arrest for repair does not necessarily contribute to radioresistance (Barbet and Carr, 1993).

An alternative model is that cells with unrepaired, or irreparable, DNA damage in G_1 are permanently removed from the cycle (Fig. 1). This may occur by induction of a prolonged arrest in some cell types, such as fibroblasts (Nagasawa and Little, 1983; Nagasawa *et al*, 1984; Di Leonardo *et al*, 1994), or by activation of an apoptotic pathway, as occurs after γ irradiation of mouse thymocytes and intestinal epithelial cells (Clarke *et al*, 1993, 1994; Lowe *et al*, 1993b). The prolonged arrest has been shown to be TP53 dependent through the use of isogenic TP53+ and TP53− normal human diploid fibroblasts (NDF) (Di Leonardo *et al*, 1994; Almasan *et al*, 1995a; Li *et al*, 1995b), Li-Fraumeni fibroblasts (Little, 1994), mouse embryo fibroblasts (MEF) (Deng *et al*, 1995) and human glioblastoma cells (Yount *et al*, 1996). The TP53+ cells undergo a dose dependent arrest in the initial G_0/G_1 phase after irradiation, generally obeying single hit kinetics. By contrast, virtually 100% of TP53− cells enter the first S phase regardless of dose (Di Leonardo *et al*, 1994; Almasan *et al*,

1995a). Arrested TP53[+] NDF exhibit several properties associated with senescence. The cells remain attached to the solid support and exclude trypan blue but become enlarged and irregularly shaped and do not divide to form viable colonies for up to 10 weeks (Di Leonardo *et al*, 1994; Li *et al*, 1995b; Yount *et al*, 1996). This is associated with a sustained increase in TP53 level, resulting in a long term five- to tenfold induction of the cyclin dependent kinase inhibitor p21 (Di Leonardo *et al*, 1994) and accumulation of hypophosphorylated RB protein (Fig. 1; Di Leonardo *et al*, 1994; Linke *et al*, 1996). In addition, $p53^{-/-}$ (Kastan *et al*, 1992), $p21^{-/-}$ (Deng *et al*, 1995) and $Rb^{-/-}$ (Slebos *et al*, 1994) MEF are all defective for the γ radiation induced G_0/G_1 arrest response. Since $p21^{-/-}$ cells appear to exhibit a partial residual arrest, it is possible that p53 dependent *trans*-activation of other factors also plays a part in this response. The permanent arrest model described here is more consistent with the generally higher radiosensitivity of p53[+] strains.

It is striking that data obtained from cell cycle analyses following γ irradiation could be interpreted to support such incompatible models as transient and permanent arrest. However, differences in experimental design and choice of cell lines can produce such ambiguity. The initial studies implicating TP53 in a transient G_1 arrest used asynchronous cells, which exhibited a temporary reduction in the S phase fraction after irradiation. This was accompanied by an increase in the G_0/G_1 fraction, which peaked approximately 16 hours after treatment. Subsequently, the S phase fraction increased to virtually normal levels (Kastan *et al*, 1991). However, interpretation of data from asynchronous cultures is difficult because of transient TP53 independent S and G_2 cell cycle delays caused by DNA damage, which can lead to partial synchronization. Delays of just a few hours in S or G_2 and subsequent re-entry of the delayed cells into cycle would produce a downstream increase in G_0/G_1 cells owing to this partial synchronization, followed by a wave of cells entering the first postirradiation S phase. Another possibility is that permanent G_0/G_1 arrest in a subset of G_1 cells experiencing irreparable DNA damage after irradiation would increase the percentage of G_1 phase cells and reduce the S phase fraction at the times those cells would normally have entered S phase. However, reproductively viable cells would continue to progress, divide and eventually restore the S phase fraction. Thus, S phase delay, G_2 phase delay or permanent G_0/G_1 arrest could lead one to conclude that a transient G_1 delay had occurred.

In both scenarios described above, second cycle events would be misinterpreted as events occurring in the cycle during which irradiation took place. G_1–S progression has been analysed more precisely by first synchronizing cells in G_0 by contact inhibition or serum deprivation and then determining cumulative S phase labelling indices using [³H]thymidine incorporation. This allows monitoring the progression of cells through G_1 and into the first post-treatment S phase. The addition of mitotic inhibitors along with [³H]thymidine immediately after treatment prevents the progression of cells into the next G_1 (see eg Nagasawa *et al*, 1995). Thus, cells in G_2/M during synchrony and cells arising from division of proliferating cells do not contribute to the G_1 fraction. These techniques allow a more detailed analysis of progression out of G_1 after treatment.

There are limitations to the radioactive labelling and colcemid arrest techniques, since they monitor progression through only one cell cycle, and radioactively labelled thymidine can directly induce TP53 and cell cycle arrest (Dover *et al*, 1994; Yeargin and Haas, 1995). These limitations can be overcome by using a procedure to quantify cell cycle progression for up to three cell cycles using continuous labelling of DNA with bromodeoxyuridine (BrdU) followed by staining with an fluorescein isothiocyanate conjugated anti-BrdU antibody (Linke *et al*, in press). Since aberrant chromosome structures can affect cell cycle progression long after the initial treatment (see eg Marder and Morgan, 1993), it is necessary to employ techniques such as this to provide a more accurate view of the cell cycle consequences of DNA damage.

Normal Human Diploid Fibroblasts with Chromosomal Damage Are Removed by TP53 Dependent and TP53 Independent Mechanisms Over Multiple Cell Cycles

Far fewer TP53$^+$ cells survive to form viable colonies than are able to escape the first G_0/G_1 after γ irradiation (Di Leonardo *et al*, 1993; Nagasawa and Little, 1983; Almasan *et al*, 1995a; Li *et al*, 1995a; Li *et al*, 1995b; Tsang *et al*, 1995). This observation suggests that some cells with misrejoined chromosomes or residual DNA damage can enter the first cycle but are eliminated in subsequent phases, perhaps after the damage is revealed by a second breakage event or additional processing. There are at least two subsets of TP53$^+$ cells that escape the initial G_0/G_1. The first subset begins entering S phase at about the same time as the untreated control and the second subset enters S phase towards the end of the entry period of the untreated control (Linke *et al*, in press). By contrast, virtually all of the cells in TP53$^-$ populations enter the first S phase without delay. The TP53$^+$ cells that escape the initial G_0/G_1 are significantly more likely to arrest in subsequent phases than TP53$^-$ cells, particularly in G_0/G_1 of subsequent cycles. Reproductive cell death in later phases appears to occur within about three cycles after irradiation for both TP53$^+$ and TP53$^-$ cells, as the number of actively cycling cells approximates the clonogenic population within this time (Linke *et al*, in press).

The fate of a cell after γ irradiation probably depends, at least in part, on the different amounts and types of DNA damage induced and where the lesions occur in the genome. Damage induction generally follows Poisson statistics, so individual cells incur broad ranges of damage (Cornforth and Goodwin, 1991). In addition, γ radiation causes a number of different lesions that appear to be processed in different ways, including single strand breaks (SSBs), double strand breaks (DSBs; consisting of two proximal SSBs), base modifications and crosslinks. Although many types of DNA damage can lead to a TP53 dependent G_1 arrest, it is generally accepted that DSBs are the primary lesions that lead to chromosomal aberrations and reproductive cell death after γ irradiation (reviewed in Ward, 1988). Several studies suggest that DSB repair has both a fast and a slow component (Sen and Hittelman, 1984; Pandita and Hittelman, 1992). Cells that repair DSBs more rapidly appear to have a lower frequency of misrejoining events and higher survival than cells with slower repair (Schwartz and Vaughan, 1989). It is probable that different types of DSBs are repaired at different rates, depending on their complexity (Ward, 1988). For example, a majority of DSBs probably consist of SSBs that are significantly offset from each other, enabling the opposite strand to serve as a template for accurate repair. Such lesions may represent sublethal damage that is repaired rapidly. On the other hand, some DSBs probably consist of closely spaced or directly opposed SSBs. These lesions, called locally multiply damaged sites (LMDS), can lead to loss of genetic information due to mis-repair, misrejoining with the end of another DSB or generation of irreparable breaks. They may represent lethal damage that undergoes slow repair (Ward, 1988). Recent data from budding yeast show that genomic location can also profoundly affect the rate at which DSBs are repaired, the fidelity of the repair process and the probability that cell death or chromosome loss will occur (Fishel and Kolodner, 1995). The cell cycle behaviour and viability of a cell may therefore be explained by the types and locations of the DSBs induced by γ irradiation.

The TP53$^+$ cells that remain irreversibly arrested in the initial G$_0$/G$_1$ after irradiation probably represent cells containing irreparable DSBs that transmit a permanent arrest signal to TP53. Irradiated cells that begin entering S phase at about the same time as the untreated controls probably incur only rapidly repaired sublethal damage, and these may comprise the majority of the clonogenic population. Since rapid repair occurs within a few hours, little or no delay would be apparent. This fraction decreases significantly with increasing dose (Linke *et al*, in press). It is likely that the number and complexity of closely spaced LMDS type DSBs increase with dose, thereby increasing the frequency of chromosome exchanges. Consistent with this prediction, cell survival at equivalent doses is enhanced at low rates of irradiation, presumably due to an increase in the ability of cells to repair lesions before additional proximal breaks and misrejoining are induced (Marchese *et al,* 1987; Nagasawa *et al,* 1992). It has also been shown that the fraction of cells without chromosomal aberrations at the first metaphase is nearly the same as the fraction that forms colonies (Cornforth and Bedford, 1987). Accordingly, the percentage of TP53$^+$ cells that enter with little or no delay roughly corresponds to the percentage survival at each dose (Linke *et al*, in press).

The second wave of irradiated cells that enter S phase may incur more complex DNA damage and may become refractory to TP53 mediated arrest. This cohort probably consists primarily of cells with LMDS type DSBs. Some DSBs may undergo misrejoining, which could induce arrest or death in a later phase through loss of an essential function or formation of a dicentric chromosome that would undergo breakage at the subsequent anaphase. Some

DSBs may also remain unrepaired or be modified in such a way that they no longer produce a damage signal detectable by the TP53 pathway. This may enable cells with persistent damage to enter the cycle after a delay. Interestingly, the RAD9 system in yeast becomes incapable of eliciting an arrest signal when low levels of DNA damage are present for an extended period, resulting in entry of cells with residual chromosome breaks into the cycle (Bennett *et al*, 1993; Sandell and Zakian, 1993). These damage adapted cells form only small, abortive colonies of inviable cells (Bennett *et al*, 1993). Similarly, small colonies with aberrant cells are observed after γ irradiation of NDF, suggesting that some progeny do not produce two viable descendants at each division (Linke SP, Clarkin KS and Di Leonardo A, unpublished). These observations suggest that the subset of cells that enter S phase later may incur more complex damage in the initial G_0/G_1, making them only a minor component of the clonogenic population.

Although virtually 100% of G_0 synchronized TP53$^-$ cells enter the first S phase after irradiation, regardless of the dose, it is important to note that they also arrest in subsequent phases in a dose dependent fashion, including G_0/G_1 (Linke *et al*, in press). These data reveal a TP53 independent mechanism of cell elimination in G_1, which is probably activated after the attempted mitosis of highly damaged chromosomes. One explanation of why these cells do not arrest in the initial G_0/G_1 is that progression through S, G_2 and/or M may be required to convert the damage to a form that can activate the TP53 independent permanent arrest mechanism. This is also probably an important component of the DNA damage response of TP53$^+$ cells that incur damage outside of G_1, escape G_1 after irradiation or acquire a type of damage that does not activate a TP53 dependent response. The increased resistance of TP53$^-$ cells to DNA damage induced apoptosis or premature senescence and their ability to tolerate repeated breakage and misrepair events are probably major sources of genetic instability. This probably contributes to the increased tumorigenicity observed in p53$^{-/-}$ mice and the increased propensity of TP53$^-$ cells to undergo tumour progression (Donehower *et al*, 1992, 1995; Donehower and Bradley, 1993; Harvey *et al*, 1993; Symonds *et al*, 1994; Ziegler *et al*, 1994).

Confluent Holding Enhances G$_1$–S Progression and Reproductive Viability Similarly in TP53$^+$ and TP53$^-$ Cells

Certain aspects of the repair of potentially lethal damage (PLD) are similar to the arrest for repair model. PLD repair occurs when irradiated cells are artificially held in G_0 before subculture. It is characterized by increased percentages of cells able to enter S phase, reduced delay in entry into S phase, reduced percentages of cells with chromosomal aberrations and enhanced survival (Little, 1969; Fornace *et al*, 1980). However, holding cells for PLD repair results in about the same degree of survival enhancement in both TP53$^+$ and TP53$^-$ cells (Linke *et al*, in press). This suggests that TP53 does not have a role in this process, either by inducing a transient arrest or through direct involvement in repair.

It is surprising that PLD repair cannot occur during G_1 progression without holding, since NDF normally take 12–18 hours to enter S phase and maximal PLD repair usually occurs within a 6 hours holding period (Little and Nagasawa, 1985). It has been suggested that unrepaired damage can become irreparable ("fixed") in cells upon re-plating (Fornace *et al*, 1980), perhaps through changes in chromatin structure as they progress into G_1 from G_0 (reviewed in Iliakis, 1988). Thus, if PLD becomes fixed in actively cycling TP53$^+$ and TP53$^-$ cells at the time of treatment, an arrest for repair would be of no use.

The TP53 Damage Sensor Fails to Elicit Cell Cycle Arrest after the G_1 Restriction Point

NDFs become refractory to γ radiation induced DNA damage in late G_1, which correlates temporally with the G_1 restriction point (Nagasawa *et al*, 1984; Di Leonardo *et al*, 1994). Similarly, overexpression of TP53 leads to arrest before or near the restriction point (Lin *et al*, 1992). The restriction point was originally defined as the time at which cells are committed to enter S phase despite the removal of growth factors or the inhibition of protein synthesis by cycloheximide (Pardee, 1989). There is no direct evidence that TP53 mediates G_1 arrest directly through inhibition of protein synthesis or growth factor signalling, but these observations suggest that the pathways may share a common downstream effector. One prime candidate is RB, which is inactivated as a repressor of E2F function by phosphorylation in late G_1 (Fig. 1; Reed, this volume; Farnham *et al*, 1993; Almasan *et al*, 1995a). Lack of RB function results in at least partial loss of both the TP53 DNA damage checkpoint and protein synthesis inhibition responsiveness (Slebos *et al*, 1994; Almasan *et al*, 1995a,b; Herrera *et al*, 1996), suggesting that RB phosphorylation may be a defining event in the G_1 restriction point.

RB phosphorylation alone cannot explain why the TP53/RB arrest pathway is unable to elicit arrest in response to DNA damage after the restriction point. Cells synchronized in late G_1 that are induced to accumulate predominantly hypophosphorylated RB by transient overexpression of TP53 or γ irradiation progress into S phase without inhibition (Harris MP, Linke SP, Clarkin KC, Shepard HM, Maneval D and Wahl GM, unpublished). This is consistent with microinjection studies indicating that RB suppression of cell cycle progression is limited to early to mid G_1 (Goodrich *et al*, 1991). This may be explained in several ways. It is possible that additional phosphorylation states of RB exist that may not be differentiable by western blot. Alternatively, modification of E2F in late G_1 may prevent rebinding of RB regardless of its phosphorylation state. Consistent with this idea, cyclin A/CDC2 complexes can phosphorylate E2F-1 (Dynlacht *et al*, 1994; Xu *et al*, 1994) and specific phosphorylation of E2F leads to dissociation of E2F from RB regardless of RB phosphorylation state (Fagan *et al*, 1994). The inability of hypophosphorylated RB to induce G_1 arrest in cells that have passed the G_1 restriction point emphasizes that traversing this checkpoint creates an irrevocable license to enter and proceed through S phase, even under conditions that can cause substantial genetic damage.

Differential Radiosensitivities between Cell Types

NDFs from presumably normal individuals display a wide range of radiosensitivities (Little *et al*, 1988). Differences also exist in the radiosensitivities of cell types within an individual. For example, normal human breast and prostate epithelial cells are significantly more resistant to γ radiation than isogenic breast fibroblasts or prostate stromal cells (Girinsky *et al*, 1995; Linke SP and Clarkin KC, unpublished). Radiosensitivities correlate well with the reparability of the damage (Cornforth and Bedford, 1985; Wlodek and Hittelman, 1988) and the rate and accuracy of DSB rejoining (Schwartz and Vaughan, 1989; Schwartz, 1992). In addition, factors such as chromatin structure and levels of free radical scavengers within a cell may dictate the severity of the damage and the likelihood that an arrest or elimination signal will be elicited. This may contribute to differences in genetic stability and cancer predisposition of different cell types and to the ability of different cell types to immortalize after an initiating lesion (see eg Shay *et al*, 1995).

DETERMINING THE TYPES AND AMOUNT OF DNA DAMAGE REQUIRED FOR TP53 DEPENDENT ARREST BY MICROINJECTION

Treatment with either UV or γ radiation results in an increase in TP53 levels and activates the signal transduction pathway leading to a G_1 arrest (Kastan *et al*, 1991; Lu and Lane, 1993). However, the primary DNA lesions caused by UV and γ radiation, pyrimidine dimers and DSBs, respectively, appear to be detected and processed by genetically separate systems (reviewed in Friedberg *et al*, 1995). It has been difficult to determine whether the inducing signal is generated directly by the lesions, by secondary products derived from them or by activation of other signal transduction cascades (Schorpp *et al*, 1984; Devary *et al*, 1992, 1993; Paulovich and Hartwell, 1995). Attempts to resolve these issues by electroporating restriction enzymes (Nelson and Kastan, 1994) have not provided definitive answers because such perturbations can themselves activate TP53 (Renzing and Lane, 1995).

Microinjection of DNA substrates into the nucleus of normal and genetically altered cells provides a powerful means of determining the types and amounts of damage required for TP53 activation. This strategy involves injecting G_0 synchronized cells with a plasmid or defined oligonucleotide containing a specific DNA lesion, along with a tracer to enable subsequent identification of the injected cells. Progression of each injected cell into S phase is determined by BrdU incorporation using immunofluorescence staining. Reasonably accurate quantification of the number of introduced lesions can be achieved by measuring the DNA concentration in the injection buffer and estimating the maximum allowable injection volume. Direct nuclear injection also allows separation of the effects of DNA damage from the collateral effects caused by radiation or drug treatments. Furthermore, appropriate design of the substrates allows analysis of DNA repair and the potential relationships between cell cycle arrest and repair.

Several considerations should be factored into interpretations of data obtained with this technique. Firstly, it is unclear whether the injected substrates acquire a chromatin conformation identical to that of resident chromosomal sequences. It is possible that some cellular factors that tightly associate with native chromatin may not associate with injected extrachromosomal substrates. For instance, cells derived from mice with severe combined immune deficiency lack the ability to join chromosomal DSBs but apparently rejoin DSBs in transfected plasmids (Harrington *et al*, 1992). Secondly, the exact number of molecules injected cannot be controlled because of unavoidable variations in the volume delivered to and retained in each injected nucleus (Minaschek *et al*, 1989). Thirdly, some substrates, such as small single stranded oligonucleotides, may be degraded rapidly or diffuse out of the nucleus, preventing them from transmitting an arrest signal. Finally, tracking the injected cells over long time intervals depends on the stability of the injection marker. However, even with such limitations, microinjection provides the only means currently available to achieve substantial control over the amount and type of damaged substrates delivered to a cell and to estimate rapidly the specific effects of DNA damage on cell cycle progression and repair.

A Single Unrepaired DSB May Be Sufficient to Induce a Prolonged Arrest in Normal Human Diploid Fibroblasts

DSBs are probably sufficient to increase TP53 levels (Nelson and Kastan, 1994). Consistent with this idea, injection of linearized plasmid with blunt ends, 5′ overhangs or 3′ overhangs induces a TP53 dependent cell cycle arrest in normal fibroblasts (Huang *et al*, 1996b). The cells fail to progress into S phase for at least 36–48 hours after injection, but they do not acquire the enlarged, irregular morphology characteristic of senescent like γ irradiated NDF (Di Leonardo *et al*, 1994). Similarly, DSBs generated by microinjection of restriction enzymes produce a prolonged arrest without morphological alterations (Huang L-C and Clarkin KC, unpublished). This raises the possibility that the senescent like morphology seen after γ irradiation may derive from radiation effects other than DNA damage or from induction of a different type of DNA lesion. Cytoplasmic injection of linearized substrates or nuclear injection of supercoiled or nicked plasmid DNA does not affect cell cycle progression (Huang *et al*, 1996b). These data demonstrate that the presence of linear DNA substrates in the nucleus triggers the TP53 dependent arrest mechanism.

The TP53 dependent arrest requires a minimum length for the injected linear substrate. Injection of approximately 1000 25 bp oligomers per cell had no effect on cell cycle progression, whereas arrest was induced after injection of the same number of 49–2900 bp molecules (Huang *et al*, 1996b). Fewer than 10 of the 2900 bp linear DNA molecules were required to arrest NDF, whereas at least 100 49 bp linear molecules were necessary. It remains to be determined whether the difference reflects the minimal size of the binding site of the DSB detection machinery or the nucleolytic degradation or diffusion of the smaller substrates from the nucleus.

Arrest is achieved when cells are injected with the DNA concentration necessary to ensure that each nucleus received at least one linearized molecule. These data are consistent with radiobiological and genetic studies, suggesting that responses measured by microinjection of linear substrates can be extended to those of true chromosomal substrates. For example, γ radiation induced arrest of NDF and normal epithelial cell strains follows single hit kinetics, implying that a single lesion, perhaps an irreparable DSB, precipitates the arrest (Di Leonardo *et al*, 1994; Almasan *et al*, 1995a). A single long lived break in yeast has also been reported to cause reproductive death (Bennett *et al*, 1996). In addition, breakage of a single dicentric chromosome in cells expressing a thermoregulated TP53 also induces a prolonged G_1 arrest in cells grown at the permissive temperature (Ishizaka *et al*, 1995). Considering that gene amplification can arise from one DSB through formation of a dicentric chromosome (Smith *et al*, 1990; Windle *et al*, 1991; Smith *et al*, 1992; Toledo *et al*, 1992; Ma *et al*, 1993) and that amplification does not occur at measurable frequencies in normal cells (Tlsty, 1990; Wright *et al*, 1990a; Livingstone *et al*, 1992; Yin *et al*, 1992), it is reasonable to infer that the TP53 dependent DNA damage sensing system detects and responds to as few as one unrepaired DSB in G_1.

TP53 function does not appear to contribute to DNA end joining capacity. A reporter gene was constructed such that linearization severs the connection between the promoter and the open reading frame. When repaired, the plasmid produces a functional gene readily scored in individual cells (Huang L-C and Clarkin KC, unpublished). No difference in reporter gene expression is observed between wild type and *p53*$^{-/-}$ MEFs cells when analysed at early passage (ie passages 1–3). However, expression was reduced two- to threefold in late passage *p53*$^{-/-}$ cells (ie passage 16). These results are consistent with γ radiation studies indicating that p53 does not participate directly in repairing DSBs (Lee *et al*, 1994; Bouffler *et al*, 1995; Nishino *et al*, 1995; Sands *et al*, 1995; Gupta *et al*, 1996) However, they suggest that passage of p53$^-$ cells in cell culture may favour the loss of other factors involved in DNA repair.

DNA Containing 30–38 Nucleotide Gaps also Induces TP53 Dependent G_1 Arrest

Since gapped DNA is a product of the reactions that repair UV induced DNA damage and since UV induces a TP53 dependent cell cycle arrest, we tested whether gaps might be sufficient to activate this arrest mechanism. The precise size of the gap generated by NER in human cells has not been determined, but data obtained in yeast suggest that it might be approximately 30 nucleotides (Huang *et al*, 1992; Carr and Hoekstra, 1995). As the supercoiled fraction of UV treated DNA does not generate a cell cycle arrest, we infer that base modification alone is insufficient to activate the TP53 dependent damage sensor and that another response generated by UV treatment may be needed. Nicks and 25 nucleotide gaps also fail to induce an arrest, whereas gaps from 38 to 2900 nucleotides long induce arrest efficiently. A 30 nucleotide gap generates an arrest in a small fraction of cells (Huang L-C, Clarkin KC and Wahl GM, unpublished). These data indicate a gap threshold of between 30 and 38 nucleotides for inducing a TP53 dependent G_1 arrest.

DNA with Mismatched Bases Induces G_1 Arrest

Errors consisting of base pair mismatches and insertion/deletion loops are often generated during replication of simple sequence repeats, and organisms spanning the phylogenetic spectrum have evolved a highly conserved system to detect and correct them (reviewed in Modrich, 1991; Kolodner, 1996). *hMSH2* and *hMLH1* are factors in the human MMR system. Inactivation of either of these genes leads to a colon cancer predisposition syndrome referred to as hereditary non-polyposis colorectal carcinoma (HNPCC) (Fishel *et al*, 1993; Bronner *et al*, 1994; Marra and Boland, 1995). The cancers of affected individuals display characteristic instability of microsatellite sequences, consisting of addition or deletion of one or a few repeat units (Aaltonen *et al*, 1993; Ionov *et al*, 1993; Peltomaki *et al*, 1993). Microsatellite instability similar to that observed in HNPCC has now been observed in a variety of sporadic cancers, including breast (Glebov *et al*, 1994; Patel *et al*, 1994; Wooster *et al*, 1994; Yee *et al*, 1994; Paulson *et al*, 1996), small cell and non-small cell lung (Merlo *et al*, 1994; Shridhar *et al*, 1994), renal (Uchida *et al*, 1994) and gastric (Rhyu *et al*, 1994). These observations have led to the speculation that defective MMR might be a frequent occurrence in many cancers. One consequence of such a defect is to increase somatic mutation frequency 100–1000-fold (Bhattacharyya *et al*, 1994; Eshleman *et al*, 1995). This, in turn, can lead to inactivation of signal transduction mechanisms such as the transforming growth factor beta pathway (Markowitz *et al*, 1995), which may explain the poor prognosis for some cancers (eg see Paulson *et al*, 1996).

Recent data indicate that TP53 can bind to insertion/deletion mismatches in vitro (Lee *et al*, 1995) and thus may also play a part in limiting the outgrowth of variants arising from replication errors. These errors should be corrected rapidly and with a low error frequency in S phase, since this is the only time in the cycle when the natural strand asymmetry created by the DNA replication process enables discrimination of the nascent strand containing the error from the parental template strand (see Friedberg *et al*, 1995, for discussion and references). If MMR occurred after DNA replication, strand polarity would no longer exist, causing such errors to be fixed into the genome approximately 50% of the time. Thus, it is not surprising that systems exist to halt cell cycle progression when replication errors are detected in G_2 (Hawn *et al*, 1995). It is unknown whether this arrest contributes to enhanced repair or to the elimination of cells with postreplication mismatches. A TP53 dependent G_1 mechanism to delay or eliminate cells with mismatches could provide a second line of defence to limit mutation in the next cell cycle. Consistent with this possibility, a proportion of the variants obtained in a selection for the replication error (RER+) phenotype had collaterally lost TP53 function (Anthoney *et al*, 1996).

Microinjected mismatches elicit variable TP53 dependent G_1 arrests. Plasmids or linear substrates with covalently closed looped ends (dumbells) containing each of the possible base pair mismatches were microinjected into cells arrested by serum deprivation to determine whether they would prevent cells from entering S phase. G/T or C/T mismatches are most efficient at inducing G_1 arrest in TP53+ NDF, since as few as 100 plasmids containing a single mismatch of either type elicit an arrest signal. Other mismatched bases either arrest a smaller proportion of injected NDF or show no effect on G_1–S progression, depending on the particular mismatch. By contrast, TP53− NDF were not affected by any of the substrates. The arrest was less efficient than that induced by DSBs in G_1, since approximately 100-fold more mismatched molecules were required and some of the arrested cells entered S phase after a delay (Huang L-C, Clarkin KC and Wahl GM, unpublished). It is unclear whether the G_1 arrest induced by some DNA mismatches enabled repair to occur or whether the cells that entered S phase had adapted to the damage and contained unrepaired lesions.

SIGNALLING THE PRESENCE OF DNA DAMAGE TO TP53

The mechanisms by which TP53 is activated by DNA lesions as diverse as mismatches and DSBs remain a mystery, especially since the detection and repair of such lesions is mediated by different proteins (Friedberg *et al*, 1995). Some possibilities of how DNA damage could activate TP53 are summarized below.

TP53 Activation by Direct Binding to DNA Lesions

In vitro, purified TP53 protein binds to the ends of short single stranded oligonucleotides (Bakalkin *et al*, 1994; Jayaraman and Prives, 1995) and to insertion/deletion mismatches (Lee *et al*, 1995) through the non-specific DNA binding domain located in the C-terminus. Binding of such DNA sequences potentiates sequence specific binding by TP53 (Jayaraman and Prives, 1995). Although microinjection of certain DNA mismatches induces a TP53 dependent arrest (Huang L-C, Clarkin KC and Wahl GM, unpublished), those most effective in vivo do not exhibit the tightest binding in vitro (Lee *et al*, 1995). The microinjection data also do not indicate whether TP53 induces arrest through direct interaction with the mismatch on the DNA or whether it requires one or more components of the mismatch detection and processing machinery. We note that injecting as many as 10 000 25 nucleotide single strand molecules into nuclei elicited no detectable arrest response, suggesting that such molecules either do not activate TP53 efficiently in tissue culture cells or are degraded too rapidly to be detected and bound by TP53 (Huang *et al*, 1996b). Since molecules with gaps exceeding 30–38 nucleotides induce arrest far more efficiently than short oligonucleotides, we infer that the gaps generated during NER are more likely to be the intracellular substrate that elicits TP53 dependent arrest induced by UV. It remains to be determined whether gap binding activates TP53 directly or whether the signal is transmitted by interaction between TP53 and other proteins (see eg Pietenpol and Vogelstein, 1993).

TP53 Activation by Kinases That Amplify Signals Elicited by DNA Damage?

Circumstantial evidence suggests that the induction of TP53 by DSBs may occur through interaction with DNA activated protein kinase (DNA-PK). Ku proteins have a high affinity for double strand ends and other abnormal DNA structures (Mimori *et al*, 1986; Gottlieb and Jackson, 1993). Ku proteins recruit the DNA-PK catalytic subunit (DNA-PK$_{cs}$) to damaged DNA and activate its kinase activity (reviewed in Weaver, this volume; Anderson and Lees, 1992). DNA-PK has been shown to phosphorylate TP53 at Ser-15 and Ser-37 in the N-terminal transcriptional activation domain (Lees-Miller *et al*, 1992; Brush *et al*, 1994). Ku binds double stranded DNA with 25 bp periodicity and exhibits helicase activity only on duplex DNA longer than 25 nucleotides (Tuteja *et al*, 1994). This may be one factor preventing double stranded oligomers smaller than 25 bp from inducing a TP53 dependent cell cycle arrest (Huang *et al*, 1996b).

Mutation of *DNA-PK$_{cs}$* underlies the immunodeficiency and radiosensitivity of *scid* mice (Biedermann *et al*, 1991; Hendrickson *et al*, 1991; Blunt *et al*, 1995; Kirchgessner *et al*, 1995; Miller *et al*, 1995). Therefore, if DNA-PK$_{cs}$/Ku relays the signal from DNA damage to TP53, then *scid* mouse fibroblasts should be defective in damage induced, TP53 mediated G$_1$ arrest. However, *scid* fibroblasts exhibit a normal arrest response to DNA breakage relative to age matched wild type fibroblasts (Huang *et al*, 1996a). Furthermore, mutations at the serine residues in TP53 that are targeted by DNA-PK$_{cs}$ have given conflicting results regarding their impact on TP53 function (Fiscella *et al*, 1993; Fuchs *et al*, 1995), and no studies have measured the effects of such mutations on the ability of TP53 to arrest cells in response to DNA damage. These results are compatible with the following interpretations. Firstly, *scid* cells may contain a small amount of residual DNA-PK$_{cs}$ that is sufficient to signal TP53. Secondly, DNA-PK$_{cs}$ may not be an upstream signal transducer for TP53. Thirdly, DNA-PK$_{cs}$ may be able to transduce signals to TP53, but other proteins may also serve this function when DNA-PK$_{cs}$ is not present.

DNA-PK is structurally related to other proteins belonging to the phosphatidylinositol 3-kinase family, including the gene mutated in ataxia telangiectasia (*ATM*) (reviewed in Rotman and Shiloh, this volume; Savitsky *et al,* 1995). Similar to TP53⁻ cells, ATM deficient cells do not arrest in G_1 after γ irradiation (Cornforth and Bedford, 1985; Dulic *et al,* 1994) or injection of linearized DNA (Huang L-C and Clarkin KC, unpublished), and TP53 induction is either absent (Kastan *et al,* 1992) or temporally delayed after γ irradiation of AT cells (Lu and Lane, 1993). The critical timing for TP53 induction relative to the restriction point (see above) might explain how delayed induction could produce a profound effect on activation of the TP53 mediated G_1 cell cycle checkpoint. The availability of *ATM* deficient mice will enable a rigorous test of the proposed pathway in which ATM relays the signal from one or more types of damage to TP53 (Barlow *et al,* 1996). If it does, it will be important to determine whether Ku proteins also associate with ATM and whether other kinases are employed as additional signal amplifiers.

TP53 Activation by Interaction with Repair Factors and/or Single Stranded DNA Binding Proteins?

TP53 may be brought to lesions through direct or indirect binding to repair factors. As one example, TP53 may detect gaps exceeding 30 nucleotides through interactions between its transcriptional activation domain and the $3' \rightarrow 5'$ helicase XPB (RAD3, ERCC3) (Wang *et al,* 1995, 1996). XPB is mutant in individuals with xeroderma pigmentosum complementation group B, a disease conferring hypersensitivity to UV radiation and a predisposition to skin cancers (Friedberg *et al,* 1995). Since XPB is a component of the TFIIH basal transcription factor, this recruitment mechanism would provide a link between transcription, damage within transcribed genes and cell cycle progression.

An alternative mechanism for TP53 activation might be through interaction with proteins that bind to gaps, such as single strand binding proteins (reviewed in Boulikas, 1996). One intriguing candidate is replication protein A (RPA), since it requires a minimum single strand length of about 30 nucleotides for binding (Kim *et al,* 1992). TP53 and RPA have been shown to interact, resulting in inhibition of DNA replication in vitro (Dutta *et al,* 1993) and co-localization with PCNA, DNA polymerase α and DNA ligase at discrete replication foci in vivo (Wilcock and Lane, 1991). Since large, gapped regions may also be generated during MMR (Friedberg *et al,* 1995) or through exonucleolytic action at DSBs, RPA and/or other single strand binding proteins may provide a common signalling mechanism for many types of DNA damage.

TP53 Activation through Structural Modification Due to Damage Induced Changes in Cellular Biochemistry?

TP53 is a zinc containing protein and its function may be altered by events that influence cellular redox potential (Milner, 1991; Hainaut and Milner,

1993a,b; Hainaut *et al,* 1995). Structural and immunological analyses demonstrate that TP53 is an extremely flexible molecule (Milner, 1995), and it has been proposed to exist in forms compatible with both cell proliferation and cell cycle arrest (Milner, 1994). Interconversion between these two forms may occur through conformational changes induced by redox variations. Such variations could be created by the stress responses generated during the induction or processing of DNA damage (Hainaut, 1995).

TP53 DEPENDENT G_0/G_1 ARREST INDUCED IN THE ABSENCE OF DNA DAMAGE

Arrest Induced by Ribonucleotide Depletion

It has been widely held that DNA damage is the exclusive upstream signal for TP53 activation, but there is now strong evidence that it can also respond to metabolic signals. Although many anticancer treatments such as γ radiation induce DNA damage directly, antimetabolites such as *n*-phosphonacetyl-L-aspartate (PALA) primarily disrupt nucleotide pools. PALA can induce DNA damage in S phase cells, presumably because of slowed replication fork progression due to depletion of nucleotide precursor pools. However, PALA can maintain G_0 synchronized TP53$^+$ NDF in the initial G_0/G_1 in the absence of replicative DNA synthesis or detectable DNA damage (Linke *et al*, 1996). By contrast, identically treated TP53$^-$ NDFs progress into and accumulate in S phase. Thus, TP53$^+$ cells probably remain undamaged in G_0/G_1, whereas TP53$^-$ cells incur DNA damage during transit through S.

Antimetabolites producing individual reductions in uridine 5′-monophosphate, cytidine 5′-triphosphate, uridine 5′-triphosphate or guanosine 5′-triphosphate all maintain G_0 synchronized TP53$^+$ human or murine fibroblasts in a G_0/early G_1 arrest (Di Leonardo *et al*, 1993; Deng *et al*, 1995; Linke *et al*, 1996). Since specific depletion of deoxyribonucleotides does not induce G_0/G_1 arrest, but rather leads to arrest in very early S phase in both TP53$^+$ and TP53$^-$ cells, we infer that the TP53 dependent G_1 arrest mechanism may be restricted to ribonucleotide depletion. This also suggests that ribonucleotide depletion does not activate TP53 by inducing DNA damage or interfering with repair of damage generated during cell synchrony, since deoxyribonucleotide pool depletion should have a more direct effect on these processes, yet cells progress into S phase during deoxyribonucleotide depletion. Taken together, these results indicate that metabolic depletion caused by ribonucleotide synthesis inhibitors, rather than DNA damage, is the major component of this TP53 dependent G_0/G_1 arrest.

The nature of the ribonucleotide depletion induced arrest is also distinctly different from that induced by ionizing γ radiation. Cells arrested in PALA resemble quiescent cells in that the arrest can be reversed by addition of the salvage metabolite uridine, the timing of entrance into S phase after release is similar to quiescent cells and the cells maintain relatively normal cellular morphologies (Fig. 1) (Linke *et al*, 1996). By contrast, γ radiation induces an apparently irreversible arrest with the gross morphological alterations associated with senescence (Di Leonardo *et al*, 1994). The differences in these responses are not expected if PALA and γ radiation induce arrest by the same mechanism.

Possible Mechanisms of Ribonucleotide Depletion Induced Arrest

The mechanism by which ribonucleotide pool reduction produces a TP53 dependent arrest remains to be elucidated. PALA reduces RNA synthesis equivalently in TP53+ and TP53− cells, yet only TP53+ cells arrest. Inhibition of general mRNA synthesis due to rNTP precursor depletion is not a likely trigger for TP53 dependent arrest since RNA polymerase II inhibitors cause a dose dependent arrest in both TP53+ and TP53− cell types (Linke *et al*, 1996). These data suggest that the ribonucleotide depletion induced arrest requires inhibition of the synthesis of specific types of RNA molecules or that other rNTP activated metabolic intermediates are involved. We do not favour the latter possibility, as we have been unable to find evidence of the involvement of intermediates such as CTP activated signal transduction molecules.

TP53 may be regulated through associations with rRNAs. It has been reported that 5.8S rRNA is covalently bound to TP53 at the CKII site (Samad and Carroll, 1991; Fontoura *et al*, 1992; Carroll B, personal communication). It has also been reported that 5S rRNA is recruited to the TP53 N-terminus through binding of the ribosomal protein L5 to MDM2 (Marechal *et al*, 1994). The importance of the CKII site for TP53 regulation or function is suggested by the conservation of the serine and adjacent three aminoacids from all species analysed thus far (Soussi *et al*, 1990). MDM2 negatively regulates TP53 function by binding to the N-terminal *trans*-activation domain, which interferes with recruitment of transcriptional co-activators (Momand *et al*, 1992; Oliner *et al*, 1993; Wu *et al*, 1993; Chen *et al*, 1994). Other indications of the importance of negative regulation of TP53 by MDM2 in vivo are the amplification of MDM2 in human tumours (Khatib *et al*, 1993; Reifenberger *et al*, 1993) and the early embryonic lethality in *Mdm2* deficient mice, which is rescued by collateral inactivation of TP53 (de Oca Luna *et al*, 1995; Jones *et al*, 1995, 1996). These observations make it tempting to speculate that association of the 5.8S and/or 5S rRNAs may provide a mechanism for informing TP53 of ribonucleotide pool depletion.

Figure 2 presents two models of how TP53 function may be regulated by association with these rRNA molecules. Figure 2A proposes that normal ribonucleotide levels enable TP53 to be retained in the cytoplasm through association with one or both of the rRNAs, possibly through binding to ribosomes or polysomes. rNTP depletion is proposed to disrupt such interactions through reduced synthesis of rRNAs, allowing TP53 to move into the nucleus and activate cell cycle inhibitors such as p21. This model does not require removal of the C-terminal 5.8S rRNA for transcriptional activation. Cytoplasmic se-

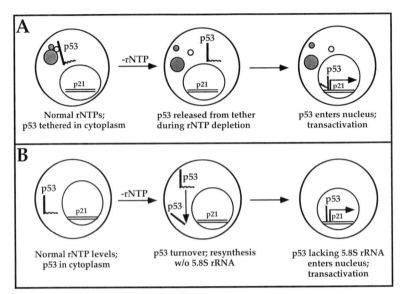

Fig. 2. Models for ribonucleotide depletion induced arrest. (A) Catch and release model. TP53 is proposed to be localized in the cytoplasm in cells growing exponentially. The mechanism of the cytoplasmic tethering may involve association with ribosomes or polysomes (indicated by small and large filled circles) through the associations of TP53 with ribosomal protein L5, 5S rRNA and/or 5.8S rRNA (squiggly line). rNTP depletion is proposed to disrupt the tether by inhibiting rRNA synthesis and promoting ribosome disaggregation. This allows TP53 to gain access to the nucleus, whereupon it would activate transcription growth inhibiting genes such as *p21*. (B) Release and catch model. This model proposes that rNTP depletion is associated with degradation of TP53 containing the 5.8S rRNA followed by resynthesis of a form lacking this rRNA. The TP53 lacking 5.8S rRNA is able to enter the nucleus and serve as an effective transcriptional regulator. Reproduced from Wahl *et al* (in press)

questration of TP53 by association with ribosomes and/or polysomes could be a very effective regulatory mechanism due to the high concentration of such structures relative to TP53. Figure 2B proposes an important role for TP53 turnover. Wild type TP53 has a half life of about 5 minutes or less under normal growth conditions (see eg Scheffner *et al*, 1990, 1993; Chowdary *et al*, 1994; Yeargin and Haas, 1995). Turnover during ribonucleotide depletion may result in the synthesis of TP53 lacking the C-terminal 5.8S rRNA. This could have several consequences, because the C-terminus is highly positively charged, binds to DNA non-specifically and is involved in forming the TP53 tetramers postulated to be the active transcriptional regulator (reviewed in Ko and Prives, 1996). TP53 containing a C-terminal 5.8S rRNA may not have access to the nucleus, may not bind DNA non-specifically and may not tetramerize because of the high density of negative charges, whereas the form lacking the 5.8S rRNA may enter the nucleus and/or form active tetramers more efficiently. Although phosphorylation of the CKII site, binding to C-terminal antibodies or deletion of the C-terminus activates TP53 for sequence specific DNA binding in cell free systems and after microinjection (Hupp *et al*, 1992,

1993), *TP53* mutations preventing modification of the CKII have also been reported to be transcriptionally active (Fiscella *et al*, 1994; Hupp and Lane, 1995). The latter observation could be explained by the model in which absence of C-terminal 5.8S rRNA is required for nuclear entry, whereas the former observations suggest involvement of a second level of control involving negative interaction of the C-terminus with the DNA binding domain (Hupp *et al*, 1995). It will be critical to determine whether mutations that prevent these interactions in the N- and C-termini affect the ability of TP53 to induce an arrest in response to ribonucleotide depletion.

A possible analogy exists between the Pho phosphate sensor in budding yeast and the TP53 ribonucleotide sensor. The Pho system enables budding yeast to respond to fluctuations in inorganic phosphate (P_i) levels. The activity of a transcriptional activator (Pho2/4) is regulated by a covalent modification using phosphate, the target of this metabolite sensing system, resulting in either a phosphorylated inactive cytoplasmic species when P_i levels are high or an active unphosphorylated nuclear entity when P_i levels are low (O'Neill *et al*, 1996).

LOSS OF TP53 FUNCTION ALLOWS, BUT DOES NOT GUARANTEE, CERTAIN TYPES OF GENETIC INSTABILITY

Additional controls exist to determine competence for cell cycle progression under conditions that do not activate the TP53 damage or ribonucleotide sensors. Loss of TP53 enables the outgrowth of PALA resistant variants but does not necessarily allow the development of other types of variants. For example, HT1080 fibrosarcoma cells can become resistant to PALA through amplification at the *CAD* locus. However, they arrest near the G_1/S boundary when challenged with the dihydrofolate reductase (DHFR) inhibitor methotrexate (MTX), which primarily depletes 3′-deoxythymidine 5′-triphosphate pools and fails to generate MTX resistant variants at a measurable frequency (Paulson TG, Almasan A, Brody L and Wahl GM, unpublished). By contrast, MTX resistant variants arise at high frequency when cells are forced to progress through S phase in the presence of PALA or other agents that induce chromosome breakage or interfere with replication fork progression (Paulson TG, Linke SP, Clarkin K, Almasan A and Wahl GM, unpublished). Other studies show that the frequencies of MTX resistance due to gene amplification, point mutation and possibly other mechanisms increase to similar degrees when amplification competent cells are passaged under conditions that interfere with S phase progression before selection (Brown *et al*, 1983; Tlsty *et al*, 1984) and that resistance to PALA or MTX can be achieved by a variety of mechanisms (Flintoff *et al*, 1984; Schaefer *et al*, 1993). Thus, controls that regulate S phase entry and progression are important for limiting the clonal heterogeneity generated by diverse mutational mechanisms.

S phase progression during metabolic challenge could affect genetic stability and gene expression by several mechanisms. There is evidence that interfering with replication fork progression can lead to breakage at fragile sites (Kuo *et al*, 1994), increasing the probability of gene amplification and translocations that may activate oncogenes (see eg Rowley, 1990a,b). It has been suggested that DNA methylation may be altered by conditions that affect replication fork progression (Baylin, 1992). Hypomethylation has been correlated with a hyperamplification phenotype (Giulotto *et al*, 1987), perhaps because of changes in gene expression or chromosome structure that increase chromosome fragility (Perry *et al*, 1992; Stopper *et al*, 1993). Hypermethylation is observed frequently during tumour progression (see eg Baylin, 1992; Makos *et al*, 1992, 1993; Vertino *et al*, 1993; Issa and Baylin, 1996) and has recently been shown to inactivate negative growth regulators such as the CDK4 inhibitors p15 and p16 (Herman *et al*, 1995, 1996; Merlo *et al*, 1995). p15/16 inactivation could lead to more severe defects in cell cycle regulation and provide new mechanisms of drug resistance due to unscheduled accumulation of hyperphosphorylated RB, activation of E2F and inappropriate expression of genes required for S phase progression, including *DHFR* and *thymidylate synthase (TS)* (see Fig. 1 and Reed, this volume). Consistent with this idea, RB deficiency in a subset of human sarcomas has been correlated with increased E2F binding activity, increased DHFR enzyme levels and resistance to MTX (Almasan *et al*, 1995a; Li *et al*, 1995c). Clearly, preventing inappropriate entry into and progression through S phase is a crucial function of normal cell cycle regulation, which contributes to minimizing many types of genetic and epigenetic variations.

CONCLUSION

The data and observations presented above suggest that TP53 and other cell cycle regulators limit the rate of large scale chromosomal alterations, more subtle point mutations and epigenetic DNA modifications by several mechanisms. Firstly, TP53 can arrest cell cycle progression in response to the types of DNA damage most commonly produced in cells undergoing tumour progression. A second mechanism involves prevention of cells from entering S phase during depletion of specific nucleotide pools. However, in some cell lines lacking such controls, the endogenous mutation rate is apparently too low to generate certain types of genetic variants at detectable levels. Thus, absence of TP53 function allows, but does not guarantee, a high intrinsic rate of genetic variation. This fact may contribute to the originally unexpected viability of p53 deficient mice (Donehower *et al*, 1992) and may help to explain why the rate at which such animals produce aberrant embryos is only about two to four times higher than p53[+] mice (Norimura *et al*, 1996). The consequences of losing p53 probably depend on the type of damage incurred, the phase of the cycle in which it occurred and the existence of other mutations that may provide the motive force for unscheduled cell cycle progression.

The function of cell cycle arrest initiated by DNA damage may depend on the availability of a suitable substrate for repair (Fig. 3). Most studies have

Damage	Optimal repair template	Optimal template present in indicated phase		
		Potential Cell Cycle Response		
		G_1	S	G_2
DSB $\quad \begin{array}{l} c_m \\ c_p \end{array}$	sister chromatid	No	Yes	Yes
		elim. cell	delay/repair	delay/repair
UV	complementary strand	Yes	Yes	Yes
		delay/repair	delay/repair	delay/repair
Mismatch	template strand	No	Yes	No
		delay/??	delay/repair	delay/??

Fig. 3. The template available for repair may affect the cell cycle response. The optimal substrates for the types of damage indicated may not be present at all cell cycle intervals and this may impact on the consequences of TP53 mediated cell cycle arrest. A DSB (LMDS type) may be difficult to repair in G_1 because the sister chromatid is not available for recombination mediated repair and the homologous chromosome may be a poor substrate (based on studies in yeast). Slow repair could allow activation of the cell elimination response described in the text. The sister chromatid present in the S or G_2 phase would provide the optimal repair substrate for NER for ensuring that recombination based DSB repair does not result in loss or rearrangement of genetic information. The appropriate repair substrate (complementary strand) is present in each phase of the cell cycle. In this case, cell cycle delay induced by the damage might augment repair independently of cell cycle phase. The structure shown for the mismatch symbolizes the absence of a substrate in G_1 (or G_2) that would enable discrimination of the parental strand from that in which the error was introduced during DNA replication. Repair of such lesions in phases other than S is likely to be error prone, unless mechanisms for strand discrimination exist that have not been identified. Microinjection analyses show that the arrest induced by substrates with mismatches in G_1 is of shorter duration than that induced by a similar substrate with a DSB, but whether repair occurred during the cell cycle delay remains to be determined (indicated by ??)

focussed on the relationship between TP53 function and the damage induced by γ radiation. As described above, cells with normal TP53 function appear to be permanently eliminated from the dividing population if they have unrepaired double strand breaks (DSBs) in G_1. Studies in yeast have shown that DSBs are repaired more efficiently after production of the sister chromatid in S or G_2 (Phipps *et al*, 1985; Kadyk and Hartwell, 1992). It is conceivable that cell elimination is the default in mammalian cells when the appropriate repair substrate is absent. This may also be the purpose of the G_2 and G_1 arrests in cells with residual mismatches. However, the existence of the complementary strand during all cell cycle intervals for cells undergoing nucleotide excision repair (NER) raises the possibility that the interactions between TP53, the NER machinery and replication factors may lead to transient arrest for repair in this instance. Direct in vivo experiments are needed to evaluate this proposal.

The low frequency of many types of mutations in TP53⁻ cells implies that in the absence of other mutations or metabolic perturbations, cells possess effective homoeostatic and damage detection/correction mechanisms to minimize such mutations. However, overexpression of oncogenes such as activated *RAS* or *MYC* (Wani *et al*, 1994; Denko *et al*, 1995; Smith *et al*, 1995a), or overexpression of S phase regulators such as E2F, induces inappropriate S phase entry in TP53⁻ cells (Johnson *et al*, 1993; DeGregori *et al*, 1995), which could contribute to chromosome instability. By contrast, overexpression of activated oncogenes in TP53⁺ cells can still produce cell cycle arrest (Yin Y and Wahl GM, unpublished) or death if growth factors are not available (Yonish-Rouach *et al*, 1991, 1993). Thus, these oncogenes may collaborate with mutant TP53 because this combination propels the cell into cycle, decreases damage induced cell death and permits replication under mutation inducing conditions. It is important to note that conditions created by tumour growth, such as hypoxia, have been associated with pyrimidine deficiency at some oxygen tensions (Palace and Lawrence, 1995) and damage induction during more severe hypoxia (Russo *et al*, 1995). The data reviewed above show that growth of TP53⁻ cells under such conditions also promotes chromosomal rearrangements and evolution of genetic variants. Furthermore, TP53 deficiency has been linked to angiogenesis (Dameron *et al*, 1994; Van Meir *et al*, 1994). Thus, TP53 deficiency could enable proliferation under conditions that both accelerate tumour cell evolution and enhance the blood supply required for additional tumour growth.

An important question for the future is whether the knowledge of TP53 function can contribute to the development of effective therapeutic strategies for TP53⁻ tumours. TP53⁻ cells have been reported to exhibit increased resistance to some chemotherapeutic agents through a lack of damage induced apoptosis (Lowe *et al*, 1993a). Furthermore, adjuvant therapy in breast tumours, which are frequently TP53⁻, can lead to the emergence of drug resistance through gene amplification (Lonn *et al*, 1996). On the other hand, TP53⁻ cells are sensitive to agents that deplete ribonucleotide pools (Linke *et al*, 1996), and one report indicates relative sensitivity to taxol (Wahl *et al*, 1996). Thus, agents such as these may be substantially more toxic to TP53⁻ cells than to normal cells. Furthermore, gene amplification is limited to cells that are functionally TP53⁻. A clue to another type of drug treatment that could be highly selective for tumour cells is suggested by the observation that gene amplification in human tumours in vivo is frequently mediated by extrachromosomal elements (Benner *et al*, 1991), which can be removed by selective entrapment within micronuclei (Von Hoff *et al*, 1992; Shimizu N and Wahl GM, unpublished). Reducing the number of amplified genes can lead to decreased tumorigenicity through induction of differentiation and/or apoptosis (Eckhardt *et al*, 1994; Shimizu *et al*, 1994). Normal cells do not undergo micronucleation at high frequency and the efficiency of micronucleation can be increased substantially in TP53⁻ cells by a variety of chemotherapeutic agents (Shimizu N and Wahl GM, unpublished). Thus, prospects for treating TP53⁻ cancers may be brighter than previously envisioned if the weaknesses in the cell cycle circuitry created by the absence of this multifunctional checkpoint protein can be exploited.

SUMMARY

TP53 serves as a key relay for signals elicited by cellular stresses arising from diverse environmental or therapeutic insults. This relay then activates a cell cycle arrest or cell death program, depending on the stimulus and cell type. The absence of TP53 function disables the cell death or arrest programmes, thereby allowing the emergence of variants with various types of genomic alterations. The data discussed focus on two different types of signals that trigger the TP53 relay system. Firstly, TP53 arrests cell cycle progression in response to the types of DNA damage most commonly detected in cells undergoing tumour progression. Secondly, TP53 is activated by specific depletion of ribonucleotide pools, which prevent cells from entering S phase under conditions that could lead to chromosome breakage. The contribution of both responses limits the emergence of genetic variants. The DNA damage induced arrest appears to be triggered by as few as one double strand break in normal human fibroblasts. Analysis of the arrest kinetics after ionizing radiation shows that TP53 activates a prolonged arrest response in cells with irreparable DNA damage and that high efficiency cell elimination is achieved by a process that can be activated over multiple cell cycles. These data indicate that the primary function of the TP53 arrest/apoptosis pathway in response to double strand break is to eliminate damaged cells from the proliferating population, not to allow additional time for lesion repair. However, it remains possible that repair of other types of damage may benefit from TP53 mediated arrest. Analyses in model genetic systems indicate that the absence of TP53 function allows, but does not ensure, a high intrinsic rate of genetic variation and that instability is increased substantially when cells proceed through S phase under inappropriate growth conditions. This implies that inactivation of TP53 function in combination with other genetic alterations, such as oncogene activation, could accelerate genomic instability and tumour progression.

Acknowledgements

We thank Kristie C Clarkin for her skilful technical assistance. We are indebted to Dr Mirit Aladjem, Dr Gretchen Jimenez and Dr Aldo Di Leonardo and to Kristie C Clarkin and Shireen Khan for thoughtful comments concerning this manuscript. We also thank Ta'Neashia Morrell for patience, good humour and skilful assistance in the preparation of the manuscript. Aspects of the work were supported by grants to GMW from the National Cancer Institute, the US Army and the G Harold and Leila Y Mathers Charitable Foundation. SPL was supported in part by funds from the HA and Mary K Chapman Charitable Trust and a National Cancer Institute training grant. TGP was supported in part by the HA and Mary K Chapman Charitable Trust and a Public Health Service genome training grant. L-c H was funded in part by a fellowship from the Charles H and Anna S Stern Foundation.

References

Aaltonen LA, Peltomaki P, Leach FS *et al* (1993) Clues to the pathogenesis of familial colorectal cancer. *Science* **260** 812-816

Agarwal M, Agarwal A, Taylor W and Stark G (1995) p53 controls both G_2/M and G_1 cell cycle checkpoints and mediates reversible growth arrest in human fibroblasts. *Proceedings of the National Academy of Sciences of the USA* **92** 8493-8497

Allen JB, Zhou Z, Siede W, Friedberg EC and Elledge SJ (1994) The SAD1/RAD53 protein kinase controls multiple checkpoints and DNA damage-induced transcription in yeast. *Genes and Development* **8** 2401-2415

Almasan A, Linke SP, Paulson TG, Huang L and Wahl GM (1995a) Genetic instability as a consequence of inappropriate entry into and progression through S-phase. *Cancer and Metastasis Reviews* **14** 59-73

Almasan A, Yin Y, Kelly R *et al* (1995b) Deficiency of retinoblastoma protein leads to inappropriate S-phase entry, activation of E2F-responsive genes, and apoptosis. *Proceedings of the National Academy of Sciences of the USA* **92** 5436-5440

Anderson CW and Lees MS (1992) The nuclear serine/threonine protein kinase DNA-PK. *Critical Reviews in Eukaryotic Gene Expression* **2** 283-314

Anthoney DA, McIlwrath AJ, Gallagher WM, Edlin AR and Brown R (1996) Microsatellite instability, apoptosis, and loss of p53 function in drug-resistant tumor cells. *Cancer Research* **56** 1374-1381

Arlett CF, Green MH, Priestley A, Harcourt SA and Mayne LV (1988) Comparative human cellular radiosensitivity. I. The effect of SV40 transformation and immortalisation on the gamma-irradiation survival of skin derived fibroblasts from normal individuals and from ataxia-telangiectasia patients and heterozygotes. *International Journal of Radiation Biology* **54** 911-928

Bakalkin G, Yakovleva T, Selivanova G *et al* (1994) p53 binds single-stranded DNA ends and catalyzes DNA renaturation and strand transfer. *Proceedings of the National Academy of Sciences of the USA* **91** 413-417

Barbet N and Carr A (1993) Fission yeast wee1 protein kinase is not required for DNA damage-dependent mitotic arrest. *Nature* **363** 824-827

Barlow C, Hirotsune S, Paylor R *et al* (1996) Atm-deficient mice: a paradigm of ataxia telangiectasia. *Cell* **86** 159-171

Baylin SB (1992) Abnormal regional hypermethylation in cancer cells. *AIDS Research and Human Retroviruses* **8** 811-820

Benner SE, Wahl GM and Von Hoff DD (1991) Double minute chromosomes and homogeneously staining regions in tumors taken directly from patients versus in human tumor cell lines. *Anti-Cancer Drugs* **2** 11-25

Bennett CB, Lewis AL, Baldwin KK and Resnick MA (1993) Lethality induced by a single site-specific double-strand break in a dispensable yeast plasmid. *Proceedings of the National Academy of Sciences of the USA* **90** 5613-5617

Bennett CB, Westmoreland TJ, Snipe JR and Resnick MA (1996) A double-strand break within a yeast artificial chromosome (YAC) containing human DNA can result in YAC loss, deletion, or cell lethality. *Molecular and Cell Biology* **16** 4414-4425

Bhattacharyya NP, Skandalis A, Ganesh A, Groden J and Meuth M (1994) Mutator phenotypes in human colorectal carcinoma cell lines. *Proceedings of the National Academy of Sciences of the USA* **91** 6319-6623

Biedermann KA, Sun JR, Giaccia AJ, Tosto LM and Brown JM (1991) scid mutation in mice confers hypersensitivity to ionizing radiation and a deficiency in DNA double-strand break repair. *Proceedings of the National Academy of Sciences of the USA* **88** 1394-1397

Blunt T, Finnie NJ, Taccioli GE *et al* (1995) Defective DNA-dependent protein kinase activity is linked to V(D)J recombination and DNA repair defects associated with the murine scid mutation. *Cell* **80** 813-823

Bouffler SD, Kemp CJ, Balmain A and Cox R (1995) Spontaneous and ionizing radiation-induced chromosomal abnormalities in p53-deficient mice. *Cancer Research* **55** 3883-3889

Boulikas T (1996) DNA lesion-recognizing proteins and the p53 connection. *Anticancer Research* **16** 225-242

Bronner CE, Baker SM, Morrison PT *et al* (1994) Mutation in the DNA mismatch repair gene homologue hMLH1 is associated with hereditary non-polyposis colon cancer. *Nature* **368** 258-261

Brown PC, Tlsty TD and Schimke RT (1983) Enhancement of methotrexate resistance and dihydrofolate reductase gene amplification by treatment of mouse 3T6 cells with hydroxyurea. *Molecular and Cell Biology* **3** 1097-1107

Brush GS, Anderson CW and Kelly TJ (1994) The DNA-activated protein kinase is required for the phosphorylation of replication protein A during simian virus 40 DNA replication. *Proceedings of the National Academy of Sciences of the USA* **91** 12520-12524

Carr AM and Hoekstra MF (1995) The cellular responses to DNA damage. *Trends in Cell Biology* **5** 32-35

Chen CY, Oliner JD, Zhan Q, Fornace AJ, Vogelstein B and Kastan MB (1994) Interactions between p53 and MDM2 in a mammalian cell cycle checkpoint pathway. *Proceedings of the National Academy of Sciences of the USA* **91** 2684-2688

Chen J, Jackson PK, Kirschner MW and Dutta A (1995) Separate domains of p21 involved in the inhibition of Cdk kinase and PCNA. *Nature* **374** 386-388

Chowdary DR, Dermody JJ, Jha KK and Ozer HL (1994) Accumulation of p53 in a mutant cell line defective in the ubiquitin pathway. *Molecular and Cell Biology* **14** 1997-2003

Clarke AR, Purdie CA, Harrison DJ *et al* (1993) Thymocyte apoptosis induced by p53-dependent and independent pathways. *Nature* **362** 849-852

Clarke AR, Gledhill S, Hooper ML, Bird CC and Wyllie AH (1994) p53 dependence of early apoptotic and proliferative responses within the mouse intestinal epithelium following gamma-irradiation. *Oncogene* **9** 1767-1773

Cleaver JE (1989) DNA repair in man. *Birth Defects* **25** 61-82

Cleaver JE (1994) It was a very good year for DNA repair. *Cell* **76** 1-4

Cornforth MN and Bedford JS (1985) On the nature of a defect in cells from individuals with ataxia-telangiectasia. *Science* **227** 1589-1591

Cornforth MN and Bedford JS (1987) A quantitative comparison of potentially lethal damage repair and the rejoining of interphase chromosome breaks in low passage normal human fibroblasts. *Radiation Research* **111** 385-405

Cornforth MN and Goodwin EH (1991) The dose-dependent fragmentation of chromatin in human fibroblasts by 3.5-MeV alpha particles from 238Pu: experimental and theoretical considerations pertaining to single-track effects. *Radiation Research* **127** 64-74

Cross SM, Sanchez CA, Morgan CA *et al* (1995) A p53-dependent mouse spindle checkpoint. *Science* **267** 1353-1356

Dameron KM, Volpert OV, Tainsky MA and Bouck N (1994) Control of angiogenesis in fibroblasts by p53 regulation of thrombospondin-1. *Science* **265** 1582-1584

DeGregori J, Leone G, Ohtani K, Miron A and Nevins JR (1995) E2F-1 accumulation bypasses a G_1 arrest resulting from the inhibition of G_1 cyclin-dependent kinase activity. *Genes and Development* **9** 2873-2887

Deng C, Zhang P, Harper JW, Elledge SJ and Leder P (1995) Mice lacking p21CIP1/WAF1 undergo normal development, but are defective in G_1 checkpoint control. *Cell* **82** 675-684

Denko N, Stringer J, Wani M and Stambrook P (1995) Mitotic and post mitotic consequences of genomic instability induced by oncogenic Ha-ras. *Somatic Cell and Molecular Genetics* **21** 241-253

de Oca Luna RM, Wagner DS and Lozano G (1995) Rescue of early embryonic lethality in mdm2-deficient mice by deletion of p53. *Nature* **378** 203-206

Devary Y, Gottlieb RA, Smeal T and Karin M (1992) The mammalian ultraviolet response is triggered by activation of Src tyrosine kinases. *Cell* **71** 1081-1091

Devary Y, Rosette C, DiDonato JA and Karin M (1993) NF-kappa B activation by ultraviolet light not dependent on a nuclear signal. *Science* **261** 1442-1445

Di Leonardo A, Linke SP, Yin Y and Wahl GM (1993) Cell cycle regulation of gene amplification. *Cold Spring Harbor Symposia on Quantitative Biology* **58** 655-667

Di Leonardo A, Linke SP, Clarkin K and Wahl GM (1994) DNA damage triggers a prolonged p53-dependent G_1 arrest and long-term induction of Cip1 in normal human fibroblasts. *Genes and Development* **8** 2540-2551

Donehower LA and Bradley A (1993) The tumor suppressor p53. *Biochimica et Biophysica Acta* **1155** 181-205

Donehower LA, Harvey M, Slagle BL *et al* (1992) Mice deficient for p53 are developmentally normal but susceptible to spontaneous tumours. *Nature* **356** 215-221

Donehower LA, Godley LA, Aldaz CM *et al* (1995) Deficiency of p53 accelerates mammary tumorigenesis in Wnt-1 transgenic mice and promotes chromosomal instability. *Genes and Development* **9** 882-895

Dover R, Jayaram Y, Patel K and Chinery R (1994) p53 expression in cultured cells following radioisotope labelling. *Journal of Cell Science* **107** 1181-1184

Dulbecco R (1989) Cancer progression: the ultimate challenge. *International Journal of Cancer* **4 (Supplement 6)** 6-9

Dulic V, Kaufmann WK, Wilson SJ *et al* (1994) p53-dependent inhibition of cyclin-dependent kinase activities in human fibroblasts during radiation-induced G_1 arrest. *Cell* **76** 1013-1023

Dutta A, Ruppert M, Aster J and Winchester E (1993) Inhibition of DNA replication factor by p53. *Nature* **365** 79-82

Dynlacht BD, Flores O, Lees JA and Harlow E (1994) Differential regulation of E2F trans-activation by cyclin/cdk2 complexes. *Genes and Development* **8** 1772-1786

Eckhardt SG, Dai A, Davidson KK, Forseth BJ, Wahl GM and Von Hoff DD (1994) Induction of differentiation in HL60 cells by the reduction of extrachromosomally amplified c-myc. *Proceedings of the National Academy of Sciences of the USA* **91** 6674-6678

Enoch T, Carr AM and Nurse P (1992) Fission yeast genes involved in coupling mitosis to completion of DNA replication. *Genes and Development* **6** 2035-2046

Eshleman JR, Lang EZ, Bowerfind GK *et al* (1995) Increased mutation rate at the hprt locus accompanies microsatellite instability in colon cancer. *Oncogene* **10** 33-37

Fagan R, Flint KJ and Jones N (1994) Phosphorylation of E2F-1 modulates its interaction with the retinoblastoma gene product and the adenoviral E4 19 kDa protein. *Cell* **78** 799-811

Farnham PJ, Slansky JE and Kollmar R (1993) The role of E2F in the mammalian cell cycle. *Biochimica et Biophysica Acta* **1155** 125-131

Fiscella M, Ullrich SJ, Zambrano N *et al* (1993) Mutation of the serine 15 phosphorylation site of human p53 reduces the ability of p53 to inhibit cell cycle progression. *Oncogene* **8** 1519-1528

Fiscella M, Zambrano N, Ullrich SJ *et al* (1994) The carboxy-terminal serine 392 phosphorylation site of human p53 is not required for wild-type activities. *Oncogene* **9** 3249-3257

Fishel R and Kolodner RD (1995) Identification of mismatch repair genes and their role in the development of cancer. *Current Current Opinion in Genetics and Development* **5** 382-395

Fishel R, Lescoe MK, Rao MR *et al* (1993) The human mutator gene homolog MSH2 and its association with hereditary nonpolyposis colon cancer. *Cell* **75** 1027-1038

Flintoff WF, Davidson SV and Siminovitch L (1984) Isolation and partial characterization of three methotrexate-resistant phenotypes from Chinese hamster ovary cells. *Somatic Cell and Molecular Genetics* **2** 245-261

Fontoura BM, Sorokina EA, David E and Carroll RB (1992) p53 is covalently linked to 5.8S rRNA. *Molecular and Cell Biology* **12** 5145-5151

Ford JM and Hanawalt PC (1995) Li-Fraumeni syndrome fibroblasts homozygous for p53 mutations are deficient in global DNA repair but exhibit normal transcription-coupled repair and enhanced UV resistance. *Proceedings of the National Academy of Sciences of the USA* **92** 8876-8880

Fornace AJ, Nagasawa H and Little JB (1980) Relationship of DNA repair to chromosome aberrations, sister-chromatid exchanges and survival during liquid-holding recovery in X-irradiated mammalian cells. *Mutation Research* **70** 323-336

Friedberg EC, Walker GC and Siede W (eds) (1995) *DNA Repair and Mutagenesis*, ASM Press, Washington, DC

Fuchs B, O'Connor D, Fallis L, Scheidtmann KH and Lu X (1995) p53 phosphorylation mutants retain transcription activity. *Oncogene* **10** 789-793

Fukasawa K, Choi T, Kuriyama R, Rulong S and Vande Woude GF (1996) Abnormal centrosome amplification in the absence of p53. *Science* **271** 1744-1747

Girinsky T, Koumenis C, Graeber TG, Peehl DM and Giaccia AJ (1995) Attenuated response of p53 and p21 in primary cultures of human prostatic epithelial cells exposed to DNA-damaging agents. *Cancer Research* **55** 3726-3731

Giulotto E, Knights C and Stark GR (1987) Hamster cells with increased rates of DNA amplification: a new phenotype. *Cell* **48** 837-845

Glebov OK, McKenzie KE, White CA and Sukumar S (1994) Frequent p53 gene mutations and novel alleles in familial breast cancer. *Cancer Research* **54** 3703-3709

Goodrich DW, Wang NP, Qian YW, Lee EY and Lee WH (1991) The retinoblastoma gene product regulates progression through the G_1 phase of the cell cycle. *Cell* **67** 293-302

Gottlieb TM and Jackson SP (1993) The DNA-dependent protein kinase: requirement for DNA ends and association with Ku antigen. *Cell* **72** 131-142

Graeber TG, Peterson JF, Tsai M, Monica K, Fornace AJ and Giaccia AJ (1994) Hypoxia induces accumulation of p53 protein, but activation of a G_1-phase checkpoint by low-oxygen conditions is independent of p53 status. *Molecular and Cell Biology* **14** 6264-6277

Graeber TG, Osmanian C, Jacks T *et al* (1996) Hypoxia-mediated selection of cells with diminished apoptotic potential in solid tumours. *Nature* **379** 88-91

Gupta N, Vij R, Haas-Kogan DA, Israel MA, Deen DF and Morgan WF (1996) Cytogenetic damage and the radiation-induced G_1-phase checkpoint. *Radiation Research* **145** 289-298

Hainaut P (1995) The tumor suppressor protein p53: a receptor to genotoxic stress that controls cell growth and survival. *Current Opinion in Oncology* **7** 76-82

Hainaut P and Milner J (1993a) Redox modulation of p53 conformation and sequence-specific DNA binding. *Cancer Research* **53** 4469-4473

Hainaut P and Milner J (1993b) A structural role for metal ions in the "wild-type" conformation of the tumor suppressor protein p53. *Cancer Research* **53** 1739-1742

Hainaut P, Rolley N, Davies M and Milner J (1995) Modulation by copper of p53 conformation and sequence-specific DNA binding: role for Cu(II)/Cu(I) redox mechanism. *Oncogene* **10** 27-32

Harrington J, Hsieh CL, Gerton J, Bosma G and Lieber MR (1992) Analysis of the defect in DNA end joining in the murine scid mutation. *Molecular and Cell Biology* **12** 4758-4768

Hartwell LH and Kastan MB (1994) Cell cycle control and cancer. *Science* **266** 1821-1828

Hartwell LH and Weinert TA (1989) Checkpoints: controls that ensure the order of cell cycle events. *Science* **246** 629-634

Harvey M, Sands A, Weiss R *et al* (1993) In vitro growth characteristics of embryo fibroblasts islolated from p53-deficient mice. *Oncogene* **8** 2457-2467

Hawn MT, Umar A, Carethers JM *et al* (1995) Evidence for a connection between the mismatch repair system and the G_2 cell cycle checkpoint. *Cancer Research* **55** 3721-3725

Hendrickson EA, Qin XQ, Bump EA, Schatz DG, Oettinger M and Weaver DT (1991) A link between double-strand break-related repair and V(D)J recombination: the scid mutation. *Proceedings of the National Academy of Sciences of the USA* **88** 4061-4065

Herman JG, Merlo A, Mao L *et al* (1995) Inactivation of the CDKN2/p16/MTS1 gene is frequently associated with aberrant DNA methylation in all common human cancers. *Cancer Research* **55** 4525-4530

Herman JG, Jen J, Merlo A and Baylin SB (1996) Hypermethylation-associated inactivation indicates a tumor suppressor role for p15INK4B. *Cancer Research* **56** 722-727

Herrera RE, Sah VP, Williams BO, Makela TP, Weinberg RA and Jacks T (1996) Altered cell cycle kinetics, gene expression, and G_1 restriction point regulation in Rb-deficient fibroblasts. *Molecular and Cell Biology* **16** 2402-2407

Hoyt MA, Totis L and Roberts BT (1991) *S. cerevisiae* genes required for cell cycle arrest in response to loss of microtubule function. *Cell* **66** 507-517

Huang JC, Svoboda DL, Reardon JT and Sancar A (1992) Human nucleotide excision nuclease removes thymine dimers from DNA by incising the 22nd phosphodiester bond 5′ and the 6th phosphodiester bond 3′ to the photodimer. *Proceedings of the National Academy of Sciences of the USA* **89** 3664-3668

Huang L-C, Clarkin KC and Wahl GM (1996a) p53 dependent cell cycle arrests are preserved in DNA-activated protein kinase deficient mouse fibroblasts. *Cancer Research* **56** 2940-2944

Huang L-C, Clarkin KC and Wahl GM (1996b) Sensitivity and selectivity of the p53-mediated DNA damage sensor. *Proceedings of the National Academy of Sciences of the USA* **93** 4827-4832

Hupp TR and Lane DP (1995) Two distinct signaling pathways activate the latent DNA binding function of p53 in a casein kinase II-independent manner. *Journal of Biological Chemistry* **270** 18165-18174

Hupp TR, Meek DW, Midgley CA and Lane DP (1992) Regulation of the specific DNA binding function of p53. *Cell* **71** 875-886

Hupp TR, Meek DW, Midgley CA and Lane DP (1993) Activation of the cryptic DNA binding function of mutant forms of p53. *Nucleic Acids Research* **21** 3167-3174

Hupp TR, Sparks A and Lane DP (1995) Small peptides activate the latent sequence-specific DNA binding function of p53. *Cell* **83** 237-245

Iliakis G (1988) Radiation-induced potentially lethal damage: DNA lesions susceptible to fixation. *International Journal of Radiation Biology and Related Studies in Physics, Chemical and Medicine* **53** 541-584

Ionov Y, Peinado MA, Malkhosyan S, Shibata D and Perucho M (1993) Ubiquitous somatic mutations in simple repeated sequences reveal a new mechanism for colonic carcinogenesis. *Nature* **363** 558-561

Ishizaka Y, Chernov M, Burns C and Stark G (1995) p53-dependent growth arrest of REF52 cells containing newly amplified DNA. *Proceedings of the National Academy of Sciences of the USA* **92** 3224-3228

Issa JP and Baylin SB (1996) Epigenetics and human disease. *Nature Medicine* **2** 281-282

Jayaraman J and Prives C (1995) Activation of p53 sequence-specific DNA binding by short single strands of DNA requires the p53 C-terminus. *Cell* **81** 1021-1029

Johnson DG, Schwarz JK, Cress WD and Nevins JR (1993) Expression of transcription factor E2F1 induces quiescent cells to enter S phase. *Nature* **365** 349-352

Jones SN, Roe AE, Donehower LA and Bradley A (1995) Rescue of embryonic lethality in Mdm2-deficient mice by absence of p53. *Nature* **378** 206-208

Jones SN, Sands AT, Hancock AR *et al* (1996) Absence of a p53-independent role for Mdm2 in control of cell growth and tumorigenesis. *Proceedings of the National Academy of Sciences of the USA* **93** 14106–14111

Kadyk LC and Hartwell LH (1992) Sister chromatids are preferred over homologs as substrates for recombinational repair in *Saccharomyces cerevisiae*. *Genetics* **132** 387-402

Kallioniemi A, Kallioniemi OP, Sudar D *et al* (1992) Comparative genomic hybridization for molecular cytogenetic analysis of solid tumors. *Science* **258** 818-821

Kallioniemi O-P, Kallioniemi A, Chen L-C *et al* (1993) DNA amplification in primary breast cancer detected by comparative genomic hybridization. *Proceedings of the American Association for Cancer Research* **34** 500

Kastan MB, Onyekwere O, Sidransky D, Vogelstein B and Craig RW (1991) Participation of p53 protein in the cellular response to DNA damage. *Cancer Research* **51** 6304-6311

Kastan MB, Zhan Q, El-Deiry WS *et al* (1992) A mammalian cell cycle checkpoint pathway utilizing p53 and GADD45 is defective in ataxia-telangiectasia. *Cell* **71** 587-597

Kaufmann WK (1995) Cell cycle checkpoints and DNA repair preserve the stability of the human genome. *Cancer and Metastasis Reviews* **14** 31-41

Khatib ZA, Matsushime H, Valentine M, Shapiro DN, Sherr CJ and Look AT (1993) Coamplification of the CDK4 gene with MDM2 and GLI in human sarcomas. *Cancer Research* **53** 5535-5541

Kim C, Snyder RO and Wold MS (1992) Binding properties of replication protein A from human and yeast cells. *Molecular and Cell Biology* **12** 3050-3059

Kirchgessner CU, Patil CK, Evans JW *et al* (1995) DNA-dependent kinase (p350) as a candidate gene for the murine SCID defect. *Science* **267** 1178-1183

Ko LJ and Prives C (1996) p53: puzzle and paradigm. *Genes and Development* **10** 1054-1072

Kolodner R (1996) Biochemistry and genetics of eukaryotic mismatch repair. *Genes and Development* **10** 1433-1442

Kuo MT, Vyas RC, Jiang LX and Hittelman WN (1994) Chromosome breakage at a major fragile site associated with P-glycoprotein gene amplification in multidrug-resistant CHO cells. *Molecular and Cell Biology* **14** 5202-5211

Lane DP (1992) p53, guardian of the genome. *Nature* **358** 15-16

Leach FS, Nicolaides NC, Papadopoulos N *et al* (1993) Mutations of a mutS homolog in hereditary nonpolyposis colorectal cancer. *Cell* **75** 1215-1225

Lee JM and Bernstein A (1993) p53 mutations increase resistance to ionizing radiation. *Proceedings of the National Academy of Sciences of the USA* **90** 5742-5746

Lee JM, Abrahamson JL, Kandel R, Donehower LA and Bernstein A (1994) Susceptibility to radiation-carcinogenesis and accumulation of chromosomal breakage in p53 deficient mice. *Oncogene* **9** 3731-3736

Lee S, Elenbaas B, Levine A and Griffith J (1995) p53 and its 14 kDa C-terminal domain recognize primary DNA damage in the form of insertion/deletion mismatches. *Cell* **81** 1013-1020

Lees-Miller S, Sakaguchi K, Ullrich SJ, Appella E and Anderson CW (1992) Human DNA-activated protein kinase phosphorylates serines 15 and 37 in the amino-terminal transactivation domain of human p53. *Molecular and Cell Biology* **12** 5041-5049

Levine AJ, Momand J and Finlay CA (1991) The p53 tumour suppressor gene. *Nature* **351** 453-456

Li CY, Nagasawa H, Dahlberg WK and Little JB (1995a) Diminished capacity for p53 in mediating a radiation-induced G_1 arrest in established human tumor cell lines. *Oncogene* **11** 1885-1892

Li CY, Nagasawa H, Tsang NM and Little JB (1995b) Radiation-induced irreversible G(0)/G(1) block is abolished in human diploid fibroblasts transfected with the human papilloma virus E6 gene: implication of the p53-Cip1/WAF1 pathway. *International Journal of of of Oncology* **6** 233-236

Li R and Murray AW (1991) Feedback control of mitosis in budding yeast. *Cell* **66** 519-531

Li W, Fan J, Hochhauser D *et al* (1995c) Lack of functional retinoblastoma protein mediates increased resistance to antimetabolites in human sarcoma cell lines. *Proceedings of the National Academy of Sciences of the USA* **92** 10436-10440

Lin D, Shields MT, Ullrich SJ, Appella E and Mercer WE (1992) Growth arrest induced by wild-type p53 protein blocks cells prior to or near the restriction point in late G_1 phase. *Proceedings of the National Academy of Sciences of the USA* **89** 9210-9214

Linke SP, Clarkin KC, Di Leonardo A, Tsou A and Wahl GM (1996) A reversible p53-dependent G_0/G_1 cell cycle arrest induced by ribonucleotide depletion in the absence of detectable DNA damage. *Genes and Development* **10** 934-947

Linke SP, Clarkin KC and Wahl GM p53 mediates permanent arrest over multiple cell cycle in response to γ radiation. *Cander Research* (in press)

Little JB (1969) Repair of sub-lethal and potentially lethal radiation damage in plateau phase cultures of human cells. *Nature* **224** 804-806

Little JB (1994) Changing views of cellular radiosensitivity. *Radiation Research* **140** 299-311

Little JB and Nagasawa H (1985) Effect of confluent holding on potentially lethal damage repair, cell cycle progression, and chromosomal aberrations in human normal and ataxia-telangiectasia fibroblasts. *Radiation Research* **101** 81-93

Little JB, Nove J, Strong LC and Nichols WW (1988) Survival of human diploid skin fibroblasts from normal individuals after X-irradiation. *International Journal of Radiation Biology* **54** 899-910

Livingstone LR, White A, Sprouse J, Livanos E, Jacks T and Tlsty TD (1992) Altered cell cycle arrest and gene amplification potential accompany loss of wild-type p53. *Cell* **70** 923-935

Loeb LA (1991) Mutator phenotype may be required for multistage carcinogenesis. *Cancer Research* **51** 3075-3079

Lonn U, Lonn S, Nilsson B and Stenkvist B (1996) Higher frequency of gene amplification in breast cancer patients who received adjuvant chemotherapy. *Cancer* **77** 107-112

Lowe SW, Ruley HE, Jacks T and Housman DE (1993a) p53-dependent apoptosis modulates the cytotoxicity of anticancer agents. *Cell* **74** 957-967

Lowe SW, Schmitt EM, Smith SW, Osborne BA and Jacks T (1993b) p53 is required for radiation-induced apoptosis in mouse thymocytes. *Nature* **362** 847-849

Lu X and Lane DP (1993) Differential induction of transcriptionally active p53 following UV or ionizing radiation: defects in chromosome instability syndromes? *Cell* **75** 765-778

Ma C, Martin S, Trask B and Hamlin JL (1993) Sister chromatid fusion initiates amplification of the dihydrofolate reductase gene in Chinese hamster cells. *Genes and Development* **7** 605-620

McClintock B (1984) The significance of responses of the genome to challenge. *Science* **226** 792-801

McIlwrath AJ, Vasey PA, Ross GM and Brown R (1994) Cell cycle arrests and radiosensitivity of human tumor cell lines: dependence on wild-type p53 for radiosensitivity. *Cancer Research* **54** 3718-3722

Makos M, Nelkin BD, Lerman MI, Latif F, Zbar B and Baylin SB (1992) Distinct hypermethylation patterns occur at altered chromosome loci in human lung and colon cancer. *Proceedings of the National Academy of Sciences of the USA* **89** 1929-1933

Makos M, Nelkin BD, Reiter RE *et al* (1993) Regional DNA hypermethylation at D17S5 precedes 17p structural changes in the progression of renal tumors. *Cancer Research* **53** 2719-2722

Malkin D, Li F, Strong LC *et al* (1990) Germ line p53 mutations in a familial syndrome of sarcomas, breast cancer and other neoplasms. *Science* **250** 1233-1238

Marchese MJ, Zaider M and Hall EJ (1987) Dose-rate effects in normal and malignant cells of human origin. *British Journal of Radiology* **60** 573-576

Marder BA and Morgan WF (1993) Delayed chromosomal instability induced by DNA damage. *Molecular and Cell Biology* **13** 6667-6677

Marechal V, Elenbaas B, Piette J, Nicolas JC and Levine AJ (1994) The ribosomal L5 protein is associated with mdm-2 and mdm-2-p53 complexes. *Molecular and Cell Biology* **14** 7414-7420

Markowitz S, Wang J, Myeroff L *et al* (1995) Inactivation of the type II TGF-beta receptor in colon cancer cells with microsatellite instability. *Science* **268** 1336-1338

Marra G and Boland CR (1995) Hereditary nonpolyposis colorectal cancer: the syndrome, the genes, and historical perspectives. *Journal of the National Cancer Institute* **87** 1114-1125

Merlo A, Mabry M, Gabrielson E, Vollmer R, Baylin SB and Sidransky D (1994) Frequent microsatellite instability in primary small cell lung cancer. *Cancer Research* **54** 2098-2101

Merlo A, Herman JG, Mao L *et al* (1995) 5′ CpG island methylation is associated with transcriptional silencing of the tumour suppressor p16/CDKN2/MTS1 in human cancers. *Nature Medicine* **1** 686-692

Miller R, Hogg J, Ozaki J, Gell D, Jackson S and Riblet R (1995) The gene for the catalytic subunit of mouse DNA-dependent protein kinase maps to the scid locus. *Proceedings of the National Academy of Sciences of the USA* **92** 10792-10795

Milner J (1991) A conformation hypothesis for the suppressor and promoter functions of p53 in cell growth control and in cancer. *Proceedings of the Royal Society of London B* **245** 139-145

Milner J (1994) Forms and functions of p53. *Seminars in Cancer Biology* **5** 211-219

Milner J (1995) Flexibility: the key to p53 function? *Trends in Biochemical Sciences* **20** 49-51

Mimori T, Hardin JA and Steitz JA (1986) Characterization of the DNA-binding protein antigen Ku recognized by autoantibodies from patients with rheumatic disorders. *Journal of Biological Chemistry* **261** 2274-2278

Minaschek G, Bereiter-Hahn J and Berthholdt G (1989) Quantitation of the volume of liquid injected into cells by means of pressure. *Experimental and Cell Research* **183** 434-442

Modrich P (1991) Mechanisms and biological effects of mismatch repair. *Annual Reviews of Genetics* **25** 229-253

Momand J, Zambetti G, Olson D, George D and Levine A (1992) The mdm-2 oncogene product forms a complex with the p53 protein and inhibits p53-mediated transactivation. *Cell* **69** 1237-1245

Mummenbrauer T, Janus F, Muller B, Wiesmuller L, Deppert W and Grosse F (1996) p53 protein exhibits 3 ′ -to-5 ′ exonuclease activity. *Cell* **85** 1089-1099

Nagasawa H and Little JB (1983) Comparison of kinetics of X-ray-induced cell killing in normal, ataxia telangiectasia and hereditary retinoblastoma fibroblasts. *Mutation Research* **109** 297-308

Nagasawa H, Robertson JB, Arundel CS and Little JB (1984) The effect of X irradiation on the progression of mouse 10T1/2 cells released from density-inhibited cultures. *Radiation Research* **97** 537-545

Nagasawa H, Little JB, Tsang NM, Saunders E, Tesmer J and Strniste GF (1992) Effect of dose rate on the survival of irradiated human skin fibroblasts. *Radiation Research* **132** 375-379

Nagasawa H, Li CY, Maki CG, Imrich AC and Little JB (1995) Relationship between radiation-induced G_1 phase arrest and p53 function in human tumor cells. *Cancer Research* **55** 1842-1846

Navas TA, Zhou Z and Elledge SJ (1995) DNA polymerase epsilon links the DNA replication machinery to the S phase checkpoint. *Cell* **80** 29-39

Nelson WG and Kastan MB (1994) DNA strand breaks: the DNA template alterations that trigger p53-dependent DNA damage response pathways. *Molecular and Cell Biology* **14** 1815-1823

Nishino H, Knoll A, Buettner VL *et al* (1995) p53 wild-type and p53 nullizygous Big Blue transgenic mice have similar frequencies and patterns of observed mutation in liver, spleen and brain. *Oncogene* **11** 263-270

Norimura T, Nomoto S, Katsuki M, Gondo Y and Kondo S (1996) p53-dependent apoptosis suppresses radiation-induced teratogenesis. *Nature Medicine* **2** 577-580

Nowell PC (1976) The clonal evolution of tumor cell populations. *Science* **194** 23-28

Nowell PC (1982) Genetic instability in cancer cells: relationship to tumor cell heterogeneity, In: Owens AH Jr, Coffey DS and Baylin SB (eds). *Tumor Cell Heterogeneity: Origins and Implications*, pp 351-365, Academic Press, New York

Oliner JD, Pietenpol JA, Thiagalingam S, Gyuris J, Kinzler KW and Vogelstein B (1993) Oncoprotein MDM2 conceals the activation domain of tumour suppressor p53. *Nature* **362** 857-860

O'Neill EM, Kaffman A, Jolly ER and O'Shea EK (1996) Regulation of PHO4 nuclear localization by the PHO80-PHO85 cyclin-CDK complex. *Science* **271** 209-212

Palace GP and Lawrence DA (1995) Nucleotide changes in oxidatively stressed lymphocytes. *Journal of Biochemical Toxicology* **10** 137-142

Pandita TK and Hittelman WN (1992) The contribution of DNA and chromosome repair deficiencies to the radiosensitivity of ataxia-telangiectasia. *Radiation Research* **131** 214-223

Pardee AB (1989) G_1 events and regulation of cell proliferation. *Science* **246** 603

Patel U, Grundfest BS, Gupta M and Banerjee S (1994) Microsatellite instabilities at five chromosomes in primary breast tumors. *Oncogene* **9** 3695-700

Paules RS, Levedakou EN, Wilson SJ *et al* (1995) Defective G_2 checkpoint function in cells from individuals with familial cancer syndromes. *Cancer Research* **55** 1763-1773

Paulovich AG and Hartwell LH (1995) A checkpoint regulates the rate of progression through S phase in *S. cerevisiae* in response to DNA damage. *Cell* **82** 841-847

Paulson TG, Wright FA, Parker BA, Russack V and Wahl GM (1996) Microsatellite instability correlates with reduced survival and poor disease prognosis in breast cancer. *Cancer Research* **56** 4021-4026

Peltomaki P, Lothe RA, Aaltonen LA *et al* (1993) Microsatellite instability is associated with tumors that characterize the hereditary non-polyposis colorectal carcinoma syndrome. *Cancer Research* **53** 5853-5855

Perry ME, Rolfe M, McIntyre P, Commane M and Stark GR (1992) Induction of gene amplification by 5-aza-2'-deoxycytidine. *Mutation Research* **276** 189-197

Phipps J, Nasim A and Miller DR (1985) Recovery, repair, and mutagenesis in *Schizosaccharomyces pombe*. *Advances in Genetics* **23** 1-72

Pietenpol J and Vogelstein B (1993) No room at the p53 inn. *Nature* **365** 17-18

Reifenberger G, Liu L, Ichimura K, Schmidt EE and Collins VP (1993) Amplification and overexpression of the MDM2 gene in a subset of human malignant gliomas without p53 mutations. *Cancer Research* **53** 2736-2739

Renzing J and Lane DP (1995) p53-dependent growth arrest following calcium phosphate-mediated transfection of murine fibroblasts. *Oncogene* **10** 1865-1868

Rhyu MG, Park WS and Meltzer SJ (1994) Microsatellite instability occurs frequently in human gastric carcinoma. *Oncogene* **9** 29-32

Rowley JD (1990a) Molecular cytogenetics: Rosetta stone for understanding cancer. *Cancer Research* **50** 3816-3825

Rowley JD (1990b) Recurring chromosome abnormalities in leukemia and lymphoma. *Seminars in Hematology* **27** 122-136

Russo CA, Weber TK, Volpe CM *et al* (1995) An anoxia inducible endonuclease and enhanced DNA breakage as contributors to genomic instability in cancer. *Cancer Research* **55** 1122-1128

Samad A and Carroll RB (1991) The tumor suppressor p53 is bound to RNA by a stable covalent linkage. *Molecular and Cell Biology* **11** 1598-1606

Sandell LL and Zakian VA (1993) Loss of a yeast telomere: arrest, recovery, and chromosome loss. *Cell* **75** 729-739

Sands AT, Suraokar MB, Sanchez A, Marth JE, Donehower LA and Bradley A (1995) p53 deficiency does not affect the accumulation of point mutations in a transgene target. *Proceedings of the National Academy of Sciences of the USA* **92** 8517-8521

Savitsky K, Bar-Shira A, Gilad S *et al* (1995) A single ataxia telangiectasia gene with a product similar to PI-3 kinase. *Science* **268** 1749-1753

Schaefer DI, Livanos EM, White AE and Tlsty TD (1993) Multiple mechanisms of N-(phosphonoacetyl)-L-aspartate drug resistance in SV40-infected precrisis human fibroblasts. *Cancer Research* **53** 4946-4951

Scheffner M, Werness BA, Huibregtse JM, Levine AJ and Howley PM (1990) The E6 oncoprotein encoded by human papillomavirus types 16 and 18 promotes the degradation of p53. *Cell* **63** 1129-1136

Scheffner M, Huibregtse JM, Vierstra RD and Howley PM (1993) The HPV-16 E6 and E6-AP complex functions as a ubiquitin-protein ligase in the ubiquitination of p53. *Cell* **75** 495-505

Schorpp M, Mallick U, Rahmsdorf H and Herrlich P (1984) UV-induced extracellular factor from human fibroblasts communicates the UV response to nonirradiated cells. *Cell* **37** 861-868

Schwartz JL (1992) The radiosensitivity of the chromosomes of the cells of human squamous cell carcinoma cell lines. *Radiation Research* **129** 96-101

Schwartz JL and Vaughan AT (1989) Association among DNA/chromosome break rejoining rates, chromatin structure alterations, and radiation sensitivity in human tumor cell lines. *Cancer Research* **49** 5054-5057

Sen P and Hittelman WN (1984) Kinetics and extent of repair of bleomycin-induced chromosome damage in quiescent normal human fibroblasts and human mononuclear blood cells. *Cancer Research* **44** 591-596

Shay JW, Tomlinson G, Piatyszek MA and Gollahon LS (1995) Spontaneous in vitro immortalization of breast epithelial cells from a Li-Fraumeni patient. *Molecular and Cell Biology* **15** 425-432

Shimizu N, Nakamura H, Kadota T *et al* (1994) Loss of amplified c-myc genes in the spontaneously differentiated HL-60 cells. *Cancer Research* **54** 3561-3567

Shridhar V, Siegfried J, Hunt J *et al* (1994) Genetic instability of microsatellite sequences in many non-small cell lung carcinomas. *Cancer Research* **54** 2084-2087

Slebos RJ, Lee MH, Plunkett BS *et al* (1994) p53-dependent G_1 arrest involves pRB-related proteins and is disrupted by the human papillomavirus 16 E7 oncoprotein. *Proceedings of the National Academy of Sciences of the USA* **91** 5320-5324

Slichenmyer WJ, Nelson WG, Slebos RJ and Kastan MB (1993) Loss of a p53-associated G_1 checkpoint does not decrease cell survival following DNA damage. *Cancer Research* **53** 4164-4168

Smith KA, Gorman PA, Stark MB, Groves RP and Stark GR (1990) Distinctive chromosomal structures are formed very early in the amplification of CAD genes in Syrian hamster cells. *Cell* **63** 1219-1227

Smith KA, Stark MB, Gorman PA and Stark GR (1992) Fusions near telomeres occur very early in the amplification of CAD genes in Syrian hamster cells. *Proceedings of the National Academy of Sciences of the USA* **89** 5427-5431

Smith ML, Chen IT, Zhan Q *et al* (1994) Interaction of the p53-regulated protein Gadd45 with proliferating cell nuclear antigen. *Science* **266** 1376-1380

Smith KA, Agarwal ML, Chernov MV *et al* (1995a) Regulation and mechanisms of gene amplification. *Philosophical Transactions of the Royal Society B* **347** 49-56

Smith ML, Chen IT, Zhan Q, O'Connor PM and Fornace AJ (1995b) Involvement of the p53 tumor suppressor in repair of u.v.-type DNA damage. *Oncogene* **10** 1053-1059

Soussi T, Caron de Fromentel C and May P (1990) Structural aspects of the p53 protein in relation to gene evolution. *Oncogene* **5** 945-952

Srivastava S, Zou Z, Pirollo K, Blattner W and Chang EH (1990) Germ-line transmission of a mutated p53 gene in a cancer-prone family with Li-Fraumeni syndrome. *Nature* **348** 747-749

Stewart N, Hicks GG, Paraskevas F and Mowat M (1995) Evidence for a second cell cycle block at G_2/M by p53. *Oncogene* **10** 109-115

Stopper H, Korber C, Schiffmann D and Caspary WJ (1993) Cell-cycle dependent micronucleus formation and mitotic disturbances induced by 5-azacytidine in mammalian cells. *Mutation Research* **300** 165-177

Sturzbecher H-W, Donzelmann B, Henning W, Knippschild U and Buchhop S (1996) p53 is linked directly to homologous recombination processes via RAD51/recA protein interaction. *EMBO Journal* **15** 1992-2002

Su LN and Little JB (1992) Transformation and radiosensitivity of human diploid skin fibroblasts transfected with SV40 T-antigen mutants defective in RB and P53 binding domains. *International Journal of Radiation Biology* **62** 461-468

Symonds H, Krall L, Remington L *et al* (1994) p53-dependent apoptosis suppresses tumor growth and progression in vivo. *Cell* **78** 703-711

Tlsty TD (1990) Normal diploid human and rodent cells lack a detectable frequency of gene amplification. *Proceedings of the National Academy of Sciences of the USA* **87** 3132-3136

Tlsty TD, Brown PC and Schimke RT (1984) UV radiation facilitates methotrexate resistance and amplification of the dihydrofolate reductase gene in cultured 3T6 mouse cells. *Molecular and Cell Biology* **4** 1050-1056

Tlsty TD, Margolin BH and Lum K (1989) Differences in the rates of gene amplification in non-tumorigenic and tumorigenic cell lines as measured by Luria-Delbruck fluctuation analysis. *Proceedings of the National Academy of Sciences of the USA* **86** 9441-9445

Toledo F, Smith KA, Buttin G and Debatisse M (1992) The evolution of the amplified adenylate deaminase 2 domains in Chinese hamster cells suggests the sequential operation of different mechanisms of DNA amplification. *Mutation Research* **276** 261-273

Tsang NM, Nagasawa H, Li C and Little JB (1995) Abrogation of p53 function by transfection of HPV16 E6 gene enhances the resistance of human diploid fibroblasts to ionizing radiation. *Oncogene* **10** 2403-2408

Tuteja N, Tuteja R, Ochem A *et al* (1994) Human DNA helicase II: a novel DNA unwinding enzyme identified as the Ku autoantigen. *EMBO Journal* **13** 4991-5001

Uchida T, Wada C, Wang C, Egawa S, Ohtani H and Koshiba K (1994) Genomic instability of microsatellite repeats and mutations of H-, K-, and N-ras, and p53 genes in renal cell carcinoma. *Cancer Research* **54** 3682-2685

Van Meir EG, Polverini PJ, Chazin VR, Su HH, de Tribolet N and Cavenee WK (1994) Release of an inhibitor of angiogenesis upon induction of wild type p53 expression in glioblastoma cells. *Nature Genetics* **8** 171-176

Vertino PM, Spillare EA, Harris CC and Baylin SB (1993) Altered chromosomal methylation patterns accompany oncogene-induced transformation of human bronchial epithelial cells. *Cancer Research* **53** 1684-1689

Visakorpi T, Kallioniemi AH, Syvanen AC *et al* (1995) Genetic changes in primary and recurrent prostate cancer by comparative genomic hybridization. *Cancer Research* **55** 342-347

Von Hoff DD, McGill JR, Forseth BJ *et al* (1992) Elimination of extrachromosomally amplified MYC genes from human tumor cells reduces their tumorigenicity. *Proceedings of the National Academy of Sciences of the USA* **89** 8165-8169

Wahl AF, Donaldson KL, Fairchild C *et al* (1996) Loss of normal p53 function confers sensitization to Taxol by increasing G_2/M arrest and apoptosis. *Nature Medicine* **2** 72-79

Wahl GM, Linke SP, Paulson TG and Huang L-C Mammalian checkpoints and genetic instability, In: Mihich E (ed). *Pezcoller Foundation Symposia*, Plenum Publishing Co, New York (in press)

Wang XW, Yeh H, Schaeffer L *et al* (1995) p53 modulation of TFIIH-associated nucleotide excision repair activity. *Nature Genetics* **10** 188-195

Wang XW, Vermeulen W, Coursen JD *et al* (1996) The XPB and XPD DNA helicases are components of the p53-mediated apoptosis pathway. *Genes and Development* **10** 1219-1232

Wani MA, Xu X and Stambrook PJ (1994) Increased methotrexate resistance and dhfr gene amplification as a consequence of induced Ha-ras expression in NIH 3T3 cells. *Cancer Research* **54** 2504-2508

Ward JF (1988) DNA damage produced by ionizing radiation in mammalian cells: identities, mechanisms of formation, and reparability. *Progress in Nucleic Acid Research and Molecular Biology*, **35** 95-125, Academic Press, New York

Weinert TA and Hartwell LH (1988) The RAD9 gene controls the cell cycle response to DNA damage in *Saccharomyces cerevisiae*. *Science* **241** 317-322

Weinert TA, Kiser GL and Hartwell LH (1994) Mitotic checkpoint genes in budding yeast and the dependence of mitosis on DNA replication and repair. *Genes and Development* **8** 652-665

Wilcock D and Lane DP (1991) Localization of p53, retinoblastoma and host replication proteins at sites of viral replication in herpes-infected cells. *Nature* **349** 429-431

Windle B, Draper BW, Yin YX, O'Gorman S and Wahl GM (1991) A central role for chromosome breakage in gene amplification, deletion formation, and amplicon integration. *Genes and Development* **5** 160-174

Wlodek D and Hittelman WN (1988) The relationship of DNA and chromosome damage to survival of synchronized X-irradiated L5178Y cells. II. Repair. *Radiation Research* **115** 566-575

Wooster R, Cleton-Jansen AM, Collins N *et al* (1994) Instability of short tandem repeats (microsatellites) in human cancers. *Nature Genetics* **6** 152-156

Wright JA, Chan AK, Choy BK, Hurta RA, McClarty GA and Tagger AY (1990a) Regulation and drug resistance mechanisms of mammalian ribonucleotide reductase, and the significance to DNA synthesis. *Biochemistry and Cell Bioligy* **68** 1364-1371

Wright JA, Smith HS, Watt FM, Hancock MC, Hudson DL and Stark GR (1990b) DNA amplification is rare in normal human cells. *Proceedings of the National Academy of Sciences of the USA* **87** 1791-1795

Wu X, Bayle JH, Olson D and Levine AJ (1993) The p53-mdm-2 autoregulatory feedback loop. *Genes and Development* **7** 1126-1132

Xu M, Sheppard KA, Peng CY, Yee AS and Piwnica WH (1994) Cyclin A/CDK2 binds directly to E2F-1 and inhibits the DNA-binding activity of E2F-1/DP-1 by phosphorylation. *Molecular and Cell Biology* **14** 8420-8431

Yeargin J and Haas M (1995) Elevated levels of wild-type p53 induced by radiolabeling of cells leads to apoptosis or sustained growth arrest. *Current Biology* **5** 423-431

Yee CJ, Roodi N, Verrier CS and Parl FF (1994) Microsatellite instability and loss of heterozygosity in breast cancer. *Cancer Research* **54** 1641-1644

Yin Y, Tainsky MA, Bischoff FZ, Strong LC and Wahl GM (1992) Wild-type p53 restores cell cycle control and inhibits gene amplification in cells with mutant p53 alleles. *Cell* **70** 937-948

Yonish-Rouach E, Resnitzky D, Lotem J, Sachs L, Kimchi A and Oren M (1991) Wild-type p53 induces apoptosis of myeloid leukaemic cells that is inhibited by interleukin-6. *Nature* **353** 345-347

Yonish-Rouach E, Grunwald D, Wilder S *et al* (1993) p53-mediated cell death: relationship to cell cycle control. *Molecular and Cell Biology* **13** 1415-1423

Yount GL, Haas KD, Vidair CA, Haas M, Dewey WC and Israel MA (1996) Cell cycle synchrony unmasks the influence of p53 function on radiosensitivity of human glioblastoma cells. *Cancer Research* **56** 500-506

Zhan Q, Carrier F and Fornace AJ (1993) Induction of cellular p53 activity by DNA-damaging agents and growth arrest. *Molecular and Cell Biology* **13** 4242-4250

Ziegler A, Jonason AS, Leffell DJ *et al* (1994) Sunburn and p53 in the onset of skin cancer. *Nature* **372** 773-776

The authors are responsible for the accuracy of the references.

Functions of the DNA Dependent Protein Kinase

SHENGFANG JIN[1,2] • SATOSHI INOUE[1] • DAVID T WEAVER[1,2]

[1]*Division of Tumour Immunology, Dana-Farber Cancer Institute, 44 Binney Street, Boston, MA 02115; and* [2]*Department of Microbiology and Molecular Genetics, Harvard Medical School, Boston, MA02115*

Introduction
Molecular biology of DNA-PK subunits
  DNA-PK$_{cs}$
  The Ku heterodimer
  Gene structure and evolutionary conservation
DNA-PK mutant cells
  Mammalian cell mutations of DNA-PK
  Drosophila mus309 mutation
  Yeast Ku mutants
  How does DNA-PK act in DSB repair?
  Telomere maintenance
Potential substrates for DNA-PK in DNA repair
  DNA-PK phosphorylation by DNA-PK
  RPA and DNA-PK
  Additional factors affecting DNA-PK dependent DNA repair pathways
DNA-PK, cell cycle checkpoints and apoptosis
  Apoptosis
  Synergistic functions of TP53 (p53) and DNA-PK
Transcriptional control by Ku/DNA-PK
Summary

INTRODUCTION

Double stranded DNA breaks are likely to be a byproduct of the routine operation of DNA metabolism as well as the result of "spontaneous" events generated by free radicals formed during radiation damage and by other agents. The ability to repair these types of lesions is therefore necessary for the overall process of maintaining chromosome stability. Eukaryotic cells have a combination of regulatory networks, programmed cell death and DNA repair to respond to broken DNA by a combined approach of interrupting cell cycle progression in concert with the mechanisms of DNA repair.

Eukaryotes have protein kinases that are activated by nucleic acids, includ-

Cancer Surveys Volume 29: *Checkpoint Controls and Cancer*
© 1997 Imperial Cancer Research Fund. 0-87969-518-8/97. $5.00 + .00

ing DNA and double stranded RNA. Kinases of this type that are localized in the nucleus may be expected to have a role in DNA and/or RNA metabolism. A DNA dependent protein kinase (DNA-PK) was originally identified in rabbit reticulocyte lysates (Walker *et al*, 1985), but an enzyme with similar properties has also been documented from mammalian cell lines and a number of marine invertebrates and insects (Lees-Miller *et al*, 1990; Carter and Anderson, 1991; Anderson and Lees-Miller, 1992; Anderson, 1993). DNA-PK is a Ser/Thr specific kinase, with a broad consensus site of SQ or TQ, although there are exceptions to this rule (Anderson and Lees-Miller, 1992; Lees-Miller *et al*, 1992; Anderson, 1993). DNA-PK is a nuclear protein of greater than 350 kDa (Carter *et al*, 1990; Lees-Miller *et al*, 1990).

With the biochemical purification of DNA-PK, it became evident that associated factors were also necessary for its activity (Lees-Miller *et al*, 1990). A number of biochemical tests suggested that a 65/81 kDa pair of proteins cofractionated with the reactivation activity. Subsequently, genetic, biochemical and immunological evidence, discussed below, has revealed that these associated proteins correspond to the Ku autoantigen heterodimer that induces the protein kinase function of DNA-PK. Therefore, active DNA-PK is composed of these three subunits, DNA-PK$_{cs}$, Ku70 and Ku80. In fact, Ku and the catalytic subunit of DNA-PK are ordinarily associated (Lees-Miller *et al*, 1990; Suwa *et al*, 1994).

Ku has had an interesting history of its own, independent of the discovery of its involvement in the DNA-PK complex. Ku was originally identified as an autoantigen in some, but not all, patients with systemic lupus erythematosus and mixed connective tissue disease (Mimori *et al*, 1981; Reeves, 1985; Mimori and Hardin, 1986). Ku is an abundant nuclear antigen (5×10^5 molecules/cell) that is a heterodimer of 70 and 80 kDa subunits. Ku binds to variations in double stranded DNA, particularly to DNA ends (Reeves, 1985; Mimori and Hardin, 1986; Mimori *et al*, 1986; deVries *et al*, 1989; Paillard and Strauss, 1991; Falzon *et al*, 1993). In addition, Ku interacts with DNA ends of differing terminal overlaps, making a strong argument for an ability to serve as a recognition molecule for DNA breaks in vivo (Rathmell and Chu, 1994a). Specific alterations in double stranded DNA, including hairpins, Y structures and single stranded DNA with secondary structure (D loops), are recognized by the Ku autoantigen (Paillard and Strauss, 1991; Falzon *et al*, 1993; Morozov *et al*, 1994). These "conformation" Ku/DNA associations appear to lack DNA sequence specificity. To detect these interactions, investigators have most frequently used electrophoretic mobility shift assays (EMSA), although DNase I footprinting also works well. An important aspect of the Ku-DNA association is that it cannot be titrated out by excess double stranded DNA, only by DNA ends or other altered conformation DNA molecules. In these tests, Ku is the most abundant DNA end binding activity. Thus, Ku-DNA complexes can be investigated almost as easily from whole cell or nuclear extracts as for the purified protein (Paillard and Strauss, 1991; Falzon *et al*, 1993; Ono *et al*, 1994; Rathmell and Chu, 1994a). ·

Several observations support the role of Ku as an activating co-factor of DNA-PK. DNA ends are an efficient substrate both for Ku binding and for DNA-PK activity. In addition, DNA-PK is activated by multiple configurations of DNA, including closed stem loop structures (hairpins) and single strand gaps as well as nicks and Y structures (Morozov *et al*, 1994). In each case, Ku is required for active DNA-PK. These various configurations may be representative of the types of DNA structures formed during DNA recombination, repair, replication and transcription. In vitro Ku-DNA binding is needed for DNA-PK activity (Gottlieb and Jackson, 1993). For Ku binding in this system, linearized plasmid DNA with DNA ends was required. Protein extracts from Ku deficient cell lines have no DNA-PK activity and addition of purified Ku is sufficient to restore DNA-PK activity (Finnie *et al*, 1995).

MOLECULAR BIOLOGY OF DNA-PK SUBUNITS

DNA-PK$_{cs}$

Determination of the gene structure of DNA-PK subunits has provided some clues as to how DNA-PK functions in the cell. The DNA-PK$_{cs}$ human gene was cloned by determining peptide sequences from purified DNA-PK$_{cs}$ and subsequent cDNA library screening (Hartley *et al*, 1995). DNA-PK$_{cs}$ is expressed as an approximately 13 kb mRNA, and the sequencing of the DNA-PK$_{cs}$ cDNAs reveals an open reading frame encoding a 460 kDa protein. Although it was appreciated that DNA-PK$_{cs}$ would probably encode a protein kinase, it was not expected that its structure might closely resemble other known genes with interesting comparative functions (see below). In fact, DNA-PK$_{cs}$ possesses a recognizable kinase homology domain, where the signature group member is the phosphatidylinositol 3 kinase (PI3-kinase; PIK) family (Fig. 1). DNA-PK contains a putative leucine zipper repeat at residues 1503–1538, with six repeats, and this motif may mediate protein-protein associations with Ku or other proteins (Hartley *et al*, 1995). An additional possibility is that an active form of DNA-PK in the nucleus would be a DNA-PK$_{cs}$ dimer. Biochemical fractionation applied thus far may not be able to resolve molecular weight differences for large protein complexes in this size range. DNA-PK$_{cs}$ contains 26 SQ motifs that are potential autoregulatory phosphorylation sites (Hartley *et al*, 1995). However, the significance of any of these sites is as yet difficult to discern. The SQ residues are spread throughout the protein and the lack of localization is uninformative about function, as is true for phosphorylation sites in many other proteins. Unfortunately, there is little additional information as yet from the primary sequence that would generate more clues as to how DNA-PK operates in the cell.

Additional proteins in the PIK group have also been discovered recently and partially characterized (reviewed in Keith and Schreiber, 1995; Zakian, 1995). These other putative kinases include the yeast proteins TOR2, TOR1,

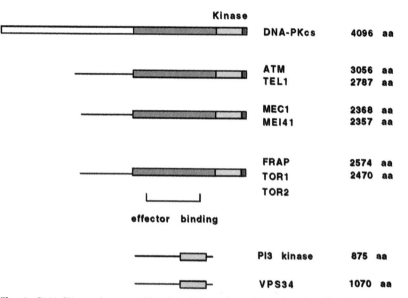

Fig. 1. DNA-PK$_{cs}$ and genes with related kinase homology domains. The DNA dependent protein kinase catalytic subunit (DNA-PK$_{cs}$) is structurally related to a family of other putative protein or lipid kinases. PI3-kinase is demonstrated to be a lipid kinase. Other potential protein kinases shown contain additional sequence similarity to DNA-PK$_{cs}$. These other family members include (ATM,TEL1), (MEC1,MEI41), (FRAP,TOR1,TOR2); groups are those that are more related to each other than to other proteins. Kinase homology domain (yellow), potential effector binding region (blue), C-terminal conserved motif (green)

DRR1, ESR1/MEC1, TEL1, the *Drosophila* MEI41 protein and the human proteins ATM, FRAP, p110α, β and γ PI3-kinase, VPS34 (Greenwell *et al*, 1995; Hari *et al*, 1995; Keith and Schreiber, 1995; Morrow *et al*, 1995; Savitsky *et al*, 1995; Zakian, 1995). Several of the related kinases are more closely aligned with the *ATM* gene than the DNA-PK$_{cs}$ open reading frame. Each of these group members also has a single kinase homology domain of approximately 40% identity with other members. The kinase homology domain is located at the C-terminal end of the protein in every case. Many of the group members are only putative protein and/or lipid kinases, since there are as yet no biochemical protein kinase activities established. In the case of MEC1 and TEL1, evidence has been presented that they may regulate the RAD53 protein kinase by direct phosphorylation (Sanchez *et al*, 1996; Sun *et al*, 1996). Aminoacids that are more N-terminal to the kinase region are more poorly conserved, although several of the group members have significant aminoacid homology in the flanking region to the kinase motif (20–22% identity) (Fig. 1). An additional motif in the C-terminal 30 residues of PIK related genes was identified (LN$_9$LN$_4$AN$_5$LN$_5$GWNP/AW/FN-COOH) (Hartley *et al*, 1995; Keith and Schreiber, 1995). The C-terminal motif is found for a subgroup of the kinases that are most related to DNA-PK. The significance of this sequence for kinase activity is not yet described.

The relatedness of the kinase homology domain and the flanking regions, with increased homologies between DNA-PK$_{cs}$ and its other closest family members, may suggest functional similarities for substrates (Fig. 1). In addition, the principal function of the two subgroups may be to act mainly as protein or lipid kinases, and this feature may be encoded by these homology differences and similarities. DNA-PK$_{cs}$ has no lipid kinase activity that is ordinarily associated with PI3-kinase (Hartley *et al*, 1995). Instead, DNA-PK protein kinase activity has been characterized for a wide variety of protein or peptide substrates in vitro (Anderson, 1993). DNA-PK is also inhibited by wortmannin with half maximal inhibition at approximately 250 nM. Wortmannin is a significantly stronger inhibitor of PI3 lipid kinase activity (5 nM) (Okada *et al*, 1994). Another DNA-PK$_{cs}$ inhibitor, OK-1035, was recently described (Take *et al*, 1995, 1996), but its effectiveness on other putative protein or lipid kinases was not shown.

A large region flanking the C-terminal kinase domain may be a general domain for the association of proteins that regulate kinase activity (Fig. 1, effector binding). The protein binding region for one of the family members, human FRAP, has been localized to this region (Keith and Schreiber, 1995). FRAP associates with a rapamycin binding protein, which mediates the effects of drugs such as FK506. Since Ku is known to be a stimulator of DNA-PK activity, this region may also be the position for Ku association by analogy to FRAP.

Many of the PIK family member proteins have been implicated in the regulation of signal transduction and/or growth control processes as well as DNA repair checkpoint pathways. *ATM* encodes the protein that is mutated in patients with ataxia telangiectasia (AT), a complex human disease consisting of cerebellar degeneration, immunodeficiency, DNA repair defects and a greatly elevated frequency of cancer (Gatti and Swift, 1985). Many of the *ATM* mutations have recently been mapped, and some are likely to be null mutations (Savitsky *et al*, 1995). All the evidence for ATM phenotypes suggests that the effects of the mutation are loss of function phenomena. The AT immunodeficiency defects are not related to direct effects on V(D)J recombination (Hsieh *et al*, 1993), unlike DNA-PK defects (see below). Instead, AT cells are deficient in a variety of responses to radiation damage, indicating a requirement of the protein in signalling DNA damage events. Included in these effects is a non-responsiveness to radiation damage in interrupting cell cycle progression (Painter and Young, 1980; Kastan *et al*, 1992). Notably, accumulation of the tumour suppressor protein, TP53, following radiation damage is retarded in AT cells (Kastan *et al*, 1992; Lu and Lane, 1993). AT cells also have an unstable chromosome content, indicating that although the gene is not essential, genome stability may depend on its functioning.

mec1 mutants show that *MEC1*, like *ATM* deficient cells, controls pathways important to DNA damage inducible cell cycle checkpoints and recombination (Allen *et al*, 1994; Kato and Ogawa, 1994; Weinert *et al*, 1994; Sanchez *et al*, 1996; Sun *et al*, 1996; Weinert *et al*, 1994). *MEC1* appears to be required for multiple cell cycle checkpoints, again like *ATM* deficient cells. It was isolated in screens for G$_2$ arrest defective mutants (Allen *et al*, 1994;

Weinert *et al*, 1994). In addition, *mec1* mutants fail to delay the duration of S phase, consistent with an S phase checkpoint defect (Paulovich and Hartwell, 1995). Several lines of evidence support the notion that *MEC1* encodes a protein kinase upstream of the RAD53 dependent pathway that transduces damage signals and replication blocks to the transcription machinery (Allen *et al*, 1994; Sanchez *et al*, 1996). Rad53 has protein kinase activity, suggesting that *MEC1* is upstream in a protein kinase cascade for the recognition of DNA damage. The Rad53 kinase is phosphorylated in response to DNA damage induction and replication blocks formed by hydroxyurea treatment (Sanchez *et al*, 1996; Sun *et al*, 1996). In mutant *mec1* strains, Rad53 protein is not phosphorylated following induction conditions. For kinase-negative point mutations of *rad53*, phosphorylation occurs to the same extent and is *MEC1* dependent, arguing that Rad53 does not phosphorylate itself (Sun *et al*, 1996).

TEL1 mutations were isolated on the basis of a telomere shortening defect, but otherwise show no significant growth or DNA repair defects (Lustig and Petes, 1986). However, *tel1Δ mec1Δ* strains are severely sensitive to ionizing radiation and show many DNA repair defects (Greenwell *et al*, 1995; Morrow *et al*, 1995). TEL1 overexpression is also able to complement *mec1* strains for DNA damage phenotypes (Sanchez *et al*, 1996). These observations suggest that *TEL1* and *MEC1* may be functionally redundant for some DNA repair functions and introduce the notion that similar effects may be observed in conjunction with DNA-PK mutations.

The Ku Heterodimer

The Ku heterodimer is ubiquitous to eukaryotes, having been identified in diverse species from *Saccharomyces cerevisiae* to mammals. By biochemical criteria in each species tested, there are two subunits of approximately 70 and 80–86 kDa in a 1:1 association. Whereas human Ku was initially characterized as an autoantigen, Ku from lower eukaryotes has been identified by functional criteria. *S cerevisiae* Ku was purified by the property of DNA end binding, based on the gel shift assays developed for the mammalian proteins. From this work, a Ku70 homologue was identified that was named HDF1 (Feldmann and Winnacker, 1993). *Drosophila melanogaster* Ku was similarly identified by binding to specific DNA sequences: the P transposable inverted repeat (Beall *et al*, 1994) or a 31 bp sequence near the transcription initiation site of *ypf1* (Jacoby and Wensink, 1994). A Ku70 homologue was independently cloned from each group and named *Irbp* (Beall *et al*, 1994) or *ypf1* (Jacoby and Wensink, 1994), respectively.

Gene Structure and Evolutionary Conservation

To date, the putative aminoacid sequences of a number of Ku open reading frames have been determined that have been useful for estimating conserved properties of the proteins (Fig. 2). Mammalian Ku70 for which there is se-

Ku70

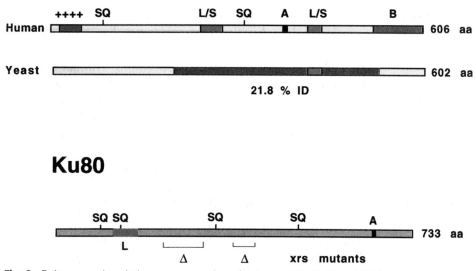

Ku80

Fig. 2. Epitopes and evolutionary conservation of Ku genes. The Ku70 and Ku80 genes are conserved among a variety of eukaryotes ranging from the yeast *S cerevisiae* to humans. Comparative analysis of Ku protein sequences reveals the regions of similarity and divergence. Human and yeast Ku protein sequences are shown. For Ku70, the mammalian Ku70 proteins contain several motifs that may be relevant to functions of Ku. The N-terminal portion contains an acidic region (++++) followed by two Leu/Ser repeats (L/S). At the C-terminal end are a related sequence to the Koonin consensus for ATPases (A = (U)n(G)XX(G)XGK[STE], where U signifies hydrophobic residues [Koonin, 1993; Cao *et al*, 1994]) and an α-helix-turn-α-helix motif that binds DNA in vitro (Chou *et al*, 1992). This may or may not be the primary DNA end binding motif considering that it is not conserved among Ku70 homologues. Yeast Ku70 is shown with the region of increased identity (ID) and a leucine (L) repeat. SQ sites of DNA-PK phosphorylation are displayed. For Ku80, human and yeast Ku80 are similar across the lengths. Position of in frame deletions from *xrs* mutations are shown

quence information available are derived from human, mouse and rat. The mammalian Ku proteins are strikingly similar, with very conservative changes at a small fraction of the aminoacids across the genes. Similarly, for Ku80 homologues, there is considerable conservation among human, mouse, rat and Chinese hamster proteins. Although the yeast and *Drosophila* Ku homologues are structurally similar to the mammalian Ku genes across their lengths, there is considerably more divergence of structure (Fig. 2). *S cerevisiae* Ku70 has reduced similarity at both N- and C-terminal portions of the human protein, with the greatest similarity in an internal region (22% identity; aminoacids 226–578 of Hdf1 and 213–556 of human Ku70) (Feldmann and Winnacker, 1993). *Drosophila* Ku70 is 27% identical and 34% similar to human Ku70 (Beall *et al*, 1994).

The Ku70 subunit may retain the ability to interact with DNA with or without Ku80, yet structural comparisons of Ku70 genes do not indicate this to be a conserved function. Full length Ku70 and a C-terminal fragment of human Ku70 bind to DNA ends by Southwestern analysis (Chou *et al*, 1992). The C-terminal portion of human Ku70 has an α-helix-turn-α-helix motif that is much like a bacterial Cro protein DNA binding region (Fig. 2). However, this motif is poorly conserved in the other Ku70 genes. Mammalian Ku70 also contains a "leucine zipper" type motif that may be important in stabilizing the Ku heterodimer and possibly for DNA-PK or other associations. Human Ku70 has a basic aminoacid region (187–214) followed by five helical turns of a Leu/Ser repetition between aminoacids 215 and 242 (Reeves and Sthoeger, 1989). Mouse Ku70 is similar, with only one fewer Leu/Ser repeat. Hdf1 contains a Leu/Ser motif of five helical repeats between aminoacids 484 and 512 but not in the same spatial region found for the mammalian proteins. Considering the diverse functions of Ku discussed below, sequence divergence between species might accommodate particular roles that are species restricted. On the other hand, considering the generality of the DNA end binding properties of Ku, significant similarities might also be expected. Because the yeast Ku70 is shorter than the mammalian proteins, the C-terminal motif for DNA binding may be lost. It will be interesting to distinguish the role of this region for DNA repair and transcription functions of Ku (discussed below).

In contrast to *HDF1* of yeast, the yeast Ku80 homologue (*ScKu80*) was identified by scanning of yeast genome databases with reduced stringency against mammalian Ku80 genes (Milne *et al*, 1996). ScKu80 is a 629 aminoacid protein with a predicted molecular mass of 74.5 kDa and thus is predicted to be smaller than mammalian Ku80 (Fig. 2). ScKu80 is 21% and 20.5% identical and 45.9% and 43.8% similar to human and mouse Ku80, respectively. As for HDF1, the homology extends the length of the protein, and the same level of similarity to the mammalian counterparts is found for ScKu80. A *Drosophila* Ku80 has not yet been reported. In the search for ScKu80, an uncharacterized *Caenorhabditis elegans* open reading frame (Genbank no. S43606, R07E5.8) was also identified. Because of the level of sequence similarity, this is very likely to be a *C elegans* Ku gene. Murine and human Ku80 putative aminoacid sequences are strikingly similar yet slightly more divergent (77% identity) than Ku70 genes. A recent report of Ku80 genes from Syrian and Chinese hamster revealed close identity with the Chinese hamster sequence, being 95.5%, 87.4% and 80.3% the same as Syrian hamster, mouse and human Ku80 proteins, respectively (Errami *et al*, 1996).

DNA-PK MUTANT CELLS

Information about the phenotypes of DNA-PK deficiencies is now available from a wide variety of cell lines and mutations in different eukaryotes for each of the three known components of the complex (Fig. 3). The evaluation of these cells has revealed new findings about the functioning of the DNA-PK

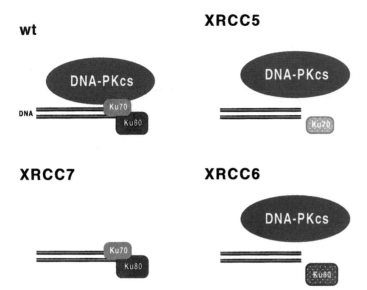

Fig. 3. Molecular defects in DNA-PK mutant cells. The mutational groups representing defects in each of the three DNA-PK subunits are shown. Wild type cells contain the three DNA-PK subunits associated with each other and with DNA. DNA-PK$_{cs}$ (protein kinase subunit) is shown in red and the Ku heterodimer (Ku70 and Ku80) in blue. *XRCC5* mutants (*xrs, sxi-3, XR-V15B* [Chinese hamster], *ku80Δ* [*S cerevisiae*]) either do not produce Ku80 or it is not stable once synthesized. In *XRCC5* mutants, Ku70 is unstable. *XRCC6* mutants are defective for Ku70 production. The *Drosophila IRBP* gene (Ku70) appears to be non-functional in *mus309* mutants. *XRCC7* mutants are deficient in DNA-PK$_{cs}$. The mouse *scid* mutation is correlated with DNA-PK$_{cs}$ deficiency (see text). Human MO59J glioblastoma cells are DNA-PK$_{cs}$ deficient. Ku is stable and binds DNA with or without DNA-PK$_{cs}$

complex. The mutant cells and mutational groups are first discussed, followed by an explanation of mutant phenotypes.

Mammalian Cell Mutations of DNA-PK

Mammalian cell mutants of DNA-PK subunits are sensitive to ionizing radiation and have errors in V(D)J recombination. The phenotypes of these mutant cell lines have been reviewed and the mutants categorized (Jackson and Jeggo, 1995; Weaver, 1995a). The sensitivity of these mutants to ionizing radiation appears to be linked to the double strand break (DSB) repair pathway in mammalian cells, since sensitivity to other agents that do not form DSBs is not usually observed. Many of the mutants have been isolated by radiation sensitivity screens, and others have been identified from spontaneous mouse mutants.

The *xrs* mutants and several other mutants in the *XRCC5* group were isolated and characterized as radiation sensitive cell lines (Jeggo, 1985a, 1985b, 1990; Jeggo and Kemp, 1983; Zdzienicka *et al*, 1988; Jeggo *et al*, 1989, 1991; Boubnov *et al*, 1995; Lee *et al*, 1995). Each of the *XRCC5* mutants are defective for DNA end binding at any appreciable levels (Getts and Stamato, 1994;

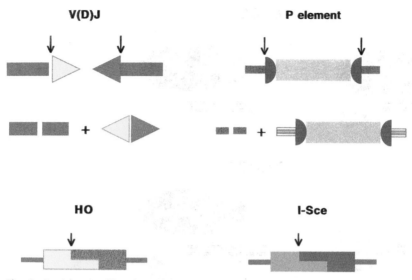

Fig. 4. Double strand break repair events. Double strand breaks are generated in eukaryotes by various means. The DNA damage created by ionizing radiation frequently leads to DSBs that are repaired using genes important for other DNA repair and recombination events (see text). In mammalian cells, a programmed recombination pathway of lymphocytes, the V(D)J recombination, involves DSBs at the borders of signal sequences (triangles) and rearrangement to produce two new DNA junctions: coding joint (green) and signal sequence joint (yellow and brown triangles). DNA-PK mutants affect the efficiency and joining events of V(D)J recombination. In *Drosophila* transposable elements such as P elements are excised from their genome position in a P transposase dependent reaction. DSBs are formed at the junctions between the P element inverted repeats (red semicircles) and genomic DNA. The genomic DSB is repaired by functions controlled by Ku. Two yeast endonucleases have been studied that will create DSBs. At the *S cerevisiae MAT* locus, an HO site is cleaved by HO endonuclease. Under circumstances where mating type switching is inactivated, HO breaks are repaired by a Ku dependent pathway. The yeast endonuclease, I-Sce I, recognizing an 18 bp site, has been engineered to be expressed in mammalian cells. I-Sce I formed DSBs in mammalian cells are repaired by Ku dependent pathways. Arrows denote cleavage sites for the enzymes for each mechanism

Rathmell and Chu, 1994a,b; Taccioli *et al,* 1994a; Boubnov *et al,* 1995). *XRCC5* mutants (*xrs, XRV15B, sxi-2, sxi-3*) are deficient in both coding and signal junction formation for V(D)J recombination (Fig. 4) (Pergola *et al,* 1993; Taccioli *et al,* 1993; Boubnov *et al,* 1995; Lee *et al,* 1995). These mutant cell lines are not yet completely characterized regarding the Ku mutations. For the *sxi-3* mutation Ku80 mRNA is not significantly produced and therefore there is very little Ku80 protein (Fig. 3). For *xrs-6,* a stable mRNA is produced, but the protein is not stable (Rathmell and Chu, 1994b). *XR-V15B* and *XR-V9B* mutants express Ku80 RNA containing internal deletions of 138 and 152 bp, respectively (Errami *et al,* 1996). Each of these mutant cell lines produces Ku80 proteins with in frame deletions of 46 and 84 aminoacids, respectively, that are located in the core middle region of the protein (Fig. 2). The consequence of these deletions may be in protein stability and/or interaction with the Ku70 subunit. For each of these cell lines, the reintroduction of Ku80 genes was sufficient to restore Ku activity, DSB repair and V(D)J recombination functions (Smider *et al*, 1994; Taccioli *et al*, 1994a; Boubnov *et al*, 1995; Errami *et al*, 1996). Additional mutagenesis strategies will be necessary to uncover more comprehensively the regions of Ku80 governing its functions.

Important to the properties of Ku, DNA-PK activity is low in *XRCC5* mutants. Addition of purified Ku to XRCC5 extracts successfully reconstitutes DNA-PK activity in vitro (Finnie *et al*, 1995). In addition, DNA-PK$_{cs}$ levels are unaltered between Ku deficient and control cell lines, arguing that DNA-PK$_{cs}$ is stable in cells in the absence of Ku (Chen *et al*, 1996).

Ku70 protein levels are apparently increased when Ku80 genes are transfected into the *XRCC5* group mutant cells (Smider *et al*, 1994; Taccioli *et al*, 1994a; Boubnov *et al*, 1995; Errami *et al*, 1996). Although Ku70 is normal in these mutants, it is functionally inactive without the Ku80 subunit, emphasizing the importance of the production of Ku as a stable heterodimer (Fig. 3).

Currently, no straightforward mammalian Ku70 mutants have been described from screens for radiation sensitivity. Null mutations in the Ku80 and Ku70 genes have been prepared by gene targeting mutagenesis in murine embryonic stem cells, and they appear to share the same molecular defects as the previously determined *XRCC5* and *XRCC6* group mutants, including deficiencies in Ku levels and DNA end binding, radiation sensitivity and V(D)J recombination coding and signal junction formation (Roth D, Hasty P, Yang Z and Alt F, unpublished).

The *scid* (severe combined immune deficient) mutational group (*XRCC7*) was originally identified by the profound immunodeficiency found in these animals (Bosma *et al*, 1983). The lack of B and T cells in *scid* is attributable to a V(D)J recombination defect (Schuler *et al*, 1986; Hendrickson *et al*, 1988, 1990,1991a; Lieber *et al*, 1988; Malynn *et al*, 1988). The *scid* recombination defect may have been caused by loss of developmental signals in the immune system, but *scid* lymphoid progenitors in culture were shown to be deficient for ongoing V(D)J recombination events. Thus, the scid factor was more directly connected to the V(D)J recombination pathway. The *scid* V(D)J recombination defect is different from that of Ku mutants. Whereas the joining of recombination signal sequence junctions is close to normal, the coding junctions form extremely inefficiently and with extensive deletions. *scid* cells are sensitive to ionizing radiation and are deficient in the same DSB repair mechanism as Ku mutant lines (Fulop and Phillips, 1990; Biedermann *et al*, 1991; Hendrickson *et al*, 1991b; Weaver, 1995b).

Because it was previously shown that the *scid* mutation could be complemented by human chromosome 8 (Kirchgessner *et al*, 1993; Komatsu *et al*, 1993; Kurimasa *et al*, 1993; Banga *et al*, 1994) and that somatic cell fusions between *scid* and Ku mutants could restore radiation resistance, it was likely that *scid* was in a unique mutational group. A radiation sensitive Chinese hamster cell line, V-3, has the same DSB repair and V(D)J recombination defects as mouse *scid* (Taccioli *et al*, 1994b). V-3 and scid somatic cell fusions do not restore radiation resistance, as if they are in the same complementation group by this test. Both scid and V-3 cells are deficient in DNA-PK activity (Blunt *et al*, 1995; Boubnov and Weaver, 1995; Peterson *et al*, 1995) but are normal for Ku and DNA end binding activity, arguing that either the Ku70 or the DNA-PK$_{cs}$ component of DNA-PK is deficient.

Several compelling findings support the connection between the product of the *scid* gene and DNA-PK$_{cs}$. Firstly, yeast artificial chromosomes containing the human DNA-PK$_{cs}$ gene are able to rescue the radiation sensitive and V(D)J recombination deficiencies of V-3 cells, in addition to rescuing DNA-PK activity (Blunt *et al*, 1995). In addition, human chromosome 8 fragments or the whole chromosome can reconstitute DNA-PK activity (see below) (Boubnov and Weaver, 1995; Peterson *et al*, 1995). Secondly, DNA-PK$_{cs}$ protein levels are reduced in scid and V-3 cells relative to control cell lines (Fig. 3) (Blunt *et al*, 1995; Kirchgessner *et al*, 1995; Peterson *et al*, 1995). Thirdly, the human *DNA-PK$_{cs}$* cDNA was used to map the position of a cross-hybridizing mouse *DNA-PK$_{cs}$* gene. Using in situ hybridization techniques, Sipley *et al* (1995) found that *DNA-PK$_{cs}$* cDNA hybridizes to an 8q11 position, which is where the complementing gene for the *XRCC7* group is located. The molecular definitions of the *scid* and V-3 mutations are not yet discerned, but they should be valuable in establishing the molecular properties of DNA-PK$_{cs}$.

Human *XRCC7* mutants have also been discriminated from the screening of tumour cell lines. Cell lines derived from a single human glioblastoma were isolated that differ significantly in their radiosensitivity. The MO59J cell line was found to be radiation sensitive, whereas its counterpart, the MO59K cell line, shows ordinary radiation resistance (Allalunis-Turner *et al*, 1993). MO59J cells are also DNA-PK deficient, since they do not express DNA-PK$_{cs}$ (Fig. 3). However, MO59J is normal for Ku and DNA end binding activity (Boubnov and Weaver, 1995; Lees-Miller *et al*, 1995) and can be restored for DNA-PK activity by addition of purified DNA-PK$_{cs}$ (Lees-Miller *et al*, 1995). MO59J is distinct from scid and V-3 cells, because there is no DNA-PK$_{cs}$, and this line may be more analogous to a null mutation. Human glioblastomas do not uniformly have DNA-PK$_{cs}$ deficiencies, and the radiosensitivity of these tumours in the clinical response is not necessarily dictated by DNA-PK (Allalunis-Turner *et al*, 1996). V(D)J recombination defects have not yet been investigated for MO59J cells.

An additional complementation group is exemplified by a single mutant cell line, XR-1, conferring radiation sensitivity, DSB repair deficiency and defects in V(D)J recombination (Stamato *et al*, 1983; Giaccia *et al*, 1985; Li Z *et al*, 1995). The *XRCC4* gene is located on human chromosome 5 (Giaccia *et al*, 1990). The V(D)J recombination defect of XR-1 cells is similar to that of Ku mutants, affecting both coding and signal junction formation (Li Z *et al*, 1995). The XR-1 mutant has a chromosomal deletion of the *XRCC4* gene and, in this regard, is a null mutation. Despite DSB repair defects in XR-1, DNA-PK activity is normal. The XRCC4 protein possesses DNA-PK phosphorylation sites, but it is not known whether these sites are functionally significant or whether XRCC4 is a phosphoprotein. Because Xrcc4 has been implicated in the same molecular pathways as DNA-PK, it is very likely that this protein will function in conjunction with DNA-PK in DNA repair in some capacity.

Drosophila mus309 **Mutation**

Drosophila Ku was discovered by purification of stimulatory activities for DNA binding of significance to the operation of P element transposase or transcriptional regulation (Beall *et al*, 1994; Jacoby and Wensink, 1994). The mechanism of P element transposition is dictated by the activity and binding of transposase at inverted repeat sites at the borders of the transposition element (Fig. 4) (Engels, 1989; Kaufman and Rio, 1992). P element excision leaves behind a DSB that must be repaired. Although this process may be DNA homology dependent, the involvement of irbp, the Ku70 homologue, in the process may indicate that end joining pathways are being used. The cloning and mapping of *irbp* demonstrated a correspondence between *irbp* and the position of *mus309* at third chromosome 86E2-3 in *Drosophila* (Beall *et al*, 1994). The *mus309* allele was identified in flies using methylmethane sulphonate (MMS) sensitivity screen (Boyd *et al*, 1976). Thus, it was suggested that the *mus309* gene was involved in DNA repair. If *mus309* was important for DSB repair, then the MMS sensitivity might correlate with other circumstances where DSBs are created. In support of a DSB repair role, it was discovered that there is a reduced survival of *mus309* males and female sterility following transposase induction (Beall and Rio, 1996). In addition, in transposase dependent P element events, excision is defective in *mus309* mutant embryos. The *mus309* MMS sensitivity, female sterility and male viability all can be complemented by introduction of genomic DNA including the *irbp* gene. Similarly, co-injection of irbp plasmid in the P element excision assay, restored the wild type properties of the excision event to *mus309* embryos. These findings offer a compelling argument for the linkage of *irbp* to *mus309*, even though *mus309* mutations have not yet been mapped.

Similar to studies with V(D)J recombination, the P element excision assay is engineered so that DNA product formation precision can be monitored in addition to the efficiency of the reaction (Fig. 4). In *mus309* embryos, excision events are reduced 30-fold from wild type levels. Of the products that can be formed, most have extended deletions in the vicinity of the transposase excision sites at the outside borders of the two 31 bp inverted repeats (Beall and Rio, 1996). The finding that an *irbp* mutant resembles Ku deficiency in V(D)J recombination in this level of molecular detail is intriguing (Fig. 4). Since irbp was originally purified as an inverted repeat binding protein, it would imply that Ku associates with P element inverted repeats under some conditions. Because of the connection between Ku and product formation in V(D)J recombination and the lack of sequence specificity of this reaction for coding joining formation, perhaps the most appropriate conclusion is that irbp binding 31 bp repeats may be functionally significant after cleavage, during the excision process rather than before cleavage. Alternatively, Ku may have a specialized function for transposition/excision, where there is a reliance on association to the inverted repeats.

Since Ku80 protein, Ku-DNA end binding and DNA-PK activity have been reported in *Drosophila*, it is likely that new alleles will be correlated with these other subunits. Additional *Drosophila* mutants have already been identified from genetic screens for P element excision or postreplication repair defects that might generally involve DNA-PK (Boyd and Setlow, 1976; Boyd *et al*, 1976, 1981). The *mus302* allele severely reduces survival of flies following P element excision (Banga *et al*, 1991) and could be a DNA-PK subunit gene.

Yeast Ku Mutants

Ku in yeast has the biochemical and genetic properties of the mammalian homologues. Null mutations have been formed by both *hdf1*Δ and *ku80*Δ. *hdf1*Δ, *ku80*Δ and *hdf1*Δ *ku80*Δ strains are viable but are totally lacking DNA end binding activity (Feldmann and Winnacker, 1993; Milne *et al*, 1996; Siede *et al*, 1996b). Knockout mutation of *ScKu80* leads to loss of DNA end binding activity (Milne *et al*, 1996). Using epitope tagged Ku80 and Hdf1, it was shown that these proteins form a heterodimer and are the DNA end binding complex of yeast (Milne *et al*, 1996). Thus, removal of either Ku subunit destabilizes the level of the other in yeast (Jin S and Weaver D, unpublished).

*Hdf1*Δ and *ku80*Δ strains have DNA repair and recombination defects. In yeast, DSB repair is ordinarily associated with homologous recombination as demonstrated by extensive genetic and molecular analysis (Petes *et al*, 1991). Mutation of the *RAD52* gene effectively eliminates the homologous recombination pathway and reduces cell survival following the formation of DSBs by ionizing radiation by several orders of magnitude. Recently, it has been shown that Ku mutants in conjunction with *rad52*Δ lead to increased radiosensitivity and MMS sensitivity (Milne *et al*, 1996; Siede *et al*, 1996). These findings indicate that Ku and Rad52 operate in separate repair pathways.

In addition to DNA damage by external agents, DSBs can be generated by the activation of site specific endonucleases in yeast (Fig. 4). The mating type switching pathway of *S cerevisiae* is initiated by a DSB at a cleavage site for the endonuclease, HO. In the presence of homologous donor DNA elsewhere on chromosome III, a gene conversion event replaces DNA flanking the HO site at the *MAT* locus with *HML* or *HMR* DNA and changes mating type. In the absence of either the Rad52 protein or homologous DNA to the cleavage site ("cassetteless" mutants), DSBs at the HO site significantly affect cell survival, and repair of these breaks resembles the end joining pathway in mammalian cells (Kramer *et al*, 1994; Moore and Haber, 1996). Mutations of Ku (*ku80*Δ, *hdf1*Δ and *ku80*Δ *hdf1*Δ strains) greatly decrease cell survival following HO breaks, indicating that Ku is important to the end joining pathway for chromosomal break repair (Milne *et al*, 1996). Cell survival is restored to normal levels by introduction of the relevant Ku genes into these strains.

In addition to chromosomal DSB repair events, Ku is also required for end joining of linearized plasmid DNA. In experiments with mammalian cells, joining of linearized plasmid DNA occurs by an end joining pathway where joints that are formed are heterogeneous and frequently show loss of terminal nucleotides closely similar to V(D)J recombination coding junctions (Weaver *et al*, 1995). In *ku80*Δ, *hdf1*Δ and *ku80*Δ *hdf1*Δ strains, plasmid end joining is reduced by 10–15-fold (Milne *et al*, 1996).

In summary, mammalian, *Drosophila* and yeast Ku complexes are important for DSB repair in several different settings. Whether *Drosophila* DNA-PK activity that has already been identified is similarly relevant is yet to be determined. In yeast, no DNA-PK activity has yet been demonstrated and no candidate genes for DNA-PK$_{cs}$ have been identified. Yeast Ku may be an interesting circumstance where DNA-PK independent functions can be explored. Additional mutant phenotypes for yeast Ku are discussed in later sections.

How Does DNA-PK Act in DSB Repair?

The molecular features of DNA-PK offer a tantalizing model that DNA-PK is a direct participant in the machinery of DNA repair. However, the elements of its involvement are still not described. Potentially, there are several alternative or additional roles for DNA-PK that may vary with cell type, cell cycle stage and possibly position on chromatin in association with other proteins. The DNA damage invoked by ionizing radiation is difficult to describe molecularly because the number and types of lesions are complex. Experimental models for repair of radiation damage have come from the use of more specific DNA lesions generated by other means. Two principal pathways that have been most intensively examined are the V(D)J recombination mechanism and induced endonuclease cutting of DNA at unique or specific sites. In both of these types of DNA breaks, the product formation steps can be discerned because of the specificity of the cleavages. Where these DSBs may differ from radiation induced DSBs is in the types of chemical modifications to DNA formed by the free radical damage. These specialized DNA changes may or may not require additional factors.

Investigations of DNA-PK mutant cell lines have mainly directed the involvement of DNA-PK and Ku to the product formation steps of V(D)J recombination. Considerable evidence has accumulated that this programmed DNA gene rearrangment mechanism is ordinarily lymphoid restricted due to the tight tissue specific regulations of the *RAG1* and *RAG2* genes (Schatz *et al*, 1992). In addition, recent molecular and biochemical assays with purified RAG1 and RAG2 proteins demonstrate that the combination of these two proteins is capable of directing the cleavage and hairpin formation steps of V(D)J recombination (McBlane *et al*, 1995; Eastman *et al*, 1996; van Gent *et al*, 1995, 1996; Steen *et al*, 1996). RAG1 and RAG2 catalyze the cleavage at borders of the recombination signal sequences (RSS) forming coding end hairpins and blunt RSS ends. These recent findings supported evidence of DNA breaks at these same positions and formation of the hairpin structures from lymphoid cells (Roth *et al*, 1992a,b, 1993; Schlissel *et al*, 1993). Because DNA-PK is not necessary for the cleavage reactions in vitro, it is likely that it will also not be used at these stages of V(D)J recombination under physiological conditions. Rather, broken DNA in hairpin and RSS end configurations with or without associated proteins is hypothesized to be the DNA template that DNA-PK binds to conduct the next steps of the reaction. Considering the exciting progress with an in vitro V(D)J recombination cleavage reaction recently, the examination of joining steps where DNA-PK is expected to be involved should be within reach.

Ku is involved in chromosomal DSB repair in mammalian cells and this pathway bears a striking resemblance to coding joint product formation of V(D)J recombination in the same cells. Recently, an alternative means to measure chromosome damage repair at specific sites has been reported (Fig. 4) (Rouet *et al*, 1994a,b; Choulika *et al*, 1995). The *S cerevisiae* endonuclease I-SceI was engineered to be translated in mammalian cells (Rouet *et al*, 1994b). Owing to the 18 bp recognition site of I-SceI, the enzyme essentially only cuts mammalian DNA where I-SceI sites are added. It was shown that DSBs formed in integrated or episomal plasmid DNA are repaired by two pathways, depending on whether or not homologous DNA sequences are present (Rouet *et al*, 1994b). In the absence of homologous DNA, the DSBs are repaired by end joining based on the molecular characterization of repair events. When an I-Sce I site was introduced into *xrs-6* (Ku deficient) cells, there was a significant decrease in the recovery of events (Liang *et al*, 1996). Thus, by analogy to V(D)J recombination, introduction of a single DSB is a lethal event in Ku deficient cells.

Ku may have some function in homologous recombination, although there is some controversy about the significance of the current findings. On the one hand, any impact on spore viability and any diminuition of cell survival after HO cleavages at MAT in *hdf1Δ* or *ku80Δ* strains have not been observed (Milne *et al*, 1996; Siede *et al*, 1996). However, in a second study, an *hdf1* deficient strain showed a marked reduction in cell survival following brief HO induction (Mages *et al*, 1996). In addition, a 10–40-fold decrease in spontaneous mitotic recombination has been noted for *hdf1Δ/hdf1Δ* diploids at three separate loci (Mages *et al*, 1996). These differences might be explained by variations in strain background. Perhaps the conditions governing meiotic and mitotic recombination are sufficiently different that Ku deficiencies may not impact both processes. Since homologous recombination can occur in several pathways (gene conversion, single strand annealing or co-ordinated with DNA replication), the involvement of Ku may be limited, whereas the role for Rad52 is very broad.

As an interesting commentary on the relationship between different recombination pathways and the ways by which recombination differences can be formed, there is now information about cells defective for the *ATM* gene.

Although AT patients are variably immune deficient, there is no evidence for the involvement of this gene in V(D)J recombination. Measurements of transiently induced V(D)J recombination in AT fibroblasts indicate that normal coding and signal joining products are formed at roughly the appropriate levels (Hsieh *et al*, 1993). On the other hand, measurements of homologous recombination in AT fibroblasts indicate large effects. Spontaneous intrachromosomal recombination was 30–200-fold higher than in normal cells (Meyn, 1993). In another study, chromosomal and extrachromosomal homologous recombination were 100- and 27-fold higher than in control cells. Since AT cells have well characterized checkpoint defects, errors in the cell recognition of spontaneous DNA damage may induce greater recombination levels.

Telomere Maintenance

The yeast gene highly related to DNA-PK$_{cs}$, *TEL1*, controls telomere maintenance, yielding chromosomes with decreased telomere length in *tel1Δ* strains (Greenwell *et al*, 1995). Ku deficient *hdf1Δ* strains also have an additional mutant phenotype that is intriguing regarding the interplay of PIK kinases in the nucleus. *hdf1Δ* strains were found to have a telomere shortening defect in a manner that was not epistatic to *tel1Δ* (Porter *et al*, 1996). This finding is consistent with a role for Ku in telomere metabolism and chromosome stability. Several models could explain the role of Ku in telomere stability. Because Ku binds to DNA ends, possibly an ordinary function of Ku would be to protect telomeres from degradation. Alternatively, Ku might affect another process that indirectly influences telomere length.

In mammalian cells, it has not yet been determined whether *ATM* or DNA-PK genes regulate the lengths of telomeres. However, one of the mutational phenotypes of AT cells is chromosome instability, which might be attributable to a telomere shortening defects. AT cells more frequently show aberrant chromosome configurations, including chromosome fusions, similar to senescing cells with shortened telomeres (Kojis *et al*, 1991; Counter *et al*, 1992). Several AT cell lines have shorter than wild type telomere lengths (Pandita *et al*, 1995).

POTENTIAL SUBSTRATES FOR DNA-PK IN DNA REPAIR

The protein kinase activity of DNA-PK may be expected to be important for functions of the complex. In this regard, identifying physiologically relevant substrates and the effects of phosphorylation of these substrates is important (Fig. 5). To date, many proteins have been demonstrated to be phosphorylated by DNA-PK in vitro. Under these artificial conditions, several general rules apply. Firstly, even for protein substrates that themselves associate with DNA there is a requirement for the intact DNA-PK complex, consisting of the kinase subunit and Ku. DNA-PK$_{cs}$ may have no activity in the absence of Ku, although the protein is stable under these conditions, as demonstrated by the presence of DNA-PK$_{cs}$ in Ku deficient cell lines. Whether Ku and DNA-PK$_{cs}$ must be co-associated before protein substrate recognition has yet to be resolved.

Secondly, activation of the complex by DNA ends or conformational changes in DNA is also required. There are several underlying issues to be investigated regarding the DNA activation of the DNA-PK complex. The cir-

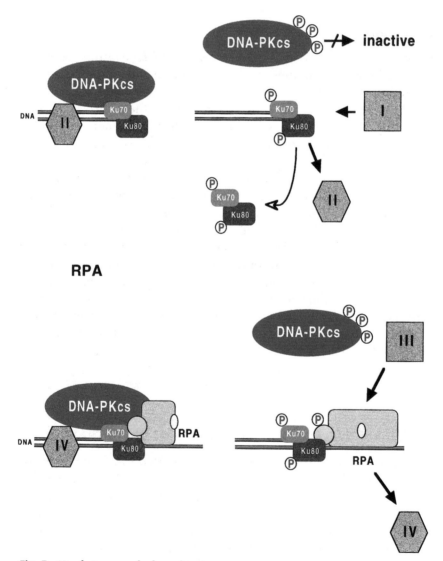

Fig. 5. (*See facing page for legend.*)

cumstances of DNA activation are restricted to conditions where Ku binds to DNA. Generally, the association with DNA ends by Ku is the most frequent, although this condition may only correspond to a rather rare circumstance in the cell—that is, the occurrence of spontaneous or induced DNA damage, where free ends are generated. Ordinarily, Ku association with DNA may occur in a non-activated state. It has been demonstrated that Ku binds to double stranded DNA and translocates along DNA as a putative helicase (Tuteja *et al*, 1994). It is not yet clear whether this type of association is sufficient to activate the protein kinase or even whether DNA-PK has helicase activity as a complex. Future experiments must distinguish pre-loaded Ku without DNA ends from Ku at DNA ends.

Several substrates for DNA-PK in vitro are themselves able to associate with DNA in either a non-specific (sequence independent) or a sequence dependent means. Transcription factor phosphorylation is dictated by the combination of a sequence specific DNA binding site with a DNA end for Ku binding (Gottlieb and Jackson, 1993). This implies that co-association is relevant to efficient phosphorylation and therefore that the DNA template serves the dual function of activation of DNA-PK and positioning. By contrast, the non-specific DNA binding complex, replication protein A (RPA), does not bind appreciably in a sequence specific manner to double stranded DNA. Instead, RPA is efficiently phosphorylated by DNA-PK, using single stranded DNA templates. In the midst of DNA replication, single strand regions of DNA where RPA associates are being replaced by replicated DNA. Although it is not yet clear whether DNA-PK associates with replication forks, a likely scenario would be that Ku association at transition configurations between single and double stranded DNA would suffice to activate the kinase.

Fig. 5. Potential mechanisms of DNA-PK action at chromosome breaks. How DNA-PK functions in DSB repair is currently unknown. One feature of importance is that the DNA-PK kinase is activated by DNA breaks. Once activated, DNA-PK phosphorylates the three DNA-PK subunits (DNA-PK$_{cs}$ in red; Ku70 and Ku80 in blue) and potentially other substrates relevant to DNA repair. P = SQ phosphorylation site. (*Upper figure*) Autophosphorylation of DNA-PK has recently been described. Autophosphorylation of DNA-PK$_{cs}$ is thought to disassociate the complex. Phosphorylated Ku may or may not be retained at DNA repair sites. Autophosphorylation of DNA-PK may be regulatory by changing the association of additional factors. (I) Proteins that may be associated to DNA repair sites by a change in the conformation/occupancy of DNA-PK subunits at DNA ends. (II) Factors that are associated prior to DNA-PK activation may themselves be activated and/or released. (*Lower figure*) Effects of RPA phosphorylation by DNA-PK at DNA repair sites. The single stranded DNA binding protein, RPA, has been implicated in many DNA repair and recombination functions. DNA-PK phosphorylates RPA in vitro and DNA-PK mutant cell lines show an altered RPA hyperphosphorylation response in vivo. RPA hyperphosphorylation occurs on the 34 kDa RPA subunit (yellow). Recent evidence indicates conformational changes in RPA following DNA association and phosphorylation by DNA-PK. As for DNA-PK subunit phosphorylation, this RPA hyperphosphorylation may change the accessibility and activity of other factors. (III) Proteins that associate with conformationally distinct RPA. (IV) Proteins that disassociate following conformational changes in RPA

DNA-PK Phosphorylation by DNA-PK

Each of the DNA-PK subunits has multiple potential phosphorylation sites for DNA-PK, raising the interesting notion that autophosphorylation of the kinase or Ku phosphorylation may be regulatory (Fig. 5, top). Ku and DNA-PK$_{cs}$ subunits of DNA-PK are phosphoryated by DNA-PK (Lees-Miller *et al*, 1990; Boubnov and Weaver, 1995). Ku can be detected as a phosphoprotein in vivo where both Ku70 and Ku80 are radioactively labelled on serine (Yaneva and Busch, 1986). The dependence of DNA-PK for Ku phosphorylation was shown by the demonstration that Ku phosphorylation is significantly reduced in DNA-PK mutants (Boubnov and Weaver, 1995). Complementation of the *scid* mutation by introduction of human chromosome 8 and human DNA-PK$_{cs}$ reconstitutes Ku phosphorylation. Thus, the human DNA-PK$_{cs}$ protein can associate with mouse Ku complexes normally. In these studies, Ku70 is phosphorylated approximately five times greater than Ku80, even though the subunits are present in a 1:1 complex.

Human Ku70 has only two DNA-PK consensus sites (Ser-51 and Ser-319), only one of which is conserved for the mouse Ku70. Therefore, of the two sites, the N-terminal site is likely to be used by DNA-PK. A simple phosphopeptide pattern is generated for Ku70 following pulldown of DNA-PK onto DNA beads that activate the kinase, suggesting that DNA-PK may act at only one phosphorylation site (Jin S and Weaver D, unpublished). Ku phosphorylation by DNA-PK also occurs in antibody complexes with the same stoichiometry. Ku80 phosphorylation is underutilized relative to Ku70 (Boubnov and Weaver, 1995). There are also more DNA-PK consensus sites (five for human Ku80), many of which are not conserved. The DNA-PK dependence of Ku phosphorylation has not yet been demonstrated in vivo. Ku is not abundantly phosphorylated in long term labelling experiments and may be rapidly dephosphorylated (Jin S and Weaver D, unpublished). No increases in Ku phosphorylation were found following intensive γ irradiation, although the number of breaks/cell are considerably limited relative to the amount of DNA-PK (Jin S and Weaver D, unpublished).

Ku phosphorylation may be relevant at the site of repair where DNA-PK is targeted, even though most DNA-PK in the cell may not be activated. Ku phosphorylation may induce conformational changes that might alter the binding affinity for DNA. However, Ku binds to DNA ends whether or not DNA-PK phosphorylation is possible. Both *scid* and MO59J cells have essentially normal levels of DNA end binding activity (Rathmell and Chu, 1994a; Boubnov and Weaver, 1995). On the other hand, DNA end binding may not be the appropriate assay for Ku phosphorylation. Perhaps Ku phosphorylation would change its mobility along DNA, affecting the DNA helicase activity of the complex (Tuteja *et al*, 1994).

A DNA dependent ATPase activity has recently been attributed to Ku (Cao *et al*, 1994). Phosphorylated and dephosphorylated Ku have intrinsically different ATPase activities in vitro. Dephosphorylated Ku has about a 350-fold reduced ATPase activity. Rephosphorylation of this dephosphorylated Ku with purified DNA-PK$_{cs}$ restores the ATPase activity. These results could imply that phosphorylation of Ku by DNA-PK alters the structure of the hetero-dimer and/or its associations with DNA. Another interpretation would be that phosphorylation changes the distribution between complete and disassociated complexes, where Ku free of DNA-PK may act as an ATPase/helicase.

Autophosphorylation is frequently used to inactivate regulatory protein kinases counteracting induction of phosphorylation cascades. The ability to shut off kinases by autoregulation allows better modulation of the inductive signal. Autophosphorylation also occurs for DNA-PK$_{cs}$ and is double stranded DNA dependent (Lees-Miller *et al*, 1990). Furthermore, DNA-PK$_{cs}$ phosphorylation is inhibitory to subsequent DNA-PK activity, supportive of a model of phosphorylation regulating the inactivation or disassociation of the complex (Fig. 5, top). Disassociation or removal of DNA-PK from a repair site may also be important for completion of repair once other repair factors are targeted, activated and/or assembled.

Experimental support for the disassociation of DNA-PK subunits by auto-phosphorylation has recently been accumulated (Chan and Lees-Miller, 1996). Purified DNA-PK is inactivated more than tenfold within 10 minutes coincidently with phosphorylation of all three DNA-PK subunits to relatively equal levels. Ku phosphorylation does not interfere with its ability to reactivate DNA-PK when more DNA-PK$_{cs}$ is added. Autophosphorylation inactivation can be eliminated by the presence of excess levels of another substrate, which may have many implications for multiprotein complexes containing DNA-PK. Because the DNA-PK complex appears to either only be formed or stabilized on DNA (Suwa *et al*, 1994; Chan and Lees-Miller, 1996), it could be evaluated whether autophosphorylation may change Ku:DNA-PK$_{cs}$ associations. Under non-phosphorylating conditions and in the presence of DNA, DNA-PK$_{cs}$ will co-precipitate with Ku using anti-Ku antibodies and Ku will co-precipitate with DNA-PK$_{cs}$ with anti-DNA-PK$_{cs}$ antibody. Phosphorylation releases the DNA-PK$_{cs}$:Ku association as monitored in the co-precipitation experiments (Chan and Lees-Miller, 1996).

An intriguing idea consistent with these results is that DNA-PK complexes may be disassociated by phosphorylation (Fig. 5, top). Whether phosphory-lated Ku can still bind to DNA remains to be determined. Phosphorylated Ku70 binds to DNA in Southwestern analysis (Chan and Lees-Miller, 1996), but this interaction may not be physiologically relevant to Ku:DNA associa-tions in repair, because the Southwestern DNA binding activity maps to a non-conserved C-terminal epitope in human Ku70. Phosphorylation of Ku could be sufficient to release DNA-PK$_{cs}$ in vivo, implying that DNA-PK$_{cs}$ independent joining steps may follow. There currently is significant precedence for dif-ferences in joining mechanism requirements between these subunits in V(D)J recombination. Disassociation of DNA-PK complexes may also alter the acces-sibility for other factors (Fig. 5, top). Two alternatives need exploration. First-ly, there may be an increased association of factors that may recognize phos-phorylated Ku, or Ku in the absence of DNA-PK$_{cs}$. Secondly, there could be a release of other proteins in conjunction with loss of DNA-PK. These mechanistic variations may become significant regarding the structures of the joints formed.

RPA and DNA-PK

A significant DNA-PK substrate may be the single stranded DNA binding protein complex, RPA, that is required for DNA synthesis (Wold and Kelly, 1988; Fairman and Stillman, 1988; Wobbe *et al*, 1987) and appears to be increasingly important for several types of DNA repair reactions (Coverley *et al*, 1991; Liu and Weaver, 1993; Carty *et al*, 1994; He *et al*, 1995). RPA 70 is a DNA binding subunit, whereas p34 and p11 have no known functions as yet, although these three subunits stably appear in a 1:1:1 complex in all eukaryotes. The 34 kDa subunit is hyperphosphorylated in S and G_2 phases of the cell cycle (Din *et al*, 1990) and in response to radiation damage (Liu and Weaver, 1993). RPA kinases that cause this high level of phosphorylation have not yet been identified, but both cyclin dependent kinases and DNA-PK may fit in this role. DNA-PK phosphorylates RPA during SV40 DNA synthesis in vitro (Brush *et al*, 1994). In addition, DNA-PK has been purified as an RPA kinase activity (Pan *et al*, 1994).

The *scid* mutation causes a defect in the RPA hyperphosphorylation following ionizing radiation, possibly meaning that DNA-PK is an inducible RPA kinase or regulates one (Boubnov and Weaver, 1995). RPA is directly phosphorylated by DNA-PK (Anderson, 1993), but the distribution of subunit phosphorylation and phosphopeptide mapping of sites has not been conducted. scid cells show restricted 34 kDa phosphorylation, where the mobility of this protein is only slightly modified, and yield underphosphorylated RPA 34 kDa that is approximately equal to cyclin dependent kinase phosphorylation alone in vitro. Thus, in the absence of DNA-PK, incomplete phosphorylation may result from the action of cyclin dependent kinases alone. Similarly, *scid* cells complemented with human chromosome 8 that contains the human DNA-PK gene were found to restore the radiation induced RPA hyperphosphorylation to *scid* cells. Thus, the 34 kDa subunit of RPA may be a true substrate for DNA-PK in vivo.

RPA hyperphosphorylaton may change how RPA complexes complete functions in DNA repair and replication by inducing conformational alterations, DNA binding properties or protein association motifs (Fig. 5, bottom). It is known that RPA interacts with TP53 (p53) such that TP53 inhibits RPA functions (Dutta *et al*, 1993), and these interactions may be phosphorylation dependent. In addition, XPA and XPG proteins of ultraviolet radiation damage repair form stable complexes with RPA in vivo (He *et al*, 1995). RPA binding to single stranded DNA induces a change in RPA structure such that DNA-PK phosphorylation is stimulated (Blackwell LJ, Borowiec JA and Mastrangelo IA, unpublished). Thus, DNA-PK activity may modify RPA complexes and change their affinity for DNA replication or DNA repair sites. An interesting model is that DNA-PK activity on RPA may induce a conformational change in RPA that leads to an altered or stabilized association with DNA. An example of how this could be relevant to repair is illustrated for DSB repair (Fig. 5, bottom). RPA binding/phosphorylation might dictate the accessibility and association of other repair co-factors just as is hypothesized for Ku.

Additional Factors Affecting DNA-PK Dependent DNA Repair Pathways

Like RPA, other factors important to the completion of DNA synthesis and recombination may be substrates for DNA-PK. As yet, no clear definition of these proteins has emerged, but experiments with *S cerevisiae* may indicate the future direction for the identification of additional genes.

Study of the ionizing radiation sensitive mutations of yeast revealed some of the differences and similarities between phenotypes illustrating common pathways of recombination and genes involved in subsets of these pathways (Petes *et al*, 1991). Recently, investigations into the relative role of recombination genes between meiotic recombination and end joining have occurred. Since end joining does not require any significant DNA homology, one notion has been that the protein machinery involved in this pathway may be totally different. On the other hand, the property of protein associations may be relevant for both pathways if these associations can occur without a requirement for DNA homology. A group of yeast genes that were previously categorized as mainly influencing homologous recombination may play an important part in the DNA end joining pathways involving DNA-PK complexes.

The *RAD50*, *MRE11* and *XRS2* genes have been implicated in a broad spectrum of DNA repair and recombination processes in yeast, such as radiation repair, formation of meiotic DSBs associated with recombination and mitotic recombination. Whereas *rad52Δ* strains have reduced levels of spontaneous mitotic recombination, *rad50*, *mre11* and *xrs2* mutant strains are hyperrecombinational (Alani *et al*, 1990; Ajimura *et al*, 1992; Ivanov *et al*, 1992; Ogawa *et al*, 1995). Rad50 is an ATP dependent double stranded DNA binding protein, containing heptad repeats that might be important to protein associations (Alani *et al*, 1989; Raymond and Kleckner, 1993). Mre11 and Xrs2 as yet have no assigned biochemical functions. Mutations in each of these three genes have a distinct phenotype in meiotic recombination events. In the presence of the *rad50S* type mutations, DSBs are formed at meiotic hot spots for recombination, but the nucleolytic degradation ordinarily associated with these breaks is not found (Sun *et al*, 1989, 1991; Cao *et al*, 1990;). These experiments would imply that the *RAD50/MRE11/XRS2* group might be important to the formation or processing of DSBs in homologous recombination events. Hyperrecombinational effects in mitotic cells could also result from an inappropriate processing of DSBs.

As discussed above for the yeast Ku mutations, in the absence of homologous recombination, HO induced DSBs at MAT need to be repaired by an end joining pathway (Kramer *et al*, 1994). In DNA end joining, the products of the *RAD50/MRE11/XRS2* group are each required for cell survival when HO breaks are formed and homologous recombination is inactivated due to the absence of donor sequences ("cassetteless") (Moore and Haber, 1996). By contrast, a number of other genes implicated in homologous recombination appear to be unnecessary for the *RAD50/MRE11/XRS2* dependent pathway. These other genes include those encoding the recA related proteins, Rad51 or Rad57, Rad54, Rad52 and the nucleotide excision endonuclease, Rad1 (Moore and Haber, 1996).

The relationship between Ku and *RAD50/MRE11/XRS2* for end joining was also evaluated by epistasis analysis for the Ku dependent DNA repair. It was observed that *ku80Δ*, *hdf1Δ* and *ku80Δ hdf1Δ* were epistatic with *rad50Δ* but not with *rad52Δ* (Milne *et al*, 1996). Thus, Ku is in the same pathway as the Rad50/Mre11/Xrs2 group and independent of homologous recombination. This DNA repair mechanism is probably analogous to end joining in mammalian cells as the ability of yeast cells to rejoin plasmid DNA ends is dependent on the same proteins. Ku and Rad50, as well as double mutants of Ku and Rad50, all show significantly reduced plasmid rejoining (Milne *et al*, 1996). These studies point towards a possible co-ordination of Ku and the Rad50/Mre11/Xrs2 group at DNA repair sites. Considering that the Rad50/Mre11/Xrs2 group may promote nucleolytic degradation, Ku may augment the accessibility of DNA ends and/or serve as a docking site for these proteins. The appreciation of the level of protein associations in this epistasis group will be valuable in the future. So far, Rad50 and Mre11 have been shown to interact by two hybrid fusion protein analysis (Johzuka and Ogawa, 1995).

The role of Rad50/Mre11/Xrs2 in mammalian repair and recombination pathways will be interesting. Since there is a extensive parallel between repair pathways in the organisms as diverse as yeast and humans, it may even be expected that these proteins would be important for the same pathways as DNA-PK, such as V(D)J recombination and radiation damage repair. A mammalian homologue of MRE11 has been reported (Petrini *et al*, 1995). Human MRE11 bears a striking similarity to the yeast homologue, although the human protein is not functional in complementing yeast *mre11* mutations.

Additional DNA-PK roles may also be revealed by study of substrates of DNA-PK identified in in vitro experiments. For example, topoisomerase II is an abundant nuclear enzyme that relaxes torsional stress by the introduction and resolution of DSBs via an enzyme-DNA intermediate in an ATP dependent reaction. In mammalian cells, topoisomerase IIa is associated with newly replicated DNA (Nelson *et al*, 1986) and a phosphoprotein with increased activity in G_2 (Saiijo *et al*, 1992). Topoisomerase II is also phosphorylated by DNA-PK in vitro (Anderson, 1993). Therefore, topoisomerase II is an interesting candidate protein to be regulated by DNA-PK, since it may be involved in many of the same functions.

Topoisomerase targeted inhibitors interfere with the DNA breakage rejoining steps of topoisomerase action, frequently stabilizing a "cleavable complex" involving the enzyme-DNA intermediate. One such inhibitor, etoposide (VP-16-23), severely diminishes topoisomerase II activity by creating these modified DSBs. Ataxia telangiectasia cells are hypersensitive to etoposide (Henner and Blazka, 1986; Smith *et al*, 1986), indicating that the *ATM* gene may be important to the surveillance mechanisms including topoisomerase II strand breaks. The effects of etoposide have been explored in DNA-PK mutants. Ku deficient cells are significantly hypersensitive to etoposide relative to parental control cells and to radiation resistant derivatives with introduced Ku80 genes (Inoue S, Lindner K, Staunton J, Liu VF and Weaver DT, unpublished). Several mutants in the *XRCC7* group that are DNA-PK$_{cs}$ deficient show no increased sensitivity to etoposide. Both murine *scid* fibroblasts and DNA-PK$_{cs}$ deficient human glioblastoma MO59J cells are no more sensitive to etoposide inhibition than isogenic wild type or other control cells (I Inoue S, Lindner K, Staunton J, Liu VF and Weaver DT, unpublished). In addition, etoposide sensitizes cells to radiation damage, particularly for cells arrested in G$_2$ (Giocanti *et al*, 1993). Thus, etoposide disruption of topoisomerase II may reveal an important difference between Ku and DNA-PK. Perhaps Ku is redundant with topoisomerase II for some essential cellular functions.

DNA-PK, CELL CYCLE CHECKPOINTS AND APOPTOSIS

As discussed elsewhere in this volume (Rotman and Shiloh, Wahl *et al* and Weinert), the protein kinases that are structurally the most related to DNA-PK$_{cs}$ control the signalling of cell cycle checkpoints induced by radiation damage (Hartwell and Kastan, 1994; Enoch and Norbury, 1995). In addition, several of these other PIK family members have significant roles in enforcing genome stability at similar checkpoints that may be triggered by spontaneous breaks and errors in the S phase DNA replication machinery. Therefore, there is some expectation that related genes such as DNA-PK may also be involved in signalling the activation of cell cycle checkpoints.

A second theme relevant to the exploration of cell cycle checkpoints is the adaptation of apoptosis, to remove damaged cells from tissue before promoting the progression of a tumorigenic pathway. Thus, the functioning of DNA repair mechanisms and the progress of apoptosis may be viewed as being diametrically opposed (Fig. 6). An example of the impetus for connecting these functions is the involvement of the tumour suppressor, TP53, in each of these pathways: cell cycle regulated checkpoints, apoptosis and frequent mutation in tumours. Thus, TP53 dependent mechanisms may co-ordinate the commitment of cells regarding life (and repair) or death. Experiments with DNA-PK mutants have clarified the role of this complex in DNA repair, cell cycle-regulated checkpoints and apoptosis. If the radiation sensitivity of DNA-PK mutants were attributable to checkpoint errors, as appears to be true for the ATM kinase deficient cells, then radiation induced cell cycle checkpoints might be disrupted in these mutants. Radiation induced cell cycle checkpoints occur in G$_1$, S, G$_2$ and G$_2$/M phases (Hartwell and Kastan, 1994).

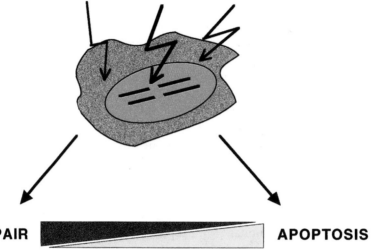

Fig. 6. Distinct mechanisms of repair and apoptosis in eukaryotes. Cells treated with ionizing radiation respond in multiple ways leading to different outcomes. For certain cell types, irradiated cells activate cell cycle regulated checkpoint mechanisms and conduct DNA repair for cell survival. In other circumstances, cells are prone to apoptosis and inactivate DNA repair in conjunction with the apoptotic pathway

DNA-PK mutants were directly examined for the operation of the TP53 dependent radiation induced cell cycle checkpoint (Inoue S, Lindner K, Staunton J, Liu VF and Weaver DT, unpublished). *scid* embryonic fibroblasts showed the same delay in entry into S phase as wild type controls, in contrast to $TP53^{-/-}$ mouse embryo fibroblast and *ATM* deficient cells, which are desensitized to the G_1/S checkpoint (Kastan *et al*, 1991, 1992; Lu and Lane, 1993). In support of the normal operation of a TP53 dependent checkpoint, *scid* cells also increase TP53 levels from 1 to 4 hours following irradiation (Inoue S, Lindner K, Staunton J, Liu VF and Weaver DT, unpublished). Ku deficient cell lines such as *sxi-3* are mutated for the endogenous TP53 protein, and therefore checkpoint functions may be altered by *TP53* mutation alone. To circumvent the problem with TP53, Ku deficient cells were transfected with an expression cassette encoding a temperature sensitive TP53 protein that has "wild type" properties at reduced temperature (33°C) and "mutant" properties at elevated temperature (37°C) (Michalovitz *et al*, 1990; Ullrich *et al*, 1992). When Ku deficient cells are tested at the "wild type" TP53 temperature, the operation of the radiation induced G_1/S checkpoint was intact and therefore did not require Ku (Inoue S, Lindner K, Staunton J, Liu VF and Weaver DT, unpublished).

As was demonstrated for G_1/S, DNA-PK mutant cell lines also retain the functioning of a G_2 stage checkpoint for radiation damage. *XRCC7* group mutants (murine *scid* and human MO59J cells) and Ku deficient cell lines each showed the activity of a delay in progression through G_2 phase following ionizing radiation (Inoue S, Lindner K, Staunton J, Liu VF and Weaver DT, unpublished). In addition, irradiated DNA-PK mutants accumulate in G_2 to a much greater extent than wild type controls. The accumulation was demonstrated to be in G_2 rather than late S phase by bromodeoxyuridine labelling experiments. It is interesting that the extended cell cycle delay occurs in G_2, because other experiments support the notion that this is also the cell cycle stage where DNA-PK is most active. S and G_2 phase cells have three- to fivefold increased DNA-PK activity relative to G_1 phase.

In yeast, Ku deficient strains have also been examined for a loss of potential cell cycle checkpoint functions. Both *hdf1Δ* and *ku80Δ* strains are viable (Feldmann and Winnacker, 1993; Milne *et al*, 1996; Siede *et al*, 1996). Both the radiation induced and hydroxyurea sensitive checkpoints that reveal functions for related PIK family kinases of yeast, such as MEC1/ESR1 and RAD3, are functioning as normal in *hdf1Δ* strains (Siede *et al*, 1996). *hdf1Δ* and *ku80Δ* strains accumulate at a large budded stage at an elevated growth temperature of 37°C (Feldmann and Winnacker, 1993). In conjunction with cell cycle arrest at elevated temperatures, *hdf1Δ* strains show increased DNA content (>4N) with time (Feldmann and Winnacker, 1993). However, the increased DNA content phenotype is not reproducible in every *hdf1Δ* strain. Disregulation of DNA synthesis by absence of Ku might be indicative of a normal involvement of Ku in the regulation of initiation or termination signals of DNA synthesis. This connection could influence the telomere shortening defect of yeast Ku mutants.

Apoptosis

Programmed cell death occurs in multistep pathways where inactivation of mechanisms preserving the cellular integrity, such as DNA repair of the genome, may be expected to be observed (Ellis *et al*, 1991). Two well studied aspects of apoptosis are proteolysis and DNA degradation. Specific proteolysis has been implicated in several contexts for apoptotic mechanisms, as demonstrated by mutational analysis in *C elegans* and *Drosophila* (Yuan *et al*, 1993). A family of cysteine proteases related to the interleukin 1β converting enzyme (ICE) have been associated with apoptosis, and a number of experiments suggest that proteolysis of key substrates drives apoptosis forward. In particular, overexpression of crmA (an ICE inhibitor) prevents the induction of apoptosis by certain ICE family members (Gagliardini *et al*, 1994). A second common feature of apoptosis is the rapid degradation of DNA upon apoptotic signalling by several inducers. DNA laddering indicates destruction of the integrity of the genome beyond repair.

If DNA repair functions are relevant to counteracting the direction of apoptosis, then elimination or modification of these activities would be expected to promote the apoptotic effect (Fig. 6). In support of this model, the functioning of several nuclear proteins implicated in DNA repair pathways has been examined. DNA-PK$_{cs}$ proteolytic cleavage occurs during the time course of apoptosis to specific sized fragments of approximately 150 and 250 kDa (Casciola-Rosen *et al*, 1995) (Inoue S, Zhang C and Weaver D, unpublished). With extended time, additional proteolysis destroys all of the DNA-PK$_{cs}$ signal. By contrast, Ku proteins are not proteolyzed by apoptotic signals in cells where DNA-PK$_{cs}$ is cleaved. In other studies, Ku is degraded during apoptosis in other cells (Ajmani *et al*, 1995). Specific Ku degradatory pathways may be found in certain cell types, such as neutrophils. The mechanism of proteolysis of DNA-PK is consistent with the timing of DNA laddering. Several features of the cleaved DNA-PK$_{cs}$ subunit are interesting. The 250 kDa fragment retains the capacity to bind to Ku, although not to the same extent as the full length DNA-PK (Inoue S and Weaver S, unpublished). In conjunction with proteolysis of DNA-PK, significantly increased levels of DNA-PK$_{cs}$ and Ku phosphorylation are also observed, indicating that the 250 kDa polypeptide has the Ku binding site and the kinase domain (see Fig. 1). These experiments indicate functional inactivation of the DNA-PK complex by a combination of proteolysis and autophosphorylation.

Synergistic Functions of TP53 (p53) and DNA-PK

TP53 and DNA-PK may have combinatorial effects on genome stability. *TP53$^{-/-}$* mice have accelerated and high frequencies of thymic lymphomas (Donehower *et al*, 1992; Jacks *et al*, 1994). Molecular events influencing the formation of these tumours may be limited to the effective operation of cell cycle checkpoints and apoptosis or might include errors in DNA repair pathways. *scid* mice that are DNA-PK deficient have relatively normal incidences of tumours. In addition, *scid* lymphocytes are eliminated by apoptotic mechanisms when the V(D)J recombination pathway fails to be completed. Thus, if DNA-PK functions primarily in DNA repair, defects in any of the DNA-PK subunits may accelerate the impact of *TP53* mutations by increasing the level of spontaneous unrepaired damage or the level of DSBs initiated but not completed during V(D)J recombination. Double mutant animals (*scid TP53$^{-/-}$*) have a significantly earlier onset of spontaneous thymic lymphomas than either *TP53$^{-/-}$* or *scid* animals (Guidos C and Jacks T, personal communication).

scid animals have an increased levels of radiation induced thymic lymphomas (Lieberman *et al*, 1992; Murphy *et al*, 1994), and radiation of neonatal *scid* mice uniformly produces these tumours (Danska *et al*, 1994). The low radiation doses that cause lymphomas in *scid* animals are possibly high enough to produce DSBs that are not properly repaired but too low to trigger apoptosis effeciently. Chromosomal translocations may form from these breaks and lead to tumorigenesis. The effects of a *scid* DNA damage defect on tumorigenesis in vivo may be masked by a strongly operating checkpoint mechanism.

The double mutant mice (*scid TP53$^{-/-}$*) are also interesting in terms of the developmental consequences of ionizing radiation. In the thymocyte maturation pathway, *scid* gene rearrangement events can be rescued by low doses of ionizing radiation (Danska *et al*, 1994; Murphy *et al*, 1994). Null mutations in *TP53* in the *scid* background rescue thymocyte development phenotypically and cause increased levels of normal gene rearrangements (Bogue *et al*, 1996). Since there does not appear to be a requirement for DNA-PK in the activation of radiation induced cell cycle checkpoint mechanisms, an alternative attractive hypothesis explaining these observations is that TP53 ordinarily activates DNA repair functions that are redundant with DNA-PK (Bogue *et al*, 1996).

TRANSCRIPTIONAL CONTROL BY Ku/DNA-PK

Several lines of evidence have emerged suggesting that DNA-PK may regulate transcription. Ku binds to particular DNA sequences of double stranded DNA internal to DNA ends and therefore in a sequence specific manner (Knuth *et al*, 1990; Dvir *et al*, 1992; Lai and Herr, 1992; Falzon *et al*, 1993; Gottlieb and Jackson, 1993; Messier *et al*, 1993; Toth *et al*, 1993; Hoff *et al*, 1994;; Roberts *et al*, 1994; Genersch *et al*, 1995). Many of these associations are interesting because they also are binding sites for known transcription factors.

The experimental motivation for associating DNA-PK functions with transcription has also come from the findings that Ku and/or DNA-PK activity is frequently derived in biochemical purifications for transcriptional complexes. Ku is frequently purified as a transcription factor based on DNA binding to transcriptionally relevant sequences. In addition, several of the in vitro substrates that are efficiently phosphorylated by DNA-PK are transcription factors (Anderson, 1993). DNA-PK phosphorylated proteins include HSP90, c-fos, c-jun, Sp1, Oct-1, c-myc, Oct-2, RNA polymerase II C-terminal fragment, TFIID, c-myc, CTF/NF-I, SRF, TP53 and several others (Bannister *et al*, 1994; Gottlieb and Jackson, 1993; Iijima *et al*, 1992; Jackson *et al*, 1990; Lees-Miller *et al*, 1990; Liu *et al*, 1993; Peterson *et al*, 1992). The significance of Ku-DNA interactions towards transcriptional regulation is yet to be explained. Association with transcription complexes alone would not be convincing towards the involvement of DNA-PK in transcriptional control. Thus, more recent evidence has been reported evaluating a direct function of DNA-PK in regulating transcription at a variety of RNA polymerase I or II promoters.

DNA-PK may function in repressing transcription of rDNA. Ku was biochemically isolated based on an rDNA transcription function. Ku acts as a negative regulator of RNA polymerase I preinitiation complexes that is overcome by the activation of an antirepressor, UBF, that is itself a transcription factor (Kuhn *et al*, 1993; Hoff *et al*, 1994). Another factor, EBF1, was purified from serum deprived cells and shown to block RNA polymerase I transcription (Niu and Jacob, 1994). In addition, RNA polymerase I from *Xenopus* is inhibited by added DNA-PK in vitro via the transcription factor Rib1/SL1 (Labhart, 1995). It appears that the repression functions occur by DNA-PK phosphorylation of the associated or co-localized transcription factors. Importantly, localization of DNA-PK activity near to RNA polymerase I initiation sites represses transcription (Kuhn *et al*, 1995; Labhart, 1995).

For RNA polymerase II dependent promoters, there is also evidence for DNA-PK involvement by negative regulation. Evidence has also been accumulated for alterations in RNA polymerase II dependent transcription functions in either the Ku deficient or DNA-PK$_{cs}$ deficient cell lines. Several retroviral long terminal repeat (LTR) promoters contain a mixture of activating and repressing elements. For mouse mammary tumour virus (MMTV), a negative regulatory element has been discovered that represses viral transcription (Lee *et al*, 1991; Giffin *et al*, 1994). In trying to understand the basis for transcriptional repression, it was discovered that Ku binds to the negative regulatory element (NRE1) to the extent that it was the principal protein purified from human Jurkat cell extracts (Giffin *et al*, 1996). Similarly, Ku binds to the site of a repressive transcription control element of the human T lymphotropic virus type I LTR (Okumura *et al*, 1995).

Ku association to NRE1 was studied by the creation of NRE1 site–DNA minicircles so that double stranded DNA association could be monitored independently of the Ku-DNA end binding properties (Giffin *et al*, 1996). Ku binds directly to NRE1 containing sequences without DNA ends. Addition of Mg^{++} facilitated translocation of Ku along the DNA, supportive of DNA helicase properties of Ku. DNA-PK was activated by NRE1 containing minicircles and phosphorylated co-localized glucocorticoid receptor and glutathione S transferase-Oct transcription factors dependent on their binding sites. Both glucocorticoid receptor and Oct1 are regulators of MMTV expression and Oct1 has been shown to be phosphorylated by DNA-PK (Anderson, 1993).

To connect DNA-PK functions with NRE1 transcriptional repression more directly, transient transfection assays were used to test the dependence on DNA-PK (Giffin *et al*, 1996). An NRE1 site in the MMTV promoter induces an approximately tenfold effect in repressing the action of glucocorticoid. In the XR-V15B Ku deficient cells, NRE1 sites had no repression activity. Complementation of these cells with Ku80 led to the re-establishment of transcriptional repression that was NRE1 dependent. Similarly, the DNA-PK deficient *scid* cells also showed a lack of repression of NRE1 dependent promoter. Therefore, Ku and DNA-PK activity have corresponding properties for transcriptional repression. Additional examples of transcriptional repression by DNA-PK continue to accumulate. Overexpression of Ku70 in rat cells suppresses the induction of HSP70 transcription (Li GC *et al*, 1995). In addition, the human parathyroid hormone gene negative regulatory elements involve the association between Ku and REF1 in a Ca^{++} responsive manner (Chung *et al*, 1996). DNA-PK may regulate a wide range of inducible promoters by the same or similar transcriptional repression mechanisms.

In support of a broad role of DNA-PK in transcriptional regulation, recent evidence indicates that DNA-PK co-purifies with high molecular weight complexes containing RNA polymerase II that were purified on the basis of transcription activity that is TFB, TFIIB and TFIIH dependent (Maldonado *et al*,

1996). Ku was present in stoichiometric amounts with RNA polymerase II and complexed even in the presence of DNase and ethidium bromide, arguing that the complex is not associated because of DNA. Because these complexes were observed to also contain additional DNA repair proteins, such as DNA polymerase ε, Rad51, RPA and replication factor C, possibly there is a coordinated transcription control/DNA repair complex to respond to DNA damage. Alternatively, more specialized subcomplexes may be utilized.

Ultraviolet radiation damage repair and RNA polymerase II transcription have been linked on the mechanistic level and by the involvement of several of the xeroderma pigmentosa complementation group genes in transcription functions (Buratowski, 1993; Feaver *et al*, 1993). For DNA-PK, transcriptional repression is likely to occur wherever DNA-PK is activated. Thus, transcriptional repression might be a normal feature of DNA damage responses (Kuhn *et al*, 1995). For RNA polymerase I transcription with high promoter activity, a temporary shutdown of potentially faulty transcription might be useful. On the other hand, DNA repair and transcriptional repression may operate essentially independently and not be functionally related.

SUMMARY

The DNA dependent protein kinase (DNA-PK) is a trimeric nuclear complex consisting of a large protein kinase and the Ku heterodimer that regulates kinase activity by its association with DNA. Recent findings have shown structural similarities between DNA-PK and a family of lipid and putative protein kinases (PIK family). DNA-PK is one of the PIK members known to be a protein kinase with clearly identified effector subunits. A broad range of observations link DNA-PK to dual roles in double strand DNA break (DSB) repair and transcription. Unlike its most closely related PIKs, DNA-PK is not required for activating cell cycle regulated DNA damage signalling mechanisms. Instead, the phenotypes and biochemical properties of DNA-PK are most consistent with functions in DSB repair and joining steps in recombination mechanisms. DNA-PK is asssociated with RNA polymerase II and RNA polymerase I transcription complexes, where it most frequently has a negative regulatory role.

References

Ajimura M, Leem S-H and Ogawa H (1992) Identification of new genes required for meiotic recombination in *Saccharomyces cerevisiae*. *Genetics* **133** 51–66

Ajmani AK, Satoh M, Reap E, Cohen PL and Reeves WH (1995) Absence of autoantigen Ku in mature human neutrophils and human promyelocytic leukemia line (HL60) cells and lymphocytes undergoing apoptosis. *Journal of Experimental Medicine* **181** 2049–2058

Alani ES, Subbiah S and Kleckner N (1989) The yeast RAD50 gene encodes a predicted 153-kd protein containing a purine nucleotide binding domain and two large heptad-repeat regions. *Genetics* **116** 541–545

Alani E, Padmore R and Kleckner N (1990) Analysis of wild-type and *rad50* mutants of yeast suggests an intimate relationship between meiotic chromosome synapsis and recombination. *Cell* **61** 419–436

Allalunis-Turner MJ, Barron GM, Day RS, Dobler KD and Mirzayans R (1993) Isolation of two cell lines from a human malignant glioma specimen differing in sensitivity to radiation and chemotherapeutic drugs. *Radiation Research* **134** 349–354

Allalunis-Turner MJ, Lintott LG, Barron GM, Day RS and Lees-Miller SP (1996) Lack of correlation between DNA-dependent protein kinase activity and tumor cell radiosensitivity. *Cancer Research* **55** 5200–5202

Allen JB, Zhou Z, Friedberg EC and Elledge SJ (1994) The SAD1/RAD53 protein kinase controls multiple checkpoints and DNA damage-induced transcription in yeast. *Genes and Development* **8** 2401–2415

Anderson CW (1993) DNA damage and the DNA-activated protein kinase. *Trends in Biochemical Sciences* **18** 433–437

Anderson CW and Lees-Miller SP (1992) The nuclear serine/threonine protein kinase DNA-PK. *Critical Reviews in Eukaryotic Gene Expression* **2** 283–314

Banga S, Velazquez A and Boyd JB (1991) P transposition in *Drosophila* provides a new tool for analyzing postreplication repair and double-strand break repair. *Mutation Research* **255** 79–88

Banga S, Hall K, Sandhu A, Weaver D and Athwal R (1994) Complementation of V(D)J recombination defect and x-ray sensitivity of *scid* mouse cells by human chromosome 8. *Mutation Research* **315** 239–247

Bannister AJ, Gottlieb T, Kouzarides T and Jackson SP (1994) c-Jun is phosphorylated by the DNA-dependent protein kinase in vitro: definition of the minimal kinase recognition motif. *Nucleic Acids Research* **21** 1289–1295

Beall EL and Rio DC (1996) *Drosophila* IRBP/Ku p70 corresponds to the mutagen-sensitive mus309 gene and is involved in P-element excision in vivo. *Genes and Development* **10** 921–933

Beall EL, Admon A and Rio DC (1994) A *Drosophila* protein homologous to the human p70 Ku autoimmune antigen interacts with the P transposable element inverted repeats. *Proceedings of the National Academy of Sciences of the USA* **91** 12681–12685

Biedermann KA, Sun J, Giaccia AJ, Tosto LM and Brown JM (1991) *scid* mutation in mice confers hypersensitivity to ionizing radiation and a deficiency in DNA double-strand break repair. *Proceedings of the National Academy of Sciences of the USA* **88** 1394–1397

Blunt T, Finnie NJ, Taccioli GE et al (1995) Defective DNA-dependent protein kinase activity is linked to V(D)J recombination and DNA repair defects associated with the murine scid mutation. *Cell* **80** 813–823

Bogue MA, Zhu C, Aguilar-Cordova E, Donehower LA and Roth DB (1996) p53 is required for both radiation-induced differentiation and rescue of V(D)J rearrangement in scid mouse thymocytes. *Genes and Development* **10** 553–565

Bosma GC, Custer RP and Bosma MJ (1983) A severe combined immunodeficiency mutation in the mouse. *Nature* **301** 527–530

Boubnov NV and Weaver DT (1995) Scid cells are deficient in Ku and RPA phosphorylation by the DNA-dependent protein kinase. *Molecular and Cellular Biology* **15** 5700–5706

Boubnov NV, Hall KT, Wills Z et al (1995) Complementation of the ionizing radiation sensitivity, DNA end binding and V(D)J recombination defects of double-strand break repair mutants by the p86 Ku autoantigen. *Proceedings of the National Academy of Sciences of the USA* **92** 890–894

Boyd JB and Setlow RB (1976) Characterization of postreplication repair in mutagen-sensitive strains of *Drosophila melanogaster*. *Genetics* **84** 507–526

Boyd JB, Golino MD, Nguyen TD and Green MM (1976) Third chromosome mutagen-sensitive mutants of *Drosophila melanogaster*. *Genetics* **84** 485–506

Boyd JB, Golino MD, Shaw KES, Osgood CJ and Green MM (1981) Third chromosome mutagen-sensitive mutants of *Drosophila melanogaster*. *Genetics* **97** 607–623

Brush GS anderson CW and Kelly TJ (1994) The DNA-activated protein kinase is required for the phosphorylation of replication protein A during simian virus 40 DNA replication. *Proceedings of the National Academy of Sciences of the USA* **91** 12520–12524

Buratowski S (1993) DNA repair and transcription: the helicase connection. *Science* **260** 37–38

Cao L, Alani E and Kleckner N (1990) A pathway for generation and processing of double-strand breaks during meiotic recombination in *S. cerevisiae*. *Cell* **61** 1089–1101

Cao QP, Pitt S, Leszyk J and Baril EF (1994) DNA-dependent ATPase from HeLa cells is related to human Ku autoantigen. *Biochemistry* **33** 8548–8557

Carter T and Anderson CW (1991) The DNA-activated protein kinase. *Progress in Molecular and Subcellular Biology* **12** 37–47

Carter TH, Vancurova IS, Lou W and DeLeon SP (1990) A DNA-activated protein kinase from HeLa cell nuclei. *Molecular and Cellular Biology* **10** 6460–6471

Carty MP, Zernik-Kobak M, McGrath S and Dixon K (1994) UV light-induced DNA synthesis arrest in HeLa cells is associated with changes in phosphorylation of human single-stranded DNA-binding protein. *EMBO Journal* **13** 2114–2123

Casciola-Rosen LA, Anhalt GJ and Rosen A (1995) DNA-dependent protein kinase is one of a subset of autoantigens specifically cleaved early during apoptosis. *Journal of Experimental Medicine* **182** 1625–1634

Chan DW and Lees-Miller SP (1996) The DNA-dependent protein kinase is inactivated by autophosphorylation of the catalytic subunit. *Journal of Biological Chemistry* **271** 8936–8941

Chen F, Peterson SR, Story MD and Chen DJ (1996) Disruption of DNA-PK in Ku80 mutant xrs-6 and the implications in DNA double-strand break repair. *Mutation Research* **362** 9–19

Chou C, Wang J, Knoth M and Reeves W (1992) Role of a major autoepitope in forming the DNA binding site of the p70(Ku) antigen. *Journal of Experimental Medicine* **175** 1677–1684

Choulika A, Perrin A, Dujon B and Nicolas J-F (1995) Induction of homologous recombination in mammalian chromosomes by using the I-SceI system of *Saccharomyces cerevisiae*. *Molecular and Cellular Biology* **15** 1963–1973

Chung U, Igarashi T, Nishishita T *et al* (1996) The interaction between Ku antigen and REF1 protein mediates negative gene regulation by extracellular calcium. *Journal of Biological Chemistry* **271** 8593–8598

Counter CM, Avilon AA, LeFeuvre CE *et al* (1992) Telomere shortening associated with chromosome instability is arrested in immortal cells which express telomerase activity. *EMBO Journal* **11** 1921–1929

Coverley D, Kenny MK, Munn M, Rupp WD, Lane DP and Wood RD (1991) Requirement for the replication protein SSB in human DNA excision repair. *Nature* **349** 538–541

Danska JS, Pflumio F, Williams CJ, Huner O, Disck JE and Guidos CJ (1994) Rescue of T cell-specific V(D)J recombination in SCID mice by DNA-damaging agents. *Science* **266** 450–455

deVries E, van Driel W, Bergsma WG, Arnberg AC and van der Vliet PC (1989) HeLa nuclear protein recognizing DNA termini and translocating on DNA forming a regular DNA-multimeric protein complex. *Journal of Molecular Biology* **208** 65–78

Din S, Brill SJ, Fairman M and Stillman B (1990) Cell-cycle-regulated phosphorylation of DNA replication factor A from human and yeast cells. *Genes and Development* **4** 968–977

Donehower L, Harvey M, Slagle B *et al* (1992) Mice deficient for p53 are developmentally normal but susceptible to spontaneous tumors. *Nature* **356** 215–221

Dutta A, Ruppert JM, Aster JC and Winchester E (1993) Inhibition of DNA replication factor RPA by p53. *Nature* **365** 79–82

Dvir A, Peterson SR, Knuth MW, Lu H and Dynan W (1992) Ku autoantigen is the regulatory component of a template-associated protein kinase that phosphorylates RNA polymerase II. *Proceedings of the National Academy of Sciences of the USA* **89** 11920–11924

Eastman QM, Leu TMJ and Schatz DG (1996) Initiation of V(D)J recombination in vitro: the 12/23 rule. *Nature* **380** 85–88

Ellis RE, Yuan J and Horvitz HR (1991) Mechanisms and functions of cell death. *Annual Review of Cell Biology* **7** 663–698

Engels WR (1989) P elements in *Drosophila*, In: Berg DE and Howe MM (eds). *Mobile DNA*, pp 437–484, American Society of Microbiology, Washington DC

Enoch T and Norbury C (1995) Cellular responses to DNA damage: cell-cycle checkpoints, apoptosis and the roles of p53 and ATM. *Trends in Biochemical Sciences* **20** 426–430

Errami A, Smider V, Rathmell WK. *et al* (1996) Ku86 defines the genetic defect and restores X-ray resistance and V(D)J recombination to complementation group 5 hamster cell mutants. *Molecular and Cellular Biology* **16** 1519–1526

Fairman MP and Stillman B (1988) Cellular factors required for multiple stages of SV40 DNA replication *in vitro*. *EMBO Journal* **7** 1211–1218

Falzon M, Fewell JW and Kuff EL (1993) EBP-80, a transcription factor closely resembling the human autoantigen Ku, recognizes single- to double-strand transitions in DNA. *Journal of Biological Chemistry* **268** 10546–10552

Feaver WJ, Svejstrup JQ, Bardwell L *et al* (1993) Dual roles of a multiprotein complex from S. cerevisiae in transcription and DNA repair. *Cell* **75** 1379–1387

Feldmann H and Winnacker EL (1993) A putative homologue of the human autoantigen Ku from *Saccharomyces cerevisiae*. *Journal of Biological Chemistry* **268** 12895–12900

Finnie NJ, Gottlieb TM, Blunt T, Jeggo PA and Jackson SP (1995) DNA-dependent protein kinase activity is absent in xrs-6 cells: implications for site-specific recombination and DNA double-strand break repair. *Proceedings of the National Academy of Sciences of the USA* **92** 320–324

Fulop GM and Phillips RA (1990) The *scid* mutation in mice causes a general defect in DNA repair. *Nature* **347** 479–482

Gagliardini V, Fernandez P, Lee RKK *et al* (1994) Prevention of vertebrate neuronal death by the crmA gene. *Science* **263** 826–828

Gatti R and Swift M (1985) *Ataxia telangiectasia: Genetics, Neuropathology and Immunology of a Degenerative Disease of Childhood*, Liss, New York

Genersch E, Eckerskorn C, Lottspeich F, Herzog C, Kuhn K and Poschl E (1995) Purification of the sequence-specific transcription factor CTCBF, involved in the control of human collagen IV genes: subunits with homology to Ku antigen. *EMBO Journal* **14** 791–800

Getts RC and Stamato TD (1994) Absence of a Ku-like DNA end binding activity in the xrs double-strand DNA repair-deficient mutant. *Journal of Biological Chemistry* **269** 15981–15984

Giaccia A, Weinstein R, Hu J and Stamato TD (1985) Cell cycle-dependent repair of double-strand DNA breaks in a gamma-ray-sensitive Chinese hamster cell. *Somatic and Cell Molecular Genetics* **11** 485–491

Giaccia AJ, Denko N, MacLaren R *et al* (1990) Human chromosome 5 complements the DNA double-strand break-repair deficiency and gamma-ray sensitivity of the XR-1 hamster variant. *American Journal of Human Genetics* **47** 459–469

Giffin W, Torrance H, Saffran H, MacLeod HL and Hache RJG (1994) Repression of mouse mammory tumor virus transcription by a transcription factor complex. *Journal of Biological Chemistry* **269** 1449–1459

Giffin W, Torrance H, Rodda DJ, Prefontaine GG, Pope L and Hache RJG (1996) Sequence-specific DNA binding by Ku autoantigen and its effects on transcription. *Nature* **380** 265–268

Giocanti N, Hennequin C, Balosso J, Mahler M and Favaudon V (1993) DNA repair and cell cycle interactions in radiation sensitization by the topoisomerase II poison etoposide. *Cancer Research* **53** 2105–2111

Gottlieb TM and Jackson SP (1993) The DNA-dependent protein kinase: requirement for DNA ends and association with Ku antigen. *Cell* **72** 131–142

Greenwell PW, Kronmal SL, Porter SE, Gassenhuber J, Obermaier B and Petes TD (1995) TEL1, a gene involved in controlling telomere length in *S. cerevisiae*, is homologous to the human ataxia telangiectasia gene. *Cell* **82** 823–829

Hari KL, Santerre A, Sekelsky JJ, McKim KS, Boyd JB and Hawley RS (1995) The mei-41 gene of *D. melanogaster* is a structural and functional homolog of the human ataxia telangiectasia gene. *Cell* **82** 815–822

Hartley KO, Gell D, Smith GCM *et al* (1995) DNA-dependent protein kinase catalytic subunit: a relative of phosphatidylinositol 3-kinase and the ataxia telangiectasia gene product. *Cell* **82** 849–856

Hartwell LH and Kastan MB (1994) Cell cycle control and cancer. *Science* **266** 1821–1828

He Z, Henricksen LA, Wold MS and Ingles CJ (1995) RPA involvement in the damage-recognition and incision steps of nucleotide excision repair. *Nature* **374** 566–569

Hendrickson EA, Schatz DG and Weaver DT (1988) The *scid* gene encodes a *trans*-acting factor that mediates the rejoining event of Ig gene rearrangement. *Genes and Development* **2** 817–829

Hendrickson EA, Schlissel MS and Weaver DT (1990) Wild-type V(D)J recombination in *scid* pre-B cells. *Molecular and Cellular Biology* **10** 5397–5407

Hendrickson EA, Liu VF and Weaver DT (1991a) Strand breaks without DNA rearrangement in V(D)J recombination. *Molecular and Cellular Biology* **11** 3155–3162

Hendrickson EA, Qin X-Q, Bump EA, Schatz DG, Oettinger M and Weaver D T (1991b) A link between double-strand break-related repair and V(D)J recombination: the *scid* mutation. *Proceedings of the National Academy of Sciences of the USA* **88** 4061–4065

Henner WD and Blazka ME (1986) Hypersensitivity of cultured ataxia telangiectasia cells to etoposide. *Journal of the National Cancer Institute* **76** 1007–1012

Hoff CM, Ghosh AK, Prabhakar BS and Jacob ST (1994) Enhancer 1 binding factor, a Ku-related protein, is a positive regulator of RNA polymerase I transcription initiation. *Proceedings of the National Academy of Sciences of the USA* **91** 762–766

Hsieh CL, Arlett C and Lieber MR (1993) V(D)J recombination in ataxia telangiectasia, Bloom's syndrome and a DNA ligase I-associated immunodeficiency disorder. *Journal of Biological Chemistry* **268** 20105–20109

Iijima S, Teraoka H, Date T and Tsukada K (1992) DNA-activated protein kinase in Raji Burkitt's lymphoma cells: phosphorylation of c-myc oncoprotein. *European Journal of Biochemistry* **206** 595–603

Ivanov EI, Korolev VG and Fabre F (1992) *XRS2*, a DNA repair gene of *Saccharomyces cerevisiae*, is needed for meiotic recombination. *Genetics* **132** 651–664

Jacks T, Remington L, Williams B *et al* (1994) Tumor spectrum analysis in p53-mutant mice. *Current Biology* **4** 1–7

Jackson SP and Jeggo PA (1995) DNA double-strand break repair and V(D)J recombination: involvement of DNA-PK. *Trends in Biochemical Sciences* **20** 412–415

Jackson SP, MacDonald JJ, Lees-Miller S and Tjian R (1990) GC box binding induces phosphorylation of Sp1 by a DNA-dependent protein kinase. *Cell* **63** 155–165

Jacoby DB and Wensink PC (1994) Yolk protein factor-1 is a *Drosophila* homolog of Ku, the DNA-binding subunit of a DNA-dependent protein kinase from humans. *Journal of Biological Chemistry* **269** 11484–11491

Jeggo PA (1985a) Genetic analysis of X-ray-sensitive mutants of the CHO cell line. *Mutation Research* **146** 265–270

Jeggo PA (1985b) X-ray sensitive mutants of Chinese hamster ovary cell line: radio-sensitivity of DNA synthesis. *Mutation Research* **145** 171–176

Jeggo P (1990) Studies on mammalian mutants defective in rejoining double-strand breaks in DNA. *Mutation Research* **239** 1–16

Jeggo PA and Kemp LM (1983) X-ray-sensitive mutants of Chinese hamster ovary cell line: isolation and cross-sensitivity to other DNA-damaging agents. *Mutation Research* **112** 313–327

Jeggo PA, Caldecott K, Pidsley S and Banks G (1989) Sensitivity of chinese hamster ovary mutants defective in DNA double strand break repair to topoisomerase II inhibitors. *Cancer Research* **49** 7057–7063

Jeggo PA, Tesmer J and Chen DJ (1991) Genetic analysis of ionising radiatiion sensitive mutants of cultured mammalian cells. *Mutation Research* **254** 125–133

Johzuka K and Ogawa H (1995) Interaction of Mre11 and Rad50: two proteins required for DNA repair and meiosis-specific double-strand break formation in *Saccharomyces cerevisiae. Genetics* **139** 1521–1532

Kastan MB, Onyekwere O, Sidransky D, Vogelstein B and Craig RW (1991) Participation of p53 protein in the cellular response to DNA damage. *Cancer Research* **51** 6304–6311

Kastan MB, Zhan Q, El-Deiry WS *et al* (1992) A mammalian cell cycle checkpoint pathway utilizing p53 and *GADD45* Is defective in ataxia-telangiectasia. *Cell* **71** 587–597

Kato R and Ogawa H (1994) An essential gene, ESR1, is required for mitotic cell growth, DNA repair and meiotic recombination in *Saccharomyces cerevisiae. Nucleic Acids Research* **22** 3104–3112

Kaufman PD and Rio DC (1992) P element transposition in vitro proceeds by a cut-and-paste mechanism and uses GTP as a cofactor. *Cell* **69** 27–39

Keith CT and Schreiber SL (1995) PIK-related kinases: DNA repair, recombination and cell cycle checkpoints. *Science* **270** 50–51

Kirchgessner C, Tosto L, Biederman K. *et al* (1993) Complementation of the radiosensitive phenotype in severe combined immunodeficient mice by human chromosome 8. *Cancer Research* **53** 6011–6016

Kirchgessner CU, Patil CK, Evans JW *et al* (1995) DNA-dependent kinase (p350) as a candidate gene for the murine SCID defect. *Science* **267** 1178–1183

Knuth M, Gunderson S, Thompson N, Straheim L and Burgess R (1990) Purification and characterization of PSE1, a transcription activating protein related to Ku and TREF that binds the proximal sequence element of the human U1 promoter. *Journal of Biological Chemistry* **265** 17911–17920

Kojis TL, Gatti RA and Sparkes RS (1991) The cytogenetics of ataxia telangiectasia. *Cancer Genetics and Cytogenetics* **56** 143–158

Komatsu K, Ohta T, Jinno Y, Niikawa N and Okumura Y (1993) Functional complementation in mouse-human radiation hybrids assigns the putative murine scid gene to the pericentric region of human chromosome 8. *Human Molecular Genetics* **2** 1031–1034

Koonin E (1993) A common set of conserved motifs in a vast variety of putative nucleic acid-dependent ATPases including MCM proteins involved in the initiation of eukaryotic DNA replication. *Nucleic Acids Research* **21** 2541–2547

Kramer KM, Brock JA, Bloom K, Moore JK and Haber JE (1994) Two different types of double-strand breaks in S. *cerevisiae* are repaired by similar *RAD52*-independent, non-homologous recombination events. *Molecular and Cellular Biology* **14** 1293–1301

Kuhn A, Stefanovsky V and Grummt I (1993) The nucleolar transcription activator UBF relieves Ku antigen-mediated repression of mouse ribosomal gene transcription. *Nucleic Acids Research* **21** 2057–2063

Kuhn A, Gottlieb TM, Jackson SP and Grummt I (1995) DNA-dependent protein kinase: a potent inhibitor of transcription by RNA polymerase I. *Genes and Development* **9** 193–203

Kurimasa A, Nagata Y, Shimizu M, Emi M, Nakamura Y and Oshimura M (1993) A human gene that restores the DNA-repair defect in scid mice is located on 8p11.1-q11.1. *Human Genetics* **93** 21–26

Labhart P (1995) DNA-dependent protein kinase specifically represses promoter-directed transcription initiation by RNA polymerase I. *Proceedings of the National Academy of Sciences of the USA* **92** 2934–2938

Lai JS and Herr W (1992) Ethidium bromide provides a simple tool for identifying genuine DNA-independent protein association. *Proceedings of the National Academy of Sciences of the USA* **89** 6958–6962

Lee JW, Moffitt PG, Morley KL and Peterson DO (1991) Multipartite structure of a negative regulatory element associated with a steroid hormone-inducible promoter. *Journal of Biological Chemistry* **266** 24101–24108

Lee SE, Pulaski CR, He DM *et al* (1995) Isolation of mammalian cell mutants that are X-ray sensitive, impaired in DNA double-strand break repair and defective for V(D)J recombination. *Mutation Research* **336** 279–291

Lees-Miller SP, Chen Y and Anderson CW (1990) Human cells contain a DNA-activated protein kinase that phosphorylates simian virus 40 T antigen, mouse p53 and the human Ku autoantigen. *Molecular and Cellular Biology* **10** 6472–6481

Lees-Miller SP, Sakaguchi K, Ullrich SJ, Appella E and Anderson CW (1992) Human DNA-activated protein kinase phosphorylates serines 15 and 37 in the amino-terminal trans-activation domain of human p53. *Molecular and Cellular Biology* **12** 5041–5049

Lees-Miller SP, Godbout R, Chan DW *et al* (1995) Absence of p350 subunit of DNA-activated protein kinase from a radiosensitive human cell line. *Science* **267** 1183–1185

Li GC, Yang SH, Kim D *et al* (1995) Suppression of heat-induced hsp70 expression by the 70-kDa subunit of the human Ku autoantigen. *Proceedings of the National Academy of Sciences of the USA* **92** 4512–4516

Li Z, Otevrel T, Gao Y *et al* (1995) The XRCC4 gene encodes a novel protein involved in DNA double-strand break repair and V(D)J recombination. *Cell* **83** 1079–1089

Liang F, Romanienko P, Weaver D, Jeggo P and Jasin M (1996) Chromosomal double-strand break repair in Ku80 deficient cells. *Proceedings of the National Academy of Sciences of the USA* **93** 8929–8933

Lieber MR, Hesse JE, Lewis S *et al* (1988) The defect in murine severe combined immune deficiency: joining of signal sequences but not coding segments in V(D)J recombination. *Cell* **55** 7–16

Lieberman M, Hansteen GA, Waller EK, Weissman IL and Sen-Majumdar A (1992) Un-expected effects of the severe combined immunodeficiency mutation on murine lym-phomagenesis. *Journal of Experimental Medicine* **176** 399–405

Liu SH, Ma JT, Yueh AY, Lees-Miller SP anderson CW and Ng S Y (1993) The carboxyl-terminal transactivation domain of human serum response factor contains DNA-activated protein kinase phosphorylation sites. *Journal of Biological Chemistry* **268** 21147–21154

Liu V and Weaver D (1993) The ionizing radiation-induced replication protein A phosphoryla-tion response differs between ataxia telangiectasia and normal human cells. *Molecular and Cell Biology* **13** 7222–7231

Lu X and Lane DP (1993) Differential induction of transcriptionally active p53 following UV or ionizing radiation: defects in chromsome instability syndromes? *Cell* **75** 765–778

Lustig AJ and Petes TD (1986) Identification of yeast mutants with altered telomere structure. *Proceedings of the National Academy of Sciences of the USA* **83** 1398–1402

McBlane JF, van Gent DC, Ramsden DA *et al* (1995) Cleavage at a V(D)J recombination signal requires only RAG1 and RAG2 proteins and occurs in two steps. *Cell* **83** 387–395

Mages GJ, Feldman HM and Winnacker E (1996) Involvement of the *Saccharomyces cerevisiae* HDF1 gene in DNA double-strand break repair and recombination. *Journal of Biological Chemistry* **271** 7910–7915

Maldonado E, Shiekhattar R, Sheldon M *et al* (1996) A human RNA polymerase II complex as-sociated with SRB and DNA-repair proteins. *Nature* **381** 86–89

Malynn BA, Blackwell TK, Fulop GM *et al* (1988) The *scid* defect affects the final step of the immunoglobulin VDJ recombinase mechanism. *Cell* **54** 453–460

Messier H, Fuller T, Mangal S *et al* (1993) p70 lupus autoantigen binds the enhancer of the T-cell receptor β-chain gene. *Proceedings of the National Academy of Sciences of the USA* **90** 2685–2689

Meyn MS (1993) High spontaneous intrachromosomal recombination rates in ataxia-telangiecta-sia. *Science* **260** 1327–1330

Michalovitz D, Halevy O and Oren M (1990) Conditional inhibition of transformation and of cell proliferation by a temperature-sensitive mutant of p53. *Cell* **62** 671–680

Milne GT, Jin S, Shannon K and Weaver DT (1996) DNA end joining is disrupted by mutation of two Ku homolog genes in *Saccharomyces cerevisiae*. *Molecular and Cellular Biology* **16** 4189–4198

Mimori T and Hardin JA (1986) Mechanism of interaction between Ku protein and DNA. *Journal of Biological Chemistry* **261** 10375–10379

Mimori T, Akizuki M, Yamagata H, Inada S, Yoshida S and Homma M (1981) Characterization of a high molecular weight acidic nuclear protein recognized by autoantibodies from patients with polymyositis-scleroderma overlap. *Journal of Clinical Investigation* **68** 611–620

Mimori T, Hardin JA and Steitz JA (1986) Characterization of the DNA-binding protein antigen Ku recognized by autoantibodies from patients with rheumatic disorders. *Journal of Biological Chemistry* **261** 2264–2278

Moore JK and Haber JE (1996) Cell cycle and genetic requirements of two pathways of non-homologous end-joining repair of double-strand breaks. *Molecular and Cellular Biology* **16** 2164–2173

Morozov VE, Falzon M anderson CW and Kuff EL (1994) DNA-dependent protein kinase is activated by nicks and larger single-stranded gaps. *Journal of Biological Chemistry* **269** 16684–16688

Morrow DM, Tagle DA, Shiloh Y, Collins FS and Hieter P (1995) TEL1, an *S. cerevisiae* homolog of the human gene mutated in ataxia telangiectasia, is functionally related to the yeast checkpoint gene MEC1. *Cell* **82** 831–840

Murphy WJ, Durum SK, Anver MR *et al* (1994) Induction of T cell differentiation and lymphomagenesis in the thymus of mice with severe combined immune deficiency (SCID). *Journal of Immunology* **153** 1004–1014

Nelson WG, Liu LF and Coffey DS (1986) Newly replicated DNA is associated with DNA topoisomerase II in cultured rat prostatic adenocarcinoma cells. *Nature* **322** 187–189

Niu H and Jacob ST (1994) Enhancer 1 binding factor (E1BF), a Ku-related protein, is a growth-regulated RNA polymerase I transcription factor: association of a repressor activity with purified E1BF from serum-deprived cells. *Proceedings of the National Academy of Sciences of the USA* **91** 9101–9105

Ogawa H, Johzuka K, Nakagawa T, Leem S-H and Hagihara AH (1995) Functions of the yeast meiotic recombination genes, MRE11 and MRE2. *Advances in Biophysics* **31** 67–76

Okada T, Sakuma L, Fukui Y, Hazeki O and Ui M (1994) Blockage of chemotactic peptide-induced stimulation of neutrophils by wortmannin as a result of selective inhibition of phosphatidylinositol 3-kinase. *Journal of Biological Chemistry* **269** 3563–3567

Okumura K, Takagi S, Sakaguchi G *et al* (1995) Autoantigen Ku protein is involved in DNA binding proteins which recognize the U5 repressive element of human T-cell leukemia virus type I long terminal repeat. *FEBS Letters* **356** 94–100

Ono M, Tucker PW and Capra JD (1994) Production and characterization of recombinant human Ku antigen. *Nucleic Acids Research* **22** 3918–3924

Paillard S and Strauss F (1991) Analysis of the mechanism of interaction of simian Ku protein with DNA. *Nucleic Acids Research* **19** 5619–5624

Painter RB and Young BR (1980) Radiosensitivity in ataxia-telangiectasia: a new explanation. *Proceedings of the National Academy of Sciences of the USA* **77** 7315–7317

Pan ZQ, Amin AA, Gibbs E, Niu HW and Hurwitz J (1994) Phosphorylation of the p34 subunit of human single-stranded DNA binding protein in cyclin A-activated G(1) extracts is catalyzed by CDK cyclin A complex and DNA-dependent protein kinase. *Proceedings of the National Academy of Sciences of the USA* **91** 8343–8347

Pandita TK, Pathak S and Geard C (1995) Chromosome end associations, telomeres and telo-merase activity in ataxia telangiectasia. *Cancer Genetics and Cytogenetics* **71** 86–93

Paulovich AG and Hartwell LH (1995) A checkpoint regulates the rate of progression through S phase in *S. cerevisiae* in response to DNA damage. *Cell* **82** 841–848

Pergola F, Zdzienicka MZ and Lieber MR (1993) V(D)J recombination in mammalian cell mutants defective in DNA double-strand break repair. *Molecular and Cellular Biology* **13** 3464–3471

Peterson SR, Dvir A anderson CW and Dynan WS (1992) DNA binding provides a signal for phosphorylation of the RNA polymerase II heptapeptide repeats. *Genes and Development* **6** 426–438

Peterson SR, Kurimasa A, Oshimura M, Dynan WS, Bradbury EM and Chen DJ (1995) Loss of the 350kD catalytic subunit of the DNA-dependent protein kinase in DNA double-strand break repair mutant mammalian cells. *Proceedings of the National Academy of Sciences of the USA* **92** 3171–3174

Petes T, Malone R and Symington L (1991) Recombination in yeast, In: Broach J, Pringle J and Jones E (eds). *The Molecular and Cellular Biology of the Yeast Saccharomyces*, pp 407–521, Cold Spring Harbor Laboratory Press, Cold Spring Harbor, New York

Petrini JHJ, Walsh M, Dimare C, Chen X-N, Korenberg JR and Weaver DT (1995) Isolation and characterization of the human MRE11 homologue. *Genomics* **29** 80–86

Porter SE, Greenwell PW, Ritchie KB and Petes TD (1996) The DNA-binding protein Hdf1p (a putative Ku homolog) is required for maintaining normal teleomere length in *Saccharomyces cerevisiae*. *Nucleic Acids Research* **24** 582–585

Rathmell WK and Chu G (1994a) A DNA end-binding factor involved in double-strand break repair and V(D)J recombination. *Molecular and Cellular Biology* **14** 4741–4748

Rathmell WK and Chu G (1994b) Involvement of the Ku autoantigen in the cellular response to DNA double-strand breaks. *Proceedings of the National Academy of Sciences of the USA* **91** 7623–7627

Raymond WE and Kleckner N (1993) Rad50 protein of *S. cerevisiae* exhibits ATP-dependent DNA binding. *Nucleic Acids Research* **21** 3851–3856

Reeves W (1985) Use of monoclonal antibodies for the characterization of novel DNA-binding proteins recognized by human autoimmune sera. *Journal of Experimental Medicine* **161** 18–39

Reeves WH and Sthoeger ZM (1989) Molecular cloning of cDNA encoding the p70 (Ku) lupus autoantigen. *Journal of Biological Chemistry* **264** 5047–5052

Roberts MR, Han Y, Fienberg A, Hunihan L and Ruddle FH (1994) A DNA-binding activity, TRAC, specific for the TRA element of the transferrin receptor gene copurifies with the Ku autoantigen. *Proceedings of the National Academy of Sciences of the USA* **91** 6354–6358

Roth DB, Menetski JP, Nakajima PB, Bosma MJ and Gellert M (1992a) V(D)J recombination: broken DNA molecules with covalently sealed (hairpin) coding ends in scid mouse thymocytes. *Cell* **70** 983–991

Roth DB, Nakajima PB, Menetski JP, Bosma MJ and Gellert M (1992b) V(D)J recombination in mouse thymocytes: double-strand breaks near T cell receptor δ rearrangement signals. *Cell* **69** 41–53

Roth DB, Zhu C and Gellert M (1993) Characterization of broken DNA molecules associated with V(D)J recombination. *Proceedings of the National Academy of Sciences of the USA* **90** 10788–10792

Rouet P, Smih F and Jasin M (1994a) Expression of a site-specific endonuclease stimulates homologous recombination in mammalian cells. *Proceedings of the National Academy of Sciences of the USA* **91** 6064–6068

Rouet P, Smith F and Jasin M (1994b) Introduction of double-strand breaks into the genome of mouse cells by expression of a rare-cutting endonuclease. *Molecular and Cellular Biology* **14** 8096–8106

Saiijo M, Ui M and Enomoto T (1992) Growth state and cell cycle dependent phosphorylation of DNA topoisomerase II in Swiss 3T3 cells. *Biochemistry* **31** 359–363

Sanchez Y, Desany BA, Jones WJ, Liu Q, Wang B and Elledge SJ (1996) Regulation of RAD53 by the ATM-like kinases MEC1 and TEL1 in yeast cell cycle checkpoint pathways. *Science* **271** 357–360

Savitsky K, Bar-Shira A, Gilad S *et al* (1995) A single ataxia telangiectasia gene with a product similar to PI-3 kinase. *Science* **268** 1749–1753

Schatz D, Oettinger M and Schlissel M (1992) V(D)J recombination: molecular biology and regulation. *Annual Reviews of Immunology* **10** 359–383

Schlissel M, Constantinescu A, Morrow T, Baxter M and Peng A (1993) Double-strand signal sequence breaks in V(D)J recombination are blunt, 5´-phosphorylated, RAG-dependent and cell cycle regulated. *Genes and Development* **7** 2520–2532

Schuler W, Weiler IJ, Schuler A *et al* (1986) Rearrangement of antigen receptor genes is defective in mice with severe combined immune deficiency. *Cell* **46** 963–972

Siede W, Friedl AA, Dianova I, Eckardt-Schupp F and Friedberg EC (1996) The *Saccharomyces cerevisiae* Ku autoantigen homologue affects radiosensitivity only in the absence of homologous recombination. *Genetics* **142** 91–102

Sipley JD, Menninger JC, Hartley KO, Ward DC, Jackson SP and Anderson CW (1995) The gene for the catalytic subunit of the human DNA-activated protein kinase maps to the site of the XRCC7 gene on chromosome 8. *Proceedings of the National Academy of Sciences of the USA* **92** 7515–7519

Smider V, Rathmell WK, Lieber MR and Chu G (1994) Restoration of X-ray resistance and V(D)J recombination in mutant cells by Ku cDNA. *Science* **266** 288–291

Smith PJ anderson CO and Watson JV (1986) Predominant role for DNA damage in etoposide-induced cytotoxicity and cell cycle perturbation in human SV40-transformed fibroblasts. *Cancer Research* **46** 5641–5645

Stamato TD, Weinstein R, Giaccia A and Mackenzie L (1983) Isolation of cell cycle-dependent gamma-ray-sensitive Chinese hamster ovary cell. *Somatic and Cell Genetics* **9** 165–173

Steen SB, Gomelsky L and Roth DB (1996) The 12/23 rule is enforced at the cleavage step of V(D)J recombination in vivo. *Genes to Cells* **1** 543–554

Sun H, Treco D, Schultes NP and Szostak JW (1989) Double-strand breaks at an initiation site for meiotic gene conversion. *Nature* **338** 87–90

Sun H, Treco D and Szostak JW (1991) Extensive 3´-overhanging, single-stranded DNA associated with the meiosis-specific double-strand breaks at the ARG4 recombination initiation site. *Cell* **64** 1155–1161

Sun Z, Fay DS, Marini F, Foiani M and Stern DF (1996) Spk1/Rad53 is regulated by Mec1-dependent protein phosphorylation in DNA replication and damage checkpoints. *Genes and Development* **10** 395–406

Suwa A, Hirakata M, Takeda Y, Jesch SA, Mimori T and Hardin JA (1994) DNA-dependent protein kinase (Ku protein-p350 complex) assembles on double-stranded DNA. *Proceedings of the National Academy of Sciences of the USA* **91** 6904–6908

Taccioli G, Gottlieb TM, Blunt T *et al* (1994a) Ku80: product of the XRCC5 gene and its role in DNA repair and V(D)J recombination. *Science* **265** 1442–1445

Taccioli GE, Rathbun G, Oltz E, Stamato T, Jeggo PA and Alt FW (1993) Impairment of V(D)J recombination in double-strand break repair mutants. *Science* **260** 207–210

Taccioli GE, Cheng H-L, Varghese AJ, Whitmore G and Alt FW (1994b) A DNA repair defect in chinese hamster ovary cells affects V(D)J recombination similarly to the murine *Scid* mutation. *Journal of Biological Chemistry* **269** 7439–7442

Take Y, Kumano M, Hamano Y *et al* (1995) OK-1035, a selective inhibitor of DNA-dependent protein kinase. *Biochemistry and Biophysics Research Communications* **215** 41–47

Take Y, Kumano M, Teraoka H, Nishimura S and Okuyama A (1996) DNA-dependent protein kinase inhibitor (OK-1035) suppresses p21 expression in HCT116 cells containing wild-type p53 induced by adriamycin. *Biochemistry and Biophysics Research Communications* **221** 207–212

Toth EC, Marusic L, Ochem A *et al* (1993) Interactions of USF and Ku antigen with a human DNA region containing a replication origin. *Nucleic Acids Research* **21** 3257–3263

Tuteja N, Tuteja R, Ochem A *et al* (1994) Human DNA helicase II: a novel DNA unwinding enzyme identified as the Ku autoantigen. *EMBO Journal* **13** 4991–5001

Ullrich SJ anderson CW, Mercer WE and Appella E (1992) The p53 tumor suppressor protein, a modulator of cell proliferation. *Journal of Biological Chemistry* **267** 15259–15262

van Gent DC, McBlane JF, Ramsden DA, Sadofsky MJ, Hesse JE and Gellert M (1995) Initiation of V(D)J recombination in a cell-free system. *Cell* **81** 925–934

van Gent DC, Mizuuchi K and Gellert M (1996) Similarities between initiation of V(D)J recombination and retroviral integration. *Science* **271** 1592–1594

Walker AI, Hunt T, Jackson RJ and Anderson CW (1985) Double-stranded DNA induces the phosphorylation of several proteins including the 90 000 Mr heat-shock protein in animal cell extracts. *EMBO Journal* **4** 139–145

Weaver DT (1995a) V(D)J recombination and double-strand break repair. *Advances in Immunology* **58** 29–85

Weaver DT (1995b) What to do at an end: DNA double-strand break repair. *Trends in Genetics* **11** 388–392

Weaver D, Boubnov N, Wills Z, Hall K and Staunton J (1995) V(D)J recombination: double-strand break repair gene products in the joining mechanism. *Annals of the New York Academy of Science* **764** 99–111

Weinert TA, Kiser GL and Hartwell LH (1994) Mitotic checkpoint genes in budding yeast and the dependence of mitosis on DNA replication and repair. *Genes and Development* **8** 652–665

Wobbe CR, Weissbach L, Borowiec JA *et al* (1987) Replication of simian virus 40 origin-containing DNA in vitro with purified proteins. *Proceedings of the National Academy of Sciences of the USA* **84** 1834–1838

Wold MS and Kelly TJ (1988) Purification and characterization of replication protein A, a cellular protein required for in vitro replication of simian virus 40 DNA. *Proceedings of the National Academy of Sciences of the USA* **85** 2523–2527

Yaneva M and Busch H (1986) A 10S particle released from deoxyribonuclease-sensitive regions of HeLa cell nuclei contains the 86-kilodalton 70-kilodalton protein complex. *Biochemistry* **25** 5057–5063

Yuan J, Shaham S, Ledoux S, Ellis HM and Horvitz HR (1993) The *C. elegans* cell death gene ced-3 encodes a protein similar to the mammalian interleukin-1β-converting enzyme. *Cell* **75** 641–652

Zakian VA (1995) ATM-related genes: what do they tell us about functions of the human gene? *Cell* **82** 685–687

Zdzienicka MZ, Tran Q, van der Schans GP and Simons JWIM (1988) Characterization of an X-ray-hypersensitive mutant of V79 Chinese hamster cells. *Mutation Research* **194** 239–249

The authors are responsible for the accuracy of the references.

Telomerase, Checkpoints and Cancer

CALVIN B HARLEY • STEVEN W SHERWOOD

Geron Corporation, 200 Constitution Drive, Menlo Park, CA 94025

INTRODUCTION

Telomeres are essential for normal chromosome structure and function and thus for long term viability of cells. However, telomeres cannot be fully replicated by the conventional DNA polymerase complex. To overcome this "end replication" problem, most eukaryotic organisms utilize telomerase, an RNA dependent DNA polymerase (Greider and Blackburn, 1985, 1996). Telomerase extends chromosome ends with a specific telomeric DNA sequence by using a portion of its integral RNA component as the template.

Interest in telomere biology has become widespread with the observations that changes in telomerase activity and telomere dynamics are correlated with senescence (the loss of replicative capacity) and immortalization (the gain of unlimited replicative potential) in mammalian cells. The importance of a mechanistic understanding of this relationship is underscored by the very high proportion of tumours that express significant telomerase activity compared with a near absence of telomerase in essentially all normal somatic tissues (Harley *et al*, 1994; Bacchetti and Counter, 1995; Greider and Blackburn, 1996; Shay and Wright, 1996), indicating that this activity is a consistent feature of tumour cells and an important potential therapeutic and diagnostic

Cancer Surveys Volume 29: *Checkpoint Controls and Cancer*
© 1997 Imperial Cancer Research Fund. 0-87969-518-8/97. $5.00 + .00

target. Additionally, telomere shortening associated with the absence of telomerase activity in most normal somatic cells is a prominent feature of cellular senescence, the loss of replicative capacity in cells derived from normal tissues (Harley *et al*, 1990; Harley, 1991; Lindsey *et al*, 1991; Allsopp *et al*, 1995). Recent evidence has conclusively linked in vitro and in vivo cellular senescence (Dimri *et al*, 1995; Campisi, 1996; Smith and Pereira-Smith, 1996). Telomeres and telomerase thus are of central importance to an understanding not only of tumorigenesis, but also of age related disorders, in which change in gene expression and decline in the ability of cells to divide contribute to tissue dysfunction.

Although a considerable body of data on telomeres and telomerase in mammalian cells has accumulated (Kipling and Cooke, 1992; de Lange, 1994; Harley and Villepontean, 1995), our understanding of basic telomere biology comes largely from biochemical and genetic characterization of telomerase and telomere associated proteins in simpler eukaryotes. Characterization of telomere associated proteins in several organisms (see Zakian, 1996; Greider, in press), the cloning of the RNA component of the telomerase RNP from *Tetrahymena*, yeast, mice and humans (Greider and Blackburn, 1989; Singer and Gottschling, 1994; Blasco *et al*, 1995; Collins *et al*, 1995; Feng *et al*, 1995) and characterization of *Tetrahymena* telomerase protein components (Collins *et al*, 1995) have provided the foundation of a molecular picture of telomere biology. Moreover, homologies between genes involved in signal transduction pathways active in cell cycle checkpoints and genes coding for proteins involved in telomere dynamics provide interesting avenues of inquiry to find out how telomeres impact cell behaviour and regulate the proliferative state of mammalian cells (Kennedy *et al*, 1995; Morrow *et al*, 1995; Sanchez *et al*, 1996).

This chapter provides a selected review of the telomere and telomerase field as it relates to tumorigenesis, with a focus on how the presence or absence of checkpoint mechanisms might link telomere loss to cell cycle exit at senescence and to proliferative crisis. In this context, stringent suppression of telomerase in most normal somatic cells and the subsequent events associated with cell mortality may have evolved as tumour suppressor mechanisms in long lived species such as humans.

TELOMERES AND TELOMERASE

Telomeres

Eukaryotic telomeres generally consist of simple arrays of tandem G rich repeats that run 5' to 3' at chromosome ends, with the C rich complementary strand sometimes recessed, forming a G rich 3' overhang. Single stranded G rich telomeric sequences can form intra- and interstrand structures, but their biological importance is unclear (Fang and Cech, 1993). Telomere sequences can vary from species to species, but a given organism has a characteristic repeat at all telomeres. In humans and all vertebrates studied, a single telo-mere sequence (TTAGGG) is found at chromosome ends, although subtelo-meric sequences vary substantially (Moyzis *et al*, 1988). Telomere containing linear plasmids transfected into human cells can truncate chromosomes and "seed" new telomere ends (Barnett *et al*, 1992; Broccoli and Cooke, 1993). Such studies have indicated that telomerase in vivo has a relatively strict specificity for the cognate TTAGGG structure, perhaps reflecting the high de-gree of specificity in binding of the major mammalian telomere associated protein (Chong *et al*, 1995).

Telomerase

Telomerase activity was first characterized biochemically in 1985 by Greider and Blackburn, using *Tetrahymena* extracts. Four years later, a very similar ac-tivity was identified in the HeLa immortal human cell line by Morin (1989). The mechanism of action of this DNA polymerase was revealed in part by the cloning of the RNA component of the *Tetrahymena* telomerase (Greider and Blackburn, 1989) and the subsequent identification of the protein components in this species (Collins *et al*, 1995). The template region of the *Tetrahymena* RNA component (5'AACCCCAAC) is complementary to 1.5 repeats of the telomere sequence and was confirmed by showing that mutations in this region led to the predicted alterations in telomere sequence when the mutated RNA was overexpressed in vivo. Mutational analysis using in vitro and in vivo reconstituted enzyme suggests that the templating RNA bases lie at the 5' end of the nine nucleotide stretch, whereas the nucleotides at the 3'end of this region serve for "alignment", allowing correct positioning of the 3'end of the telomeric DNA at the beginning of each elongation cycle (Greider, in press). Thus, telomerase is capable of processive cycling between rounds of telomere polymerization (elongation) and translocation in which the enzyme complex remains associated with the DNA (Fig. 1). The RNA components of *Saccharomyces cerevisiae* (Singer and Gottschling, 1994) and *Kluyveromyces lactis* yeast telomerase (McEachern and Blackburn, 1995) have been identi-fied, and several candidates for the protein components or associated factors have been proposed (Lundblad and Szostak, 1989; Lin and Zakian, 1995; Steiner *et al*, 1996; Zakian, 1996).

The human and mouse telomerase RNA components (hTR and mTR) have also been cloned (Blasco *et al*, 1995; Feng *et al*, 1995), and each encodes the predicted template region complementary to approximately 1.5 repeats of the vertebrate TTAGGG telomeric sequence. Identification of the cDNAs was confirmed by mutating the template sequence in a genomic copy of the gene and demonstrating that telomerase isolated from cells transfected with the altered gene synthesized the corresponding mutant telomere sequence. HeLa cells transfected with antisense vectors to hTR had reduced telomerase activity and gradually lost telomeric DNA, ultimately showing proliferative crisis (Feng *et al*, 1995). Although not all cell lines stably transfected with the antisense

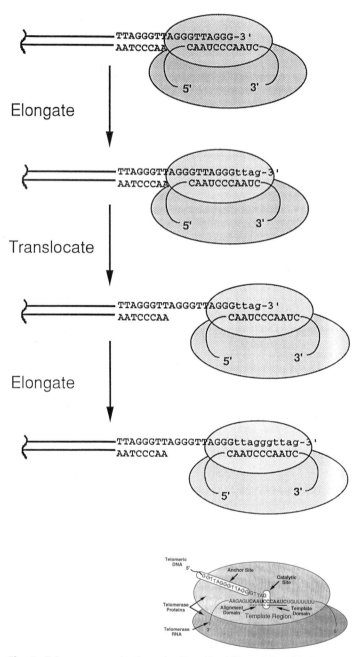

Fig. 1. Telomerase mechanism of action. (Top) The relationship between telomerase protein subunits (shaded ovals) and telomerase RNA (curved strand showing the human template region, CAAUCCCAAUC). (Bottom) Presumptive primer:template configuration at the catalytic site of human telomerase just following a translocation event. The template region contains an alignment domain which, together with the protein:DNA anchor site, positions the terminal bases of the telomeric DNA such that the catalytic site is ready to add dNTPs in the sequence dictated by the template domain

vectors showed telomere loss and cell crisis, and among those that did some cells survived the population crisis, these cells showed near wild type telomerase activity, suggesting that they were refractory to inhibition by the antisense hTR vector (Feng *et al*, 1995; Wang S-S, personal communication). Yeast with altered or no telomerase RNA component also generally show proliferative crisis following telomere loss, with a low frequency of survivors. In this case, however, it appears that an alternative mechanism may be involved in telomere maintenance in the surviving cells (McEachern and Blackburn, 1996).

The Link to Ageing and Cancer

In general, non-transformed human cells either do not express telomerase activity or express it a very low levels, as determined by a highly sensitive polymerase chain reaction (PCR) based assay (Kim *et al*, 1994). Continued division of primary cells in the absence of telomerase activity results in progressive telomere shortening, and ultimately cells exit the cell cycle expressing a unique phenotype, that of replicative senescence (Harley *et al*, 1990; Hastie *et al*, 1990; Lindsey *et al*, 1991; Vaziri *et al*, 1993; Allsopp and Harley, 1995; Chang and Harley, 1995). This point in the proliferative life of the cells, the Hayflick limit, represents a stable cell cycle state in which non-proliferating cells can remain viable and metabolically active for long periods (Goldstein, 1990; Campisi, 1996; Smith and Pereira-Smith, 1996). Senescence is not cell death but rather appears to be a cell cycle checkpoint arrest associated with the first critically short telomeres. Accurate measurement of the length of continuous TTAGGG repeats at the ends of chromosomes is very difficult. Thus, terminal restriction fragment (TRF) length measurements are typically performed with Southern blot analysis. Comparison of the rate of loss of TRF and telomere signal suggests that TRFs may have approximately 3–5 kbp of non-TTAGGG DNA. Since the average TRF at senescence is typically 5–7 kbp, and there is significant variation in telomere length from chromosome to chromosome (Moyzis *et al*, 1988; Henderson *et al*, 1996), it is likely that one or more chromosomes will have critically short telomeres at senescence.

Presenescent cells can be driven into a period of extended proliferation, bypassing the Hayflick phenomenon, by oncoproteins of viral origin, such as simian virus 40 (SV40) T, human papillomavirus (HPV) E6/E7 and adenovirus E1A/E1B (Shay *et al*, 1991; Counter *et al*, 1992; Klingelhutz *et al*, 1993). These proteins override the proliferative checkpoints operating during senescence, allowing the extended lifespan phenotype to be expressed. Because telomerase is not directly activated by these proteins, continued telomere shortening occurs during extended proliferation, bringing chromosomes eventually to a second critical length, typically in the 2–4 kbp range (Fig. 2). Rather than entering a stable non-proliferative state, cells at this stage enter

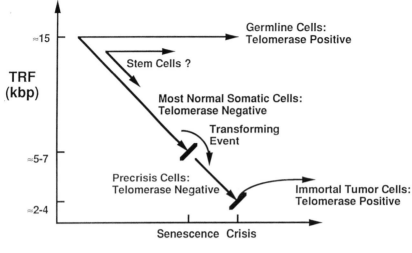

Replicative Age

Fig. 2. Telomere and telomerase hypothesis of cell ageing and immortalization. Reproductive cells and certain normal somatic cells express telomerase activity. During development, telomerase is repressed in most somatic cells, and telomeres shorten with cell division both in vitro and in vivo. Telomere loss is a marker of cell ageing and is thought to signal senescent growth arrest when one or more telomeres reaches a functionally critical size. Transformation events such as inactivation of tumour suppressor genes allow cells to bypass senescence but do not themselves activate telomerase, so cells continue to lose telomeres until the end of their extended lifespan is reached (crisis). At crisis, telomeres may be critically short on many chromosomes, causing massive genomic instability and cell death. Cells that survive crisis generally have active telomerase and stable telomeres

proliferative crisis. TRF length at crisis ranges between 2 kbp and 4 kbp (Counter *et al*, 1992, 1994), suggesting that essentially all chromosomes have lost functional telomeres. In contrast to senescence, crisis represents a highly unstable cellular state involving extensive chromosome structural instability and cell death, presumably by apoptosis (Pandita *et al*, 1995). Just as senescence represents the "end" of proliferative lifespan, crisis represents the "end" of extended lifespan. Following crisis, proliferation is maintained only in rare $(1 \text{ in } 10^{-7})$ immortalized cells in which additional growth deregulating mutations occur. These surviving cells are proliferatively metastable, corresponding to the immortal phenotype, and continued proliferation commonly involves reexpression of telomerase, which maintains telomere stability and function (Counter *et al*, 1992; Shay *et al*, 1993).

The number of tumours and normal tissue samples for which telomerase data are available is growing rapidly. It was only in 1994 that the first report of specific activation of telomerase in a human tumour (ovarian) was reported (Counter *et al*, 1994). In that study, the "conventional" non-PCR based telomerase assay was used to characterize telomerase activity in ascites fluid cells from ten patients with ovarian cancer. Separation of tumour cells from normal cells in the ascites fluid showed that telomerase was associated with the

TABLE 1. Telomerase activity in normal and malignant tissues*

Tissue	No positive/ No tested (%)
Reproductive tissues	
ovaries and testes	4/4 (100%)
Normal somatic tissues	
breast, liver, ascites	0/20 (0%)
peripheral blood mononuclear cells	
1000 cells	0/124 (0%)
10000 cells	55/124 (44%)
100000 cells	124/124 (100%)
Benign and premalignant tissues	
prostate PIN3, gastric adenoma, breast fibroadenoma, hepatitis, cirrhosis	31/56 (55%)
BPH, intestinal metaplasia, anaplastic astrocytoma, benign meningioma	7/76 (9%)
colon polyps, colon adenoma, ganglioneuroma, leiomyoma, miscellaneous breast disorders	0/134 (0%)
Malignant tissues	
neuroblastoma, colon, skin, breast, uterine, ovarian	350/374 (94%)
lung, prostate, gastric hepatic	247/296 (83%)
brain, renal, hematological	161/225 (72%)
Tissue adjacent to tumours	
head and neck, Wilms'	8/22 (35%)
prostate, breast, lung, gastric	11/220 (5%)
renal neuroblastoma	0/68 (0%)

*Adapted from Shay and Wright (1996)

tumour compartment only and that the tumour cells had short, stable telomeres in contrast to the normal cells, which had long telomeres that presumably decreased slowly with age as demonstrated in other studies. Telomerase could not be detected in normal peripheral blood leucocytes in that study with the less sensitive conventional assay, nor could telomerase activity be detected in ovarian epithelial cells. Since that report, thousands of tumour biopsies have been studied representing over 20 different human tumour types, including the ten major tumours in the USA. In a recent review covering only some of these reported studies, Shay and Wright (1996) summarize observations on telomerase activity in 198 normal tissue samples, 266 benign and premalignant samples and 1205 malignant and adjacent samples (Table 1). The overwhelming impression is that telomerase activity is highly specific for reproductive and malignant tissues and is nearly universally absent in normal somatic tissues and tissues adjacent to tumour material. Analyses of cultured normal cell strains and immortal cell lines have generated similar data (Kim *et al*, 1994). Moreover, in recent months, qualitatively similar studies have been reported using reverse transcription-PCR and in situ analysis of hTR, in which the total abundance and cell specific presence of the RNA component of telomerase in normal and tumour tissues are investigated.

In addition to the data using antisense expression vectors to inhibit telomerase and drive HeLa cells into crisis following apparent critical telomere loss (Feng *et al*, 1995), there has been another recent experimental test of the causal link between telomerase and cell immortality. Wright *et al* (1996) demonstrated that telomerase positive immortal cells grown exposed to TTAGGG containing oligonucleotides significantly elongated their telomeres over time in culture. The mechanism is unknown, but growth of telomeres does not occur in telomerase negative cells. The authors then fused telomerase negative normal cells containing very long telomeres to telomerase positive cells with short telomeres that were pre-exposed to either the G rich oligonucleotide or control oligonucleotides. In these studies, fusion of mortal and immortal cells dramatically inhibited telomerase activity and led to a finite lifespan of the hybrid. Consistent with a causal link between telomere length, telomerase activity and replicative capacity, the hybrids arising from the immortal parent with longer telomeres had significantly longer lifespans than hybrids arising from the immortal parents with short telomeres.

Caveats

Gradual loss of telomeric DNA with replicative age in the absence of telomerase activity is now an accepted characteristic of most normal somatic cells and tissues. Similarly, maintenance of generally short telomeres in the presence of telomerase is the norm for most tumour cells and tissues. However, this simple model does account for all the data. In our view, the exceptions point to important cases where new avenues of basic and applied research should be explored, but they do not detract significantly from the general model. The major exceptions to the model are reviewed below.

Telomere loss as a function of replicative ageing in culture or as a function of donor age in vivo has been documented in a variety of human cells and tissues that undergo significant self renewal, including candidate haemopoietic stem cells, epidermis, peripheral blood lymphocytes and gut mucosa (Hastie *et al*, 1990; Vaziri *et al*, 1993, 1994; Taylor *et al*, 1995). However, weak telomerase activity has now been detected with the highly sensitive PCR based assay in all of these tissues, including liver (Tahara H *et al*, 1995; Taylor *et al*, 1995; Chiu *et al*, 1996; Bodnar *et al*, 1996). Given that telomeric DNA is lost with age in vitro and in vivo in these cells and tissues, the biological significance of this activity is uncertain. However, it is intriguing that telomerase activation in at least some of these normal cells appears to be transient and associated only with the early commitment of stem or progenitor cells to significant clonal expansion (Chiu *et al*, 1996; Bodnar *et al*, 1996). Bodnar *et al* (1996) carefully measured telomere length throughout the lifespan of stimulated T cells and noticed a correlation between high telomerase activity early in the clonal expansion and a reduced rate of telomere loss. Thus, it is possible that transient telomerase expression extends cell lifespan in certain tissues that require frequent and extensive rounds of cell proliferation. Since the quiescent stem cells and the more committed exponentially growing cells have little if any detectable telomerase, it is unlikely that telomerase inhibition would have a significant biological effect on these tissues.

Although the vast majority of all transformed and immortalized cell lines and tumours express telomerase, some do not. There are several possible interpretations of these data, ranging from suggestions that telomerase is there but not detected for trivial reasons (eg transient bursts of activity or cryptic inhibitors of the in vitro assay) to suggestions that there are prevalent alternative mechanisms for telomere maintenance. A comprehensive description of the biology is probably complex. Some telomerase negative "immortal cells" may turn out to be mortal, but some may display recombination type pathways for telomere maintenance. Some telomerase negative tumours may spontaneously regress, as is seen in stage IVS neuroblastoma (Hiyama *et al*, 1995), but it is also possible that biopsies occasionally miss tumour tissue with intact or measurable telomerase. In addition, we cannot at this point rule out the possibility that some tumours will not require telomerase activation to be lethal, either because they have sufficient replicative capacity without telomerase or because they utilize a telomerase independent mechanism of telomere maintenance. However, the high degree of association between telomerase activity and tumour progression in humans, and the fact that all available evidence still indicates that this enzyme is the major pathway for a critical genetic function of immortal cells, makes telomerase an extremely attractive target for both cancer diagnostics and therapy.

A controversial aspect of the telomere model for ageing and cancer is the significant difference between humans and rodents in telomere structure and telomerase regulation. Compared with humans, most common laboratory strains of mice and other rodents have relatively large arrays of TTAGGG or TTAGGG like repeats near chromosomal ends, and these are hypervariable (Kipling and Cooke, 1990; Starling *et al*, 1990). Thus, it has been difficult to assess the significance of reports that telomere loss does not occur as a function of age in mice, since loss of true telomeric DNA may be obscured by large subtelomeric arrays of TTAGGG or degenerate TTAGGG sequences. It is also possible that the importance of telomere loss and replicative senescence is relatively minor in mice compared with humans.

Senescence in mice is complicated by the relatively high rate of spontaneous transformation and immortalization of murine cells, coupled with little distinction in some reports among growth adaptation to tissue culture conditions, cell senescence and crisis. Thus, the phenomenon observed about 5–20 population doublings after the primary culture of cells is established from murine embryonic fibroblasts has been variously called "senescence" or "crisis", with the population surviving this period termed "immortal". However, these cells are telomerase positive from the start, and it is very likely that the senescence/crisis phenomenon is simply cell selection or adaptation to growth conditions in vitro. Prowse and Greider (1995) demonstrated that adult *Mus spretus* fibroblasts put into culture are telomerase negative initially, and only after approximately 60 population doublings with gradual telomere loss does the culture undergo a subtle change in growth kinetics and a gradual conversion to a telomerase positive population with stable telomeres. The involvement of telomerase activation in tumours in mice is also complex (Chadeneau *et al*, 1995; Blasco *et al*, 1996). Unlike the situation in humans, telomerase activity can readily be detected in a variety of normal rodent tissues in vivo (Chadeneau *et al*, 1995; Prowse and Greider, 1995). Cultured rodent cells in general display cell cycle checkpoint controls that differ significantly from cultured human cells (Schimke *et al*, 1991). These observations may account for the high frequency of spontaneous immortalization of rodent cells in vitro and the extremely high frequency of cancer in mice in vivo on a per cell, per year basis (reviewed in Harley *et al*, 1994). However, even though telomerase activity and mTR expression are both upregulated in general as cells progress from the normal to a preoplastic state to overt malignancy, existing correlations are not as tight in the rodent system as they are in human studies. A strong case can be made for a tight relationship in mice between telomerase expression and growth rate, rather than telomerase expression and cell immortalization. It is clear that further work is needed to understand fully the significance, if any, of telomere dynamics and telomerase expression during ageing and tumorigenesis in mice.

CHECKPOINTS AND POSSIBLE SIGNALLING PATHWAYS

Cytogenetic Changes Occurring at Senescence and Crisis

A variety of cytogenetic abnormalities, including telomere associations, are prominent during cellular senescence and crisis. In the case of cellular senescence, polyploidy increases consistently throughout the proliferative lifespan, whereas the frequency of chromosomal aberrations (per cell) remains relatively low until the end of the proliferative lifespan, when there is a dramatic increase in aberration frequency (Sherwood *et al*, 1988). This temporal correlation of loss of proliferative capacity (ie cell cycle exit) with increased chromosomal instability would be expected if a "damage" signal generated by these aberrations and/or by the process or events giving rise to the aberrations (eg critically shortened telomeres) was transduced into cell cycle exit. From this perspective, senescence is an adaptive cellular response.

Oncoprotein driven proliferation of cells "beyond" senescence does not require telomerase activation, and so, during extended lifespan, telomeres continue to shorten (Counter *et al*, 1992). A variety of data broadly suggest that in human cells, the second critical TRF length is encountered when the average TRF is 2–4 kbp. Because TRF is averaged over the entire set of chromosomes, telomere shortening at crisis may on any given chromosome extend into subtelomeric regions where TTAGGG repeats become interspersed and degenerate. Cytogenetic data showing extensive between telomere associations and a high frequency of true dicentric (telomere fusion) chromosomes demonstrate the presence of "sticky" chromosome ends in cells undergoing crisis, consistent with loss of protective telomere sequence (Pandita *et al*, 1995).

Cell Cycle Phenomenology of Senescence

The cell cycle behaviour of cells at senescence corresponds to the cell cycle "engine" (Murray, 1992) receiving a "stop" signal. The events of crisis probably reflect a similar process, except that cells are unable to exit cell cycle progression because checkpoint gene products are inactivated by oncoprotein action or tumour suppressor gene mutation (see Apoptosis, below). Although polyploidy represents a significant feature of both senescence and crisis, senescent cells most frequently arrest with euploid G_1 DNA content. The nucleocytoplasmic ratio of senescent G_1 cells is much smaller than that of quiescent proliferative G_1 cells (Sherwood *et al*, 1988). Although reported activity or level of specific gene products (mRNA, protein) are sometimes inconsistent, the general picture is that replicatively senescent cells are arrested in late G_1, perhaps corresponding to the "restriction point". In contrast to quiescence, senescent cell cycle arrest is irreversible in the absence of viral oncoproteins, and from the population, perspective tissues age as populations of differentiated cells "semistochastically" lose the capacity to traverse a cell cycle successfully. Changes in the relative proportion of senescent cells in a tissue may contribute to in vivo (organismal) ageing.

Senescence specific gene expression has been examined for a variety of proteins including "housekeeping" genes, transcription factors and cell cycle regulatory activities. Greatest attention has been given to changes in gene transcription (Rittling *et al*, 1986; Goldstein, 1990; Seshardi and Campisi, 1990; Warner *et al*, 1992; Jazwinski *et al*, 1995; Vojta and Barrett, 1995). Although senescence related change in promoter binding of presumed regulatory proteins cells has been shown for the *PAI1* gene, demonstrating that constitutive high level expression of the gene may directly reflect de-regulated gene transcription (West, 1994), senescent cell specific gene expression cascades or regulatory elements have yet to be described (Seshardi and Campisi, 1990; Vojta and Barrett, 1995). An analysis of steady state RNA levels using differential display demonstrated 23 genes specifically expressed in senescent human fibroblasts and 19 expressed at levels similar in quiescent cells (Linskens *et al*, 1995). Surprisingly, genes expressed at similar levels in quiescent and senescent cells following serum stimulation include several that are associated with cellular proliferation, among them *JUN, MYC, MYB, JUNB* and several unidentified cloned sequences. Epidermal and platelet derived growth factor receptors in senescent cells are unaltered in ligand binding and autophosphorylation, although alterations in downstream signal transducing kinase and/or phosphatase activities suggest that mitogenic signal transduction is impaired in senescent cells (Carlin *et al*, 1983; Phillips *et al*, 1983; Cristofalo *et al*, 1989; Afshari *et al*, 1993; Garfinkel *et al*, 1996). Insulin like growth factor 1 receptor expression appears to be downregulated in senescent cells (Ferber *et al*, 1993). In contrast to these genes, *FOS*, an early gene in the mitogenic cascade, is downregulated in senescent cells (Seshardi and Campisi, 1990). Concomitant downregulation of *AP1* transcriptional activity in senescent cells may reflect altered stoichiometry of *FOS*, which is downregulated, and *JUN*, which is not (Riabowol *et al*, 1992). The *JUN-FOS* relationship in senescence may be particularly useful in defining active/inactive signal transduction pathways as related to homoeostatic potential of cells in vivo.

Some late G_1 events, for example, expression of *HRAS* and the ornithine decarboxylase gene (*ODC*), occur in senescent cells in a manner similar to that of serum stimulated quiescent cells (Chang and Chen, 1988). S phase genes, however, are not expressed during quiescence or senescence, including cyclin A, the large subunit of DNA polymerase α, thymidylate synthetase, thymidine kinase and dihydrofolate reductase. However, changes in gene regulation occur posttanscriptionally as well as transcriptionally. *ODC* mRNA levels are similar in serum stimulated quiescent and senescent cells (Seshardi and Campisi, 1990), whereas enzyme activity is downregulated in senescence (Chang and Chen, 1988). Elongation factor EF1 shows a similar disparate change in mRNA and protein activity (Giordano *et al*, 1989). *TP53* message and protein levels require further examination.

The cyclins D1 and E are stably elevated in senescent cells, indicating either that these genes are deregulated or that cell cycle arrest occurs after the point at which these proteins normally begin to accumulate during normal cell cycle progression (Stein *et al*, 1991; Won *et al*, 1992; Lucibello *et al*, 1993; Afshari *et al*, 1993; Dulic *et al*, 1993; Fukami *et al*, 1995). These proteins exist in complexes with specific cyclin dependent kinases (CDKs), which themselves do not appear to be elevated. Although cyclin/CDK complexes increase in abundance during senescence, activating changes in phosphorylation do not occur and the kinase activity of the complexes is inhibited. A key downstream consequence of inactive G_1 cyclin/CDK is the failure of RB phosphorylation and the consequent release of active transcription factor E2F (Stein *et al*, 1990; Futreal and Barrett, 1991). Failure of transcriptional activation of S phase genes probably results from maintained hypophosphorylation of RB, presumably acting in concert with *AP1*. Thus, both transcriptional and posttranscriptional changes in gene expression are correlated with exit from cell cycle progression at senescence, when telomere shortening has proceeded as far as TRFs.

The telomere hypothesis of senescence postulates that signals arising from shortened telomeres are transduced into cell cycle arrest. A critical goal in understanding telomere function is to establish the signalling pathways from chromosomes in general, and telomeres in particular, to cell cycle regulatory processes. Although the nature of the signals and pathways of transduction from telomeres remains speculative, recent data showing the importance of phosphatidylinositol kinases (PIK) in chromosome stability and telomere maintenance provide interesting clues to possible pathways and mechanisms between chromosomes, telomeres and the cell cycle.

ATM

Ataxia-telangectasia (AT) is an autosomal recessive disease cytogenetically characterized by chromosome instability, including a very high frequency of telomere fusions (Kojis *et al*, 1991). The identification of the gene involved in AT, *ATM*, as a phosphatidylinositol-3 kinase suggests the significance of signal transduction pathways in chromosome stability (Savitsky *et al* 1995; Keith and Schreiber, 1996). More compelling still, *ATM* shares significant homology with yeast genes involved in DNA damage sensing checkpoint mechanisms, including *MEC1* (Kato and Ogawa, 1994; Weinert *et al*, 1994), *TOR1* and *TOR2* (Heitman *et al*, 1991; Cafferkey *et al*, 1993; Kunz *et al*, 1993; Brown *et al*, 1994) and TEL1, a protein implicated in telomere dynamics in yeast (Greenwall *et al* 1995; Morrow *et al*, 1995). Yeast telomeres are involved in checkpoint recognition of broken chromosomes (Sandell and Zakian, 1993), but *TOR1* and *TOR2* and *MEC*, which act at a G_1 checkpoint, do not show telomere loss as a consequence of mutational inactivation. Potential involvement of transcriptional changes in AT is suggested by correction of radiation sensitivity in AT fibroblasts by mutations in the gene for IκB-a, an inhibitor of the transcriptional activator NF-κB, an observation linking activities of *ATM* related genes and transcriptional changes associated with alterations in cell cycle progression (Jung *et al*, 1995). The involvement of *TEL1* and *MEC1* in controlling the DNA damage checkpoint kinase RAD53 in yeast further demonstrates the importance of the interactions of these genes in regulating the cell cycle (Sanchez *et al*, 1996). Taken together, these observations indicate that a family of related genes with signal transduction activities are involved in maintenance of chromosome stability, including telomeres acting perhaps to signal chromosome state to transcriptional regulators and the cell cycle machinery. The growing evidence that genes coding for PIK kinases have an important role in chromosome stability and that the cessation of cell cycle progression in senescent cells involves changes in kinase/phosphatase activities points to pathways by which chromatin/chromosome structural or functional organization is transmitted to the cell cycle "engine".

TP53

As a prototypical gene involved in response to DNA damage, the tumour suppressor gene *TP53* has served as a model for transcriptional regulators transducing DNA damage signals in senescent cells. Proliferative human diploid fibroblasts exposed to gamma irradiation show elevated *TP53* levels, cyclin E/CDK2 accumulation and cell cycle arrest at the G_1/S boundary (Dulic *et al*, 1993; DiLeonardo *et al*, 1994). Cyclin E/CDK2 complexes are not activated, however, and downstream phosphorylation of RB does not occur. In this respect, the phenomenology of cell cycle arrest in "damaged" actively proliferating mortal cells resembles that of senescent cells.

The level and activity state of *TP53* in senescent cells have recently been reported, and although there are alterations of *TP53* in cellular senescence, the data do not directly demonstrate that *TP53* is an effector of cell cycle arrest in senescent cells. TP53 levels (mRNA and protein) have been reported to increase (Kulju and Lehman, 1995) or remain unchanged (Atadja *et al*, 1995) in senescent fibroblasts. In the latter study, increased DNA binding and transcriptional activity of *TP53* in senescent cells suggests that posttranslational changes in *TP53* may be involved in regulating senescent gene expression. Viral oncoproteins that extend proliferative lifespan—SV40 T antigen, HPV E6 and adenovirus E1A—all bind to TP53. Radiation damage in proliferative diploid fibroblasts induces TP53 expression and long term senescence like cell cycle arrest (DiLeonardo *et al*, 1994), and senescence or senescence like states can be induced by oxidative damage, suggesting that stable proliferative arrest can be a response to cellular damage (Chen and Ames, 1994; von Zglinicki *et al*, 1996). The status of TP53 and cyclin E/CDK2 complexes has not, however, been determined in these situations.

p21

An additional role for *TP53* in cell cycle exit may be as transcriptional activator of the CDK inhibitor p21 (WAF 1, CIP 1). p21 is a nuclear protein that binds CDK/cyclin–proliferating cell nuclear antigen (PCNA) complexes and is expressed at high levels in quiescent, senescent and differentiated (ie noncycling) cells (Noda *et al*, 1994; Tahara T *et al*, 1995). p21 is an inhibitor of cyclin/CDK kinases as well as of the DNA replication factor PCNA, and thus its activities and regulation would appear to be central in blocking progression into S phase. The importance of p21 in controlling proliferation has been shown by cell growth inhibitory and tumour suppressive activity when the protein is overexpressed in tumour cells (Noda *et al*, 1994) and by the re-entry into the cell cycle of senescent cells in which p21 expression is inhibited by antisense RNA (Nakanishi *et al*, 1995). Association of p21 expression with quiescence as well as senescence indicates that it is not an effector of "irreversible" cell cycle exit. Although a causal connection between TP53 and p21 expression has been suggested by some studies, TP53 independent expression has been reported for differentiating HL-60 cells (Jiang *et al*, 1994) which, combined with reports of a complex pattern of expression in SV40 transformed fibroblasts, indicates that p21 can be regulated independently of TP53 function (Tahara T *et al*, 1995). Although p21 appears to represent an important general mediator of cell cycle arrest, the pathway by which its expression is upregulated is unclear. Relatively unimpaired viability, growth and differentiation of tissues in p21$^{-/-}$ mice indicate that the pathways in which p21 acts are redundant (Deng *et al*, 1995).

CDC25

Cyclin D1/CDK and cyclin E/CDK complexes accumulate in senescent cells but do not undergo an activating change in phosphorylation state. Mitotic transit in proliferating cells depends on the balance of kinase phosphatase activities, of which CDC25 and CDI1 are examples. Identification of CDC25A as a phosphatase critical in regulating transit into S phase demonstrates that a similar mechanism operates in G_1 as well (Gyuris *et al* 1993; Hoffman *et al* 1994; Jinno *et al* 1994).

CDC25A is itself phosphorylated by cyclin E/CDK in an autoamplifying feedback loop similar to processes at the G_2-M checkpoint (Hoffman *et al*, 1994). CDI1 is another mammalian phosphatase, which interacts with cyclin/CDK complexes early in the cell cycle (Gyuris *et al*, 1993). CDC25 is frequently overexpressed in breast tumours, suggesting that deregulated expression can be involved in changes in proliferative potential (Galaktionov *et al* 1995). CDC25A interacts with RAF1 and 14-3-3 proteins, suggesting a role in mitogenic signal transduction cascades significant in controlling proliferation (Conklin *et al*, 1995). Evidence for phosphatase inhibitors (sodium vanadate and okadaic acid) to stimulate DNA synthesis in senescent cells (Afshari *et al*, 1994) suggests a central role for phosphatase activity in cell cycle exit occurring at senescence. Thus, CDC25 phosphatase activity may be central to regulation of early cell cycle transitions and represent one element of the transduction pathway by which signals are transduced to the cell cycle. The status of CDC25A in senescent cells has not been reported.

Apoptosis

The role of cell death in the in vitro senescence of primary human cell populations has yet to be completely defined (Jazwinski *et al*, 1995). Clearly, senescent cells in culture remain functional and fully viable for long periods after ceasing to proliferate, and senescence represents a state of relative physiological stability. Proliferating mortal cells undergo apoptosis induced by nutrient deprivation (Wang, 1995) or following a period of cytostasis induced in proliferating mortal cells by antiproliferative agents, for example, G_1/S arrest induced by the DNA synthesis inhibitor aphidicolin. Thus, mortal cells are competent to carry out apoptosis while proliferating and will do so when arrested in cell cycle progression by drugs, with many of the cell cycle regulatory molecules in a state similar or identical to that seen in senescent cells (eg elevated hypophosphorylated cyclin E/CDK2 complexes). In general, however, non-cycling quiescent or senescent cells are less sensitive to apoptotic inducing treatments than are proliferating cells, whether primary or immortal.

Changes in sensitivity to the induction of apoptosis in senescent cells reflect in part the restriction in the spectrum of potential inducers but may also reflect changes in apoptotic pathways accompanying stable cell cycle exit. The absence of apoptosis during cell cycle arrest in quiescent and senescent cells suggests that a key element of cell cycle exit is a change in the regulation/expression of the apoptotic program. This has been reported to reflect expression of *BCL2* which in senescent WI-38 cells is not downregulated (Wang, 1995). The correlation of *p21* expression with cell cycle exit in quiescent and senescent cells and reduced sensitivity of these cells to apoptosis have also been suggested to indicate that *p21* is a survival factor or a marker for survival pathways, at least in the context of the proapoptotic activity of *TP53* (Xu *et al*,

1995). Overexpression of p21 in tumour cells induces cell cycle arrest at G_1/S but does not induce apoptosis (Katayose *et al*, 1995), suggesting that p21 is directly or indirectly antiapoptotic (Xu *et al*, 1995). This interesting possibility needs further examination.

Crisis is poorly described with respect to cell cycle and apoptoptic events, but it is likely that extensive cell death occurring at this time is apoptotic in nature. Cytogenetic instability is a hallmark of cells undergoing crisis, and it has recently been suggested that telomeric associations (and other chromosomal instabilities) occurring during crisis are a manifestation of apoptosis (Pandita *et al*, 1995). Although a causal relationship between cytogenetic and apoptotic events of crisis has yet to be firmly established, both clearly represent elements of a "syndrome" of cellular pathologies which occurs coincidentally with telomere shortening to the second critical length. Apoptosis during extended lifespan, and particularly at crisis, may arise from conflicting proliferative and antiproliferative signals, the hypothesized role of telomeres being to contribute antiproliferative signals.

SUMMARY

Telomere dynamics and changes in telomerase activity are consistent elements of cellular alterations associated with changes in proliferative state. In particular, the highly specific correlations and early causal relationships between telomere loss in the absence of telomerase activity and replicative senescence or crisis, on the one hand, and telomerase reactivation and cell immortality, on the other, point to a new and important paradigm in the complementary fields of ageing and cancer.

Although the signalling pathways between telomeres and transcriptional and cell cycle machinery remain undefined, recently described homologies between telomeric proteins and lipid/protein kinase activities important in chromosome stability provide evidence for the existence of pathways transducing signals originating in chromosome structure to cell cycle regulatory processes. Similarities between cell cycle arrest at senescence and the response of mortal cells to DNA/oxidative damage suggest overlap in the signal transduction mechanisms culminating in irreversible and stable cell cycle arrest.

The feasibility of targeting telomeres/telomerase as a strategy for antiproliferative therapeutics has been shown in studies in yeast, in which mutations in specific telomere associated genes result in delayed cell death. Similarly, antisense oligonucleotide inhibition of telomerase activity in human tumour cells (HeLa) results in delayed cell death. The mechanism of cell death and possible escape from this fate require further study. In human cells, however, it would seem reasonable to predict that in these circumstances, apoptosis is induced in the vast majority of cells either directly in response to a DNA damage signal arising from critically shortened telomeres or as a secondary consequence of genetic instability.

References

Afshari CA, Vojta PJ, Annab LA, Futreal A, Willard TB and Barrett JC (1993) Investigation of the role of G_1/S cell cycle mediators in cellular senescence. *Experimental Cell Research* **209** 231–237

Allsopp RC and Harley CB (1995) Evidence for a critical telomere length in senescent human fibroblasts. *Experimental Cell Research* **219** 130–136

Allsopp RC, Chang E, Kashefi-Aazam M *et al* (1995) Telomere shortening is associated with cell division in vitro and in vivo. *Experimental Cell Research* **220** 194–200

Atadja P, Wong H, Garkavtsev, Veillette C and Riabowol K (1995) Increased activity of p53 in senescing fibroblasts. *Proceedings of National Academy of Sciences of the USA* **92** 8348–8352

Bacchetti S and Counter CM (1995) Telomeres and telomerase in human cancer. *International Journal of Oncology* **7** 423–432

Barnett MA, Buckle VJ, Evans EP *et al* (1992) Telomere directed fragmentation of mammalian chromosomes. *Nucleic Acids Research* **21** 27–36

Blasco MA, Funk W, Villeponteau B and Greider CW (1995) Functional characterization and developmental regulation of mouse telomerase RNA. *Science* **269** 1267–1270

Blasco MA, Rizen M, Greider CW and Hanahan D (1996) Differential regulation of telomerase activity and telomerase RNA during multi-stage tumorigenesis. *Nature Genetics* **12** 200–204

Bodnar A, Kim NW, Effros RB and Chiu C-P (1996) Mechanism of telomerase induction during T cell activation. *Experimental Cell Research* **228** 58–64

Broccoli D and Cooke H (1993) Aging, healing, and the metabolism of telomeres. *American Journal of Human Genetics* **52** 657–660

Brown EJ, Albers MW, Shin TB *et al* (1994) A mammalian protein targeted by G_1-arresting rapamycin-receptor complex. *Nature* **369** 756–758

Cafferkey R, Young PR, McLaughlin MM *et al* (1993) Dominant negative missense mutations in a novel yeast protein related to mammalian phosphitidylinositol 3-kinase and VP-S34 abrogate rapamycin cytotoxicity. *Molecular Cell Biology* **13** 6012–6023

Campisi J (1996) Replicative senescence: an old lives' tale? *Cell* **84** 497–500

Carlin C, Phillips PD, Knowles BB and Cristofalo VJ (1983) Diminished in-vitro kinase activity of the EGF receptor in senescent human fibroblasts. *Nature* **306** 617–620

Chadeneau C, Siegel P, Harley CB, Muller WJ and Bacchetti S (1995) Telomerase activity in normal and malignant murine tissues. *Oncogene* **11** 893–898

Chang E and Harley CB (1995) Telomere length and replicative aging in human vascular tissues. *Proceedings of the National Academy of Sciences of the USA* **92** 11190–11194

Chang ZF and Chen KY (1988) Regulation of ornithine decarboxylase and other cell cycle-dependent genes during senescence of IMR-90 human diploid fibroblasts. *Journal of Biological Chemistry* **263** 11431

Chen Q and Ames BN (1994) Senescence-like growth arrest induced by hydrogen peroxide in human diploid fibroblast F65 cells. *Proceedings of the National Academy of Sciences of the USA* **91** 4130–4134

Chiu C-P, Dragowska W, Kim NW *et al* (1996) Differential expression of telomerase activity in hematopoietic progenitors from adult human bone marrow. *Stem Cells* **14** 239–248

Chong L, Steensel B, Broccoli D *et al* (1995) A human telomeric protein. *Science* **270** 1663–1667

Collins K, Kobayashi R and Greider CW (1995) Purification of tetrahymena telomerase and cloning of genes encoding the two protein components of the enzyme. *Cell* **81** 677–686

Conklin DS, Galaktinov K and Beach D (1995) 14-3-3 proteins associate with cdc25 phosphatases. *Proceedings of the National Academy of Sciences of the USA.* **92** 7892–7896

Counter CM, Avilion AA, LeFeuvre CE *et al* (1992) Telomere shortening associated with chromosome instability is arrested in immortal cells which express telomerase activity. *EMBO Journal* **11** 1921–1929

Counter CM, Hirte HW, Bacchetti S and Harley CB (1994) Telomerase activity in human ovarian carcinoma. *Proceedings of the National Academy of Sciences of the USA* **91** 2900–2904

Cristofalo VJ, Phillips PD, Sorger T and Gerhard G (1989) Alterations in the responsiveness of senescent cells to growth factors. *Journal of Gerontology* **44** 55–62

De Lange T (1994) Activation of telomerase in human tumor. *Proceedings of the National Academy of Sciences of the USA* **91** 2882–2885

Deng C, Zhang P, Harper JW, Elledge SJ and Leder P (1995) Mice lacking p21 undergo normal development, but are defective in G_1 checkpoint control *Cell* **82** 675–684

Di Leonardo A, Linke SP, Clarkin K and Wahl GM (1994) DNA damage triggers prolonged p53-dependent G_1 arrest and long-term induction of Cip 1 in normal human fibroblasts. *Genes and Development* **8** 2540–2551

Dimri GP, Lee X, Basile G *et al* (1995) A biomarker identifies senescent human cells in culture and in aging skin in vivo. *Proceedings of the National Academy of Sciences of the USA* **92** 9363–9367

Dulic V, Drullinger LF, Lees E, Reed SI and Stein GH (1993) Altered regulation of G_1 cyclins in senescent human diploid fibroblasts: accumulation of inactive cyclin E-cdk2 and cyclin D1-cdk2 complexes. *Proceedings of the National Academy of Sciences of the USA* **90** 11034–11038

Fang G and Cech TR (1993) Characterization of a G-quartet formation reaction promoted by the beta--subunit of the *Oxytricha* telomere binding protein. *Biochemistry* **32** 11646–11657

Feng J, Funk WD, Wang S-S *et al* (1995) The RNA component of human telomerase. *Science* **269** 1236 – 1241

Ferber AC, Chang C, Sell A *et al* (1993) Failure of senescent human fibroblasts to express the insulin-like growth factor gene IGF-1. *Journal of Biological Chemistry* **268** 17883–17888

Fukami J, Anno K, Ueda K, Takaashi T and Ide T (1995) Enhanced expression of cyclin D1 in senescent human fibroblasts. *Mechanisms of Ageing and Development* **81** 139–157

Futreal PA and Barrett JC (1991) Failure of senescent cells to phosphorylate the RB protein. *Oncogene* **6** 1109–1113

Galaktionov K, Lee AK, Eckstein J *et al* (1995) CDC25 phosphatases as potential human oncogenes. *Science* **269** 1575–1577

Garfinkel S, Hu X, Prudovsky IA *et al* (1996) FGF-1-dependent proliferative and migratory responses are impaired in senescent human umbilical vein endothelial cells and correlate with the inability to signal tyrosine phosphorylation of fibroblast growth factor receptor substrates. *Journal of Cell Biology* **134(3)** 783–791

Giordano T, Kleinsek D and Foster DN (1989) Increase in abundance of a transcript hybridizing to elongation factor 1 alpha during cellular senescence and quiescence. *Experimental Gerontology* **24** 501–513

Goldstein S (1990) Replicative senescence: the human fibroblast comes of age. *Science* **249** 1129–1133

Greenwall PW, Kronmal SL, Porter SE, Gassenhuber J, Obermaier B and Petes TD (1995) TEL1, a gene involved in controlling telomere lkength in *S. cerevisiae* is homologous to the ataxia telangectasia gene. *Cell* **82** 841–852

Greider CW Telomere length regulation. *Annual Review Biochemistry* (in press)

Greider CW and Blackburn EH (1985) Identification of a specific telomere terminal transferase enzyme with two kinds of primer specificity. *Cell* **51** 405–413

Greider CW and Blackburn EH (1989) A telomeric sequence in the RNA of tetrahymena telomerase required for telomere repeat synthesis. *Nature* **337** 331–337

Greider CW and Blackburn EH (1996) Telomeres, telomerase and cancer. *Scientific American* **274** 92–97

Gyuris J, Golemis E, Chertkov H and Brent R (1993) Cdi1, a human G_1 and S phase protein phosphatase that associates with cdk2. *Cell* **75** 791–803

Harley CB (1991) Telomere loss: mitotic clock or genetic time bomb? *Mutation Research* **256** 271–282

Harley CB, Futcher AB and Greider CW (1990) Telomeres shorten during ageing of human fibroblasts. *Nature* **345** 458–460

Harley CB, Kim NW, Prowse KR *et al* (1994) Telomerase, cell immortality, and cancer. *Cold Spring Harbor Symposia on Quantitative Biology* **59** 1–9

Harley CB and Villeponteau B (1995) Telomeres and telomerase in aging and cancer. *Current Opinion in Genetics and Development* **5** 249–255

Hastie ND, Dempster M, Dunlop MG, Thompson AM, Green DK and Allshire RC (1990) Telomere reduction in human colorectal carcinoma and with ageing. *Nature* **346** 866–868

Heitman J, Movva NR and Hall MN (1991) Targets for cell cycle arrest by the immunosuppressant rapamycin in yeast. *Science* **253** 905–909

Henderson S, Allsopp R, Spector D, Wang S-S and Harley C (1996) In situ analysis of changes in telomere size during replicative aging and cell transformation. *Journal of Cell Biology* **134** 1–12

Hiyama E, Hiyama K, Yokoyama T, Mitsuura Y, Piatyszek MA and Shay JW (1995) Correlating telomerase activity levels with human neuroblastoma outcomes. *Nature Medicine* **1** 249–255

Hoffmann I, Draetta G and Karsenti E (1994) Activation of the phosphatase activity of human cdc25a by a cdc2-cyclin E dependent phosphorylation at the G_1/S transition. *EMBO Journal* **13(18)** 4302–4310

Jazwinski SM, Howard BH and Nayak RK (1995) Cell cycle progression, aging and cell death. *Journal of Gerontology* **50A** B1–B8

Jiang H, Lin J, Su Z-z, Collart FR, Huberman E and Fisher PB (1994) Induction of differentiation in human promyelocytic HL-60 leukemia cells activates p21, WAF1/Cip1, expression in the absence of p53. *Oncogene* **9** 3397–3406

Jinno S, Suto K, Nagata A *et al* (1994) Cdc25a is a novel phosphatase functioning early in the cell cycle. *EMBO Journal* **13(7)** 1549–1556

Jung M, Zhang Y, Lee S and Dritschilo A (1995) Correction of radiation senstivity in ataxia telangectasia cells by a truncated IkB-alpha. *Science* **268** 1619–1621

Katayose D, Wersto R, Cowan KH and Seth P (1995) Effects of a recombinant adenovirus expressing WAF1/Cip1 on cell growth, cell cycle and apoptosis. *Cell Growth and Differentiation.* **6** 1207–1212

Kato R and Ogawa H (1994) An essential gene, ESR-1, is required for mitotic cell growth, DNA repair and meiotic recombination in *Saccharomyces cerevisiae*. *Nucleic Acids Research* **23** 3104–3112

Keith CT and Schreiber SL (1996) PIK-related kinases: DNA repair, recombination and cell cycle checkpoints. *Science* **270** 50–51

Kennedy BK, Austriaco NR Jr, Zhang J and Guarente L (1995) Mutation in the silencing gene SIR4 can delay aging in S *cerevisiae*. *Cell* **80** 485–496

Kim NW, Piatyszek MA, Prowse KT *et al* (1994) Specific association of human telomerase activity with immortal cells and cancer. *Science* **266** 2011–2014

Kipling D and Cooke HJ (1990) Hypervariable ultra-long telomeres in mice. *Nature* **347** 400–402

Kipling D and Cooke HJ (1992) Beginning or end? Telomere structure, genetics and biology. *Human Molecular Genetics* **1** 3–6

Klingelhutz AJ, Barber SA, Smith PP, Dyer K and McDougall JK (1993) Restoration of telomeres in human papillomavirus-immortalized human anogenital epithelial cells. *Molecular and Cellular Biology* **144** 961–969

Kojis TL, Gatti RA and Sparkes RS (1991) The cytogenetics of ataxia telangectasia. *Cancer Genetics and Cytogenetics* **56** 143–156

Kulju KS and Lehman L (1995) Increased p53 protein associated with aging in human diploid fibroblasts. *Experimental Cell Research* **21** 336–345

Kunz J, Henriquez R, Schneider U, Deuter-Reinhard H, Movva NR and Hall MN (1993) Target for rapamycin in yeast TOR2 is an essential phosphitidylinositol kinase homolg required for G_1 progression. *Cell* **73** 585–596

Lin J-J and Zakian VA (1995) An in vitro assay for *Saccharomyces* telomerase requires EST1. *Cell* **81** 1127–1135

Lindsey J, McGill NI, Lindsey LA, Green DK and Cooke HJ (1991) In vivo loss of telomeric repeats with age humans. *Mutation Research* **256** 45–48

Linskens MHK, Feng J, Andrews WH *et al* (1995) Cataloging altered gene expression in young and senescent cells using enhanced differential display. *Nucleic Acids Research* **23** 3244–3251

Lucibello FC, Sewing A, Brusselbach S, Burger C and Muller R (1993) Deregulation of cyclins D1 and E and suppression of cdk2 and cdk4 in senescent human fibroblasts. *Journal of Cell Science* **105** 123–133

Lundblad V and Szostak JW (1989) A mutant with defect in telomere elongation leads to senescence in yeast. *Cell* **57** 633–643

McEachern MJ and Blackburn EH (1995) Runaway telomere elongation caused by telomerase RNA gene mutations. *Nature* **376** 403 – 409

McEachern MJ and Blackburn EH (1996) Cap-prevented recombination between terminal telomeric repeat arrays (telomere CPR) maintains telomeres in *Kluyveromyces lactis* lacking telomerase. *Genes and Development* **10** 1822–1834

Morin GB (1989) The human telomere terminal transferase enzyme is a ribonucleoprotein that synthesizes TTAGGG repeats. *Cell* **59** 521–529

Morrow DM, Tagle DA, Shiloh Y, Collins FS and Hieter P (1995) TEL1, an *S. cerevisiae* homolog of the human gene mutated in ataxia telangiectasia, is functionally related to the yeast checkpoint gene MEC1. *Cell* **82** 831–840

Moyzis RK, Buckingham JM, Cram LS *et al* (1988) A highly conserved repetitive DNA sequence (TTAGGG)n, present at the telomeres of human chromosomes. *Proceedings of the National Academy of Sciences of the USA* **85** 6622–6626

Murray AW (1992) Creative blocks: cell cycle checkpoints and feedback controls. *Nature* **359** 599–604

Nakanishi M, Adami GR, Robetoyre RS *et al*(1995) Exit from G0 and entry into the cell cycle of cells expressing p21 antisense RNA. *Proceedings of the National Academy of Sciences of the USA* **92** 4352–4356

Noda A, Ning Y, Venable SF, Pereira-Smith O and Smith JR (1994) Cloning of senescent cell-derived inhibitors of DNA synthesis using an expression screen. *Exerimental Cell Research* **211** 90–98

Pandita TK, Pathak S and Geard CR (1995) Chromosome and associations, telomeres and telomerase activity in ataxia telangiectasia cells. *Cytogenetics and Cell Genetics* **71** 86–93

Phillips PD, Kuhnle E and Cristofalo VJ (1983) [125] EGF binding ability is stable throughout the replicative lifespan of WI-38 cells. *Journal of Cell Physiology* **114** 311–316

Prowse KR and Greider CW (1995) Developmental and tissue specific regulation of mouse telomerase and telomere length. *Proceedings of the National Academy of Sciences of the USA* **92** 4818–4822

Riabowol K, Schiff J and Gilman MZ (1992) Transcription factor AP-1 activity is required for initiation of DNA synthesis and is lost during cellular aging. *Proceedings of the National Academy of Sciences of the USA* **89** 157–161

Rittling SR, Brooks CM, Cristofalo VJ and Baserga R (1986) Expression of cell-cycle dependent genes in young and senescent WI-38 cells. *Proceedings of the National Academy of Sciences of the USA* **83** 3316–3320

Sanchez Y, Desnay BA, Jones WJ, Liu Q, Wang B and Elledge SJ (1996) Regulation of RAD53 by the ATM-like kinases MEC1 and TEL1 in yeast cell cycle checkpoint pathways. *Science* **271** 357–360

Sandell LL and Zakian VA (1993) Loss of a yeast telomere: arrest, recovery, and chromosome loss. *Cell* **75** 729–739

Savitsky K, Sfez S, Tagle DA *et al* (1995) The complete sequence of the coding region of the ATM gene reveals simlarity to cell cycle regulators in different species. *Human Molecular Genetics* **4** 2025–2032

Schimke RT, Kung AL, Rush DF and Sherwood SW (1991) Differences in mitotic control among mammalian cell lines. *Cold Spring Harbor Symposia on Quantitative Biology* **56** 417–425

Shay JW and Wright WE (1996) Telomerase activity in human cancer. *Current Opinion in Oncology* **8** 66–71

Shay JW, Pereira-Smith OM and Wright WE (1991) A role for both RB and p53 in the regulation of human cellular senescence. *Experimental Cell Research* **196** 33–39

Seshardi T and Campisi J (1990) Repression of c-fos transcription and altered genetic program in senescent human fibroblasts. *Science* **247** 205–209

Shay JW, Wright WE, Brasiskyte D and Van DerHaegen BA (1993) E6 of human papillomavirus type 16 can overcome the M1 stage of immortalization in human mammary epithelial cells but not in human fibroblasts. *Oncogene* **8** 1407–1413

Sherwood SW, Rush DR, Ellsworth JL and Schimke RT (1988) Defining cellular senescence in IMR-90 cells. A flow cytometric analysis. *Proceedings of the National Academy of Sciences of the USA* **85** 9086–9090

Singer MS and Gottschling DE (1994) TLC1: template RNA component of *Saccharomyces cerevisiae* telomerase. *Science* **266** 404–409

Smith JR and Pereira-Smith OM (1996) Replicative senescence: implications for in vivo aging and tumor suppression. *Science* **273** 63–67

Starling JA, Maule J, Hastie ND and Allshire RC (1990) Extensive telomere repeat arrays in mouse are hypervariable. *Nucleic Acids Research* **18** 6881–6888

Stein GH, Beeson M and Gordon L (1990) Failure to phosphorylate the retinoblastoma gene product in senescent human fibroblasts. *Science* **249** 666–669.

Stein GH, Drullinger LF, Robetory RS, Pereira-Smith O and Smith J (1991) Senescent cells fail to express cdc2, cycA and cycB in response to mitogen stimulation. *Proceedings of the National Academy of Sciences of the USA* **889** 11012–11016

Steiner BR, Hidaka K and Futcher B (1996) Association of the Est1 protein with telomerase activity in yeast. *Proceedings of the National Academy of Sciences of the USA* **93** 2817–2821

Tahara H, Nakanishi T, Kitamoto M *et al* (1995) Telomerase activity in human liver tissues: comparison between chronic liver disease and hepatocellular carcinomas. *Cancer Research* **55** 2734–2736

Tahara T, Sato E, Noda A and Ide T (1995) Increase in expression level of p21 with increasing division age in both normal and SV-40 transformed human fibroblasts. *Oncogene* **10** 835–840

Taylor RS, Ramirez RD, Ogoshi M, Chaffins M, Piatyszek MA and Shay JW (1995) Detection of telomerase activity in malignant and nonmalignant skin conditions. *Journal of Investigative Dermatology* **106** 759–765

Vaziri H, Schachter F, Uchida I *et al* (1993) Loss of telomeric DNA during aging of normal and trisomy 21 human lymphocytes. *American Journal of Human Genetics* **52** 661–667

Vaziri H, Dragowska W, Allsopp RC, Thomas TE, Harley CB and Lansdorp PM (1994) Evidence for a mitotic clock in human hematopoietic stem cells: loss of telomeric DNA with age. *Proceedings of the National Academy of Sciences of the USA* **91** 9857–9860

Vojta PJ and Barrett JC (1995) Genetic analysis of cellular senescence. *Biochimica et Biophysica Acta* **1242** 29–41

von Zglinicki T, Saretzki G, Docke W and Lotze C (1995) Mild hyperoxia shortens telomeres and inhibits proliferation of fibroblasts: a model for senescence? *Experimental Cell Research* **220** 186–193

Wang E (1995) Senescent human fibroblasts resist programmed cell death, and failure to suppress bcl2 is involved. *Cancer Research* **55** 2284–2292

Warner HR, Campisi J, Cristofalo VJ *et al* (1992) Control of cell proliferation in senescent cells. *Journal of Gerontology* **47(6)** B185–B189

Weinert TA, Kiser GL and Hartwell LH (1994) Mitotic checkpoint genes in budding yeast and the dependence of mitosis on DNA replication and repair. *Genes and Development* **8** 652–665

West MD (1994) The cellular and molecular biology of skin aging. *Archives of Dermatology* **130** 87–95

Won K-A, Xiong Y, Beach D and Gilman ML (1992) Growth-regulated expression of D-type cyclin genes in human diploid fibroblasts. *Proceedings of the National Academy of Sciences of the USA* **89** 9910–9914

Wright WE and Shay JW (1992) Telomere positional effects and the regulation of cellular senescence. *Trends In Genetics* **8** 193–197

Wright WE, Brasiskyte D, Piatyszek MA and Shay JW (1996) Experimental elongation of telomeres extends the lifespan of immortal x normal cell hybrids. *EMBO Journal* **15** 1734–1741

Xu C, Meikrantz W, Schlegel R and Sager R (1995) The human papilloma virus 16E6 gene sensitizes human mammary epithelial cells to apoptosis induced by DNA damage. *Proceedings of the National Academy of Sciences of the USA* **92** 7829–7833

Zakian V (1996) Telomere functions: lessons from yeast. *Trends in Cell Biology* **6** 29–33

The authors are responsible for the accuracy of the references.

The *ATM* Gene and Protein: Possible Roles in Genome Surveillance, Checkpoint Controls and Cellular Defence against Oxidative Stress

GALIT ROTMAN • YOSEF SHILOH

Department of Human Genetics, Sackler School of Medicine, Tel Aviv University, Ramat Aviv 69978

Ataxia-telangiectasia: a defect in a pleiotropic physiological function
    Clinical and laboratory characteristics of AT
    Cellular characteristics of AT
The AT gene (*ATM*) and its protein product
Summary

ATAXIA-TELANGIECTASIA: A DEFECT IN A PLEIOTROPIC PHYSIOLOGICAL FUNCTION

The human genetic disorder ataxia-telangiectasia (AT) is a prime example of a defect in a single gene affecting numerous cellular systems (reviewed in Sedgwick and Boder, 1991). The clinical and cellular manifestations of this disease include abnormalities in the development and function of various tissues, premature ageing, cancer predisposition, genome instability, defective cellular responses to certain DNA damaging agents and aberrant cell cycle checkpoint controls. Extensive studies of the cellular phenotype of AT and the recent cloning of the responsible gene, *ATM*, have provided initial clues to the nature of the elusive physiological junction linking all of these phenomena to a single protein.

Clinical and Laboratory Characteristics of AT

Progressive ataxia that develops into a general neuromotor dysfunction is the early and most profound sign of AT (reviewed in Harnden 1994; Shiloh, 1995; Lavin and Shiloh, in press). The predominant neuropathological defect is cerebellar degeneration involving primarily the Purkinje cells, followed at a later age by degenerative changes in other parts of the central nervous system. The second external hallmark is telangiectases—dilated blood vessels in the conjunctivae and sometimes in the facial skin. Recurrent infections, primarily sinopulmonary, affect up to 80% of AT patients and reflect an array of immune defects spanning both cellular and humoral systems. Hypogammaglobulin-aemia is manifested as varying reductions in immunoglobulin A, G_2 and E levels, and cellular immunity defects are expressed by poor response to mitogens and diminished number of circulating lymphocytes. A striking developmental abnormality associated with the immune system is the absence or severe degeneration of the thymus. Elevated levels of two oncofetal proteins, alpha fetoprotein and carcinoembryonic antigen, indicate a possible developmental abnormality in the liver (Lavin, 1993). Acute cancer predisposition is another major feature of AT. Patients are predisposed mainly to a range of lymphoid malignancies of B cell and T cell origin, but at later ages, a variety of solid tumours may appear (Hecht and Hecht, 1990; Peterson et al, 1992). AT heterozygotes were reported to be more susceptible to various malignancies, particularly breast cancer in women (Swift et al, 1991; Easton, 1994). Other clinical manifestations of AT include premature ageing, a high incidence of female hypogonadism, growth retardation, delayed puberty and, less commonly, insulin resistant diabetes.

Cellular Characteristics of AT

The major cellular characteristics of AT are chromosomal instability and radio-sensitivity (reviewed in McKinnon, 1987; Taylor et al, 1989, 1994; Gatti, 1991). An increased rate of chromosomal breakage is found in various types of cells from AT patients. Lymphocytes occasionally exhibit clonal expansion of trans-locations involving primarily the sites of the immune system genes; these translocations have been implicated in the generation of lymphoid malignancies observed in AT patients (reviewed in Kojis et al, 1991).

The acute sensitivity of AT patients to ionizing radiation was first noted as severe reaction to radiotherapy (Morgan et al, 1968). Cultured AT cells were found to be hypersensitive to the cytotoxic and clastogenic effects of this radiation, as well as to a variety of radiomimetic chemicals that induce oxidative stress (reviewed in Lehmann, 1982; Shiloh et al, 1985a; McKinnon, 1987). Cells from AT heterozygotes show a moderate degree of sensitivity to the same agents (Shiloh et al, 1982; Arlett and Priestly, 1985; Paterson et al, 1985).

DNA repair mechanisms have been repeatedly found to be normal in AT cells. However, extensive investigation of the defective response of these cells to DNA damaging agents revealed significant abnormalities in the G_1/S, S and G_2/M checkpoints (Nagasawa and Little, 1983; Smith et al, 1985; Beamish and Lavin, 1994; Beamish et al, 1994; Lavin et al, 1994). The arrest of cells at the G_1/S checkpoint, to allow repair of radiation induced damage, is less pronounced in AT cells. A defect in an S phase checkpoint is evident as reduced inhibition of DNA synthesis by ionizing irradiation, known as radioresistant DNA synthesis, which is characteristic of AT cells (Houldsworth and Lavin, 1980; Painter and Young, 1980). There are two separate and opposing abnormal G_2 responses of AT cells: a shorter G_2 delay in cells irradiated at this phase, and a greater delay of G_2 in cells irradiated at earlier stages of the cell cycle. The latter probably reflects the accumulation of cells with severe unrepaired DNA damage, which does not allow progression through mitosis (Scott *et al*, 1994; Paules *et al*, 1995).

Radiation induced cell cycle arrest at the G_1/S checkpoint is mediated by a signal transduction system involving a rise in cellular levels and/or activation of the TP53 protein, which activates the transcription of several genes (recently reviewed in Cox and Lane, 1995; Bates and Vousden, 1996). The best studied of these genes encodes the p21 protein (also called CIP1 or WAF1), which acts as an inhibitor of various cyclin dependent kinases, preferentially affecting cell cycle progression through the G_1/S checkpoint (Gu *et al*, 1993; Harper *et al*, 1993, 1995; Dulic *et al*, 1994; El Deiry *et al*, 1994; Harper and Elledge, 1996). p21 also inhibits the essential DNA replication factor proliferating cell nuclear antigen (PCNA) (Waga *et al*, 1994), a subunit of the DNA polymerase δ enzyme complex, which appears to be involved in both DNA replication and DNA repair. The phosphorylation of the retinoblastoma gene product (RB) by cyclin dependent kinases is inhibited following TP53 dependent p21 expression (reviewed in Picksley and Lane, 1994; White, 1994; Hinds, 1995; Kouzarides, 1995). The hypophosphorylated form of RB sequesters the transcription factor E2F, thereby inhibiting progression through G_1 by preventing expression of E2F responsive genes, whose products are required for entry into S phase (reviewed in Muller, 1995).

Another gene transcriptionally activated by TP53 is *GADD45* (Kastan *et al*, 1992; Zhan *et al*, 1994). The product of this gene has been shown to interact with p21 and PCNA (Smith *et al*, 1994; Chen *et al*, 1995; Hall *et al*, 1995; Kearsey *et al*, 1995) and may have a role in co-ordinating various aspects of DNA repair and replication.

The *MDM2* gene is also induced by TP53 following irradiation (Momand *et al*, 1992; Barak *et al*, 1993; Oliner *et al*, 1993; Price and Park, 1994), and its protein seems to inactivate TP53 transcriptional activity as part of a negative regulatory system that terminates the cellular response to DNA damage (Wu *et al*, 1993; Chen *et al*, 1995). However, MDM2 not only releases a proliferative block by silencing TP53, but also augments proliferation by interacting with RB and inhibiting its growth suppressive function (Xiao *et al*, 1995) and by activating the transcription factors E2F1 and DP1 which are involved in S phase progression (Martin *et al*, 1995). TP53 also participates in a G_2/M checkpoint (Agarwal *et al*, 1995; Stewart *et al*, 1995) and a mitotic checkpoint (Cross *et al*, 1995). A possible S phase block may also be mediated by TP53, leading either to inhibition of PCNA by p21 and GADD45 or to direct inhibition by TP53 of the DNA replication protein A (RPA; Dutta *et al*, 1993).

AT cells show a reduced and delayed activation of TP53 following treatment with ionizing radiation (Kastan *et al*, 1992; Lu and Lane, 1993; Khanna and Lavin, 1993). Several radiation induced TP53 dependent signal transduction pathways were found defective in AT. Among these defects is reduced or delayed induction of p21, GADD45 and MDM2 by radiation (Kastan *et al*,

1992; Canman *et al*, 1994; Price and Park, 1994; Khanna *et al*, 1995; Beamish *et al*, 1996). No inhibition of cyclin E and cyclin A associated kinase activity was observed in AT cells following irradiation, in agreement with the delayed p21 response (Khanna *et al*, 1995). This lack of inhibition was also reflected in accumulation of the hyperphosphorylated (inactive) form of RB in these cells (Khanna *et al*, 1995). These results agree with the observed failure of AT cells to delay progression from G_1 to S phase after irradiation. Lack of postirradiation inhibition of cyclin dependent kinase (CDK) activities in S phase and G_2 phase was also noted in AT cells, which correlated with lack of significant increase in the amount of CDK associated p21 (Beamish *et al*, 1996). These results suggest that the defective TP53/p21 mediated response to ionizing radiation in AT cells is indeed related to the abnormalities in cell cycle checkpoints.

A recent report showed that the S phase checkpoint, manifested as radiation induced inhibition of DNA synthesis in normal cells, is not mediated by TP53 but by a calmodulin dependent signal transduction pathway (Mirzayans *et al*, 1995). Malfunction in this pathway was observed in AT fibroblasts, and deficiencies in intracellular Ca^{++} homoeostatic mechanisms involving calmodulin were suggested to be the cause of the RDS phenomena observed in AT cells (Mirzayans *et al*, 1995).

Defective regulation of RPA, another protein implicated in cell cycle progression, was observed in AT cells (Liu and Weaver, 1993). RPA is a multifunctional single strand binding protein complex required in DNA replication, recombination and repair (reviewed in Stillman *et al*, 1992). The phosphorylation of the p34 subunit of this holoenzyme is cell cycle regulated (Din *et al*, 1990; Dutta and Stillman, 1992) and is rapidly induced in response to ionizing (Liu and Weaver, 1993) and ultraviolet (UV) radiation (Carty *et al*, 1994). RPA may regulate the transcription of a number of genes involved in DNA repair and DNA metabolism (Singh and Samson, 1995). An apparent delay or decrease in the accumulation of the radiation dependent phosphorylated form of the RPA p34 subunit was observed in AT cells (Liu and Weaver, 1993).

THE AT GENE (*ATM*) AND ITS PROTEIN PRODUCT

The AT locus was first mapped to chromosome 11q22–23 by Gatti *et al* (1988). Extensive physical mapping and repeated linkage studies by a number of groups narrowed the search interval for the AT gene to less than 1 Mb of DNA (McConville *et al*, 1994; Rotman *et al*, 1994; Lange *et al*, 1995). The gene was finally identified and cloned in our laboratory and designated *ATM* (AT, mutated) (Savitsky *et al*, 1995a,b). The *ATM* gene occupies approximately 150 kb of genomic DNA and is transcribed into a large transcript of about 13 kb, representing 66 exons (Savitsky *et al*, 1995b; Uziel *et al*, 1996). Most *ATM* mutations identified in AT patients are expected to inactivate the ATM protein completely by truncation or large deletions (Savitsky *et al*, 1995a; Byrd *et al*, 1996; Gilad *et al*, 1996; Telatar *et al*, 1996).

The open reading frame of the *ATM* transcript encodes a protein of 3056 aminoacids (Savitsky *et al*, 1995b) with an expected molecular mass of 350 kDa. The ATM protein belongs to an expanding family of large proteins involved in cell cycle progression and checkpoint response to DNA damage (Savitsky *et al*, 1995b; reviewed in Jackson, 1995; Keith and Schreiber, 1995; Lavin *et al*, 1995; Zakian 1995). The C-termini of all members of this family are highly conserved and show sequence similarity to the catalytic domain of phosphatidylinositol (PI) 3-kinases (reviewed in Carpenter and Cantley, 1996).

This family of PI3-kinase like proteins can be divided into several subgroups. One subgroup, defined by sequence and functional similarities, includes Tor1p and Tor2p of *Saccharomyces cerevisiae* (Cafferkey *et al*, 1993; Kunz *et al*, 1993; Helliwell *et al*, 1994) and their mammalian homologues, RAFT1 and FRAP (Brown *et al*, 1994; Sabatini *et al*, 1994; Sabers *et al*, 1995). These proteins were identified as the targets of the immunosuppressant rapamycin and are involved in a pathway that leads to progression through the G_1 phase of the cell cycle. However, ATM is more closely related to another subgroup of this family—one that includes the cell cycle checkpoint proteins Mec1p (Esr1p) of *S cerevisiae*, rad3 of *Schizosaccharomyces pombe*, mei-41 of *Drosophila melanogaster* and the recently identified FRP1 of humans (Kato and Ogawa, 1994; Hari *et al*, 1995; Cimprich *et al*, 1996).

S cerevisiae Tel1p, which shows the highest homology with ATM, is involved in maintenance of telomere length rather than cell cycle control (Lustig and Petes, 1986; Greenwell *et al*, 1995). AT cells, like *tel1* mutants, have shortened telomeres (Pandita *et al*, 1995). Tel1p and the cell cycle checkpoint protein Mec1p probably share some overlapping functions, since overexpression of TEL1 partially restores the checkpoint function of *mec1* mutants (Morrow *et al*, 1995), and *mec1/tel1* double mutants are synergistically sensitive to ionizing and UV radiation and hydroxyurea. The double mutants, like AT cells, display sensitivity to radiomimetic drugs to which neither mutant alone is sensitive (Morrow *et al*, 1995).

Another member of the PI3-kinase like protein family is the catalytic subunit of DNA dependent protein kinase (DNA-PK$_{cs}$), which also contains the PI3-kinase like C-terminal domain (Hartley *et al*, 1995). The holoenzyme has an additional component, the Ku antigen, that probably acts as a sensor of DNA breaks and activates the serine/threonine protein kinase activity of the catalytic subunit, which subsequently phosphorylates various DNA bound substrates (reviewed in Jeggo *et al*, 1995; Anderson and Carter, 1996). SCID mice are defective in DNA-PK$_{cs}$ and display hypersensitivity to ionizing radiation, defective double strand break repair and V(D)J recombination, which result in severe immunodeficiency (Bosma and Carrol, 1991).

Despite similarities with the PI3-kinase catalytic domain, no member of the ATM related family of proteins has proved to possess a lipid kinase activity; several of these enzymes were, however, found to contain intrinsic protein kinase activity (reviewed in Hunter, 1995). DNA-PK, for example, is a genuine serine/threonine protein kinase that phosphorylates numerous targets in vitro (Anderson, 1993). FRAP can autophosphorylate on serine in vitro, and the p70 S6 kinase (Proud, 1996) is a candidate target for its kinase activity (Brown EJ et al, 1995). Human FRP1 (also designated ATR) has been shown to have an associated protein kinase activity in vitro (Keegan et al, 1996).

It is very probable that ATM, like other members of this family, is a protein kinase, suggesting a role for this protein in signal transduction systems. The predominance of the checkpoint defects in the cellular phenotype in AT and the similarity of ATM to Mec1p, rad3 and mei-41 suggest ATM involvement in activation of cell cycle checkpoints in response to DNA damaging agents. On the other hand, telomere shortening in AT cells and yeast *tel1* mutants argues for a role in telomere maintenance as well. Another human member of the PI3-kinase like protein family, DNA-PK$_{cs}$, is probably not involved in either cell cycle regulation or telomere maintenance but rather in DNA double strand break repair and site specific recombination. Taken together, these observations are commonly viewed as representing a possible role for ATM in genome surveillance, leading to activation of several signal transduction pathways in response to specific types of DNA damage.

The genomic instability and cancer proneness observed in AT may be a direct consequence of the defective cell cycle checkpoints, since loss of the TP53 dependent response is associated with genome instability and tumorigenesis (Hartwell, 1992; Hartwell and Kastan, 1994; Tlsty et al, 1995). There is general consensus in the literature that the sensitivity of AT cells to ionizing radiation cytotoxicity is also due to the defective cell cycle response to DNA damage. However, as pointed out previously (Mirzayans and Paterson, 1991; Murnane and Kapp, 1993; Murnane and Schwartz, 1993), several lines of evidence suggest that the hypersensitivity of AT cells to the cytotoxic effects of ionizing radiation may not result directly from their inefficient repair of DNA damage and aberrant cell cycle response to radiation: (a) unlike control cells, AT cell cultures do not recover from potentially lethal damage when maintained in a non-proliferating state for several hours after irradiation (Cox et al, 1981; Arlett and Priestly, 1983; Little and Nagasawa, 1985); (b) upon lethal doses of radiation, normal cells divide several times before death, whereas the majority of AT cells undergo cell death while arrested in G$_2$, before their first postirradiation mitoses and before any segregation of DNA damage (Beamish and Lavin, 1994); (c) radiation sensitivity and abnormal cell cycle delay in AT are separable phenotypes (Lehmann et al, 1986; Kapp and Painter, 1989; Komatsu et al, 1989; Mirzayans and Paterson, 1991; Sasaki and Taylor, 1994; Ziv et al, 1995); and (d) loss of the TP53 dependent cell cycle arrest of normal cells does not decrease their postirradiation survival (Slichenmyer et al, 1993).

The sensitivity of AT cells to ionizing radiation could be due to failure of a more general cellular response to this agent. The inducible transcription factor NF-κB stands out as a central co-ordinating regulator of the response to ionizing radiation. A broad spectrum of mostly pathogenic conditions induce NF-κB, leading to subsequent activation of multiple target genes involved in cellular defence mechanisms to such pathogens (reviewed in Baeuerle and Henkel, 1994; Siebenlist *et al*, 1994; Thanos and Maniatis, 1995). Unlike most other transcription factors, NF-κB is not normally located in the nucleus and is not bound to DNA. Before stimulation, NF-κB resides in the cytoplasm and is associated with its inhibitor, IκB, which retains it in the cytoplasm in the absence of activating signals (reviewed in Siebenlist *et al*, 1994; Israel, 1995; Thanos and Maniatis, 1995). IκB-α is one member of this family of proteins that has been characterized in great detail. Activation of NF-κB requires a signal induced phosphorylation of IκB-α, which is followed by complete proteolysis of the inhibitor. An autoregulatory pathway exists between NF-κB and IκB-α: although cytoplasmic IκB-α is rapidly degraded in response to activating stimuli, de novo synthesis of IκB-α is promoted by NF-κB transcriptional activation of the IκB-α promoter. Newly synthesized IκB-α can enter the nucleus and bind the activated NF-κB, thereby restoring its inhibited state (Brown *et al*, 1993; Sun *et al*, 1993; Chiao *et al*, 1994). This feedback inhibition may ensure a transient response once the initiating event fades, an essential feature for the regulation of genes whose functions may be harmful when escaping tight control. The inducible phosphorylation sites of IκB-α were mapped to two serine residues at the N-terminus of this protein (Brown K *et al*, 1995; DiDonato *et al*, 1996). Several protein kinases were shown to phosphorylate IκB-α in vitro (Liou and Baltimore, 1993; Israel, 1995). However, the physiologically relevant kinase(s) responsible for the phosphorylation of IκB-α remains to be identified.

Cells exposed to ionizing radiation and various radiomimetic drugs increase the production of reactive oxygen intermediates (ROIs), including hydroxyl radicals, superoxide anions and hydrogen peroxide, leading to a state of oxidative stress (reviewed in Vuillaume, 1987). It has been suggested that ROIs act as a common messenger for various NF-κB inducing agents, since a variety of antioxidants suppress NF-κB activation by such inducers (Schreck *et al*, 1992). Support for this notion also comes from the increased levels of ROIs observed in response to several NF-κB inducing agents (reviewed in Schreck *et al*, 1992; Siebenlist *et al*, 1994). It is noteworthy that induction of DNA damage by ROIs is not a prerequisite for activation of NF-κB.

AT cells are sensitive to a variety of ROIs generators, such as ionizing radiation, bleomycin, neocarzinostatin, streptonigrin and adriamycin (Taylor *et al*, 1975; Morris *et al*, 1983; Shiloh *et al*, 1983, 1985a). In addition, AT cells are more sensitive to the direct addition of hydrogen peroxide to cell cultures (Shiloh *et al*, 1983; Yi *et al*, 1990) or to treatment with xanthine/xanthine oxidase, an enzyme system that generates superoxide and hydrogen peroxide (Ward *et al*, 1994). Some of these agents, such as ionizing radiation and hydrogen peroxide, are potent inducers of NF-κB (Brach *et al*, 1991; Schreck *et al*, 1992; Siebenlist *et al*, 1994). The others are chemicals known to produce increased levels of intracellular hydrogen peroxide, which activates NF-κB at low concentrations (Schreck *et al*, 1992). AT cells have also shown hypersensitivity to the tumour promoter phorbol-2-myristate-3-acetate (PMA) (Shiloh *et al*, 1985b), another potent NF-κB inducer (Meyer *et al*, 1991). Contrary to the agents mentioned above, PMA is not directly involved in the release of radicals, but its clastogenic effects involve the formation of superoxide anions as intermediates (Emerit and Cerutti, 1982).

NF-κB shows constitutive activation in AT cells and no induction in response to ionizing radiation (Jung *et al*, 1995). Although the IκB-α transcript was highly expressed in these cells, as expected from the constitutive activation of NF-κB, no difference in the protein levels of the IκB-α inhibitor or the p65 and p50 subunits of NF-κB was observed between AT and normal cells (Jung *et al*, 1995; Siddoo-Atwal *et al*, 1996). Constitutive activation of the interferon-β induction pathway has been observed in AT cells and probably resulted from the constitutive activation of NF-κB in these cells (Siddoo-Atwal *et al*, 1996). Lavin and co-workers reported the constitutive presence of a DNA binding protein in the nuclei of AT cells, which translocates from the cytoplasm to the nucleus upon irradiation of normal cells (Singh and Lavin, 1990; Teale *et al*, 1993). A similar finding was reported regarding the pattern of p34^{CDC2} proteins in AT cells, which resembled the pattern observed in normal fibroblasts upon exposure to ionizing radiation (Paules *et al*, 1995). These observations suggest a continuous state of oxidative stress in AT cells.

AT lymphoblasts exhibit defective postirradiation activation of the stress activated protein kinase (SAP-kinase; Shafman *et al*, 1995). SAP kinases (also known as JUN kinases or JNKs) constitute a family of serine/threonine kinases that phosphorylate the transcription factor JUN (Sanchez *et al*, 1994), an early response gene to radiation exposure (Sherman *et al*, 1990). SAP kinases are stimulated by diverse agents such as tumour necrosis factor, UV light, protein synthesis inhibitors and ionizing radiation (Kyriakis *et al*, 1994; Shafman *et al*, 1995). However, only the signal transduction pathway that induces SAP kinase activity in response to ionizing radiation is defective in AT cells.

The hypersensitivity of AT cells to ionizing radiation and the chemical agents mentioned above has traditionally been explained as a defective response to DNA damage. However, the recent findings described above support the hypothesis that AT cells are defective in a more general cellular response to oxidative stress (Fig. 1). The constitutive activation of specific functions in AT cells, which resembles that of irradiated cells, can be explained by a deficiency in the detoxification of ROIs, in particular hydrogen peroxide, which is produced by various NF-κB inducers including ionizing radiation and might serve as the intracellular common messenger of NF-κB activation (Schreck *et al*, 1992). This idea gains support from studies showing that although AT cells have normal levels of glutathione, a major component of the cellular system that detoxifies hydrogen peroxide, its resynthesis upon a depleting challenge, such as radiation or xenobiotic compounds, is deficient in these cells (Meredith and Dodson, 1987). In addition, reduced levels of catalase activity, a scavanger of hydrogen peroxide, were detected in AT cells

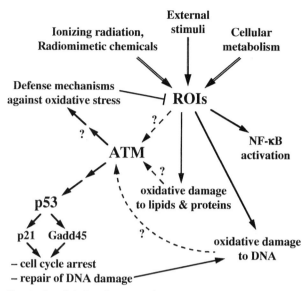

Fig. 1. Possible involvement of ATM in cellular defence mechanisms against oxidative stress

(Vuillaume, 1987). It should be noted, however, that these results were regarded as controversial (Sheridan and Huang, 1979; Dean and Jaspers, 1988).

This hypothesis offers an attractive alternative explanation not only for the radiosensitivity of AT cells, but also for other aspects of the pleiotropic AT phenotype that are not easily explained by defective cell cycle checkpoints and processing of DNA damage.

Abnormal NF-κB signalling in cells of the immune system of AT patients could explain several aspects of their immunodeficiency, since NF-κB has an important role in the development and function of the immune system. A vast array of defence proteins, whose genes are regulated by NF-κB in cells of the immune system, act as a rapid response to invading agents. These proteins include various cytokines, growth factors, immunoreceptors and adhesion molecules involved in the recruitment of circulating monocytes and lymphocytes (reviewed in Muller *et al*, 1993; Baeuerle and Henkel, 1994).

AT cells undergo programmed cell death (apoptosis) upon exposure to lower doses of radiation or radiomimetic drugs than those needed to induce appreciable apoptosis in control cells (Meyn *et al*, 1994). These observations were attributed to a TP53 mediated pathway that is induced by DNA damage and leads to apoptosis (Meyn *et al*, 1994; Meyn, 1995). However, a defective mechanism in the detoxification of ROIs could also offer a valid explanation for the higher triggering of apoptotic death in AT cells, since apoptosis may also be triggered by oxidative stress (Buttke and Sandstrom, 1994). ROIs production was suggested to serve as an early signal in neuronal apoptosis upon deprivation of nerve growth factor (Greenlund *et al*, 1995).

Oxidative stress is also involved in neurological diseases (reviewed in Beal,

1995). A possible relationship between NF-κB, ROIs and neurodegenerative disorders has been suggested (Kaltschmidt *et al*, 1993) and could account for the most devastating aspect of AT, the progressive neuromotor dysfunction that results from degeneration of Purkinje cells. This differential neurodegeneration might involve the action of a novel neurotransmitter, nitric oxide (NO). NO is produced in neurons by neuronal nitric oxide synthase (nNOS), which is present in all granule and basket cells of the cerebellum. Purkinje cells do not produce NO but receive it via synapses from the surrounding basket and granule cells (reviewed in Jaffrey and Snyder, 1995).

NO reacts with superoxide anion with very high affinity, resulting in the formation of the powerful oxidant, peroxynitrite anion. There is a critical balance between NO and superoxide in cells: a rise in the concentration of the latter increases the prevalence of oxidation reactions of NO, leading to toxicity if the cellular defences are overwhelmed (Beckman and Crow, 1993). A recent study showed that NO mediated cytotoxicity was dependent on hydrogen peroxide rather than superoxide. A compound with glutathione peroxidase like activity was able to protect cells against NO toxicity, whereas depletion of endogenous glutathione levels resulted in increased susceptibility to NO mediated toxicity (Farias-Eisner *et al*, 1996). Thus, the toxicity of NO originates predominantly from its oxidized metabolites, including peroxynitrite and other derived species (Lipton *et al*, 1993). The levels of ROIs and NO, combined with the antioxidant protective mechanisms, play a major part in determining whether NO will be defensive or injury producing. Depletion of glutathione, or the loss of other antioxidant defence mechanisms, could permit a rise in the endogenous levels of superoxide and hydrogen peroxide and enhance the toxicity of NO. Reactive nitrogen and oxygen species may act in concert, via such a route, to inactivate key metabolic enzymes and cause lipid peroxidation and DNA strand breaks, resulting eventually in irreversible cellular injury and death (reviewed in Schmidt and Walter, 1994; Gross and Wolin, 1995).

The degeneration of Purkinje cells in AT could thus be due to a defect in cellular anti-oxidative mechanisms. A contribution of NO to the pathogenesis of other neurodegenerative disorders, such as Alzheimer's disease, Huntington's disease, AIDS dementia and amyotrophic lateral sclerosis (ALS), has been suggested (Dawson and Dawson, 1996). In Alzheimer's disease and Huntington's disease, for example, the nNOS producing neurons survive, but the cells immediately surrounding them, which are the target of NO, die. NOS producing neurons are also resistant to NO neurotoxicity in vascular stroke, suggesting the presence of protective factors that render these cells resistant to the toxic NO environment they create. This could explain the specific loss of neurons that do not produce NO but receive it in relatively large amounts from their environment, such as the Purkinje cells in AT.

ALS is characterized by degeneration of large motor neuron nuclei in the spinal cord, which are surrounded by NO producing interneurons. Familial ALS is associated with dominantly inherited mutations in the gene encoding copper-zinc superoxide dismutase (CuZnSOD; Rosen *et al*, 1993). These mutations were suggested to result in a gain of function effect, where the mutated CuZnSOD catalyzes oxidation of cellular substrates by hydrogen peroxide, peroxynitrite or other oxidants (Beckman *et al*, 1993; Wiedau-Pazos *et al*, 1996). In a neuronal cell culture model, the mutant CuZnSOD enhanced apoptosis by serum withdrawal, whereas the wild type enzyme inhibited this apoptotic death (Wiedau-Pazos *et al*, 1996). Thus, the neurodegeneration in familial ALS may be a consequence of increased apoptosis of motor neurons due to excessive oxidation of critical cellular targets by the mutated form of CuZnSOD.

Further indication of the possible involvement of NO in the pathogenesis of AT is supplied by the finding that a rise in intracellular levels of NO results in TP53 accumulation. In addition, TP53 mediates transrepression of inducible NOS (iNOS) gene expression, creating a negative feedback loop between TP53 accumulation and NO formation (Forrester *et al*, 1996). Transcription of iNOS is activated by NF-κB (Spink *et al*, 1995). On the other hand, NO can inhibit NF-κB activation (De Caterina *et al*, 1995), suggesting a negative feedback mechanism.

The premature ageing and enhanced in vitro cellular senescence observed in AT could also result from higher levels of oxidative stress. The age related increase in amounts of oxidized proteins and unrepaired DNA damage has long been thought to be due to progressive oxidative damage inflicted on cellular macromolecules (Stadtman, 1992; Beal, 1995; Martin *et al*, 1996). A deficiency in the cellular anti-oxidative defence mechanisms could speed up this natural process and result in premature ageing.

During evolution, multiple defence mechanisms evolved to protect cells from the toxic effects of ROIs. These include small antioxidant molecules (such as glutathione), various ROIs detoxifying enzymes and enzymes that repair oxidative damage in lipids, proteins and DNA (Yu, 1994). ATM could be a component in a regulatory system that activates these defence mechanisms against oxidative stress. This protein might be involved in sensing an increase in ROIs and subsequently recruiting the proper inducible transcription factors, such as NF-κB and TP53, which translate the stress signal into an altered pattern of gene expression. The genes activated by these transcription factors may be charged with protecting a wide variety of essential molecules in the cell from oxidative damage, arresting the cell cycle to allow repair of oxidative damage inflicted on the DNA, and recruiting the DNA repair machinery.

The strikingly pleiotropic nature of AT represents all these aspects. It has traditionally been difficult to tie the clinical, pathological and cellular manifestations of this disease to a single function. This function might become clearer by adding new information about the delicate regulatory networks operating in cells under stress to what is known about the nature of the ATM protein, a signal transduction protein kinase with potentially multiple targets. The common denominator underlying all AT defects might be increased oxidative stress.

SUMMARY

The autosomal recessive disorder ataxia-telangiectasia (AT) is highly pleio-tropic. It is characterized by gradual loss of Purkinje cells in the cerebellum, leading to progressive neuromotor deterioration, immunodeficiency, developmental defects in specific tissues, profound predisposition to malignancy and acute sensitivity to ionizing radiation. AT cells show chromosomal instability, premature senesence, radiosensitivity and defects in cell cycle checkpoints activated by ionizing radiation. Several radiation induced pathways that regulate the cell cycle seem to be defective in AT cells, at least one of which is mediated by TP53. Extensive characterization of the cellular defects of AT cells, together with the recent isolation of the *ATM* gene, has provided some insight into the possible physiological roles of the ATM protein. Several lines of evidence, including the nature of the agents that elicit the hypersensitivity of AT cells, point to the possibility of a defect in the response to damage induced by oxidative stress, which affects various cellular macromolecules. The ATM protein might have a role in activating defence mechanisms against oxidative stress. This hypothesis broadens the previous concept of the AT defect and explains several aspects of the AT phenotype that cannot be accounted for by defective processing of DNA damage.

Acknowledgements

We are indebted to Dr Yinon Ben-Neriah, Dr Ayala Hochman, Dr Yoram Groner, Dr Hagop Youssoufian and Dr Carl Anderson for productive discussions. Studies at the authors' laboratory were supported by the AT Children's Project, the AT Medical Research Foundation, the Thomas Appeal (AT Medical Research Trust), the United States-Israel Binational Science Foundation and the National Institute of Neurological Disorders and Stroke (NS31763).

References

Agarwal ML, Agarwal A, Taylor WR and Stark GR (1995) p53 controls both the G2/M and the G1 cell cycle checkpoints and mediates reversible growth arrest in human fibroblasts. *Proceedings of the National Academy of Sciences of the USA* **92** 8493–8497

Anderson CW (1993) DNA damage and the DNA-activated protein kinase. *Trends in Biochemical Sciences* **18** 433–437

Anderson CW and Carter TH (1996) The DNA-activated protein kinase—DNA-PK. *Current Topics in Microbiology and Immunology* **217** 91–111

Arlett CF and Priestley A (1983) Defective recovery from potentially lethal damage in some human fibroblast cell strains. *International Journal of Radiation Biology* **43** 157–167

Arlett CF and Priestley A (1985) An assessment of the radiosensitivity of ataxia-telangiectasia heterozygotes, In: Gatti RA and Swift M (eds). *Ataxia-Telangiectasia: Genetics, Neuropathology and Immunology of a Degenerative Disease of Childhood*, pp 101–109, Alan R Liss, New York

Baeuerle PA and Henkel T (1994) Function and activation of NF-κB in the immune system. *Annual Review of Immunology* **12** 141–179

Barak Y, Juven T, Haffner R and Oren M (1993) mdm2 expression is induced by wild type p53 activity. *EMBO Journal* **12** 461–468

Bates S and Vousden KH (1996) p53 in signaling checkpoint arrest or apoptosis. *Current Opinion in Genetics and Development* **6** 12–19

Beal MF (1995) Aging, energy, and oxidative stress in neurodegenerative diseases. *Annals of Neurology* **38** 357–366

Beamish H and Lavin MF (1994) Radiosensitivity in ataxia-telangiectasia: anomalies in radiation-induced cell cycle delay. *International Journal of Radiation Biology* **65** 175–184

Beamish H, Khanna KK and Lavin MF (1994) Ionizing radiation and cell cycle progression in ataxia-telangiectasia. *Radiation Research* **138** 130–133

Beamish H, Williams R, Chen P and Lavin MF (1996) Defect in multiple cell cycle checkpoints in ataxia-telangiectasia post-irradiation. *Journal of Biological Chemistry* **271** 20786–20793

Beckman JS and Crow JP (1993) Pathological implications of nitric oxide, superoxide and peroxynitrite formation. *Biochemical Society Transactions* **21** 330–334

Beckman JS, Carson M, Smith CD and Koppenol WH (1993) ALS, SOD and peroxynitrite. *Nature* **364** 584

Bosma MJ and Carroll AM (1991) The SCID mouse mutant: definition, characterization, and potential uses. *Annual Review of Immunology* **9** 323–350

Brach MA, Hass R, Sherman ML, Gunji H, Weichselbaum R and Kufe D (1991) Ionizing radiation induces expression and binding activity of the nuclear factor κB. *Journal of Clinical Investigation* **88** 691–695

Brown EJ, Albers MW, Shin TB *et al* (1994) A mammalian protein targeted by G1-arresting rapamycin-receptor complex. *Nature* **369** 756–758

Brown EJ, Beal PA, Keith CT, Chen J, Shin TB and Schreiber SL (1995) Control of p70 S6 kinase by kinase activity of FRAP in vivo. *Nature* **377** 441–446

Brown K, Park S, Kanno T, Franzoso G and Siebenlist U (1993) Mutual regulation of the transcriptional activator NF-κB and its inhibitor, IκB-α. *Proceedings of the National Academy of Sciences of the USA* **90** 2532–2536

Brown K, Gerstberger S, Carlson L, Franzoso G and Siebenlist U (1995) Control of IκB-α proteolysis by site-specific signal-induced phosphorylation. *Science* **267** 1485–1488

Buttke TM and Sandstrom PA (1994) Oxidative stress as a mediator of apoptosis. *Immunology Today* **15** 7–10

Byrd PJ, McConville CM, Cooper P *et al* (1996) Mutations revealed by sequencing the 5′ half of the gene for ataxia-telangiectasia. *Human Molecular Genetics* **5** 145–149

Cafferkey R, Young PR, McLaughlin MM *et al* (1993) Dominant missense mutations in a novel yeast protein related to mammalian phosphatidylinositol 3-kinase and VPS34 abrogate rapamycin cytotoxicity. *Molecular and Cellular Biology* **13** 6012–6023

Canman CE, Wolff AC, Chen CY, Fornace AJ and Kastan MB (1994) The p53-dependent G1 cell cycle checkpoint pathway and ataxia-telangiectasia. *Cancer Research* **54** 5054–5058

Carpenter CL and Cantley LC (1996) Phosphoinositide kinases. *Current Opinion in Cell Biology* **8** 153–158

Chen IT, Smith ML, O'Connor PM and Fornace AJ (1995) Direct interaction of Gadd45 with PCNA and evidence for competitive interaction of Gadd45 and p21$^{Waf1/Cip1}$ with PCNA. *Oncogene* **11** 1931–1937

Chen J, Lin J and Levine AJ (1995) Regulation of transcription functions of the p53 tumor suppressor by the mdm-2 oncogene. *Molecular Medicine* **1** 142–152

Chiao PJ, Miyamoto S and Verma IM (1994) Autoregulation of IκBα activity. *Proceedings of the National Academy of Sciences of the USA* **91** 28–32

Cimprich KA, Shin TB, Keith CT and Schreiber SL (1996) cDNA cloning and gene mapping of a candidate human cell cycle checkpoint protein. *Proceedings of the National Academy of Sciences of the USA* **93** 2850–2855

Cox LS and Lane DP (1995) Tumour suppressors, kinases and clamps: how p53 regulates the cell cycle in response to DNA damage. *BioEssays* **17** 501–508

Cox R, Masson WK, Weichselbaum RR, Nove J and Little JB (1981) The repair of potentially lethal damage in X-irradiated cultures of normal and ataxia-telangiectasia human fibroblasts. *International Journal of Radiation Biology* **39** 357–365

Cross SM, Sanchez CA, Morgan CA *et al* (1995) A p53-dependent mouse spindle checkpoint. *Science* **267** 1353–1356

Dawson VL and Dawson TM (1996) Nitric oxide in neuronal degeneration. *Proceedings of the Society for Experimental Biology and Medicine* **211** 33–40

Dean SW and Jaspers NGJ (1988) Impaired glutathione biosynthesis in cultured ataxia-telangiectasia cells. *Cancer Research* **48** 5374–5376

De Caterina R, Libby P, Peng HB *et al* (1995) Nitric oxide decreases cytokine-induced endothelial activation. *Journal of Clinical Investigation* **96** 60–68

DiDonato J, Mercurio F, Rosette C *et al* (1996) Mapping of the inducible IκB phosphorylation sites that signal its ubiquitination and degradation. *Molecular and Cellular Biology* **16** 1295–1304

Din S, Brill SJ, Fairman MP and Stillman B (1990) Cell-cycle-regulated phosphorylation of DNA replication factor A from human and yeast cells. *Genes and Development* **4** 968–977

Dulic V, Kaufmann WK, Wilson SJ *et al* (1994) p53-dependent inhibition of cyclin-dependent kinase activities in human fibroblasts during radiation-induced G1 arrest. *Cell* **76** 1013–1023

Dutta A and Stillman B (1992) cdc2 family kinases phosphorylate a human cell DNA replication factor, RPA, and activate DNA replication. *EMBO Journal* **11** 2189–2199

Dutta A, Ruppert JM, Aster JC and Winchester E (1993) Inhibition of DNA replication factor RPA by p53. *Nature* **365** 79–82

Easton DF (1994) Cancer risks in A-T heterozygotes. *International Journal of Radiation Biology* **66** S177–S182

El-Deiry WS, Harper JW, O'Connor PM *et al* (1994) WAF1/CIP1 is induced in p53-mediated G1 arrest and apoptosis. *Cancer Research* **54** 1169–1174

Emerit I and Cerutti PA (1982) Tumor promoter phorbol 12-myristate 13-acetate induces a clastogenic factor in human lymphocytes. *Proceedings of the National Academy of Sciences of the USA* **79** 7509–7513

Farias-Eisner R, Chaudhuri G, Aeberhard E and Fukuto JM (1996) The chemistry and tumoricidal activity of nitric oxide/hydrogen peroxide and the implications to cell resistance/susceptibility. *Journal of Biological Chemistry* **271** 6144–6151

Forrester K, Ambs S, Lupold SE *et al* (1996) Nitric oxide-induced p53 accumulation and regulation of inducible nitric oxide synthase expression by wild-type p53. *Proceedings of the National Academy of Sciences of the USA* **93** 2442–2447

Gatti RA (1991) Localizing the genes for ataxia-telangiectasia: a human model for inherited cancer susceptibility. *Advances in Cancer Research* **56** 77–104

Gatti RA, Berkel I, Boder E *et al* (1988) Localization of an ataxia-telangiectasia gene to chromosome 11q22–23. *Nature* **336** 577–580

Gilad S, Khosravi R, Shkedy D *et al* (1996) Predominance of null mutations in ataxia-telangiectasia. *Human Molecular Genetics* **5** 433–439

Greenlund LJS, Deckwerth TL and Johnson EM (1995) Superoxide dismutase delays neuronal apoptosis: a role for reactive oxygen species in programmed neuronal death. *Neuron* **14** 303–315

Greenwell PW, Kronmal SL, Porter SE, Gassenhuber J, Obermaier B and Petes TD (1995) TEL1, a gene involved in controlling telomere length in *S. cerevisiae*, is homologous to the human ataxia-telangiectasia gene. *Cell* **82** 823–829

Gross SS and Wolin MS (1995) Nitric oxide: pathophysiological mechanisms. *Annual Review of Physiology* **57** 737–769

Gu Y, Turck CW and Morgan DO (1993) Inhibition of CDK2 activity in vivo by an associated 20K regulatory subunit. *Nature* **366** 707–710

Hall PA, Kearsey JM, Coates PJ, Norman DG, Warbrick E and Cox LS (1995) Characterisation of the interaction between PCNA and Gadd45. *Oncogene* **10** 2427–2433

Hari KL, Santerre A, Sekelsky JJ, McKim KS, Boyd JB and Hawley RS (1995) The mei-41 gene of *D. melanogaster* is a structural and functional homolog of the human ataxia-telangiectasia gene. *Cell* **82** 815–821

Harnden DG (1994) The nature of ataxia-telangiectasia: problems and perspectives. *International Journal of Radiation Biology* **66** (**Supplement**) S13–S19

Harper JW and Elledge SJ (1996) Cdk inhibitors in development and cancer. *Current Opinion in Genetics and Development* **6** 56–64

Harper JW, Adami GR, Wei N, Keyomarsi K and Elledge SJ (1993) The p21 Cdk-interacting protein Cip1 is a potent inhibitor of G1 cyclin-dependent kinases. *Cell* **75** 805–816

Harper JW, Elledge SJ, Keyomarsi K *et al* (1995) Inhibition of cyclin-dependent kinases by p21. *Molecular Biology of the Cell* **6** 387–400

Hartley KO, Gell D, Smith GCM *et al* (1995) DNA-dependent protein kinase catalytic subunit: a relative of phosphatidylinositol 3-kinase and the ataxia-telangiectasia gene product. *Cell* **82** 849–856

Hartwell L (1992) Defects in a cell cycle checkpoint may be responsible for the genomic instability of cancer cells. *Cell* **71** 543–546

Hartwell LH and Kastan MB (1994) Cell cycle control and cancer. *Science* **266** 1821–1828

Hecht F and Hecht BK (1990) Cancer in ataxia-telangiectasia patients. *Cancer Genetics and Cytogenetics* **46** 9–19

Helliwell SB, Wagner P, Kunz J, Deuter-Reinhard M, Henriquez R and Hall MN (1994) TOR1 and TOR2 are structurally and functionally similar but not identical phosphatidylinositol kinase homologues in yeast. *Molecular Biology of the Cell* **5** 105–118

Hinds PW (1995) The retinoblastoma tumor suppressor protein. *Current Opinion in Genetics and Development* **5** 79–83

Houldsworth J and Lavin MF (1980) Effect of ionizing radiation on DNA synthesis in ataxia-telangiectasia cells. *Nuclecic Acids Research* **8** 3709–3720

Hunter T (1995) When is a lipid kinase not a lipid kinase? When it is a protein kinase. *Cell* **83** 1–4

Israel A (1995) A role for phosphorylation and degradation in the control of NF-κB activity. *Trends in Genetics* **11** 203–205

Jackson SP (1995) Ataxia-telangiectasia at the crossroads. *Current Biology* **5** 1210–1212

Jaffrey SR and Snyder SH (1995) Nitric oxide: a neural messenger. *Annual Review in Cell Biology* **11** 417–440

Jeggo PA, Taccioli GE and Jackson SP (1995) Menage a trois: double strand break repair, V(D)J recombination and DNA-PK. *BioEssays* **17** 949–957

Jung M, Zhang Y, Lee S and Dritschilo A (1995) Correction of radiation sensitivity in ataxia-telangiectasia cells by a truncated IκB-α. *Science* **268** 1619–1621

Kaltschmidt B, Baeuerle PA and Kaltschmidt C (1993) Potential involvement of the transcription factor NF-κB in neurological disorders. *Molecular Aspects of Medicine* **14** 171–190

Kapp LN and Painter RB (1989) Stable radioresistance in ataxia-telangiectasia cells containing DNA from normal human cells. *International Journal of Radiation Biology* **56** 667–675

Kastan MB, Zhan Q, El-Deiry WS *et al* (1992) A mammalian cell cycle checkpoint pathway utilizing p53 and GADD45 is defective in ataxia-telangiectasia. *Cell* **71** 587–597

Kato R and Ogawa H (1994) An essential gene, ESR1, is required for mitotic cell growth, DNA repair and meiotic recombination in *Saccharomyces cerevisiae*. *Nucleic Acids Research* **22** 3104–3112

Kearsey JM, Coates PJ, Prescott AR, Warbrick E and Hall PA (1995) Gadd45 is a nuclear cell cycle regulated protein which interacts with p21^Cip1. *Oncogene* **11** 1675–1683

Keegan KS, Holzman DA, Plug, AW *et al* (1996) The Atr and Atm protein kinases associate with different sites along meiotically pairing chromosomes. *Genes and Development* **10** 2423–2437

Keith CT and Schreiber SL (1995) PIK-related kinases: DNA repair, recombination, and cell cycle checkpoints. *Science* **270** 50–51

Khanna KK and Lavin MF (1993) Ionizing radiation and UV induction of p53 protein by different pathways in ataxia-telangiectasia cells. *Oncogene* **8** 3307–3312

Khanna KK, Beamish H, Yan J *et al* (1995) Nature of G1/S cell cycle checkpoint defect in ataxia-telangiectasia. *Oncogene* **11** 609–618

Kojis TL, Gatti RA and Sparkes RS (1991) The cytogenetics of ataxia-telangiectasia. *Cancer Genetics and Cytogenetics* **56** 143–156

Komatsu K, Okumura Y, Kodama S, Yoshida M and Miller RC (1989) Lack of correlation between radiosensitivity and inhibition of DNA synthesis in hybrids (A-T × HeLa). *International Journal of Radiation Biology* **56** 863–867

Kouzarides T (1995) Functions of pRb and p53: what's the connection? *Trends in Cell Biology* **5** 448–450

Kunz J, Henriquez R, Schneider U, Deuter-Reinhard M, Movva NR and Hall MN (1993) Target of rapamycin in yeast, TOR2, is an essential phosphatidylinositol kinase homolog required for G1 progression. *Cell* **73** 585–596

Kyriakis JM, Banerjee P, Nikolakaki E *et al* (1994) The stress-activated protein kinase subfamily of c-Jun kinases. *Nature* **369** 156–160

Lange E, Borresen AL, Chen X *et al* (1995) Localization of an ataxia-telangiectasia gene to an approximately 500-kb interval on chromosome 11q23.1: linkage analysis of 176 families by an international consortium. *American Journal of Human Genetics* **57** 112–119

Lavin MF (1993) Biochemical defects in ataxia-telangiectasia, In: Gatti RA and Painter RB (eds). *Ataxia-telangiectasia*, pp 235–255, NATO ASI Series, vol H77, Springer-Verlag, Berlin

Lavin MF, Khanna KK, Beamish H, Teale B, Hobson K and Watters D (1994) Defect in radiation signal transduction in ataxia-telangiectasia. *International Journal of Radiation Biology* **66** S151–S156

Lavin MF, Khanna KK, Beamish H, Spring K, Watters D and Shiloh Y (1995) Relationship of the ataxia-telangiectasia protein ATM to phosphoinositide 3-kinase. *Trends in Biochemical Sciences* **20** 382–383

Lavin MF and Shiloh Y Ataxia-telangiectasia. In: Ochs HD, Smith CI and Puck J (eds). *Primary Immunodeficiency Diseases, A Molecular and Genetic Approach*. Oxford University Press, Oxford (in press)

Lehmann AR (1982) The cellular and molecular responses of ataxia-telangiectasia cells to DNA damage, In: Bridges BA and Harnden DG (eds). *Ataxia-telangiectasia—A Cellular and Molecular Link between Cancer, Neuropathology and Immune Deficiency*, pp 83–102, Wiley and Sons, Chichester

Lehmann AR, Arlett CF, Burke JF, Green MHL, James MR and Lowe JE (1986) A derivative of an ataxia-telangiectasia (A-T) cell line with normal radiosensitivity but A-T like inhibition of DNA synthesis. *International Journal of Radiation Biology* **49** 639–643

Liou HC and Baltimore D (1995) Regulation of the NF-κB/rel transcription factor and IκB inhibitor system. *Current Opinion in Cell Biology* **5** 477–487

Lipton SA, Choi YB, Pan ZH *et al* (1993) A redox-based mechanism for the neuroprotective and neurodestructive effects of nitric oxide and related nitroso-compounds. *Nature* **364** 626–632

Little JB and Nagasawa H (1985) Effect of confluent holding on potentially lethal damage repair, cell cycle progression and chromosomal aberrations in human normal and ataxia-telangiectasia fibroblasts. *Radiation Research* **101** 81–93

Liu VF and Weaver DT (1993) The ionizing radiation-induced replication protein A phosphorylation response differs between ataxia-telangiectasia and normal human cells. *Molecular and Cellular Biology* **13** 7222–7231

Lu X and Lane DP (1993) Differential induction of transcriptionally active p53 following UV or ionizing radiation: defects in chromosome instability syndromes? *Cell* **75** 765–778

Lustig AJ and Petes TD (1986) Identification of yeast mutants with altered telomere structure. *Proceedings of the National Academy of Sciences of the USA* **83** 1398–1402

Martin GM, Austad SN and Johnson TE (1996) Genetic analysis of ageing: role of oxidative damage and environmental stresses. *Nature Genetics* **13** 25–34

Martin K, Trouche D, Hagemeler C, Sorensen TS, LaThange NB and Kouzarides T (1995) Stimulation of E2F1/DP1 transcriptional activity by MDM2 oncoprotein. *Nature* **375** 691–694

McConville CM, Byrd PJ, Ambrose HJ and Taylor AMR (1994) Genetic and physical mapping of the ataxia-telangiectasia locus on chromosome 11q22–23. *International Journal of Radiation Biology* **66** S45–S56

McKinnon PJ (1987) Ataxia-telangiectasia: an inherited disorder of ionizing-radiation sensitivity in man. *Human Genetics* **75** 197–208

Meredith MJ and Dodson ML (1987) Impaired glutathione biosynthesis in cultured human ataxia-telangiectasia cells. *Cancer Research* **47** 4576–4581

Meyer R, Hatada EN, Hohmann HP *et al* (1991) Cloning of the DNA-binding subunit of human nuclear factor κB: the level of its mRNA is strongly regulated by phorbol ester or tumor necrosis factor α. *Proceedings of the National Academy of Sciences of the USA* **88** 966–970

Meyn MS (1995) Ataxia-telangiectasia and cellular responses to DNA damage. *Cancer Research* **55** 5991–6001

Meyn MS, Strasfeld L and Allen C (1994) Testing the role of p53 in the expression of genetic instability and apoptosis in ataxia-telangiectasia. *International Journal of Radiation Biology* **66** S141–S149

Mirzayans R and Paterson MC (1991) Lack of correlation between hypersensitivity to cell killing and impaired inhibition of DNA synthesis in ataxia-telangiectasia fibroblasts treated with 4-nitroquinoline 1-oxide. *Carcinogenesis* **12** 19–24

Mirzayans R, Famulski KS, Enns L, Fraser M and Paterson MC (1995) Characterization of the signal transduction pathway mediating γ ray-induced inhibition of DNA synthesis in human cells: indirect evidence for involvement of calmodulin but not protein kinase C nor p53. *Oncogene* **11** 1597–1605

Momand J, Zambetti GP, Olson DC, George D and Levine AJ (1992) The mdm-2 oncogene product forms a complex with the p53 protein and inhibits p53-mediated transactivation. *Cell* **69** 1237–1245

Morgan JL, Holcomb TM and Morrisey RW (1968) Radiation reaction in ataxia-telangiectasia. *American Journal of Diseases of Children* **116** 557–558

Morris C, Mohamed R and Lavin MF (1983) DNA replication and repair in ataxia-telangiectasia cells exposed to bleomycin. *Mutation Research* **112** 67–74

Morrow DW, Tagle DA, Shiloh Y, Collins FS and Hieter P (1995) TEL1, an *S. cerevisiae* homolog of the human gene mutated in ataxia-telangiectasia, is functionally related to the yeast checkpoint gene MEC1. *Cell* **82** 831–840

Muller JM, Ziegler-Heitbrock HWL and Baeuerle PA (1993) Nuclear factor kappa B, a mediator of lipopolysaccharide effects. *Immunobiology* **187** 233–256

Muller R (1995) Transcriptional regulation during the mammalian cell cycle. *Trends in Genetics* **11** 173–178

Murnane JP and Kapp LN (1993) A critical look at the association of human genetic syndromes with sensitivity to ionizing radiation. *Seminars in Cancer Biology* **4** 93–104

Murnane JP and Schwartz JL (1993) Cell checkpoint and radiosensitivity. *Nature* **365** 22

Nagasawa H and Little JB (1983) Comparison of kinetics of X-ray induced cell killing in normal, ataxia-telangiectasia and hereditary retinoblastoma fibroblasts. *Mutation Research* **109** 297–308

Oliner JD, Pietenpol JA, Thiagalingam S, Gyuris J, Kinzler KW and Vogelstein B (1993) Oncoprotein MDM2 conceals the activation domain of tumour suppressor p53. *Nature* **362** 857–860

Painter RB and Young BR (1980) Radiosensitivity in ataxia-telangiectasia: a new explanation. *Proceedings of the National Academy of Sciences of the USA* **77** 7315–7317

Pandita TK, Pathak S and Geard C (1995) Chromosome end associations, telomeres and telomerase activity in ataxia-telangiectasia cells. *Cytogenetics and Cell Genetics* **71** 86–93

Paterson MC, MacFarlane SJ, Gentner N and Smith BP (1985) Cellular hypersensitivity to chronic γ-radiation in cultured fibroblasts from ataxia-telangiectasia heterozygotes, In: Gatti RA and Swift M (eds). *Ataxia-telangiectasia: Genetics, Neuropathology and Immunology of a Degenerative Disease of Childhood*, pp 73–87, Alan R Liss, New York

Paules RS, Levedakou EN, Wilson SJ et al (1995) Defective G2 checkpoint function in cells from individuals with familial cancer syndromes. *Cancer Research* **55** 1763–1773

Peterson RDA, Funkhouser JD, Tuck-Muller CM and Gatti RA (1992) Cancer susceptibility in ataxia-telangiectasia. *Leukemia* **6 (Supplement 1)** 8–13

Picksley SM and Lane DP (1994) p53 and Rb: their cellular roles. *Current Opinion in Cell Biology* **6** 853–858

Price BD and Park SJ (1994) DNA damage increases the levels of MDM2 messenger RNA in wtp53 human cells. *Cancer Research* **54** 896–899

Proud CG (1996) p70 S6 kinase: an enigma with variations. *Trends in Biochemical Sciences* **21** 181–185

Rosen DR, Siddique T, Patterson D et al (1993) Mutations in Cu/Zn superoxide dismutase gene are associated with familial amyotrophic lateral sclerosis. *Nature* **362** 59–62

Rotman G, Savitsky K, Vanagaite L et al (1994) Physical and genetic mapping at the ATA/ATC locus on chromosome 11q22–23. *International Journal of Radiation Biology* **66** S63–S66

Sabatini DM, Erdjument-Bromage H, Lui M, Tempst P and Snyder SH (1994) RAFT1: a mammalian protein that binds to FKBP12 in a rapamycin-dependent fashion and is homologous to yeast TORs. *Cell* **78** 35–43

Sabers CJ, Martin MM, Brunn GJ et al (1995) Isolation of a protein target of the FKBP12-rapamycin complex in mammalian cells. *Journal of Biological Chemistry* **270** 815–822

Sanchez I, Hughes RT, Mayer BJ et al (1994) Role of SAPK/ERK kinase-1 in the stress-activated pathway regulating transcription factor c-jun. *Nature* **372** 794–798

Sasaki MS and Taylor AMR (1994) Dissociation between radioresistant DNA replication and chromosomal radiosensitivity in ataxia-telangiectasia cells. *Mutation Research* **307** 107–113

Savitsky K, Bar-Shira A, Gilad S et al (1995a) A single ataxia-telangiectasia gene with a product similar to PI-3 kinase. *Science* **268** 1749–1753

Savitsky K, Sfez S, Tagle DA et al (1995b) The complete sequence of the coding region of the ATM gene reveals similarity to cell cycle regulators in different species. *Human Molecular Genetics* **4** 2025–2032

Schmidt HHW and Walter U (1994) NO at work. *Cell* **78** 919–925

Schreck R, Albermann K and Baeuerle PA (1992) Nuclear factor κB: an oxidative stress-responsive transcription factor of eukaryotic cells. *Free Radical Research Communications* **17** 221–237

Scott D, Spreadborough AR and Roberts SA (1994) Radiation-induced G2 delay and spontaneous chromosome aberrations in ataxia-telangiectasia homozygotes and heterozygotes. *International Journal of Radiation Biology* **66** S157–S163

Sedgwick RP and Boder E (1991) Ataxia-telangiectasia, In: Vinken PJ, Bruyn GW and Klawans HL (eds). *Handbook of Clinical Neurology*, vol 16, pp 347–423, Elsevier, Amsterdam

Shafman TD, Saleem A, Kyriakis J, Weichselbaum R, Kharbanda S and Kufe DW (1995) Defective induction of stress-activated protein kinase activity in ataxia-telangiectasia cells exposed to ionizing radiation. *Cancer Research* **55** 3242–3245

Sheridan RB and Huang PC (1979) Superoxide dismutase and catalase activities in ataxia-telangiectasia and normal fibroblast cell extracts. *Mutation Research* **61** 381–386

Sherman ML, Datta R, Hallahan DE, Weichselbaum RR and Kufe DW (1990) Ionizing radiation regulates expression of the c-jun protooncogene. *Proceedings of the National Academy of Sciences of the USA* **87** 5663–5666

Shiloh Y (1995) Ataxia-telangiectasia: closer to unraveling the mystery. *European Journal of Human Genetics* **3** 116–138

Shiloh Y, Tabor E and Becker Y (1982) The response of ataxia-telangiectasia homozygous and heterozygous skin fibroblasts to neocarzinostatin. *Carcinogenesis* **3** 815–820

Shiloh Y, Tabor E and Becker Y (1983) Abnormal response of ataxia-telangiectasia cells to agents that break the deoxyribose moiety of DNA via a targeted free radical mechanism. *Carcinogenesis* **4** 1317–1322

Shiloh Y, Tabor E and Becker Y (1985a) In vitro phenotype of ataxia-telangiectasia (AT) fibroblast strains: clues to the nature of the "AT DNA lesion" and the molecular defect in AT. In: Gatti RA and Swift M (eds). *Ataxia-telangiectasia: Genetics, Neuropathology and Immunology of a Degenerative Disease of Childhood*, pp 111–121, Alan R Liss, New York

Shiloh Y, Tabor E and Becker Y (1985b) Cell from patients with ataxia-telangiectasia are abnormally sensitive to the cytotoxic effect of a tumor promoter, phorbol-12-myristate-13-acetate. *Mutation Research* **149** 283–286

Siddoo-Atwal C, Haas AL and Rosin MP (1996) Elevation of interferon β-inducible proteins in ataxia-telangiectasia cells. *Cancer Research* **56** 443–447

Siebenlist U, Franzoso G and Brown K (1994) Structure, regulation and function of NF-κB. *Annual Review of Cell Biology* **10** 405–455

Singh SP and Lavin MF (1990) DNA-binding protein activated by gamma radiation in human cells. *Molecular and Cellular Biology* **10** 5279–5285

Singh KK and Samson L (1995) Replication protein A binds to regulatory elements in yeast DNA repair and DNA metabolism genes. *Proceedings of the National Academy of Sciences of the USA* **92** 4907–4911

Slichenmyer WJ, Nelson WG, Slebos RJ and Kastan MB (1993) Loss of a p53-associated G1 checkpoint does not decrease cell survival following DNA damage. *Cancer Research* **53** 4164–4168

Smith ML, Chen IT, Zhan Q *et al* (1994) Interaction of the p53-regulated protein Gadd45 with proliferating cell nuclear antigen. *Science* **266** 1376–1380

Smith PJ, Anderson CO and Watson JV (1985) Abnormal retention of X-irradiation ataxia-telangiectasia fibroblasts in G2 phase of the cell cycle: cellular RNA content, chromatin stability and the effect of 3-amino-benzamide. *International Journal of Radiation Biology* **47** 701–712

Spink J, Cohen J and Evans TJ (1995) The cytokine responsive vascular smooth muscle cell enhancer of inducible nitric oxide synthase. *Journal of Biological Chemistry* **270** 29541–29547

Stadtman ER (1992) Protein oxidation and aging. *Science* **257** 1220–1224

Stewart N, Hicks GG, Paraskevas F and Mowat M (1995) Evidence for a second cell cycle block at G2/M by p53. *Oncogene* **10** 109–115

Stillman B, Bell SP, Dutta A and Marahrens Y (1992) DNA replication and the cell cycle, In: Marsh J (ed). *Regulation of the Eukaryotic Cell Cycle*, Ciba Foundation Symposium 170, pp 147–160, Wiley, Chichester

Sun SC, Ganchi PA, Ballard DW and Greene WC (1993) NF-κB controls expression of inhibitor IκBI: evidence for an inducible autoregulatory pathway. *Science* **259** 1912–1915

Swift M, Morrell D, Massey RB and Chase CL (1991) Incidence of cancer in 161 families affected by ataxia-telangiectasia. *New England Journal of Medicine* **325** 1831–1836

Taylor AMR, Harnden DG, Arlett CF *et al* (1975) Ataxia-telangiectasia: a human mutation with abnormal radiation sensitivity. *Nature* **258** 427–429

Taylor AMR, Metcalfe JA and McConville C (1989) Increased radiosensitivity and the basic defect in ataxia-telangiectasia. *International Journal of Radiation Biology* **56** 677–684

Taylor AMR, Byrd PJ, McConville CM and Thacker S (1994) Genetic and cellular features of ataxia-telangiectasia. *International Journal of Radiation Biology* **65** 65–70

Teale B, Khanna KK, Singh SP and Lavin MF (1993) Radiation-activated DNA-binding protein constitutively present in ataxia-telangiectasia nuclei. *Journal of Biological Chemistry* **268** 22450–22455

Telatar M, Wang Z, Udar N *et al* (1996) Ataxia-telangiectasia: mutations in cDNA detected by protein truncation screening. *American Journal of Human Genetics* **59** 40–44

Thanos D and Maniatis T (1995) NF-κB: a lesson in family values. *Cell* **80** 529–532

Tlsty TD, Briot A, Gualberto A *et al* (1995) Genomic instability and cancer. *Mutation Research* **337** 1–7

Uziel T, Savitsky K, Platzer M *et al* (1996) Genomic organization of the ATM gene. *Genomics* **33** 317–320

Vuillaume M. (1987) Reduced oxygen species, mutation, induction and cancer initiation. *Mutation Research* **186** 43–72

Waga S, Hannon GJ, Beach D and Stillman B (1994) The p21 inhibitor of cyclin-dependent kinases controls DNA replication by interaction with PCNA. *Nature* **369** 574–578

Ward AJ, Olive PL, Burr AH and Rosin MP (1994) Response of fibroblast cultures from ataxia-telangiectasia patients to reactive oxygen species generated during inflammatory reactions. *Environmental and Molecular Mutagenesis* **24** 103–111

White E (1994) p53, guardian of Rb. *Nature* **371** 21–22

Wiedau-Pazos M, Goto JJ, Rabizadeh S *et al* (1996) Altered reactivity of superoxide dismutase in familial amyotrophic lateral sclerosis. *Science* **271** 515–518

Wu X, Bayle JH, Olson D and Levine AJ (1993) The p53-mdm-2 autoregulatory feedback loop. *Genes and Development* **7** 1126–1132

Xiao ZX, Chen J, Levine AJ *et al* (1995) Interaction between the retinoblastoma protein and the oncoprotein MDM2. *Nature* **375** 694–698

Yi M, Rosin MP and Anderson CK (1990) Response of fibroblast cultures from ataxia-telangiectasia patients to oxidative stress. *Cancer Letters* **54** 43–50

Yu BP (1994) Cellular defenses against damage from reactive oxygen species. *Physiological Reviews* **74** 139–162

Zakian VA (1995) ATM-related genes: what do they tell us about functions of the human gene? *Cell* **82** 685–687

Zhan Q, Bae I, Kastan MB and Fornace AJ (1994) The p53-dependent γ-ray response of GADD45. *Cancer Research* **54** 2755–2760

Ziv Y, Bar-Shira A, Jorgensen TJ *et al* (1995) Human cDNA clones that modify radiomimetic sensitivity of ataxia-telangiectasia (group A) cells. *Somatic Cell and Molecular Genetics* **21** 99–111

The authors are responsible for the accuracy of the references.

Apoptosis and Cancer Mechanisms

HUICHIN PAN • CHAOYING YIN • TERRY VAN DYKE

Department of Biochemistry and Biophysics, Lineberger Comprehensive Cancer Center, University of North Carolina at Chapel Hill, Chapel Hill, NC

Introduction
Overview of apoptosis
The BCL2 family
The TP53 tumour suppressor
Animal models establish possible mechanisms of cancer development
    Apoptosis suppression and predisposition to tumorigenesis
    Apoptosis suppression and tumour progression
    TP53 independent apoptosis
Current challenges and future directions
Summary

INTRODUCTION

For nearly two decades, studies in cancer research focussed on identifying genes that act as positive and negative regulators of cell growth. Only relatively recently was it recognized that the regulation of cell death (apoptosis) is also an important modulator of tumorigenesis. At least two genes linked to human cancers, *BCL2* and *TP53*, have been shown to regulate apoptosis. The correlation between apoptosis modulating genes and human tumours raises an important question as to how dysregulation of apoptosis contributes to neoplastic transformation and malignant cell growth. Cell culture studies have clearly demonstrated that TP53 can induce, and BCL2 suppress, apoptosis in response to various stimuli. Extension of findings in genetically tractable organisms, such as *Caenorhabditis elegans* and *Drosophila melanogaster*, to vertebrate systems has expanded the roster of genes with a role in apoptosis and has begun to provide the foundation for elucidating a common cellular pathway. Studies of mammalian viruses, which possess mechanisms for both inducing and evading apoptosis, have also extended our understanding of this process. On the basis of such findings, several animal models have been developed that begin to address the role of apoptosis regulation in tumorigenesis. This chapter discusses these early models and the insights that have been derived from them.

Cancer Surveys Volume 29: *Checkpoint Controls and Cancer*
© 1997 Imperial Cancer Research Fund. 0-87969-518-8/97. $5.00 + .00

OVERVIEW OF APOPTOSIS

Apoptosis, or programmed cell death, was first described as a unique form of cell death distinct from necrosis (Wyllie, 1980). It is encoded intrinsically by a "suicide" genetic program that is triggered when cells are exposed to certain intracellular or extracellular stimuli. Apoptotic cells undergo characteristic morphological changes, including chromatin condensation, membrane blebbing, loss of a nuclear envelope and fragmentation of the dying cells into several membrane bound compartments called "apoptotic bodies" (Wyllie, 1980; Arends and Wyllie, 1991). Apoptosis is distinct from necrosis in several ways. Unlike necrosis, apoptosis is generally not associated with inflammation. Apoptotic cells are somehow recognized and engulfed by neighbouring cells and macrophages, leaving the tissue with minimal damage. Gene transcription and protein synthesis are often required for cells undergoing apoptosis but not necrosis, indicating an active role for newly synthesized gene products. In addition, apoptosis is usually associated with the activation of a specific Ca^{++} dependent endonuclease that cleaves DNA between nucleosomes, leading to chromosomal DNA degradation first into large fragments (50–300 kb) and then into small multimers of 180 base pairs commonly referred to as "DNA ladders" (Wyllie, 1980; Walker *et al*, 1993).

Apoptosis is now widely recognized as the major mode of cell death observed during development throughout the animal kingdom. In vertebrate systems, the two currently best studied examples occur in the central nervous system where apoptosis plays an essential part in facilitating the establishment of effective synaptic networks (Oppenheim, 1991), and in the lymphoid system, where autoreactive T cells are deleted (Rothenberg, 1992). In addition to its role in development, apoptosis is also important for normal tissue turnover, because it allows for the precise regulation of cell numbers. It is now clear that apoptosis also serves as a defence mechanism, eliminating potentially dangerous cells, such as virus infected cells and cells exposed to toxins or other adverse environmental conditions (Thompson *et al*, 1995; Meyn *et al*, 1996). Primarily through the study of animal tumorigenesis models, it has become clear that the modulation of apoptosis levels can play an important part in the genesis of a tumour, although precisely how a reduction in apoptosis contributes to tumour growth at both the biological and molecular levels is not yet known (see below).

The molecular pathways involved in apoptosis and its regulation are now being studied intensively, and progress has been steady. There is increasing evidence that apoptosis occurs through a central death pathway that has been conserved in evolution from nematodes to mammals. Known details of this pathway and how it was derived have been reviewed elsewhere (Korsmeyer, 1995; Fraser and Evan, 1996; White, 1996) and are summarized here. Genetic analyses in the nematode *C. elegans* have led to the isolation of genes that promote cell death (eg the *ced-3* and *ced-4* genes) or suppress cell death (the *ced-9* gene). The *ced-3* and *ced-9* genes are homologous to mammalian gene families, interleukin 1β converting enzyme (ICE) and *BCL2*, respectively. ICE belongs to a family of cysteine proteases that appear to act in a cascade as principal effectors of apoptosis, with the cleavage of the ultimate substrates (largely unknown) resulting in orderly cellular demise. Ectopic expression of several ICE like proteases induces apoptosis in mammalian cells (Miura *et al*, 1993; Fraser and Evan, 1996). Viral proteins that inhibit ICE like proteases, such as CrmA (poxvirus) or p35 (baculovirus) (Ray *et al*, 1992; Komiyama *et al*, 1994; Bump *et al*, 1995), in addition to chemical inhibitors (Nicholson *et al*, 1995), also suppress apoptosis. Since p35 inhibits cell death in insects, nematodes and mammals (Rabizadeh *et al*, 1993; Hay *et al*, 1994; Sugimoto *et al*, 1994), this process appears to be conserved in evolution. Moreover, such studies suggest that the ICE like enzymes are either shared among different signalling pathways or act downstream from a convergence point. The mechanism of BCL2 action is less well understood (see below). However, whatever its function, it has been conserved during evolution in that human BCL2 can function to complement a *ced-9* defect in *C. elegans* (Vaux *et al*, 1992).

Clearly, there is much yet to be learned about the molecular mechanisms involved in apoptosis. However, extensive research on two genes found to be mutated in certain human cancers, *BCL2* and the tumour suppressor *TP53*, have provided a foundation for exploring the involvement of apoptosis in tumorigenesis. Several animal models developed thus far focus on *BCL2* (and its relatives) and *TP53* and were guided by in vitro and cultured cell studies summarized below.

THE BCL2 FAMILY

The *BCL2* gene was isolated from a common human follicular B cell lymphoma, where a chromosomal t(14;18) translocation moved the gene into juxtaposition with the immunoglobulin (IG) heavy chain transcription enhancer, resulting in overproduction of *BCL2* mRNA and protein (Tsujimoto *et al*, 1985). At the time of its isolation, most oncogenes had been associated with facilitating cell proliferation. Thus, uncovering the biological function of BCL2 proved pivotal in conceptualizing the possible mechanisms of tumorigenesis. Rather than inducing aberrant proliferation, *BCL2* was shown to extend the lifespan of B cells (Wagner *et al*, 1993). Extended survival was due to suppression of apoptosis (Hockenberry *et al*, 1990) and resulted in a predisposition to B cell lymphoma (Wagner *et al*, 1993) (see below). Involvement of *bcl-2* in lymphoid cell homoeostasis was confirmed both in gain of function transgenic mice (McDonnell *et al*, 1989; Strasser *et al*, 1990), where bcl-2 expression protected thymocytes from apoptosis induced by various stimuli (Sentman *et al*, 1991) and in *bcl-2* deficient mice, in which abundant lymphocyte death occurs (Veis *et al*, 1993; Kamada *et al*, 1995). The BCL2 protein is associated with mitochondrial endoplasmic reticular and nuclear membranes (Hockenberry *et al*, 1993; Krajewski *et al*, 1994). However, its function therein is not known. It appears to act as an antioxidant in preventing lipid membrane peroxidation

(Hockenberry *et al*, 1993), which may provide some clues for exploring its mechanism.

It is now clear that BCL2 is a member of a large family of related proteins that share conserved regions designated BCL2 homologous (BH) 1 and 2 domains (Korsmeyer, 1995; White, 1996). Several BCL2 related proteins have been shown to regulate apoptosis, either as repressors or promoters. Like BCL2, the long form of BCL-x (BCL-xL) functions as an apoptosis antagonist (Chittenden *et al*, 1995), whereas BAX (Oltvai *et al*, 1993), BAD (Yang *et al*, 1995), BAK (Boise *et al*, 1993) and BCL-xS (Chittenden *et al*, 1995) each promotes apoptosis. Members of the BCL2 family can form both homo- and heterodimers, and it has been suggested that the ratio of the apoptosis promoting relative to the apoptosis inhibiting member determines survival (Oltvai *et al*, 1993). However, apparent exceptions to this mechanism have been demonstrated (Knudson *et al*, 1995), and the mechanism(s) by which these proteins function under physiological conditions is(are) not yet fully understood. The large family of BCL2 related genes raises the possibility that the role of a given member could depend on cell type and/or differentiation state. For example, bcl-2 confers resistance to various death signals in mature lymphocytes, whereas bcl-xL is required to maintain the lifespan of immature counterparts (Motoyama *et al*, 1995). Moreover, bcl-xL's function is critical for embryonic development, since in its absence mice die around embryonic day 13 due to extensive apoptotic cell death in postmitotic immature neurons in brain, spinal cord and dorsal root ganglia (Motoyama *et al*, 1995). By contrast, bcl-2 deficient embryos develop normally but display growth retardation and early mortality postnatally (Kamada *et al*, 1995).

THE TP53 TUMOUR SUPPRESSOR

Mutations in the *TP53* gene appear to be the most common genetic aberration present in human cancers and have been detected in most tumour classes (Nigro *et al*, 1989; Hollstein *et al*, 1991; Levine *et al*, 1991). Two different biological responses that could impact tumorigenesis, cell growth arrest and apoptosis, have been firmly associated with wild type TP53 function (Haffner and Oren, 1995; Kastan *et al*, 1995; Lee and Bernstein, 1995). The factors that determine which response is elicited are not understood, although the cell type and nature of the stimulus appear to play a part. TP53 has several biochemical activities, including sequence specific DNA binding and transcription activation (Kern *et al*, 1991; Farmer *et al*, 1992; Zambetti *et al*, 1992), interaction with general transcription factors (Chen *et al*, 1993; Truant *et al*, 1993) and transcriptional repression (Seto *et al*, 1992), binding to non-specific single stranded (ss) DNA ends (Haffner and Oren, 1995) and to small ssDNA loops (Lee *et al*, 1995) and interaction with repair components of the TFIIH complex (Wang *et al*, 1995b). On the basis of crystallographic data (Cho *et al*, 1994) and biochemical studies (Haffner and Oren, 1995), the TP53 aminoacid residues most frequently mutated in human cancer (Greenblatt *et al*, 1994; Prives, 1994) alter or destroy the TP53 sequence specific DNA binding function, indicating a role for TP53 regulated genes in tumour suppression. Candidate genes harbouring TP53 responsive elements include those encoding GADD45 (a DNA damage induced protein of unknown function) (Kastan *et al*, 1992), MDM2 (a cellular protein that binds to and regulates TP53) (Wu *et al*, 1993), cyclin G, p21 (Cip1/Sdi1/Waf1; a cyclin dependent kinase inhibitor) (El-Deiry *et al*, 1993; Harper *et al*, 1993; Xiong *et al*, 1993) and BAX (Miyashita *et al*, 1994; Miyashita and Reed, 1995). Of these, only BAX has an established role in apoptosis thus far (see above). Studies showing that the TP53 *trans*-activation function either is necessary (Sabbatini *et al*, 1995; Friedlander *et al*, 1996) or is dispensable (Caelles *et al*, 1994; Haupt *et al*, 1995) for the induction of apoptosis have been reported, indicating that multiple TP53 functions could be involved in this process. Further details on TP53 and its functions can be found in several reviews (Perry and Levine, 1993; Prives and Manfredi, 1993; Clarke, 1995; Kastan *et al*, 1995; Ko and Prives, 1996).

TP53 was first shown to induce apoptosis in certain cultured tumour cell lines (Yonish-Rouach *et al*, 1991; Shaw *et al*, 1992) but has since been shown to be required for apoptosis in response to a variety of stimuli. For example, normal mouse embryonic fibroblasts undergo apoptosis in response to expression of the adenovirus early region 1A oncogene (*E1A*), a process that is defective in TP53 deficient cells (Lowe and Ruley, 1993; Debbas and White, 1993). In addition to *E1A*, other viral oncogenes reported to induce TP53 dependent apoptosis include an N-terminal fragment of simian virus 40 (*SV40*) T antigen (*T-Ag*) (Symonds *et al*, 1994), human papillomavirus (HPV) *E7* (Pan and Griep, 1994) and hepatitis B virus X gene (*HBx*) (Wang *et al*, 1995a). The specific stimulus that induces TP53 dependent apoptosis in cells expressing these viral genes has not been clearly identified, although aberrant cell proliferation has been postulated to be involved. This prediction is based on the involvement of RB inactivation by some viral proteins, such as E7, E1A and T-Ag (DeCaprio *et al*, 1988; Whyte *et al*, 1988; Dyson *et al*, 1989), as well as the observation that upregulation of the transcription factor E2F1 can induce TP53 dependent apoptosis (Qin *et al*, 1994; Wu and Levine, 1994).

Studies of viral genes have also identified a class of genes that inhibit apoptosis. Among these are adenovirus *E1B 19K*, HPV-16 *E6*, poxvirus *CrmA* and baculovirus *p35*. The E1B 19K protein blocks the TP53 dependent apoptosis induced by E1A, a function that can also be provided by BCL2 (Chiou *et al*, 1994a,b). Indeed, the E1B 19K protein shares regions of homology with BCL2 and can interact with BAX (Han *et al*, 1996). The E6 protein is known to target TP53 for degradation (Scheffner *et al*, 1990, 1993) and the CrmA and p35 proteins inhibit ICE like proteases (Ray *et al*, 1992; Komiyama *et al*, 1994; Bump *et al*, 1995).

Several non-viral stimuli have been shown to induce TP53 dependent apoptosis. For example, TP53 is required for irradiation induced apoptosis of mouse thymocytes (Clarke *et al*, 1993; Lowe *et al*, 1993b) (see below). Cells treated with various chemotherapeutic agents also undergo apoptosis in a TP53 dependent fashion (Lowe *et al*, 1993a). Furthermore, oxygen deprivation (hypoxia) induces TP53 dependent apoptosis (Graeber *et al*, 1996). Thus, in considering the development or treatment of tumours, each of these conditions—viral gene expression, irradiation, chemotherapy, oxygen status—would be predicted to have a different impact on tumour cells depending on the TP53 status (see below).

Exactly how TP53 modulates apoptosis is unknown. However, indirect evidence suggests that at least in some cases, TP53 dependent apoptosis may operate through pathway(s) also involving BCL2 related genes. A temperature sensitive TP53 can induce a temperature dependent decrease of Bcl-2 expression and a simultaneous increase in Bax expression in murine leukaemia cells (Miyashita *et al*, 1994; Han *et al*, 1996). Co-transfection assays have also shown that TP53 can activate expression via the TP53 responsive elements of the *Bax* promoter (Miyashita and Reed, 1995). In support of the hypothesis that TP53 can regulate BAX, the level of *BAX* transcripts is reduced in many tissues of TP53 deficient mice in comparison with normal mice (Miyashita *et al*, 1994).

ANIMAL MODELS ESTABLISH POSSIBLE MECHANISMS OF CANCER DEVELOPMENT

Apoptosis Suppression and Predisposition to Tumorigenesis

Discovery of BCL2's function as an apoptosis repressor provided the first clear evidence indicating the importance of apoptosis regulation in tumorigenesis. Transgenic mice bearing a *BCL2-IG* mini gene designed to mimic the human t(14:18) translocation express BCL2 highly in B cells (McDonnell *et al*, 1989). As with the human counterpart, this leads to the accumulation of small resting B cells. After 1–2 years, benign hyperplasia progresses clonally to high grade malignant large B cell lymphoma (McDonnell and Korsmeyer, 1991), consistent with the long 10–16 year delay observed in human follicular lymphoma. The extended latency and monoclonality of lymphoma in the *BCL2-IG* mice indicate that BCL2 expression alone is insufficient for tumorigenesis and that additional transforming events are required. Indeed, half of the tumours carry a rearranged *Myc* gene and consequently overexpress *Myc* (McDonnell and Korsmeyer, 1991).

These results suggest that the inhibition of apoptosis, in this case by dysregulation of BCL2, can initiate multistep tumorigenesis as shown in model 1 (Fig. 1). In this model, the first genetic mutation causes either overexpression of an oncogene or inactivation of a tumour suppressor gene, leading to a reduction in apoptosis. The resulting prolonged cellular lifespan does not cause transformation directly, but it does increase the probability that these cells will acquire stochastic mutations, some of which will promote aberrant cell growth and result in selective clonal outgrowth.

This model predicts that tumours would be accelerated by genetically engineering multiple mutations into the target cells. Indeed, mice harbouring both the *BCL2-IG* transgene and a *Myc* transgene also under IG control (*Eμ-myc*) developed malignant lymphoma at a higher rate than mice of either monotransgenic strain (Table 1). The mean latency shifts to 4 weeks, compared with 5 months in *Eμ-myc* mice and more than 1 year in *BCL2-IG* mice (Strasser *et al*, 1990). Similarly, human lymphomas overexpressing both BCL2 and MYC, resulting from t(14;18) and t(8;14) translocations, respectively, develop highly aggressive malignant tumours, whereas single translocation lymphomas emerge slowly and/or sporadically (Hsu *et al*, 1995).

Accelerated tumorigenesis in the doubly *Bcl-2/Myc* transgenic mice indicates a multistep conversion to malignant status, with Bcl-2 extending the survival and Myc driving the proliferation of lymphocytes. Since lymphocytes expressing only the *Myc* transgene are also predisposed to tumorigenesis (Schmidt *et al*, 1988), it is possible that in the natural progression of a B cell tumour, induction of BCL2 expression would be a selected alteration subsequent to MYC induction as outlined in model 2 (Fig. 1). Consistent with this view, overexpression of MYC can induce apoptosis (Evan *et al*, 1992) inhibitable by Bcl-2 (Bissonnette *et al*, 1992), along with driving proliferation (Horning and Rosenberg, 1984). This model is discussed further below.

Promotion of cell survival also appears to play a part in T cell tumorigenesis in p53 deficient mice. About 70% of such mice develop T cell lymphoma; a smaller percentage develop a variety of sarcomas (Donehower *et al*, 1992). This high incidence of T cell tumours indicates that p53 has a critical role in producing and/or maintaining a normal T cell population such that p53 inactivation predisposes mice to this tumour type. Analysis of thymocyte apoptosis induced by a variety of stimuli has shown that p53 is required for irradiation induced thymocyte apoptosis but is dispensable for apoptosis induced in vitro by other common stimuli, such as glucocorticoids and T cell receptor engagement (Clarke *et al*, 1992; Lowe *et al*, 1993b). Furthermore, apoptosis during thymocyte self tolerance induction (clonal deletion) proceeds normally in transgenic mice in which thymocyte p53 is inactivated by T-Ag (McCarthy *et al*, 1994). These studies indicate that a p53 dependent surveillance pathway leads to the death of cells with accumulated mutations. Although the natural DNA damage signal that induces this pathway has not been identified, one possibility is that p53 dependent apoptosis is required to guard against mutations resulting from T cell receptor gene recombination events. The results of recent studies in which severe combined immunodeficiency (*scid*) mice were crossed with *p53* null mice suggest that this is the case (Guidos *et al*, 1996; Nacht *et al*, 1996). The *scid* mice carry a defective DNA dependent protein kinase (DNA-PK) and cannot properly undergo DNA end joining during VDJ recombination in B and T cells. Consequently, these cells fail to mature normally. A combined deficiency in DNA-PK and p53 promotes T cell development, and T cell tumours develop earlier than in *p53* null mice. Thus, p53 apparently acts as a checkpoint in early T cell development such that inactiva-

TABLE 1. Animal models for apoptosis roles in tumourigenesis

Model/Strain	Protein	Tissue/cell type studied	Additional genotype modification	Tumour-related phenotype	Reference
Bcl-2-Ig (T)	human Bcl-2	B cells	–	lymphoid hyperplasia; malignant B-cell lymphoma, t_{50} = 1–2 y	McDonnel and Korsmeyer (1991)
			Eμ-myc (T)	malignant lymphoma, t_{50} = 4 wk	Marin et al (1995)
			p53⁻/⁻ (H)	lymphoma	Marin et al (1994)
TgT$_{121}$ (T)	SV40 TAg fragment; inactivates RB proteins	brain CPE	–	high proliferation, high apoptosis, slow brain tumor, t_{50} = 26 wk	Symonds et al (1994)
			p53⁺/⁻ (H)	progression to focal aggressive tumour t_{50} = 11 wk, focal low apoptosis	
			p53⁻/⁻ (H)	aggressive brain tumour, t_{50} = 3.5 wk, low apoptosis	
			DN p53 (T)	aggressive brain tumour, t_{50} = 7 wk low apoptosis	Bowman et al (1996)
			bax⁺/⁻ (H)	intermediate brain tumour growth, t_{50} = 18 wk	Yin et al (1997)
			bax⁻/⁻ (H)	fast brain tumour growth, t_{50} = 8.5 wk, intermediate apoptosis	

Model (T/H)	Gene/oncoprotein	Tissue	Cross	Phenotype	Reference
αAE6TTL/E7 (T)	HPV16 E7 inactivates RB proteins	lens	–	microphthalmia, high proliferation, high apoptosis	Pan and Griep (1994)
αAE6/E7TTL (T) (E6 expressed)			p53$^{-/-}$ (H)	lens tumour, reduced apoptosis; lens tumour, intermediate apoptosis; p53 independent apoptosis demonstrated	Pan and Griep (1995)
IRBP-E7 (T)	HPV16 E7	retina	–	retinal degeneration retinoblastoma	Howes et al (1994)
RIP-Tag (T)	TAg	pancrease β cell	p53$^{-/-}$ (H); –; IGF-2$^{-/-}$ (H); RIP-bcl-X$_L$ (T)	hyperplasia, 1–2% solid tumour; tumour size diminished, apoptosis increased; tumour expanded, apoptosis diminished	Folkman et al (1989); Christofori et al (1994); Naik et al (1996)
p53$^{-/-}$ (H)	p53	thymocytes	–	thymoma[1]; inhibition of DNA damage apoptosis	Donehower et al (1992); Clarke et al (1993); Lowe et al (1993b)
bax$^{-/-}$ (H)	Bax	B/T-cells	–	lymphoid hyperplasia	Knudson et al (1995)
E1A+Ha-ras in MEF/nude mice		fibroblasts	p53$^{-/-}$ (H)	high apoptosis in hypoxic region; low apoptosis in hypoxic region	Graeber et al (1996)

CPE = choroid plexus epithelium; DN = dominant negative; E1A = adenovirus early region 1 A; Eμ, Ig = immunoglobulin heavy chain enhancer; H = homologous recombination, gene disruption; HPV = human papillomavirus; IRPB = interphotoreceptor retinoid binding protein promoter; MEF = mouse embryo fibroblasts; RIP = rat insulin promoter; t_{50} = the time at which 50% of the animals had died; T = transgene; TTL = translation termination linker;
[1]Sarcomas also induced, but no information on relation to apoptosis.

I. Initiation

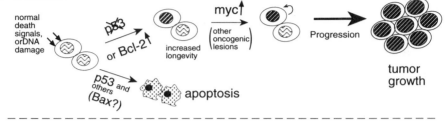

II. Progression

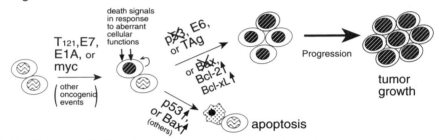

Fig. 1. Models for contribution of apoptosis attenuation to tumorigenesis. (I) Attenuation of apoptosis predisposes cells to tumorigenesis. This model is based on the observation of long latency for some lymphomas and that TP53 is critical for checkpoint function in response to DNA damage. Dysregulation of an oncogene, such as *BCL2*, or inactivation of a tumour suppressor, such as *TP53*, makes a cell irresponsive to normal death signals or to DNA damage. Additional events are required for transformation to malignancy. In the case of DNA damage, the mutation frequency is increased, theoretically increasing the probability for such secondary events. (II) Selective pressure exists for attenuation of apoptosis during tumorigenesis, leading to tumour progression. This model is based on the observation that *TP53* mutation is often a late event in the progression of many human cancers and that overexpression of BCL2 accelerates tumour growth with no effects on proliferation rate. A key difference from the first model is the existence of oncogenic events before apoptosis impairment. Both models have TP53 dependent and independent pathways

tion of p53 promotes the survival of cells with genetic mutations, increasing the probability of their oncogenic progression (Fig. 1, model 1). Indeed, as with the *BCL2-IG* mice, additional aberrations are required for T cell tumour development, since tumours arise clonally in spite of p53 inactivation throughout the thymocyte population (McCarthy *et al*, 1994).

The *BCL2* related gene, *BAX*, can promote apoptosis and can also be transcriptionally regulated by p53 (Miyashita and Reed, 1995), indicating the possibility that BAX is an effector of TP53 dependent apoptosis. However, thymocytes from BAX deficient mice are not resistant to irradiation induced apoptosis as are *p53* null thymocytes (Knudson *et al*, 1995). Furthermore, expression of a *Bax* transgene in *p53* deficient thymocytes does not restore the apoptosis response to irradiation (Brady *et al*, 1996). Bax expression in $p53^{+/+}$ thymocytes, however, accelerates apoptosis in response to a variety of stimuli, including irradiation (Brady *et al*, 1996). Hence, Bax is neither necessary nor sufficient for p53 dependent thymocyte apoptosis, although it could be involved in the process. Inactivation of *Bax* in B cells appears not to be equivalent to Bcl-2 expression in that cell type. Although B and T cell hyperplasias occur in Bax deficient mice, these have not yet progressed to malignancy (Knudson *et al*, 1995), as occurs with the *BCL2-IG* mice (McDonnell *et al*, 1989).

Much of the work indicating a predisposition to tumorigenesis via apoptosis inhibition (model 1, Fig. 1) has been focussed on lymphocytes. Lymphocytes comprise a population of cells that are normally regulated by substantial apoptosis. Furthermore, the fact that gene rearrangement is an ongoing process in these cells could make them more susceptible to DNA damage. Hence, interference with apoptosis may have a greater impact on lymphocytes than on other cell types that do not undergo extensive apoptosis during normal differentiation. It will be important in the future to determine whether and how a predisposition to cancer occurs in other cell types upon apoptosis inhibition. Neurons constitute another population of cells in which apoptosis regulation has a key role in normal development. In this regard, future studies on tumorigenesis of this cell type should indicate whether there could be a general role for apoptosis attenuation in tumorigenesis of cells with a dependency on apoptosis for normal function.

With respect to apoptosis in response to toxic or DNA damage exposure, it is interesting that *TP53* mutations appear to arise early in the development of human cancers of the skin (Ziegler *et al*, 1994) and oesophagus (Parenti *et al*, 1995; Schneider *et al*, 1996). This observation fulfils the predictions of model 1 (Fig. 1), suggesting the possibility that cell types exposed to such toxicity may be susceptible to tumour initiation upon perturbation of the apoptotic response.

Apoptosis Suppression and Tumour Progression

Recent studies using viral oncoproteins as tools in transgenic mouse models have also clearly indicated a role for apoptosis suppression in tumour progression (Fig. 1, model 2). In several systems, expression of viral oncoproteins that disrupt cell cycle regulation by inactivating the RB family proteins also induces apoptosis. In each case examined thus far, a large fraction of the apoptosis is dependent on functional p53, although the induction of p53 independent apoptosis has also been clearly established (Pan and Griep, 1995; Naik *et al*, 1996) (see below). In these experimental systems, tumour progression can be induced by inactivation of p53 and concomitant reduction of apoptosis. Thus far, transgenic mouse models examining three distinct cell types have shown similar results (see Table 1). Expression in lens epithelium of either the HPV-16 E7 oncoprotein (Pan and Griep, 1994, 1995) or a truncated T-Ag (Fromm *et al*, 1994)—both of which inactivate pRb family proteins—resulted in microphthalmia, an atrophy of lens tissue. This condition was characterized by abnormal proliferation along with abundant apoptosis in the developing lens. When crossed with mice expressing HPV-16 E6 (which causes the destruction of p53) or with *p53* null mice, the level of apoptosis in the lens was diminished and tumours emerged (Pan and Griep, 1994, 1995). Similar results were obtained in the retina, wherein expression of HPV-16 E7 was targeted with the interphotoreceptor retinoid binding protein (IRBP) promoter (Howes *et al*, 1994). Expression of E7 alone caused retinal degeneration through photoreceptor cell death, but expression of E7 in a *p53* null background resulted in retinoblastoma. These data suggest that inhibition of apoptosis contributes to tumorigenesis in E7 positive cells.

A transgenic model of tumour progression has been generated in our laboratory in which the expression of T antigen or mutant derivatives is directed to the choroid plexus epithelium (CPE), a normally non-dividing brain epithelial cell of neurectoderm derivation (Symonds *et al*, 1994). As with the eye, this tissue can be readily followed using in situ techniques and has proved a powerful system for deciphering the molecular determinants of tumorigenesis and their resulting biological effects. Expression of wild type T antigen in CPE results in aggressive non-clonal tumour growth (Chen and Van Dyke, 1991), whereas a truncated T antigen (T_{121}), which inactivates the pRB proteins but not p53, induces very slow growing tumours (Saenz Robles *et al*, 1994). This indicates that p53 inactivation in these cells contributes to tumour progression rather than initiation. Indeed, tumours grow aggressively in a p53 null background, confirming that T_{121} induced tumour growth is retarded by wild type p53 (Symonds *et al*, 1994). This role is distinct from that in T cells, where p53 inactivation predisposes thymocytes to tumorigenesis (see above), further indicating the importance of cell type specificity in determining the mechanism of tumour suppression.

Analysis of fast growing versus slow growing CPE tumours in situ indicates that the critical difference is the percentage of cells undergoing apoptosis. Whereas T_{121} is able to induce proliferation at rates similar to that induced by wild type T antigen (Saenz Robles *et al*, 1994), substantial apoptosis is also induced (Symonds *et al*, 1994). The apoptosis is reduced by more than 90% when T_{121} is expressed in the absence of p53 (in transgenic (Tg)$T_{121}p53^{-/-}$ mice and in mice expressing wild type T antigen). A similar result is produced in doubly transgenic mice that co-express T_{121} and a dominant negative fragment of p53 (p53DD) in the presence of wild type p53 (Bowman *et al*, 1996).

We have recently used the TgT_{121} model to begin to define the pathway by which p53 induces apoptosis. By expressing T_{121} in a Bax deficient background, we have shown that Bax is a an effector of p53 mediated apoptosis in this tissue (Yin *et al*, 1997). The apoptotic index in the CPE of Tg$T_{121}Bax^{-/-}$ mice is reduced by about 50%, indicating that a large fraction of the p53 dependent apoptosis is mediated through Bax. Since the tumour growth rate is also intermediate compared with that of Tg$T_{121}TP53^{-/-}$ mice, these data further strengthen the argument that tumour growth rate inversely correlates with the level of apoptosis.

Perhaps the system that most closely models the evolution of human cancer is the study of tumorigenesis in mice heterozygous for *p53*. This is because tumours begin to develop while p53 is functional, such that selective pressure for $p53$ inactivation can be assessed. For example, do tumours of $TgT_{121}p53^{+/-}$ mice progress to an aggressive state with the functional loss of p53? Importantly, in $TgT_{121}p53^{+/-}$ mice, tumours do progress from slow hyperplastic growth characteristic of $TgT_{121}p53^{+/+}$ mice to focally aggressive growth (Symonds *et al*, 1994). The emergence of aggressive tumour nodules can be observed in the brains of 100% of these mice within 2 weeks of death at around 11 weeks, by which time the tumour mass has grown dramatically. (When both $p53$ alleles are intact, mice live substantially longer, to an average of 26 weeks.) The wild type $p53$ allele is often lost in the tumours that emerge in $TgT_{121}p53^{+/-}$ mice, indicating that there is selective pressure during tumour growth for the inactivation of p53 and thus cell death. Indeed, apoptosis is high in morphologically hyperplastic tissue and low in the adjacent tumour nodules. This reproducible mouse model provides a powerful tool for further understanding the molecular and biological steps of tumour progression.

Evidence for similar selective pressure to inactivate p53 in oncogenically transformed cells under conditions of hypoxia has also recently been reported (Graeber *et al*, 1996). Since hypoxia is a condition often associated with tumour growth, the fact that p53 dependent apoptosis is triggered in low oxygen is particularly important and could explain why $p53$ is so frequently mutated in solid tumours. Mouse embryo fibroblasts (MEFs) transformed with adenovirus E1A and activated Ha-ras die by a p53 dependent mechanism under hypoxic, but not aerobic, conditions in culture (Graeber *et al*, 1996). Moreover, when $p53^{+/+}$ and $p53^{-/-}$ transformed cells are mixed, the $p53^{-/-}$ cells selectively overtake the culture when grown under low oxygen, but not when grown aerobically. Finally, when these transformed MEFs are injected into athymic nude mice, tumours that develop contain hypoxic regions that demonstrate p53 dependent apoptosis. As predicted from the cultured cell experiments, tumours derived from $p53^{+/+}$ cells show a correlation between hypoxic regions and areas of high apoptosis, whereas tumours derived from $p53^{-/-}$ cells demonstrate lower levels of apoptosis in hypoxic regions and no difference in aerobic regions (Graeber *et al*, 1996).

Overall, the animal studies described here have provided critical in vivo evidence that p53 dependent apoptosis, occurring in response to oncogenic events and also to physiological changes within developing tumours, can be a critical regulator of tumorigenesis and may explain the observation that $p53$ mutation often occurs late in the development of human cancer. A critical difference in comparison to the model proposed for thymocytes is that p53 dependent apoptosis is summoned subsequent to oncogenic transformation, rather than initially to DNA damaging agents (Fig. 1). Further studies carried out in multiple specific cell types will be required to determine the extent to which these mechanisms differ at the molecular level and with regard to their overall impact in tumorigenesis. In addition, several questions remain to be explored concerning the tumour suppression model. For example, what is the molecular mechanism by which an "oncogenic stimulus" triggers p53 dependent apoptosis in these models? Does cell cycle dysregulation via pRb inactivation and E2F activation have a role? What are the essential downstream effectors of p53 dependent apoptosis? To what extent does loss of a p53 dependent DNA damage checkpoint impact on tumour progression subsequent to apoptosis reduction? Does loss of p53 dependent apoptosis in response to inappropriate cell cycle activity have a role in tumorigenesis of other cell types? The animal models described here will serve as powerful systems with which to address these important issues.

TP53 Independent Apoptosis

Clearly, several normal cell death mechanisms do not involve TP53. In thymocytes, the induction of apoptosis by glucocorticoids or T cell receptor engagement is TP53 independent (Clarke *et al*, 1993; Lowe *et al*, 1993b). Other examples include apoptosis of prostatic glandular cells in androgen ablated animals (Berges *et al*, 1993) and peripheral lymphoid cells (Strasser *et al*, 1994). In addition, much of the apoptosis occurring during normal vertebrate development does not require p53, because most *p53* null embryos develop normally (Donehower *et al*, 1992; Jacks *et al*, 1994).

Although many in vitro and in vivo tumorigenesis studies have focussed on p53 dependent apoptosis, there is also evidence that p53 independent apoptosis is involved in tumour suppression. In the lens system described above, a single oncogenic stimulus, expression of E7, can trigger both p53 dependent and p53 independent apoptosis in the same cell type simultaneously (Pan and Griep, 1995). At an early developmental stage (E13.5), the molecular pathways to apoptosis are almost completely dependent on p53, similar to what is seen in the lenses of *Rb/p53* doubly null embryos (Morgenbesser *et al*, 1994). However, as the lens grows in late embryogenesis and in postnatal life, E7 induced apoptosis relies increasingly on a p53 independent pathway(s). E6 inhibits both p53 dependent and p53 independent apoptosis, resulting in subsequent tumour formation. p53 independent apoptosis also accounts for about 10% of the apoptosis induced by T_{121} in CPE (Symonds *et al*, 1994). Since the molecular mechanism(s) for the p53 independent apoptosis observed in these systems is not yet understood, it is unclear whether reduction in p53 independent apoptosis is as effective as p53 dependent apoptosis in tumorigenesis.

In a transgenic model of pancreatic β cell tumour progression, however, apoptosis is not dependent on p53 and modulates up and down during various defined stages of tumour development (Naik *et al*, 1996). The tumours are induced by targeting wild type T antigen to pancreatic β cells with the rat insulin promoter (RIP) (Christofori *et al*, 1994). Studies on these mice (RIP-Tag2) have revealed a molecular and biological basis for several events during this multistage tumorigenesis. In these cells, expression of T antigen is insufficient for tumour development (Teitelman *et al*, 1988), and dysregulated proliferation is accompanied with p53 independent apoptosis (Naik *et al*, 1996). However, in all mice, 1–2% of the islets progress to carcinoma. During this progression, the transition from hyperplasia to an angiogenic state results in an increase in apoptosis, but the final progression to carcinoma correlates with a reduction in apoptosis (Naik *et al*, 1996). Insulin like growth factor (IGF2) acts as a survival factor during the early stages of tumorigenesis in these cells, since apoptosis is increased and tumour size is decreased when RIP-Tag mice are intercrossed into an *Igf-2* null background (Christofori *et al*, 1994). The rate of islet cell proliferation, however, is unaffected. Finally, member(s) of the *Bcl-2* family may be involved in the reduction of apoptosis observed in the transition from angiogenic progenitor to carcinoma. Bcl-xL is upregulated in these tumours, and mice doubly transgenic for RIP-Tag and RIP-bcl-xL exhibit a reduction in apoptosis and an increase in tumour incidence (Naik *et al*, 1996). A recent report of T antigen induced mammary tumorigenesis also demonstrates suppression of p53 independent apoptosis in the transition from preneoplasia to carcinoma (Shibata *et al*, 1996). These studies emphasize the importance of apoptosis suppression in tumour progression and further demonstrate that additional pathway(s) independent of p53 act in tumour suppression.

CURRENT CHALLENGES AND FUTURE DIRECTIONS

Although animal models have demonstrated possible roles for apoptosis modulation in cancer development, we do not yet know the extent to which these mechanisms translate to the genesis of human cancer. Relatively few correlative studies have been carried out because of the difficulty in following tumour progression in humans. However, the animal studies should provide specific predictions and thus guidance for the examination of human samples.

The animal studies of apoptosis and tumour development have also made several clear predictions in the area of cancer therapy (Lowe, 1995). For example, the susceptibility of cancer cells to anticancer drugs and irradiation lies in their ability to undergo apoptosis. Embryonic fibroblasts transformed by co-expression of adenovirus E1A and activated Ha-*ras* oncogenes form tumours when injected subcutaneously into nude mice regardless of their p53 status. The tumours expressing p53 contain a high proportion of apoptotic cells and typically regress after treatment with γ radiation or adriamycin. By contrast, p53 deficient tumours continue to enlarge and contain few apoptotic cells after treatment with the same regimens (Lowe *et al*, 1994). In addition, in mice bearing preneoplastic mammary lesions, radiation induced apoptosis is manifested in lesions expressing normal p53 but not in those that are p53 deficient (Meyn *et al*, 1996). Thus, in the future, tumour genotyping may provide important guidance as to which mode of therapy may be most effective for treatment of a given tumour.

The mouse models also indicate the potential efficacy of introducing apoptosis inducing genes, such as *p53* or *bax*, for cancer therapy. Genetic therapy methods are being developed which allow for high efficiency gene delivery, predominantly using viral vectors (Martin and Green, 1994; Avalosse *et al*, 1995; Neubauer *et al*, 1996; Vos *et al*, 1995; Roth, 1996; Bartlett *et al*, 1995). However, the fact that viruses harbour several genes effective in inhibiting apoptosis cautions against using viral vectors that retain uncharacterized viral genes. Nevertheless, initial experiments on *p53* gene therapy seem promising.

Adenovirus or retrovirus mediated transfer of the wild type *TP53* gene has been tested in cell lines derived from human tumours, including glioma (Gjerset *et al*, 1995; Gomez-Manzano *et al*, 1996), prostate (Srivastava *et al*, 1995), breast (Seth *et al*, 1996; Zhang *et al*, 1996), ovarian (Mujoo *et al*, 1996), small cell lung carcinoma (Gjerset *et al*, 1995) and squamous cell carcinoma of the head and neck (Liu *et al*, 1995). The restoration of wild type TP53 in several cancer cell lines induces rapid cell death via apoptosis and, where tested, also sensitizes the cells to radiation and cisplatin with correlation of increased apoptosis. The first clinical trial using this approach on non-small cell lung cancer was recently reported (Roth *et al*, 1996). Retroviral mediated transfer of wild type *TP53* resulted in increased apoptosis in post-treatment biopsies compared with those taken before treatment, and tumour regression was noted in three of nine patients. Although these studies were only phase I clinical trials to assess toxicity (which was absent) and are preliminary with regard to efficacy, they do show promise for this approach.

Although the mouse models show that induction of apoptosis in tumours should provide a mechanism by which to slow tumour growth, they also caution that this approach is theoretically not particularly feasible. For example, in $TgT_{121}p53^{+/-}$ mice, there is selective pressure for growth of cells lacking functional p53 such that 100% of the mice exhibit tumour progression, leading to rapid tumour growth. Presumably, since the tumours develop focally, only a single cell had to lose *p53* to develop into an aggressive tumour mass. Thus, theoretically, delivery of an apoptosis inducing gene would have to reach 100% of the cells to stop tumour growth. Perhaps combined therapies and the development of chemotherapeutic agents that will induce apoptosis in cells lacking the therapeutic gene will be warranted.

Overall, the animal models described here should provide valuable experimental systems for establishing which genes are most effective at tumour suppression, for optimizing somatic delivery methods and for assessing the effectiveness of apoptosis inducing vectors against cancer growth in designing new regimens for cancer therapy.

SUMMARY

For nearly two decades, studies in the cancer research field focussed on identifying genes that act as positive and negative regulators of cell growth. Only relatively recently was it recognized that the regulation of cell death (apoptosis) is also an important modulator of tumorigenesis. At least two genes linked to human cancers, *BCL2* and *TP53*, have been shown to regulate apoptosis. The correlation between apoptosis modulating genes and human tumours raises an important question as to how dysregulation of apoptosis contributes to neoplastic transformation and malignant cell growth. Cell culture studies have clearly demonstrated that *TP53* can induce and *BCL2* can suppress apoptosis in response to various stimuli. Studies of mammalian viruses, which possess mechanisms for both inducing and evading apoptosis, have also extended our understanding of this process. On the basis of such findings, several animal models have been developed which begin to address the role of apoptosis regulation in tumorigenesis. This chapter discusses those animal models, focussing on *bcl-2* (and its relatives) and *p53*.

Acknowledgements

The National Cancer Institute is gratefully acknowledged for continued support of our research program. We also thank John Kim, Leah Akins and Stacey Bridge for help with references.

References

Arends MJ and Wyllie AH (1991) Apoptosis: mechanisms and roles in pathology. *International Reviews of Experimental Pathology* **32** 223–254

Avalosse B, Dupont F and Burny A (1995) Gene therapy for cancer. *Current Opinion in Oncology* **7** 94–100

Bartlett JS, Quattrocchi KB and Samulski RJ (1995) The development of adeno-associated virus as a vector for cancer gene therapy, In: Sobol RE and Scanlon KJ (eds). *The Internet Book of Gene Therapy Cancer Therapeutics*, pp 27–39, Appleton and Lange, Stamford, Connecticut

Berges RR, Furuya Y, Remington L, English HF, Jacks T and Isaacs JT (1993) Cell proliferation, DNA repair and p53 function are not required for programmed death of prostatic glandular cells induced by androgen ablation. *Proceedings of the National Academy of Sciences of the USA* **90** 8910–8914

Bissonnette RP, Exheverri F, Mahboubi A and Green DR (1992) Apoptotic cell death induced by c-myc is inhibited by bcl-2. *Nature* **359** 552–554

Boise LH, Gonzalez-Garcia M, Postema CE *et al* (1993) bcl-x, a bcl-2-related gene that functions as a dominant regulator of apoptotic death. *Cell* **74** 597–608

Bowman T, Symonds H, Gu LY, Yin CY, Oren M and Van Dyke T (1996) Tissue-specific inactivation of p53 tumor suppression in the mouse. *Genes and Development* **10** 826–835

Brady HJM, Salomons GS, Bobeldijk RC and Berns AJM (1996) T cells from bax transgenic mice show accelerated apoptosis in response to stimuli but do not show restored DNA damage-induced cell death in the absence of p53. *EMBO Journal* **15** 1221–1230

Bump NJ, Hackett M, Hugunin M *et al* (1995) Inhibition of ICE family proteases by baculovirus anti-apoptotic protein p53. *Science* **269** 1885–1888

Caelles C, Helmberg A and Karin M (1994) p53-dependent apoptosis in the absence of transcriptional activation of p53-target genes. *Nature* **370** 220–223

Chen J and Van Dyke T (1991) Uniform cell-autonomous tumorigenesis of the choroid plexus by papovavirus large T antigens. *Molecular and Cellular Biology* **11** 5968–5976

Chen X, Farmer G, Zhu H, Prywes R and Prives C (1993) Cooperative DNA binding of p53 with TFIID (TBP): a possible mechanism for transcriptional activation. *Genes and Development* **7** 1837–1849

Chiou S, Rao L and White E (1994a) Bcl-2 blocks p53-dependent apoptosis. *Molecular and Cellular Biology* **14** 2556–2563

Chiou S, Tseng C, Rao L and White E (1994b) Functional complementation of the adenovirus E1B 19-kilodalton protein with bcl-2 in the inhibition of apoptosis in infected cells. *Journal of Virology* **68** 6553–6566

Chittenden T, Harrington EA, O'Connor R *et al* (1995) Induction of apoptosis by the Bcl-2 homologue Bak. *Nature* **733** 736

Cho Y, Gorina S, Jeffrey P and Pavletich N (1994) Crystal structure of a p53 tumor suppressor-DNA complex: understanding tumorigenic mutations. *Science* **265** 346–355

Christofori G, Naik P and Hanahan D (1994) A second signal supplied by insulin-like growth factor II in oncogene-induced tumorigenesis. *Nature* **369** 414–418

Clarke AR (1995) Murine models of neoplasia: functional analysis of the tumor suppressor genes Rb-1 and p53. *Cancer and Metastasis Reviews* **14** 125–148

Clarke AR, Maandag ER, van Roon M *et al* (1992) Requirement for a functional Rb-1 gene in murine development. *Nature* **359** 328–330

Clarke AR, Purdie CA, Harrison DJ *et al* (1993) Thymocyte apoptosis induced by p53-dependent and independent pathways. *Nature* **362** 849–852

Debbas M and White E (1993) Wild-type p53 mediates apoptosis by E1A, which is inhibited by E1B. *Genes and Development* **7** 546–554

DeCaprio JA, Ludlow JW, Figge J *et al* (1988) SV40 large tumor antigen forms a specific complex with the product of the retinoblastoma susceptibility gene. *Cell* **54** 275–283

Donehower LA, Harvey M, Slagle BL *et al* (1992) Mice deficient for p53 are developmentally normal but are susceptible to spontaneous tumours. *Nature* **356** 215–221

Dyson N, Howley PM, Munger K and Harlow E (1989) The human papilloma virus-16 E7 oncoprotein is able to bind to the retinoblastoma gene product. *Science* **243** 934–937

El-Deiry WS, Tokino T, Velculescu VE *et al* (1993) WAF1, a potential mediator of p53 tumor suppression. *Cell* **75** 817–825

Evan GI, Wyllie AH, Gilbert CS *et al* (1992) Induction of apoptosis in fibroblasts by *c-myc* protein. *Cell* **69** 119–128

Farmer G, Bargonetti J, Zhu H, Friedman P, Prywes R and Prives C (1992) Wild-type p53 activates transcription *in vitro*. *Nature* **358** 83–86

Folkman J, Watson K, Ingber D and Hanahan D (1989) Induction of angiogenesis during the transition from hyperplasia to neoplasia. *Nature* **339** 58–61

Fraser A and Evan G (1996) A license to kill. *Cell* **85** 781–784

Friedlander P, Haupt Y, Prives C and Oren M (1996) A mutant p53 that discriminates between p53-responsive genes cannot induce apoptosis. *Molecular and Cellular Biology* **16** 4961–4971

Fromm L, Shawlot W, Gunning K, Butel J and Overbeek P (1994) The retinoblastoma protein binding region of simian virus 40 large T antigen alters cell cycle regulation in lenses of transgenic mice. *Molecular and Cellular Biology* **14** 6743–6754

Gjerset RA, Turla ST, Sobol RE *et al* (1995) Use of wild-type p53 to achieve complete treatment sensitization of tumor cells expressing endogenous mutant p53. *Molecular Carcinogenesis* **14** 275–285

Gomez-Manzano C, Fueyo J, Kyritsis AP *et al* (1996) Adenovirus-mediated transfer of the p53 gene produces rapid and generalized death of human glioma cells via apoptosis. *Cancer Research* **56** 694–699

Graeber TG, Osmanian C, Jacks T *et al* (1996) Hypoxia-mediated selection of cells with diminished apoptotic potential in solid tumours. *Nature* **379** 88–91

Greenblatt MS, Bennett WP, Hollstein M and Harris CC (1994) Mutations in the p53 tumor suppressor gene: clues to cancer etiology and molecular pathogenesis. *Cancer Research* **54** 4855–4878

Guidos CJ, Williams CJ, Grandal I, Knowles G, Guang MTF and Danska FS (1996) V(D)J recombination activates a p53-dependent DNA damage checkpoint in *scid* lymphocyte precursors. *Genes and Development* **10** 2038–2054

Haffner R and Oren M (1995) Biochemical properties and biological effects of p53. *Current Opinion in Genetics and Development* **5** 84–90

Han JH, Sabbatini P, Perez D, Rao L, Modha D and White E (1996) The E1B 19K protein blocks apoptosis by interacting with and inhibiting the p53-inducible and death-promoting Bax protein. *Genes and Development* **10** 461–477

Harper JW, Adami GR, Wei N, Keyomars K and Elledge SJ (1993) The p21 Cdk-interacting protein Cip1 is a potent inhibitor of G1 cyclin-dependent kinases. *Cell* **75** 805–816

Haupt Y, Rowan S, Shaulian E, Vousden KH and Oren M (1995) Induction of apoptosis in HeLa cells by trans-activation-deficient p53. *Genes and Development* **9** 2170–2183

Hay BA, Wolff T and Rubin GM (1994) Expression of baculovirus p35 prevents cell death in *Drosophila*. *Development* **120** 2121–2129

Hockenberry D, Nunez G, Milliman C, Schreiber RD and Korsmeyer SJ (1990) Bcl-2 is an inner mitochondrial membrane protein that blocks programmed cell death. *Nature* **348** 334–336

Hockenberry D, Oltvai Z, Yin X, Milliman C and Korsmeyer S (1993) Bcl-2 functions in an antioxidant pathway to prevent apoptosis. *Cell* **75** 241–251

Hollstein M, Sidransky D, Vogelstein B and Harris CC (1991) p53 mutations in human cancers. *Science* **253** 49–53

Horning SJ and Rosenberg SA (1984) The natural history of initially untreated low-grade non-Hodgkin's lymphomas. *New England Journal of Medicine* **311** 1471–1475

Howes KA, Ransom N, Papermaster DS, Lasudry JGH, Albert DM and Windle JJ (1994) Apoptosis or retinoblastoma: alternative fates of photoreceptors expressing the HPV-16 E7 gene in the presence or absence of p53. *Genes and Development* **8** 1300–1310

Hsu B, Marin MC and McDonnell TJ (1995) Cell death regulation during multistep lymphomagenesis. *Cancer Letter* **94** 17–23

Jacks T, Remington L, Williams B *et al* (1994) Tumor spectrum analysis in p53-mutant mice. *Current Biology* **4** 1–7

Kamada S, Shinto AA, Tsujimura Y *et al* (1995) Bcl-2 deficiency in mice leads to pleiotropic abnormalities: accelerated lymphoid cell death in the thymus and spleen, polycystic kidney, hair hypopigmentation, and distorted small intestine. *Cancer Research* **55** 354–359

Kastan MB, Zhan Q, El-Deiry WS *et al* (1992) A mammalian cell cycle checkpoint pathway utilizing p53 and GADD45 is defective in ataxia-telangiectasia. *Cell* **71** 587–597

Kastan MB, Canman CE and Leonard CJ (1995) p53, cell cycle control and apoptosis: implications for cancer. *Cancer and Metastasis Reviews* **14** 3–15

Kern SE, Kinzler DT, Bruskin A, Jarosz D, Riedman P, Prives C and Vogelstein B (1991) Identification of p53 as a sequence-specific DNA binding protein. *Science* **252** 1708–1711

Knudson CM, Tung KSK, Tourtellotte WG, Brown GAJ and Korsmeyer SJ (1995) Bax-deficient mice with lymphoid hyperplasia and male germ cell death. *Science* **270** 96–99

Ko LJ and Prives C (1996) p53: puzzle and paradigm. *Genes and Development* **10** 1054–1072

Komiyama T, Ray CA, Pickup DJ *et al* (1994) Inhibition of interleukin-1b converting enzyme by the cowpox virus serpin CrmA. *Journal of Biological Chemistry* **269** 19331–19337

Korsmeyer SJ (1995) Regulators of cell death. *Trends in Genetics* **11** 101–105

Krajewski S, Tanaka S, Sckibler MJ, Fenton W and Reed JC (1994) Investigations of the subcellular distribution of the BCL-2 oncoprotein: residence in the nuclear envelope, endoplasmic reticulum and outer mitochondrial membranes. *Cancer Research* **53** 4701–4714

Lee JM and Bernstein A (1995) Apoptosis, cancer and the p53 tumor suppressor gene. *Cancer and Metastasis Reviews* **14** 149–161

Lee S, Elenbaas B, Levine A and Griffith J (1995) p53 and its 14 kDa c-terminal domain recognize primary DNA damage in the form of insertion/deletion mismatches. *Cell* **81** 1013–1020

Levine AJ, Momand J and Finlay CA (1991) The p53 tumour suppressor gene. *Nature* **351** 453–

456

Liu TJ, el-Naggar AK, McDonnell TJ *et al* (1995) Apoptosis induction mediated by wild-type p53 adenoviral gene transfer in squamous cell carcinoma of the head and neck. *Cancer Research* **55** 3117–3122

Lowe SW (1995) Cancer therapy and p53. *Current Opinion in Oncology* **7** 547–553

Lowe SW and Ruley HE (1993) Stabilization of the p53 tumor suppressor is induced by adenovirus 5 E1A and accompanies apoptosis. *Genes and Development* **7** 535–545

Lowe SW, Ruley HE, Jacks T and Housman DE (1993a) p53-dependent apoptosis modulates the cytotoxicity of anticancer agents. *Genes and Development* **7** 535–545

Lowe SW, Schmitt EM, Smith SW, Osborne BA and Jacks T (1993b) p53 is required for radiation-induced apoptosis in mouse thymocytes. *Nature* **362** 847–849

Lowe SW, Bodis S, McClatchey A *et al* (1994) p53 status and the efficacy of cancer therapy in vivo. *Science* **266** 807–810

McCarthy SA, Symonds HS and Van Dyke T (1994) Regulation of apoptosis in transgenic mice by SV40 T antigen-mediated inactivation of p53. *Proceedings of the National Academy of Sciences of the USA* **91** 3979–3983

McDonnell T and Korsmeyer S (1991) Progression from lymphoid hyperplasia to high-grade malignant lymphoma in mice transgenic for the t(14; 18) *Nature* **349** 254–256

McDonnell TJ, Deane N, Platt FM, Nunez G, Jaeger U, McKearn JP and Korsmeyer SJ (1989) bcl-2-immunoglobulin transgenic mice demonstrate extended B cell survival and follicular lymphoproliferation. *Cell* **57** 79–88

Marin MC, Hsu B, Meyn RE, Donehower LA, El Naggar A and McDonnell TJ (1994) Evidence that p53 and bcl-2 are regulators of a common cell death pathway important for in vivo lymphomagenesis. *Oncogene* **9** 3107–3112

Marin MC, Hsu B, Stephens LC, Brisbay S and McDonnell TJ (1995) The functional basis of c-myc and bcl-2 complementation during multistep lymphomagenesis in vivo. *Experimental Cell Research* **217** 240–247

Martin S and Green D (1994) Apoptosis as a goal of cancer therapy. *Current Opinion in Oncology* **6** 616–621

Meyn RE, Stephens LC, Mason KA and Medina D (1996) Radiation-induced apoptosis in normal and pre-neoplastic mammary glands *in vivo*: significance of gland differenciation and p53 status. *International Journal of Cancer* **65** 466–472

Miura M, Zhu H, Rotello R, Hartwieg EA and Yuan J (1993) Induction of apoptosis in fibroblasts by IL-1b converting enzyme, a mammalian homolog of the *C. elegans* cell death. *Cell* **75** 653–660

Miyashita T and Reed JC (1995) Tumor suppressor p53 is a direct transcriptional activator of the human *bax* gene. *Cell* **80** 293–299

Miyashita T, Krajewski S, Krajewska *et al* (1994) Tumor suppressor p53 is a regulator of bcl-2 and *bax* gene expression in vitro and in vivo. *Oncogene* **9** 1799–1805

Morgenbesser S, Williams B, Jacks T and DePinho R (1994) p-53-dependent apoptosis produced by Rb-deficiency in the developing mouse lens. *Nature* **371** 72–74

Motoyama N, Wang F, Roth KA *et al* (1995) Massive cell death of immature hematopoietic cells and neurons in Bcl-x-deficient mice. *Science* **267** 1506–1510

Mujoo K, Maneval SC, Anderson SC and Gutterman JU (1996) Adenoviral-mediated p53 tumor suppressor gene therapy of human ovarian carcinoma. *Oncogene* **12** 1617–1623

Nacht M, Strasser A, Chan Y *et al* (1996) Mutations in the *p53* and *scid* genes cooperate in tumorigenesis. *Genes and Development* **10** 2055–2065

Naik P, Karrim J and Hanahan D (1996) The rise and fall of apoptosis during multistage tumorigenesis: down-modulation contributes to tumor progression from angiogenic progenitors. *Genes And Development* **10** 2105–2117

Neubauer A, Thiede C, Huhn D and Wittig B (1996) p53 and induction of apoptosis as a target for anticancer therapy. *Leukemia* **10** S2–S4

Nicholson DW, Ali A, Thornberry NA *et al* (1995) Identification and inhibition of the

ICE/CED-3 protease necessary for mammalian apoptosis. *Nature* **376** 37–43

Nigro JM, Baker SJ, Preisinger AC *et al* (1989) Mutations in the p53 gene occur in diverse human tumor types. *Nature* **342** 705–708

Oltvai Z, Milliman C and Korsmeyer S (1993) Bcl-2 heterodimerizes in vivo with a conserved homolog, bax, that accelerates programmed cell death. *Cell* **74** 609–619

Oppenheim RW (1991) Cell death during development of the nervous system. *Annual Reviews of Neuroscience* **14** 453–501

Pan H and Griep AE (1994) Altered cell cycle regulation in the lens of HPV-16 E6 or E7 transgenic mice: implications for tumor suppressor gene function in development. *Genes and Development* **8** 1285–1299

Pan H and Griep AE (1995) Temporally distinct patterns of p53-dependent and p53-independent apoptosis during mouse lens development. *Genes and Development* **9** 2157–2169

Parenti AR, Rugge M, Frizzera E *et al* (1995) p53 overexpression in the multistep process of esophageal carcinogenesis. *American Journal of Surgical Pathology* **19** 1418–1422

Perry ME and Levine AJ (1993) Tumor-suppressor p53 and the cell cycle. *Current Opinion in Genetics and Development* **3** 50–54

Prives C (1994) How loops, β sheets and α helices help us to understand p53. *Cell* **78** 543–546

Prives C and Manfredi JJ (1993) The p53 tumor suppressor protein: meeting review. *Genes and Development* **7** 529–534

Qin XQ, Livingston DM, Kaelin WG and Adams PD (1994) Deregulated transcription factor E2F-1 expression leads to S-phase entry and p53-mediatied apoptosis. *Proceedings of the National Academy of Sciences of the USA* **91** 10918–10922

Rabizadeh S, LaCount DJ, Friesen PD and Bredesen DE (1993) Expression of baculovirus p35 inhibits mammalian neural cell death. *Journal of Neurochemisty* **61** 2318–2321

Ray CA, Black RA, Kronheim SR *et al* (1992) Viral inhibition of inflammation: cowpox virus encodes an inhibitor of the interleukin-1b converting enzyme. *Cell* **69** 597–604

Roth JA (1996) Gene replacement strategies for cancer. *Israeli Journal of Medical Science* **32** 89–94

Roth JA, Nguyen D, Lawrence DD *et al* (1996) Retrovirus-mediated wild-type p53 gene transfer to tumors of patients with lung cancer. *Nature Medicine* **2** 985–991

Rothenberg EV (1992) The development of functionally responsive T cells. *Advances in Immunology* **51** 85–214

Sabbatini P, Lin J, Levine AJ and White E (1995) Essential role for p53-mediated transcription in E1A-induced apoptosis. *Genes and Development* **9** 2184–2192

Saenz Robles MT, Symonds H, Chen J and Van Dyke T (1994) Induction versus progression of brain tumor development: differential functions for the pRB- and p53-targeting domains of SV40 T antigen. *Molecular and Cellular Biology* **14** 2686–2698

Scheffner M, Werness BA, Huibregtse JM, Levine AJ and Howley PM (1990) The E6 oncoprotein encoded by human papillomavirus types 16 and 18 promotes the degradation of p53. *Cell* **63** 1129–1136

Scheffner M, Huibregtse JM, Vierstra RD and Howley PM (1993) The HPV-16 E6 and E6-AP complex function as a ubiquitin–protein ligase in the ubiquitination of p53. *Cell* **75** 495–505

Schmidt EV, Pattengale PK, Weir L and Leder P (1988) Transgenic mice bearing the human c-myc gene activated by an immunoglobulin enhancer: a pre-B-cell lymphoma model. *Proceedings of the National Academy of Sciences of the USA* **85** 6047–6051

Schneider PM, Casson AG, Levin B *et al* (1996) Mutations of p53 in Barrett's esophagus and Barrett's cancer: a prospective study of ninety-eight cases. *Journal of Thoracic and Cardiovascular Surgery* **111** 323–331

Sentman CL, Shutter JR, Hockenbery D, Kanagawa O and Korsmeyer SJ (1991) Bcl-2 inhibits multiple forms of apoptosis but not negative selection in thymocytes. *Cell* **67** 879–888

Seth P, Brinkmann U, Schwartz GN, Katayose D, Gress R, Pastan Ira and Cowan K (1996) Adenovirus-mediated gene transfer to human breast tumor cells: an approach for cancer gene therapy and bone marrow purging. *Cancer Research* **56** 1346–1351

Seto E, Usheva A, Zambetti GP *et al* (1992) Wild-type p53 binds to the TATA-binding protein and represses transcription. *Proceedings of the National Academy of Sciences of the USA* **89** 12028–12032

Shaw P, Bovey R, Tardy S, Sahli R, Sordat B and Costa J (1992) Induction of apoptosis by wild-type p53 in a human colon tumor-derived cell line. *Proceedings of the National Academy of Sciences of the USA* **89** 4495–4499

Shibata MA, Maroulakou IG, Jorcyk CL, Gold LG, Ward JM and Green JE (1996) p53-independent apoptosis during mammary tumor progression in C3(1)/SV40 large T antigen transgenic mice: suppression of apoptosis during the transition from preneoplasia to carcinoma. *Cancer Research* **56** 2998–3003

Srivastava S, Katayose D, Tong YA *et al* (1995) Recombinant adenovirus vector expressing wild-type p53 is a potent inhibitor of prostate cancer cell proliferation. *Urology* **46** 843–848

Strasser A, Harris AW, Bath ML and Cory S (1990) Novel primitive lymphoid tumours induced in transgenic mice by cooperation between myc and bcl-2. *Nature* **348** 331–333

Strasser A, Harris AW, Jacks T and Cory S (1994) DNA damage can induce apoptosis in proliferating lymphoid cells via p53-independent mechanisms inhibitable by bcl-2. *Cell* **79** 329–339

Sugimoto A, Friesen PD and Rothman JH (1994) Baculvirus p35 prevents developmentally programmed cell death and rescues a *ced-9* mutant in the nematode *Caenorhabditis elegans*. *EMBO Journal* **13** 2023–2028

Symonds H, Krall L, Remington L *et al* (1994) p53-dependent apoptosis suppresses tumor growth and progression in vivo. *Cell* **78** 703–711

Teitelman G, Alpert S and Hanahan D (1988) Proliferation, senescence and neoplastic progression of beta cells in hyperplasic pancreatic islets. *Cell* **52** 97–105

Thompson TC, Park SH, Timme TL *et al* (1995) Loss of p53 function leads to metastasis in ras+myc-initiated mouse prostate cancer. *Oncogene* **10** 869–879

Truant R, Xiao H, Ingles J and Greenblatt J (1993) Direct interaction between the transcriptional activation domain of human p53 and the TATA box-binding protein. *Journal of Biological Chemistry* **268** 2284–2287

Tsujimoto Y, Gorham J, Cossman J, Jaffe E and Croce CM (1985) The t(14;18) chromosome translocations involved in B-cell neoplasms result from mistakes in VDJ joining. *Science* **229** 1390–1393

Vaux D, Weissman I and Kim S (1992) Prevention of programmed cell death in *Caenorgabitis elegans* by human bcl-2. *Science* **258** 1955–1957

Veis D, Sorenson C, Shutter J and Korsmeyer S (1993) Bcl-2-deficient mice demonstrate fulminant lymphoid apoptosis, polycystic kidneys and hypopigmented hair. *Cell* **75** 229–240

Vos JM (1995) *Viruses in Human Gene Therapy*, Carolina Academic Press, Durham, North Carolina

Wagner AJ, Small MB and Hay N (1993) Myc-mediated apoptosis is blocked by ectopic expression of Bcl-2. *Molecular and Cellular Biology* **13** 2432–2440

Walker PR, Kokileva JL and Sikorska M (1993) Detection of the initial stages of DNA fragmentation in apoptosis . *Biotechniques* **15** 1032–1040

Wang XW, Gibson MK, Vermeulen W *et al* (1995a) Abrogation of p53-induced apoptosis by the hepatitis B virus X gene. *Cancer Research* **55** 6012–6016

Wang XW, Yeh H, Schaeffer L *et al* (1995b) p53 modulation of TFIIH-associated nucleotide excision repair activity. *Nature Genetics* **10** 188–195

White E (1996) Life, death and the pursuit of apoptosis. *Genes and Development* **10** 1–15

Whyte P, Buchkovich K, Horowitz JM *et al* (1988) Association between an oncogene and an anti-oncogene: the adenovirus E1A proteins bind to the retinoblastoma gene product. *Nature* **334** 124–129

Wu X and Levine AJ (1994) p53 and E2F-1 cooperate to mediate apoptosis. *Proceedings of the National Academy of Sciences of the USA* **91** 3602–3606

Wu X, Bayle JH, Olson D and Levine AJ (1993) The p53-mdm-2 autoregulatory feedback loop. *Genes and Development* **7** 1126–1132

Wyllie AH (1980) Glucocorticoid-induced thymocyte apoptosis is associated with endogenous endonuclease activation. *Nature* **284** 555–556

Xiong Y, Hannon GJ, Zhang H, Casso D, Kobayashl R and Beach D (1993) p21 is a universal inhibitor of cyclin kinases. *Nature* **366** 701–704

Yang E, Zha E, Jockel J, Boise LH, Thompson CB and Korsmeyer SJ (1995) Bad, a heterodimeric partner of Bcl-xL and Bcl-2, displaces Bax and promotes cell death. *Cell* **80** 285–291

Yin C, Knudson CM, Korsmeyers JS and Van Dyke T (1997) Bax suppresses tumorigenesis and stimulates apoptosis in vivo. *Nature* **385** 637–640

Yonish-Rouach E, Resnitzky D, Lotem J, Sachs L, Kimchi A and Oren M (1991) Wild-type p53 induces apoptosis of myeloid leukaemic cells that is inhibited by interleukin-6. *Nature* **352** 345–340

Zambetti GP, Bargonetti J, Walker K, Prives C and Levine AJ (1992) Wild-type p53 mediates positive regulation of gene expression through a specific DNA sequence element. *Genes and Development* **6** 1143–1152

Zhang JF, Hu C, Geng Y, Selm J, Klein B, Orazi A and Taylor MW (1996) Treatment of human breast cancer xenograft with an adenovirus vector containing an interferon gene results in rapid progression due to viral oncolysis and gene therapy. *Proceedings of the National Academy of Sciences of the USA* **93** 4513–4518

Ziegler A, Jonason AS, Leffell DJ *et al* (1994) Sunburn and p53 in the onset of skin cancer. *Nature* **372** 773–776

The authors are responsible for the accuracy of the references.

Genetic Instability in Animal Tumorigenesis Models

LARRY A DONEHOWER

Division of Molecular Virology, Baylor College of Medicine, Houston, Texas 77030

Introduction
    Genes that affect genetic instability
    Transgenic animal models in assessment of genetic instability and cancer
Animal tumorigenesis models for genes that maintain genetic stability
    Tumour suppressors with cell cycle regulatory roles
    Cell cycle mediators and inhibitors
    DNA repair genes
    Telomerase and telomeres
Discussion
Summary

INTRODUCTION

Many of the alterations that take place in the progression of a normal cell to a cancer cell have a genetic origin. Initiating mutations can be inherited or they can be somatic in origin, but it is evident that multiple genetic lesions are required for a cell to be fully malignant (Fearon and Vogelstein, 1990). Each successive mutation may provide a growth advantage to the evolving precancerous cell (Nowell, 1976). Mutations in cellular proto-oncogenes that constitutively activate a growth signal transduction pathway would be likely to provide a direct proliferative advantage to the cell. Likewise, inactivation of a tumour suppressor gene or cell cycle inhibitor could provide a growth stimulus. Finally, mutational activation or inactivation of genes that regulate cell death could provide a survival advantage to a nascent tumour cell clone even in the absence of a direct proliferative effect.

Other mutations that may arise in an evolving neoplastic cell are those that do not provide a direct growth or survival advantage but provide an indirect growth advantage through the relaxation of controls on the cell's genetic stability (Nowell, 1976; Tlsty *et al*, 1995). A mutation that increases genetic instability is likely to result in subsequent oncogenic lesions that do provide a direct growth advantage to the cell. The very high incidence of karyotypic abnormalities in human and animal cancers argues for the importance of genetic instability in the cancer cell evolution process. Although the association between tumour cell progression and genetic instability has been known for many years, only recently has it been possible to identify and isolate many of the genes that maintain genetic stability. Inherited or somatic mutations in these genes have been correlated with cancer in a number of cases, some of which are discussed in this chapter.

Genes That Affect Genetic Instability

Genetic instability can encompass multiple mechanisms, ranging from instability at the nucleotide level to that of entire chromosome complements. Some human tumours may exhibit a mutator phenotype, with greatly elevated rates of point mutations or replication errors (Loeb, 1991; Eshleman and Markowitz, 1995). Alternatively, many human tumours have displayed polyploidy, aneuploidy and abnormal chromosome structures, including translocations and double minute chromosomes. Genes in tumours can be amplified to high copy number or be deleted through loss of an entire chromosome or subchromosomal fragment (Stark, 1993; Tlsty et al, 1995). Abnormal recombination rates may be elevated in some tumours and this may be a mechanism for activating certain oncogenic pathways (Kirsch et al, 1994). The many potential mechanisms of genetic instability have been previously summarized (Hartwell 1992; Almasan et al, 1995a; Smith and Fornace, 1995; Tlsty et al, 1995; Wahl et al, this volume), and thus this chapter focuses on those cancer and genetic instability associated genes for which animal models currently exist.

The most obvious category of genes expected to influence genetic instability during tumorigenesis are the genes that mediate DNA repair. Defects in DNA repair genes would be expected to result in increased overall mutation rates and consequently a higher likelihood of oncogenic mutations (Loeb, 1991). In fact, many inherited DNA repair deficiencies in humans are associated with increased cancer susceptibility. Xeroderma pigmentosum (nucleotide excision repair defect), Fanconi's anaemia (crosslink repair) and Lynch syndrome (mismatch repair) are examples of DNA repair syndromes accompanied by elevated risk of cancer (dos Santos et al, 1994; Hanawalt, 1994; Kolodner, 1995). Moreover, some spontaneously arising tumours have mutations in DNA repair genes, and these tumours exhibit higher mutation rates than other types of tumours without such mutations (Eshleman and Markowitz, 1995).

Genes that regulate cell cycle progression may also have a role in preserving genetic stability. As direct mediators of the cell cycle, the cyclins and cyclin dependent kinases are obvious candidates. Overexpression of cyclins or constitutive activation of cyclin dependent kinases might be expected to result in accelerated cell cycle progression and unresponsiveness to external growth control signals and cell cycle checkpoints (Hunter and Pines, 1994). Loss of responsiveness to G_1, S and G_2/M checkpoints could result in various types of genetic instability, including mutations occurring as a result of premature DNA replication (G_1 checkpoint loss) or chromosome abnormalities occurring as a result of failure of proper segregation of chromosomes during mitosis (G_2/M checkpoint loss) (Hartwell and Kastan, 1994; Hartwell *et al*, 1994).

Cyclin dependent kinase (CDK) inhibitors have also been shown to be important players in cell cycle regulation (Sherr and Roberts, 1995; Harper and Elledge, 1996). In response to DNA damage or inhibitory growth signals, these proteins may be the direct mediators of cell cycle checkpoint control. By binding to the cyclin CDK complexes, these molecules interfere with the critical kinase function of the CDK that mediates cell cycle progression. That at least some of these cyclin dependent kinase inhibitors have a key role in cancer is underscored by the finding that the *p16INK4a* CDK inhibitor gene is lost in a number of human tumour types (Hirama and Koeffler, 1995; Shapiro and Rollins, 1996).

Among the classical tumour suppressor genes, the *RB* and *TP53* genes encode important regulators of the cell cycle. These genes are frequently mutated and lost in human and animal cancers (Weinberg, 1991; Knudson, 1993; Greenblatt *et al*, 1994). The RB protein directly mediates a G_1 arrest function, perhaps in part through its binding in the unphosphorylated form to members of the E2F transcriptional factor family (Bagchi *et al*, 1991; Chellappan *et al*, 1991). When RB is phosphorylated by cyclin/CDK complexes, E2F proteins are released and are free to activate transcriptionally important S phase genes, allowing S phase entry (Hatakeyama *et al*, 1994; Weinberg, 1995). Loss of RB results in accelerated proliferation in several in vivo models (Howes *et al*, 1994; Lee *et al*, 1994; Morgenbesser *et al*, 1994; Pan and Griep, 1994; Symonds *et al*, 1994), and there is evidence that loss of RB increases genetic instability (White *et al*, 1994; Almasan *et al*, 1995a).

The TP53 protein may be the checkpoint control protein par excellence, since G_1 and G_2/mitotic checkpoint functions have been ascribed to it (Kuerbitz *et al*, 1992; Cross *et al*, 1995). Following DNA damage or inappropriate growth signals, TP53 levels increase and mediate either G_1 arrest or apoptosis (Kastan *et al*, 1995; Bates and Vousden, 1996). The G_1 arrest function is mediated in part by the transcriptional activation of the p21CIP1/WAF1 CDK inhibitor (El-Deiry *et al*, 1993; Harper *et al*, 1993; Dulic *et al*, 1994). In several studies, TP53 has been implicated in a G_2/M or mitotic checkpoint (Agarwal *et al*, 1995; Cross *et al*, 1995; Stewart *et al*, 1995). It is now well established that the absence of TP53 in a normal cell or a tumour cell has significant adverse effects on genetic stability (Almasan *et al*, 1995a; Smith and Fornace, 1995). Polyploidy, aneuploidy, gene amplification and increased rates of recombination have been noted in TP53 deficient cells (Livingstone *et al*, 1992; Yin *et al*, 1992; Meyn *et al*, 1994; Cross *et al*, 1995).

Finally, an important aspect of preservation of genetic stability must be the maintenance of intact chromosomal ends, the telomeres. Uncapped chromosomes are susceptible to degradation and fusion and may activate DNA damage checkpoints (Harley *et al*, 1994; de Lange, 1995). Telomerase is the critical enzyme that can maintain the long tandem arrays of short repeats that make up the telomeres. In human cells, the telomeres become progressively shortened with increasing numbers of cell divisions, and it has been hypothesized that the chromosomal instability and gene amplification observed during tumour cell progression may be due in part to shortened telomeres, which are more vulnerable to forming telomere associated dicentric chromosomes (de Lange, 1995). Tumour cells that are able to increase telomerase expression levels could restore normal telomere ends, which is consistent with the observation that late stage human tumour cells usually have high levels of telomerase activity (Harley *et al*, 1994; Kim *et al*, 1994; de Lange, 1995).

Transgenic Animal Models in Assessment of Genetic Instability and Cancer

The ability to introduce oncogenes stably into the germline of a mouse has provided opportunities to study the oncogenic role of the added gene in the context of an intact animal. Ectopic expression of the oncogene often results in the formation of hyperplasia, followed by focal emergence of malignant tumours (Hanahan 1989; Christofori and Hanahan, 1994). The usual delayed tumour appearance indicates that additional oncogenic steps, aside from expression of the oncogene, must occur to induce a tumour. The multistep genesis of tumours in these models is analogous to the stages observed in many human tumour types. Each stage of the multistage process is directly accessible to the investigator, since the tumours usually progress in a predictable fashion. As each step in tumour progression is likely to be accompanied by genetic changes, it may be possible to identify some of these alterations both at the karyotypic level and ultimately at the level of individual genes. This is particularly useful for studies of the role of genetic instability, where its relative contribution to each stage of tumour progression can be assessed.

The development of gene targeting techniques in mouse embryonic stem cells has added a further dimension in the development of animal tumour models. With the addition of this powerful new technology, genes can be inactivated in the germline of mice (Capecchi, 1989; Frohman and Martin, 1989). This approach has been particularly useful in assessing the oncogenic role of tumour suppressor genes and genes involved in DNA repair. In some instances, these mouse tumour suppressor and DNA repair deficient mice may mimic a known inherited human cancer predisposition. In addition, the crossing of mice heterozygous for a mutated allele may reveal the developmental importance of the targeted gene if normally functioning null animals are not obtained (Jones *et al*, 1995). If null animals survive at least until mid-gestation, then it is a simple matter to derive embryonic fibroblasts that can be used for important functional assays on the role of the targeted gene, including its ability to maintain genetic stability.

The ability to obtain primary null cells provides a critical advantage in molecular functional studies. Previously, studies of tumour suppressor gene function relied heavily on the introduction of the tumour suppressor gene into tumour cells lacking it. The mutational complexity of tumour cells may complicate the interpretation of results, particularly in the realm of genetic instability studies, where karyotypically intact tumour lines are scarce. In theory, primary cells derived from null mice should be different from cells of wild type littermates only in the single targeted gene. This single difference should also extend to in vivo tumorigenesis studies. However, since wild type mice rarely develop tumours before two years of age (and often the tumours are of a different type from those observed in the knockout mouse), it is difficult to compare the tumorigenesis process in the presence and absence of the targeted gene. In the analysis of the role of a particular targeted gene during tumorigenesis, the appropriate controls are essential, and we and others have addressed this issue by crossing knockout mice with transgenic mice susceptible to a single tumour type (Howes *et al*, 1994; Symonds *et al*, 1994; Donehower *et al*, 1995). The transgenic offspring of these crosses containing or lacking the targeted gene of interest can be compared for mechanistic aspects of tumorigenesis, including genetic instability. Usually, the tumours observed in these offspring will be of the single type initiated by the transgene, further simplifying mechanistic analyses.

ANIMAL TUMORIGENESIS MODELS FOR GENES THAT MAINTAIN GENETIC STABILITY

In this section, I describe some of the genes implicated in maintenance of genetic stability for which animal models have been generated. Since new models are continually being generated, the genes and models described here should be considered representative but not exhaustive. The focus will be primarily on mice generated by gene targeting that develop accelerated spontaneous or carcinogen induced tumours and that are likely to have increased genetic instability during tumour formation. For each gene, I briefly discuss its known role in maintaining genetic stability. Then, additional insights provided by experiments using the relevant animal model(s) are addressed.

Tumour Suppressors with Cell Cycle Regulatory Roles

TP53

The *TP53* tumour suppressor gene is mutated or lost in almost half of all human cancers, indicating that functional inactivation of this gene has a central role in the progression of many human tumours (Greenblatt *et al*, 1994). In addition to the role of *TP53* in somatic tumours, inherited mutations in the germline *TP53* gene have been shown to cause a familial cancer predisposition, Li-Fraumeni syndrome (Malkin *et al*, 1990; Srivastava *et al*, 1990).

Since the discovery of the TP53 protein in 1979, an extensive series of molecular studies has gradually revealed TP53 as a negative regulator of cell growth that appears to act at multiple levels (Ko and Prives, 1996). It functions as a transcriptional activator and binds to defined nucleotide sequences within a number of genes that have growth regulatory activity (Raycroft *et al*, 1990; El-Deiry *et al*, 1992, 1993; Farmer *et al*, 1992). Kastan and others have shown that TP53 levels are greatly increased following DNA damage and that this activation of TP53 results in late G_1 arrest (Kuerbitz *et al*, 1992; Lu and Lane, 1993). At approximately the same time, the Oren lab showed that in some contexts, upregulation of TP53 could induce apoptosis (Yonish-Rouach *et al*, 1991). These results led Lane to propose the "guardian of the genome" hypothesis, which argued that the central role of TP53 may be to respond to DNA damage either by arresting the cell in G_1 or by initiating a programmed cell death pathway (Lane, 1992). Presumably, the G_1 arrest would allow time for the damage to be repaired before entry into S phase. Alternatively, the cells with DNA damage could be killed through apoptosis. Either strategy would fulfil the role of protecting the cell from replicating damaged DNA templates. The implication of this model is that cells with inactivated TP53 would be more prone to various forms of genetic instability, ranging from point mutations to chromosomal abnormalities.

The testing of the "guardian of the genome" hypothesis has been facilitated by the development of both transgenic mice containing a mutant *TP53* transgene and *p53* knockout mice. Since some mutant forms of TP53 have a dominant negative effect on wild type TP53, it might be expected that overexpression of a dominant negative mutant form of TP53 would have oncogenic effects in the mouse. In fact, *TP53* transgenic mice overexpressing a codon 135 alanine to valine mutant TP53 do develop a variety of tumours with some frequency (30% by 18 months of age) (Lavigueur *et al*, 1989). At least four groups have generated *TP53* knockout mice by standard gene targeting methods (Donehower *et al*, 1992; Tsukada *et al*, 1993; Jacks *et al*, 1994; Purdie *et al*, 1994) Null *p53* mice are for the most part developmentally viable but have a very rapid onset of tumours, primarily of lymphoid origin. Unlike null *p53* mice, all of which succumb to tumours by 10 months of age, the *p53* heterozygous mice have a more delayed tumour onset (50% by 18 months) and acquire both lymphomas and sarcomas of various types (Jacks *et al*, 1994; Purdie *et al*, 1994; Donehower *et al*, 1995)

The availability of primary cells from *p53* null animals was exploited by several groups to show that cells missing *p53* were more prone to genetic instability. The Wahl and Tlsty labs, following up on initial studies using diploid fibroblast cultures from patients with Li-Fraumeni syndrome, showed that human fibroblasts and mouse embryo fibroblasts missing *p53* lacked the normal G_1 arrest checkpoint when challenged with the uridine biosynthesis inhibitor n-(phosphonacetyl)-l-aspartate (PALA) (Livingstone *et al*, 1992; Yin *et al*, 1992). In addition, the *p53* deficient cells were able to undergo PALA selected gene amplification much more readily than cells containing intact *p53*. Both Bischoff *et al* (1990) and Livingstone *et al* (1992) correlated high levels of aneuploidy with loss or absence of intact *TP53* in cultured fibroblasts. These initial studies on karyotypic instability in *p53* null cells were confirmed and extended by us to show a strong correlation between genomic instability and in vitro immortalization and transformation (Harvey *et al*, 1993; Yahanda *et al*, 1995). In other studies, it has been shown that cells with inactivated TP53 were significantly more prone to homologous recombination events (Meyn *et al*, 1994)

An aspect of the "guardian of the genome" hypothesis is that the damage induced, TP53 mediated G_1 arrest may allow sufficient time for DNA repair to take place before S phase entry. This indirect genetic protection related to cell cycle delay may be supplemented by a direct role in DNA repair. A number of lines of evidence support this, including the ability of TP53 to preferentially bind free DNA ends, single stranded DNA, short mismatched loops and radiation damaged DNA and to reanneal DNA strands (Oberosler *et al*, 1993; Bakalkin *et al*, 1994; Lee *et al*, 1995; Reed *et al*, 1995). Studies have suggested that TP53 may bind to DNA repair associated proteins such as ERCC3, RPA, XPB and XPD (Dutta *et al*, 1993; Wang XW *et al*, 1994, 1995) and co-localize with them in the cell to regions of DNA repair (Coates *et al*, 1995) and that loss of functional TP53 inhibits some aspects of DNA repair (Ford and Hanawalt, 1995; Smith *et al*, 1995).

It has been postulated that TP53 recognizes and binds to DNA lesions and then recruits other DNA repair proteins to the site to effect repair (Lee *et al*, 1995). Therefore, cells lacking functional TP53 might be expected to show higher rates of point mutations in response to mutagenic agents that activate TP53. At least one study has suggested an increased point mutation rate in the absence of functional TP53 following ultraviolet (UV) irradiation (Havre *et al*, 1995). We have used the *p53* deficient mice with a transgenic *lacI* marker to estimate point mutation rates (nucleotide substitutions, small deletions and insertions) in normal tissue, tumour tissue and primary cells treated with mutagens (Sands *et al*, 1995b). Following treatment with a mutagen known to induce p53, cells lacking *p53* did not exhibit a higher rate of mutation than cells containing *p53*, nor did tumours from *p53* deficient mice show higher mutation frequencies than tissue from normal mice. We could find no evidence for a mutator phenotype in the absence of *p53*, which included a failure to find increased microsatellite instability in tumours from *p53* deficient mice (Choi J and Donehower L, unpublished). Other investigators have also failed to find decreases in DNA repair efficiency in the absence of p53 in cells in culture and in in vitro assays (Ishizaki *et al*,1994; Sancar, 1995; Leveillard *et al*, 1996) Further studies are clearly needed to resolve whether *p53* loss actually does increase genetic instability at the nucleotide level.

Although p53 mediates G_1 arrest, the rapid loss of diploidy in cultured *p53* null cells is actually more consistent with abnormal segregation of chromosomes during mitosis. In fact, several groups have reported evidence of a G_2/M block mediated by p53 (Agarwal *et al*, 1995; Cross *et al*, 1995; Stewart *et al*, 1995). Cross *et al* (1995) showed that, in cultured fibroblasts from *p53* null embryos exposed to spindle inhibitors, the cells underwent multiple rounds of DNA synthesis without completing chromosome segregation. Normal pancreatic tissue from intact *p53* null mice also showed a much higher proportion of tetraploid cells than pancreatic cells from normal control mice. Further evidence that p53 is implicated in a mitotic spindle checkpoint was provided by Fukasawa *et al* (1996), who demonstrated that early passage embryo fibroblasts from *p53* null mice had increased proportions of cells with abnormally duplicated centrosomes that could be generated within a single cell cycle. The abnormally amplified centrosomes reduced mitotic fidelity and resulted in unequal segregation of chromosomes, providing a possible mechanism for the increased aneuploidy seen even in early passage *p53* null mouse embryo fibroblasts (Harvey *et al*, 1993).

An important question is the relationship of this genetic instability mediated by *TP53* loss to cancer formation. Is genetic instability a driving force in tumour initiation and progression or is it an epiphenomenon relatively unimportant in the neoplastic process? Several groups have now shown that the loss of *TP53* in a nascent tumour is often accompanied by more aggressive growth, primarily as a result of attenuated apoptosis, indicating one important mechanism by which *TP53* mutation could accelerate tumour progression (Howes *et al*, 1994; Pan and Griep, 1994; Symonds *et al*, 1994). A question we have attempted to answer, using the *p53* deficient mice, is whether tumour progression induced by *p53* loss is affected by increased chromosomal instability. Using two different models, *p53* deficient mice and bitransgenic *TP53* deficient *WNT-1* transgenic mice, we showed that tumours missing or losing *TP53* had significantly higher levels of chromosomal instability than tumours that retained wild type *TP53* (Donehower *et al*, 1995; Shi Y, Prakel D and Donehower L, unpublished). Our data further suggested that *TP53* loss and its accompanying genomic instability does not affect tumour initiation, but does affect tumour growth and progression. Those tumours lacking *p53* increased in volume more rapidly than tumours with wild type *p53*, and this increased volume was correlated with increased proliferation rates and higher levels of genomic instability (Jones J and Donehower L, unpublished). In these models, tumour cell apoptosis was very low and not dependent on *p53* status (Jacks T, personal communication; Donehower L, unpublished). Thus, our data suggest that genomic instability promoted by p53 loss does have a role in tumour progression, probably by accelerating the rate at which other oncogenic lesions arise.

RB

The retinoblastoma gene (*RB*) encodes a prototypical tumour suppressor and was first identified through its involvement in an inherited predisposition to childhood retinoblastoma (Knudson, 1971). In familial retinoblastoma cases, the remaining wild type *RB* allele is invariably lost or mutated, indicating that "hits" in both alleles of this gene are required for tumour progression to occur (Cavenee *et al*, 1985). The *RB* gene is mutated or lost in a number of human tumour types, providing further evidence of its importance in the neoplastic process (Hamel *et al*, 1993).

Whereas TP53 protein stands one or more steps removed from direct control of the cell cycle, RB protein is more immediately involved in controlling cell cycle progression as sort of a G_1 gatekeeper (Hatakeyama *et al*, 1994; Weinberg, 1995). In early and mid G_1, the proliferating cell integrates a variety of positive and negative growth signals. As the cell arrives at the restriction or R point of the cell cycle in late G_1, a decision is made whether to proceed into S phase or to remain arrested in late G_1. The RB protein may be an important mediator of the R point. RB undergoes a conversion from a hypophosphorylated to a hyperphosphorylated form several hours before the end of G_1 (Hatakeyama *et al*, 1994). The cell cycle restrictive form of RB is hypophosphorylated and effectively binds members of the E2F transcription factor family, which may function as important transcriptional activators of S phase genes (Bagchi *et al*, 1991; Chellappan *et al*, 1991). When hyperphosphorylated, the RB protein loses its ability to restrict progression of the cell through the R point and into S phase. The cell cycle dependent phosphorylation of RB is mediated by cyclin/CDK complexes, in particular those complexes containing the cyclin D and cyclin E proteins (Hatakeyama *et al*, 1994; Sherr *et al*, 1994; Weinberg, 1995).

The loss of this important RB-mediated gatekeeper function would be likely to have important growth promoting effects in a cell. This is supported by the frequent loss of both *RB* alleles in tumours and the preferential complexing (and presumed functional inactivation) by viral oncoproteins to the hypophosphorylated forms of RB in cells transformed by DNA tumour viruses (Ludlow *et al*, 1989; Imai *et al*, 1991). The generation of *RB* deficient mice has been instrumental in furthering our understanding of the role of *RB* in development, cell cycle control and tumorigenesis. Unlike *p53* deficient mice, *RB* null mice die during day 14–16 of embryogenesis from defects in erythrogenesis and neurogenesis (Clarke *et al*, 1992; Jacks *et al*, 1992; Lee *et al*, 1992). The mice heterozygous for a defective *RB* allele develop tumours, and these are invariably pituitary adenomas rather than the retinoblastomas seen in children (Jacks *et al*, 1992; Hu *et* al, 1994). In virtually every heterozygous tumour, the remaining *RB* allele is lost (Hu *et al*, 1994). The relatively late lethality of *RB* null embryos has allowed the derivation and characterization of null embryonic fibroblasts. *RB* null fibroblasts are smaller and have a shorter G_1 phase, perhaps due in part to derepression of cyclin E, a late G_1 cyclin that assists in cell cycle progression (Herrera *et al*, 1996). Two E2F target genes showed elevated mRNA and protein levels, consistent with the evidence that E2F responsive S phase genes are activated in the absence of *RB* despite the presence of RB-related proteins p107 and p130 (Almasan *et al*, 1995b).

By introducing high risk human papillomavirus oncoproteins E6 (which inactivates TP53) and E7 (which inactivates RB) into human fibroblasts and epithelial cells, at least three groups have assessed the role of RB (and TP53) in maintaining genetic stability (Reznikoff *et al*, 1994; White *et al*, 1994; Almasan *et al*, 1995a,b). Human cell types lacking functional TP53 (E6 expressing cells) after long term passaging show high levels of chromosome abnormalities, including multiple rearrangements, telomeric associations, aneuploidy and gains and losses of chromosomal DNA copy number, consistent with previous studies. By contrast, cells without functioning RB (E7 expressing cells) maintain relatively diploid karyotypes even after prolonged passaging (Reznikoff *et al*, 1994). In response to the metabolic inhibitor PALA, TP53 deficient human cells were relatively efficient at forming PALA resistant colonies through amplification of the carbamoyl phosphate synthetase 2/ aspartate transcarbamylase/dihydro-orotase (*CAD*) gene. However, RB deficient cells were inefficient at forming PALA resistant colonies, and those that did form did not directly amplify the *CAD* gene but contained multiple copies of the chromosome containing the *CAD* gene (White *et al*, 1994). Since RB deficient cells were able to form significantly more colonies than normal cells, RB loss clearly has some ability to induce genomic instability, although not to the degree to which absence of *TP53* promotes such instability. Cells lacking functional RB did show efficient amplification of the dihydrofolate reductase (*DHFR*) gene when incubated with methotrexate (Almasan *et al*, 1995a). The above results argue that RB does have a role in maintaining genetic stability, presumably by preventing premature entry into S phase under growth compromising conditions (Almasan *et al*, 1995a).

Cell Cycle Mediators and Inhibitors

Cyclin D1

Cyclin D is one of the more intensively studied G_1 cyclins, in part because it has been shown to be the target of translocations and amplification in a number of human tumour types including breast carcinomas (Lammie *et al*, 1991; Motokura *et al*, 1991; Jiang *et al*, 1992; Hunter and Pines, 1994). Its overexpression seems to be oncogenic in nature and has been associated with a poorer prognosis compared with tumours without overexpression (Jares *et al*, 1994). When the cyclin D1 gene is targeted to the mouse mammary gland as a transgene, it induces mammary hyperplasia and carcinomas (although the carcinomas usually arise after 1 year of age) (Wang TC *et al*, 1994). A number of in vitro studies have shown that overexpression of cyclin D1 results in earlier phosphorylation of RB protein and accelerated progression through G_1 (Jiang *et al*, 1993; Resnitzky *et al*, 1994; Quelle *et al*, 1993). In addition, others have shown that ectopic expression of cyclin D1 can enhance transformation parameters and increase gene amplification frequencies (Asano *et al*, 1995; Zhou *et al*, 1996). Cyclin D1 overexpressing cells can produce more PALA resistant colonies with amplified *CAD* genes than cells producing normal levels of cyclin D1, thus demonstrating that some cyclins may play an important part in disrupting genomic stability when expressed inappropriately.

p21CIP1/WAF1

In 1993, several groups identified a novel CDK inhibitor of 21 kDa (El-Deiry *et al*, 1993,; Gu *et al*, 1993; Harper *et al*, 1993; Xiong *et al*, 1993). This inhibitor, p21CIP1/WAF1, was found to be associated with inactive cyclin E/ CDK2 complexes that normally mediate G_1 to S phase transition, indicating that the primary role of p21CIP1/WAF1 is to effect G_1 arrest (El Deiry *et al*, 1994; Dulic *et al*, 1994). Overexpression of p21CIP1/WAF1 in proliferating cells does indeed result in G_1 arrest (Harper *et al*, 1995). Vogelstein and colleagues (El-Deiry *et al*, 1993) demonstrated that p21CIP1/WAF1 expression is transcriptionally activated by overexpression of TP53, providing a potential mechanism through which DNA damage induced TP53 can mediate G_1 arrest. Basal p21CIP1/WAF1 expression is independent of TP53, but increased expression following γ-irradiation is TP53 dependent (Macleod *et al*, 1995; Parker *et al*, 1995). p21CIP1/WAF1 induction has been observed in cell lines undergoing induction of differentiation or senescence (Noda *et al*, 1994; Halevy *et al*, 1995; Parker *et al*, 1995). It has also been shown to bind proliferating cell nuclear antigen and may directly inhibit DNA replication (Flores-Rozas *et al*, 1994; Waga *et al*, 1994). Thus, p21CIP1/WAF1 may have multiple roles in reducing cell cycle progression at critical junctures, such as following DNA damage or during terminal differentiation.

To explore further the biological role of p21CIP1/WAF1, two groups have generated p21CIP1/WAF1 null mice or embryonic stem cells (Brugarolas *et al*, 1995; Deng *et al*, 1995). p21CIP1/WAF1 null mice are developmentally viable, arguing that p21CIP1/WAF1, although associated with terminally differentiating cells, is not absolutely required for terminal differentiation. These mice, which lack a cell cycle inhibitor, do not develop tumours up to 7 months of age (Deng *et al*, 1995). Nevertheless, cells from p21CIP1/WAF1 null mice do have a defective G_1 arrest function in response to DNA damage and incubation with PALA (Deng *et al*, 1995). p21CIP1/WAF1 null cells also have normal apoptotic responses following irradiation and a normal mitotic spindle checkpoint, indicating that these TP53 associated functions are not mediated by p21CIP1/WAF1 (Brugarolas *et al*, 1995; Deng *et al*, 1995). Not surprisingly, since they have intact mitotic spindle checkpoints, long term cultures of p21CIP1/WAF1 null embryo fibroblasts do not exhibit the high levels of polyploidy, tetraploidy and aneuploidy seen in passaged *TP53* null cells. Given that loss of G_1 checkpoint control in the presence of PALA is associated with increased rates of gene amplification in *TP53* null cells, it will be interesting to see whether *p21CIP1/WAF1* null cells also can amplify genes more readily than normal cells.

p16INK4a

Another CDK inhibitor, p16INK4a, was first observed as a CDK4 associated protein in human cells (Xiong *et al*, 1993b). It was demonstrated to be a specific inhibitor of the CD4-6/cyclin D complexes (Serrano *et al*, 1993; Sherr and Roberts, 1995). Since one of the critical substrates of the CDK4-6/cyclin D kinases is the RB protein, p16INK4a probably has an important role in assisting RB induced G_1 arrest by blocking CDK4-6/cyclin D phosphorylation of RB. Overexpression of p16INK4a in cells in culture results in G_1 arrest in the presence of functional RB but not in its absence, arguing that the primary role of p16INK4a is indeed to mediate G_1 arrest through the RB pathway (Lukas *et al*, 1995; Medema *et al*, 1995)

The oncogenic importance of p16INK4a in human cancer was underscored by the finding that the *INK4a* gene maps to a chromosomal region frequently mutated in human tumours (Kamb *et al*, 1994; Nobori *et al*, 1994; Okamoto *et al*, 1994). Point mutations and small deletions of p16INK4a occur in spontaneously arising oesophageal carcinomas, biliary tract cancers and pancreatic adenocarcinomas, and in families with inherited predisposition to melanoma and pancreatic adenocarcinomas (Pollock *et al*, 1996). Thus, p16INK4a has many of the hallmarks of a classic tumour suppressor.

To study the role of p16INK4a in tumour suppression more directly, Serrano *et al* (1996) have generated a p16INK4a knockout mouse. The *p16INK4a* null mice develop relatively normally but develop tumours at an early age, and fibroblasts from these mice have accelerated growth kinetics and susceptibility to transformation by the Ha-*ras* oncogene. In addition, these mice show greatly accelerated incidence of tumorigenesis in the presence of carcinogens. The apparent ability of the embryo fibroblasts from *p16INK4a* null mice to immortalize more readily during long term passage than wild type counterparts suggests an increase in genetic instability in the *p16INK4a* deficient cells.

p27KIP1

The p27KIP1 protein inhibits a wide range of cyclin/CDK complexes in vitro, and its overexpression blocks progression through G_1 (Hunter and Pines, 1994). The growth inhibitor transforming growth factor beta, which inhibits RB phosphorylation, apparently activates p27 and stimulates its binding to the appropriate G_1 cyclin/CDK complexes (Polyak *et al*, 1994). *p27KIP1* deficient mice have been generated by two groups, and the null mice, although viable, have some interesting developmental abnormalities (Fero *et al*, 1996; Kiyokawa *et al*, 1996; Nakayama *et al*, 1996). They are larger than their wild type littermates, and all organs are larger than normal, although the spleen and thymus are the most enlarged. The null mice exhibit intermediate lobe pituitary hyperplasia, but malignant pituitary tumours or other types of tumours have not been observed. This is consistent with the absence of mutations in the *p27KIP1* gene in human tumours (reviewed in Harper and Elledge, 1996). The lack of tumours in the null animals suggests that p27KIP1 will not have a major role in preserving genetic stability.

DNA Repair Genes

Nucleotide Excision Repair (XPA and XPC)

The xeroderma pigmentosum group A gene (*XPA*) is a prototypical member of a group of DNA repair genes involved in nucleotide excision repair (NER). Humans with deficiencies in one of the *XP* genes typically exhibit defective nucleotide excision repair, resulting in cellular hypermutability and elevated rates of skin cancer following exposure to sunlight (Hanawalt, 1994). The *XPA* gene product recognizes and binds damaged nucleotides and then recruits the other members of the nucleotide excision repair apparatus (Lehmann, 1995).

Since the relationship between defective nucleotide excision repair, genetic instability and cancer is clear-cut in humans, it was expected that the mouse models for xeroderma pigmentosum should also be susceptible to cancer. In fact, *XPA* null mice generated by two groups elegantly validated the human syndrome, since they showed no obvious physical abnormalities but their cells were defective in nucleotide excision repair and more sensitive to mutagens including UV radiation (De Vries *et al*, 1996; Nakane *et al*, 1996). The *XPA* null mice were highly susceptible to UV induced and dimethylbenz[a]anthracene induced skin carcinogenesis compared with normal mice.

The *XPC* gene is also involved in nucleotide excision repair, but the exact molecular role of the XP protein in the repair process remains unclear (Lehmann, 1995). However, human patients defective in this complementation group exhibit the classic sensitivity to UV induced skin cancer. Mice deficient in *XPC*, as expected, are sensitive to UV induced skin carcinogenesis, and cells from these mice display higher sensitivity to UV irradiation than wild type cells (Sands *et al*, 1995a).

Mismatch Repair (MSH2 and PMS2)

The *MSH2* gene in humans was identified as a homologue of *Escherichia coli* and yeast mismatch repair genes which directly recognize and bind to small mismatched regions of DNA that may arise following DNA damage, replication or recombination (Kolodner, 1995). In conjunction with other proteins in the mismatch repair pathway, MSH2 serves to maintain genetic stability at the nucleotide level, preventing point mutations and small deletions and insertions. The clinical importance of the mismatch repair pathway in humans was dramatically demonstrated when individuals with hereditary non-polyposis colorectal cancer (HNPCC), who are predisposed to the early development of proximal colon tumours, were shown to have germline mutations either in *MSH2* or in one of the other known mismatch repair genes *MLH1, PMS1* or *PMS2* (Fishel *et al*, 1993; Leach *et al*, 1993; Nicolaides *et al*, 1994; Papadopoulos *et al*, 1994). In tumours from these patients, mismatch repair is lost, probably through loss of the remaining wild type allele. Many spontaneously arising tumour types also acquire mutations in mismatch repair genes, suggesting that loss of genetic stability in cells through absence of functional mismatch repair may be an important factor in tumorigenesis (Eshleman and Markowitz, 1995).

The recent development of *MSH2* deficient mice has confirmed the importance of *MSH2* function both in preserving genetic stability and in suppressing tumours. *MSH2* null mice are developmentally viable and develop tumours (primarily lymphomas) as early as 2 months of age (de Wind *et al*, 1995) As expected, cell extracts from $MSH2^{-/-}$ cells show loss of binding to mismatched DNA templates in vitro, and microsatellite DNA sequences in the *MSH2* null cells show increased variation in size, a hallmark of cells lacking functional mismatch repair. In addition, homologous recombination in *MSH2* null cells was up to 50-fold more efficient than in normal cells when constructs with only 0.6% sequence divergence were tested in homologous recombination assays. Thus, MSH2 null cells appear to have lost this heterology dependent suppression of recombination. de Wind *et al* (1995) hypothesize that decreased stringency in recombination in cells without *MSH2* may lead to an increased rate of aberrant recombination events, which in turn results in activation of oncogenes or inactivation of tumour suppressors and, ultimately, cancer. The likely increase in point mutation events as a loss of mismatch repair is expected to affect cancer associated genes, and this has been directly observed in colon carcinomas in HNPCC patients, who were found to have frameshifts in repeated nucleotides or G→A transitions in the *TP53* or *APC* genes in their tumours (Lazar *et al*, 1994). These particular types of point mutations are signature mutation types in cells lacking functional mismatch repair.

The *PMS2* gene is another mismatch repair gene and is the yeast and mammalian homologue of *E coli mutL*, which appears to affect mismatch recognition by mutS and enhance the endonuclease function of mutH, another *E coli* mismatch repair protein (Kolodner, 1995). PMS2 mutations are inherited in some HNPCC families, and the colon, endometrial, ovarian and stomach cancers observed in the family members show microsatellite DNA sequence instability (Nicolaides *et al*, 1994).

The *PMS2* null mice derived by Baker *et al* (1995) are viable and susceptible to sarcoma and lymphoma development. Tumours from the *PMS2* deficient mice consistently display microsatellite instability, arguing that instability associated genetic events played a significant part in the tumorigenesis process. The *PMS2* null males are infertile and produce only abnormal spermatozoa, which may be caused by the abnormal synapsis observed in the spermatogonia of these animals, but whether this synaptic failure syndrome has a role in predisposition to cancer of somatic cells is unclear.

Telomerase and Telomeres

One component of genetic stability is the maintenence of normal chromosome ends (telomeres). In genetically unstable cells, a frequently observed karyotypic abnormality is the formation of dicentric chromosomes resulting from telomere associations. Recently, the relationship between telomeres and cancer has received a good deal of attention. Part of the attention is based on two important observations derived from the long term passaging of human cells in culture. Firstly, the length of telomeres decreases proportionally to the number of cell divisions in culture. Secondly, telomerase (the enzyme capable of elongating telomeres) is rarely detected in normal human cells (except for germ cells) but is frequently detected in human tumour cells (Harley *et al*,

1994; de Lange, 1995). This has led to the formulation of a telomere hypothesis, which postulates that as a somatic cell divides in the absence of active telomerase, the gradual shortening of telomeres limits the number of mitoses that a cell can undergo before exposing uncapped chromosome ends (Harley *et al,* 1994). At this point, a DNA damage checkpoint would be activated and the cell arrested. In those cells in which both DNA damage checkpoint control is lost and telomerase activity is activated, the chromosome ends can be maintained and the cells can divide indefinitely. The evidence supporting this model is mostly correlative, and whether telomerase activation is a prerequisite for tumour formation has yet to be established.

Mouse tumour models may provide a more accessible way to explore the role of telomeres in tumour formation. However, their usefulness is complicated by the fact that *Mus musculus* telomeres are very long (20–50 kb), so that telomere shortening is difficult to assess (Kipling and Cooke, 1990). In addition, there is preliminary evidence that mouse telomerase is weakly repressed in normal somatic tissues (Chadeneau *et al,* 1995; Prowse and Greider, 1995). Several groups have explored the telomeres and telomerase in mouse tumour models. Chadeneau *et al* (1995) showed in a mouse mammary tumour model that whereas mammary tumours showed high levels of telomerase activity, detectable, albeit lower, levels of activity could also be observed in normal tissue. Blasco *et al* (1996) demonstrated in two mouse tumour models that telomerase activity could not be detected in early stage hyperplastic tissues but could be detected in most, but not all, late stage carcinomas and metastases. Although the results clearly demonstrate that telomerase is activated to higher levels in the late stages of tumour progression, the observation of late stage tumours with no detectable telomerase suggests that telomerase activation is not an absolute prerequisite for tumour evolution. A study by Broccoli *et al* (1996) on a mouse mammary tumour model failed to detect any evidence of telomere shortening in tumours but showed 10–20-fold increases in telomerase levels compared with normal and hyperplastic mammary tissues. This suggests that telomerase is activated in tumours but that such activation is unlikely to be linked to telomere shortening or loss. Given the differences between the mouse and human tumour systems cited above, it will be interesting to determine what factors activate telomerase and whether such activation is an important event in human tumour progression. Another question is whether telomerase activation and telomere shortening play a part in enhancement of genomic instability in tumours.

DISCUSSION

The animal models presented here provide further compelling evidence of the link between loss of genetic stability and cancer. The models (except the telomerase models) fall into two major categories: those that develop cancer because of defects in cell cycle control and those that acquire tumours because of DNA repair dysfunction. The fact that DNA repair deficient mice are more susceptible to spontaneous or mutagen induced cancers is not surprising. Clearly, increased mutation rates in the somatic cells will increase the likelihood that one cell will acquire the necessary combination of lesions to become oncogenic. Perhaps most satisfying was the UV induced skin cancer susceptibility in the XP models, replicating what is observed in XP patients quite well. By contrast, the mismatch repair mutant mice, although developing lymphoid and mesenchymal tumours, rarely if ever developed the hallmark HNPCC intestinal carcinomas. This inability to replicate inherited human cancer predisposition syndromes in tumour spectra seems to be a recurring theme for knockout tumour models (Harlow, 1992). It underscores the fact that genes may operate differently in given tissues of mice and men. Even so, such differences should not obscure the value of the knockout mice for molecular functional studies as well as tumour biology studies.

The mice defective for various cell cycle regulators showed varying susceptibilities to cancer. At one end of the spectrum are the *TP53* null mice, which show very rapid tumour development. This may be due to the fact that not only are these mice lacking G_1 and mitotic checkpoint controls, but they are also partially defective in apoptotic function. The CDK inhibitor knockout mice (*p21CIP1/WAF1* and *p27KIP1*) showed few or no aggressive tumours, whereas the *p16INK4a* deficient mice were quite susceptible to tumours. Such a result is actually consistent with the observations that *p21CIP1/WAF1* and *p27KIP1* mutations are rarely observed in human tumours, whereas *16INK4a* mutations are relatively common. Since all three CDK inhibitors regulate G_1 arrest, why such a stark contrast in tumour phenotype? Elledge *et al* (1996) have postulated that *p16INK4a* loss is tumorigenic because it specifically targets cyclin D dependent kinases but not cyclin E dependent kinases. Thus, RB may be inactivated, but in the absence of full cyclin E activation, an abnormal S phase results. This in turn is hypothesized to lead to increased genetic instability and the gradual accumulation of oncogenic lesions. Testing of genetic instability in *p16INK4a* deficient cells may corroborate this hypothesis.

SUMMARY

In this review I have attempted to a describe some of the recent mouse tumour models and their impact on our understanding of cancer aetiology. The focus has been on cell cycle regulatory genes and DNA repair genes which are likely to affect cancer development at least in part through genetic instability mechanisms. The cell cycle regulatory genes classified as tumour suppressors, TP53 and RB, maintain genomic stability and inhibit cancer through their roles in preserving cell cycle checkpoints. The cell cycle inhibitors have variable effects on cancer prevention, and their role in preserving genetic stability remains largely unexplored. The DNA repair gene models described here show the most direct connection between genetic instability and cancer, even in the absence of demonstrable cell cycle effects. It should be clear that the development of mice deficient in cell cycle control or DNA repair will provide useful tools for studying the interplay of these processes with genetic instability and cancer. Important new insights into the mechanisms of cancer initiation and progression are likely to come increasingly from such models in the coming years.

Acknowledgements

This is not intended to be an exhaustive review and I apologize to any individuals who feel they should have been cited but were not. I thank Wade Harper and Jeff Jones for editorial comments and I am grateful for the support of the National Cancer Institute, the US Army Breast Cancer Program and the Council for Tobacco Research.

References

Agarwal ML, Agarwal A, Taylor WR and Stark GR (1995) p53 controls both the G2/M and the G1 cell cycle checkpoints and mediates reversible growth arrest in human fibroblasts. *Proceedings of the National Academy of Sciences of the USA* **92** 8493–8497

Almasan A, Linke SP, Paulson TG, Huang L-C and Wahl G (1995a) Genetic instability as a consequence of inappropriate entry into and progression through S phase. *Cancer and Metastasis Reviews* **14** 59–73

Almasan A, Yin Y, Kelly RE *et al* (1995b) Deficiency of retinoblastoma protein leads to inappropriate S-phase entry, activation of E2F-responsive genes and apoptosis. *Proceedings of the National Academy of Sciences of the United States of America* **92** 5436–5440

Asano K, Sakamoto H, Sasaki H *et al* (1995) Tumorigenicity and gene amplification potentials of cyclin D1-overexpressing NIH3T3 cells. *Biochemical and Biophysical Research Communications* **217** 1169–1176

Bagchi S, Weinmann R and Raychaudhuri P (1991) The retinoblastoma protein copurifies with E2F-1, an E1A-regulated inhibitor of the transcription factor E2F. *Cell* **65** 1063–1072

Bakalkin G, Yakovleva T, Selivanova G *et al* (1994) p53 binds single-stranded DNA ends and catalyzes DNA renaturation and strand transfer. *Proceedings of the National Academy of Sciences of the USA* **91** 413–417

Baker SM, Bronner CE, Zhang L *et al* (1995) Male mice defective in the DNA mismatch repair gene PMS2 exhibit abnormal chromosome synapsis in meiosis. *Cell* **82** 309–319

Bates S and Vousden KH (1996) p53 in signaling checkpoint arrest or apoptosis. *Current Opinion in Genetics and Development* **6** 12–19

Bischoff FZ, Yim SO, Pathak S *et al* (1990) Spontaneous abnormalities in normal fibroblasts from patients with Li-Fraumeni cancer syndrome: aneuploidy and immortalization. *Cancer Research* **50** 7979–7984

Blasco MA, Rizen M, Greider CW and Hanahan D (1996) Differential regulation of telomerase activity and telomerase RNA during multi-stage tumorigenesis. *Nature Genetics* **12** 200–204

Broccoli D, Godley LA, Donehower LA, Varmus HE and de Lange T (1996) Telomerase activation in mouse mammary tumors: lack of detectable telomere shortening and evidence for regulation of telomerase RNA with cell proliferation. *Molecular and Cellular Biology* **16** 3765–3772

Brugarolas, J, Chandrasekaran, C, Gordon JI *et al* (1995) Radiation-induced cell cycle arrest compromised by p21 deficiency. *Nature* **377** 552–557

Capecchi MR (1989) Altering the genome by homologous recombination. *Science* **244** 1288–1292

Cavenee WK, Hansen MF, Nordenskjold M *et al* (1985) Genetic origin of mutations predisposing to retinoblastoma. *Science* **228** 501–503

Chadeneau C, Siegel P, Harley CB, Muller WJ and Bacchetti S (1995) Telomerase activity in normal and malignant murine tissues. *Oncogene* **11** 893–898

Chellappan SP, Hiebert S, Mudryj M, Horowitz JM and Nevins JR (1991) The E2F transcription factor is a cellular target for the RB protein. *Cell* **65** 1053–1062

Christofori G and Hanahan D (1994) Molecular dissection of multi-stage tumorigenesis in transgenic mice. *Seminars in Cancer Biology* **5** 3–12

Clarke AR, Maandag ER, van Roon M *et al* (1992) Requirement for a functional *Rb-1* gene in murine development. *Nature* **359** 328–330

Coates PJ, Save V, Ansari B and Hall PA (1995) Demonstration of DNA damage/repair in individual cells using in situ end labelling: association of p53 with sites of DNA damage. *Journal of Pathology* **176** 19–26

Cross SM, Sanchez CA, Morgan CA *et al* (1995) A p53-dependent mouse spindle checkpoint. *Science* **267** 1353–1356

de Lange T (1995) Telomere dynamics and genome instability in human cancer, In: Blackburn EH and Greider CW (eds). *Telomeres*, pp. 265–293, Cold Spring Harbor Laboratory Press, Cold Spring Harbor, New York

Deng C, Zhang P, Harper JW, Elledge SJ and Leder P (1995) Mice lacking p21CIP1/WAF1CIP1/WAF1 undergo normal development, but are defective in G1 checkpoint control. *Cell* **82** 675–684

De Vries A, Van Oostrom CTM, Hofhuis FMA *et al* (1995) Increased susceptibility to ultraviolet-B and carcinogens of mice lacking the DNA excision repair gene XPA. *Nature* **377** 169–173

de Wind N, Dekker, Berns A, Radman M and te Riele H (1995) Inactivation of the mouse *Msh2* gene results in mismatch repair deficiency, methylation toleration tolerance, hyper-recombination and predisposition to cancer. *Cell* **82** 321–330

Donehower LA, Harvey M, Slagle BL *et al* (1992) Mice deficient for p53 are developmentally normal but susceptible to spontaneous tumours. *Nature* **356** 215–221

Donehower LA, Godley LA, Aldaz CM *et al* (1995) Deficiency of p53 accelerates mammary tumorigenesis in *Wnt-1* transgenic mice and promotes chromosomal instability. *Genes and Development* **9** 882–895

Donehower LA, Harvey M, Vogel H *et al* (1995) Effects of genetic background on tumorigenesis in p53-deficient mice. *Molecular Carcinogenesis* **14** 16–22

dos Santos CC, Gavish H and Buchwald M (1994) Fanconi anemia revisited: old ideas and new advances. *Stem Cells* **12** 142–153

Dulic, V, Kaufmann WK, Wilson SJ *et al* (1994) p53-dependent inhibition of cyclin-dependent kinase activities in human fibroblasts during radiation-induced G1 arrest. *Cell* **76** 1013–1023

Dutta A, Ruppert JM, Aster JC and Winchester E (1993) Inhibition of DNA replication factor RPA by p53. *Nature* **365** 79–82

El-Deiry WS, Kern SE, Pietenpol JA, Kinzler KW and Vogelstein B (1992) Definition of a consensus binding site for p53. *Nature Genetics* **1** 45–49

El-Deiry WS, Tokino T, Velculescu VE *et al* (1993) WAF1, a potential mediator of p53 tumor suppression. *Cell* **76** 817–825

El-Deiry WS, Harper JW, O'Connor PM *et al* (1994) WAF1/CIP1 is induced in p53 mediated G1 arrest and apoptosis. *Cancer Research* **4** 1169–1174

Elledge SJ, Winston J and Harper JW (1996) A question of balance: the role of cyclin-kinase inhibitors in development and tumorigenesis.*Trends in Cell Biology* **6** 388–392

Eshleman JR and Markowitz SD (1995) Microsatellite instability in inherited and sporadic neoplasms. *Current Opinion in Oncology* **7** 83–89

Farmer G, Bargonetti J, Zhu H, Friedman P, Prywes R and Prives C (1992) Wild-type p53 activates transcription in vitro. *Nature* **358** 83–86

Fearon ER and Vogelstein B (1990) A genetic model for colorectal tumorigenesis. *Cell* **61** 759–767

Fishel R, Lescoe MK, Rao MRS *et al* (1993) The human mutator gene homolog *MSH2* and its association with hereditary nonpolyposis colon cancer. *Cell* **75** 1027–1038

Flores-Rozas H, Kelman Z, Dean F *et al* (1994) Cdk-interacting protein 1 directly binds with PCNA and inhibits replication catalyzed by the DNA polymerase d holoenzyme. *Proceedings of the National Academy of Sciences of the USA* **91** 8655–8659

Ford JM and Hanawalt PC (1995) Li-Fraumeni syndrome fibroblasts homozygous for p53 mutations are deficient in globals DNA repair but exhibit normal transcription-coupled repair and enhanced UV resistance. *Proceedings of the National Academy of Sciences of the USA* **92** 8876–8880

Frohman MA and Martin GR (1989) Cut, paste and save: new approaches to altering specific genes in mice. *Cell* **56** 145–147

Fukasawa K, Choi T, Kuriyama R, Rulong S and Van de Woude GF (1996) Abnormal centrosome amplification in the absence of p53. *Science* **271** 1744–1747

Greenblatt MS, Bennett W, Hollstein M and Harris CC (1994) Mutations in the p53 tumor suppressor gene: clues to cancer etiology and molecular pathogenesis. *Cancer Research* **54** 4855–4878

Gu Y, Turck CW and Morgan DO (1993) Inhibition of CDK2 activity in vivo by an associated 20K regulatory subunit. *Nature* **366** 707–710

Halevy O, Novitch BG, Spicer DB *et al* (1995) Correlation of terminal cell cycle arrest of skeletal muscle with induction of p21CIP1/WAF1 by MyoD. *Science* **267** 1018–1021

Hamel PA, Phillips RA, Muncaster M and Gallie BL (1993) Speculations on the roles of RB1 in tissue-specific differentiation, tumor initiation and tumor progression. *FASEB Journal* **7** 846–854

Hanahan D (1989) Transgenic mice as probes into complex systems. *Science* **246** 1265–1275

Hanawalt P (1994) Transcription-coupled repair and human disease. *Science* **266** 1957–1960

Harley CB, Kim NW, Prowse KR *et al* (1994) Telomerase, cell immortality and cancer. *Cold Spring Harbor Symposia on Quantitative Biology* **59** 307–315

Harlow E (1992) For our eyes only. *Nature* **359** 270–271

Harper JW and Elledge SJ (1996) Cdk inhibitors in development and cancer *Current Opinions in Genetics and Development* **6** 56–64

Harper JW, Adami GR, Wei N, Keyomarsi K and Elledge SJ (1993) The p21CIP1/WAF1 Cdk-interacting protein Cip1 is a potent inhibitor of G1 cyclin-dependent kinases. *Cell* **75** 805–816

Harper JW, Elledge SJ, Keyomarsi K *et al* (1995) Inhibition of cyclin-dependent kinases by p21CIP1/WAF1. *Molecular Biology of the Cell* **6** 387–400

Hartwell L (1992) Defects in a cell cycle checkpoint may be responsible for the genomic instability of cancer cells. *Cell* **71** 543–546

Hartwell L and Kastan M (1994) Cell cycle control and cancer. *Science* **266** 1821–1828

Hartwell L, Weinert T, Kadyk L and Garvik B (1994) Cell cycle checkpoints, genomic integrity and cancer. *Cold Spring Harbor Symposia on Quantitative Biology* **59** 259–263

Harvey M, Sands AT, Weiss RS *et al* (1993) In vitro growth characteristics of embryo fibroblasts isolated from p53-deficient mice. *Oncogene* **8** 2457–2467

Hatekeyama M, Herrera RA, Makela T, Dowdy SF, Jacks T and Weinberg RA (1994) The cancer cell and the cell cycle clock. *Cold Spring Harbor Symposia on Quantitative Biology* **59** 1–10

Havre PA, Yuan J, Hedrick L, Chu KR and Glazer PM (1993) p53 inactivation by HPV16 E6 results in increased mutagenesis in human cells. *Cancer Research* **55** 4420–4424

Herrera RE, Sah VP, Williams BO, Makela TP, Weinberg RA and Jacks T (1996) Altered cell cycle kinetics, gene expression and G1 restriction point regulation in Rb-deficient fibroblasts. *Molecular and Cellular Biology* **16** 2402–2407

Hirama T and Koeffler HP (1995) Role of the cyclin-dependent kinase inhibitors in the development of cancer. *Blood* **86** 841–854

Howes KA, Ransom LN, Papermaster DS, Lasudry JGH, Albert DM, and Windle JJ (1994) Apoptosis or retinoblastoma—alternative fates of photoreceptors expressing the HPV-16 E7 gene in the presence or absence of p53. *Genes and Development* **8** 1300–1310

Hu N, Gutsmann A, Herbert DC, Bradley A, Lee WH and Lee EYHP (1994) Heterozygous Rb-1 delta/+ mice are predisposed to tumors of the pituitary gland with a nearly complete penetrance. *Oncogene* **9** 1021–1027

Hunter T and Pines J (1994) Cyclins and cancer II: cyclin D and CDK inhibitors come of age. *Cell* **79** 573–582

Imai Y, Matsushima Y, Sugimura T and Terada M (1991) Purification and characterization of human papillomavirus type 16 E7 protein with preferential binding capacity to the underphosphorylated form of retinoblastoma gene product. *Journal of Virology* **65** 4966

Ishizaki K, Ejima T, Matsunaga R et al (1994) Increased UV-induced SCEs but normal repair of DNA damage in p53-deficient mouse cells. *International Journal of Cancer* **57** 254–257

Jacks TA, Fazeli E, Schmitt E, Bronson RT, Goodell M and Weinberg RA (1992) Effects of an RB mutation in the mouse. *Nature* **359** 295–300

Jacks T, Remington L, Williams BO *et al* (1994) Tumor spectrum analysis in p53-mutant mice. *Current Biology* **4** 1–7

Jares P, Fernandez PL, Campo E *et al* (1994) The PRAD-1/cyclin D1 gene amplification correlates with messenger RNA overexpression and tumor progression in human laryngeal carcinomas. *Cancer Research* **54** 4813–4817

Jiang W, Kahn SM, Tomita N, Zhang YJ, Lu SH and Weinstein IB (1992) Amplification and expression of the human cyclin D gene in esophageal cancer. *Cancer Research* **52** 2980–2983

Jiang W, Kahn SM, Zhou P, Zhang Y-J *et al* (1993) Overexpression of cyclin D1 in rat fibroblasts causes abnormalities in growth control, cell cycle progression and gene expression. *Oncogene* **8** 3447–3457

Jones SN, Donehower LA and Bradley A (1995) Analysis of tumor suppressor genes using transgenic mice. *Methods: A Companion to Methods in Enzymology* **8** 247–258

Kamb A, Gruis NA, Weaver-Feldhaus J *et al* (1994) A cell cycle regulator potentially involved in genesis of many tumor types. *Science* **264** 436–440

Kastan MB, Canman CE and Leonard CJ (1995) p53, cell cycle control and apoptosis: implications for cancer. *Cancer and Metastasis Reviews* **14** 3–15

Kim NW, Piatyszek, Prowse KR *et al* (1994) Specific association of human telomerase activity with immortal cells and cancer. *Science* **266** 2011–2015

Kipling D and Cooke HJ (1990) Hypervariable ultra-long telomeres in mice. *Nature* **347** 347–402

Kirsch IR, Abdallah JM, Bertness VL *et al* (1994) Lymphocyte-specific genetic instability and cancer. *Cold Spring Harbor Symposia on Quantitative Biology* **59** 287–295

Kiyokawa H, Kineman RD, Mahova-Todorova KO *et al* (1996) Enhanced growth of mice lacking the cyclin-dependent kinase inhibitor function of p27Kip1 . *Cell* **85** 721–732

Knudson AG (1971) Mutation and cancer: statistical study of retinoblastoma. *Proceedings of the National Academy of Sciences of the USA* **68** 820–823

Knudson AG (1993) Antioncogenes and human cancer. *Proceedings of the National Academy of Sciences of the USA* **90** 10914–10921

Ko LJ and Prives C (1996) p53: puzzle and paradigm. *Genes and Development* **10** 1054–1072

Kolodner RD (1995) Mismatch repair: mechanisms and relationship to cancer susceptibility. *Trends in Biochemical Sciences* **20** 397–401

Kuerbitz SJ, Plunkett BS, Walsh WV and Kastan MB (1992) Wild-type p53 is a cell cycle checkpoint determinant following irradiation. *Proceedings of the National Academy of Sciences of the USA* **89** 7491–7495

Lammie GA, Fantl V, Smith R *et al* (1991) D11S128, a putative oncogene on chromosome

11q13 is amplified and expressed in squamous cell and mammary carcinomas and linked to BCL-1. *Oncogene* **6** 439–444

Lane DP (1992) p53, guardian of the genome. *Nature* **358** 15–16

Lavigueur A, Maltby V, Mock D, Rossant J, Pawson T and Bernstein A (1989) High incidence of lung, bone and lymphoid tumors in transgenic mice overexpressing mutant alleles of the p53 oncogene. *Molecular and Cellular Biology* **9** 3882–3991

Lazar V, Grandjouan S, Bognel C *et al* (1994) Accumulation of multiple mutations in tumour suppressor genes during colorectal tumorigenesis in HNPCC patients. *Human Molecular Genetics* **3** 2257–2260

Leach FS, Nicolaides NC, Papadopoulos N *et al* (1993) Mutations of a mutS homolog in hereditary nonpolyposis colorectal cancer. *Cell* **75** 1215–1225

Lee EYHP, Chang CY, Hu NP *et al* (1992) Mice deficient for RB are nonviable and show defects in neurogenesis and haematopoiesis. *Nature* **359** 288–294

Lee EYHP, Hu NP, Yuan SSF *et al* (1994) Dual roles of the retinoblastoma in cell cycle regulation and neuron differentiation. *Genes and Development* **8** 2008–2021

Lee S, Elenbaas B, Levine A and Griffith J (1995) p53 and its 14 kD C-terminal domain recognize primary DNA damage in the form of insertion/deletion mismatches. *Cell* **81** 1013–1020

Lehmann AR (1995) Nucleotide excision repair and the link with transcription. *Trends in Biochemical Sciences* **20** 402–405

Leveillard T, Andera L, Bissonnette N *et al* (1996) Functional interactions between p53 and the TFIIH complex are affected by tumour-associated mutations. *EMBO Journal* **15** 1615–1624

Livingstone LR, White A, Sprouse J, Livanos E, Jacks T and Tlsty TD (1992) Altered cell cycle arrest and gene amplification potential accompany loss of wild type p53. *Cell* **70** 605–620

Loeb LA (1991) Mutator phenotype may be required for multistage carcinogenesis *Cancer Research* **51** 3075–3079

Lu X and Lane DP (1993) Differential induction of transcriptionally active p53 following UV or ionizing radiation: defects in chromosome instability syndromes? *Cell* **75** 765–778

Ludlow JW, DeCaprio JA, Huang CM, Lee WH, Paucha E and Livingston D (1989) SV40 large T antigen binds preferentially to an underphosphorylated member of the retinoblastoma susceptibility gene product family. *Cell* **56** 57–65

Lukas J, Parry D, Aagaard L *et al* (1995) Retinoblastoma-protein-dependent cell cycle inhibition by the tumour suppressor p16. *Nature* **375** 503–506

Macleod KF, Sherry N, Hannon G *et al* (1995) p53-dependent and independent expression of p21CIP1/WAF1 during cell growth, differentiation and DNA damage. *Genes and Development* **9** 935–944

Malkin D, Li FP, Strong LC *et al* (1990) Germ line p53 mutations in a familial syndrome of breast cancer, sarcomas and other neoplasms. *Science* **250** 1233–1238

Medema RH, Herrera RE, Lam F and Weinberg RA (1995) Growth suppression by p16ink4 requires functional retinoblastoma protein. *Proceedings of the National Academy of Sciences of the USA* **92** 6289–6293

Meyn MS, Strasfeld L and Allen C (1994) Testing the role of p53 in the expression of genetic instability and apoptosis in ataxia telangiectasia. *International Journal of Radiation Biology* **66** **(Supplement 6)** S141–S149

Morgenbesser SD, Williams BO, Jacks T and DePinho RA (1994) p53-dependent apoptosis produced by Rb-deficiency in the developing mouse lens. *Nature* **371** 72–74

Motokura T, Bloom T, Kim HG *et al* (1991) A novel cyclin encoded by a bcl-linked candidate oncogene. *Nature* **350** 512–515

Nakane H, Takeuchi S, Yuba S *et al* (1995) High incidence of ultraviolet-B- or chemical-carcinogen-induced skin tumours in mice lacking the xeroderma pigmentosum group A gene. *Nature* **377** 165 168

Nakayama K, Ishida N, Shirane M *et al* (1996) Mice lacking p27Kip1 display increased body size, multiple organ hyperplasia, retinal dysplasia, and pituitary tumors. *Cell* **85** 707–720

Nicolaides NC, Papadopoulos N, Liu B *et al* (1994) Mutations of two PMS homologues in hereditary nonpolyposis colon cancer. *Nature* **371** 75–80

Nobori T, Miura K, Wu DJ, Lois A, Takabayashi K and Carson DA (1994) Deletions of the cyclin-dependent kinase-4 inhibitor gene in multiple human cancers. *Nature* **368** 753–756

Noda A, Ning Y, Venable SF, Pereira-Smith OM and Smith JR (1994) Cloning of cell-derived inhibitory of DNA synthesis using an expression screen. *Experimental Cell Research* **211** 90–98

Nowell PC (1976) The clonal evolution of tumor cell populations. *Science* **194** 23–28

Oberosler P, Hloch P, Ramsperger U and Stahl H (1993) p53-catalyzed annealing of complementary single-stranded nucleic acids. *EMBO Journal* **12** 2389–2396

Okamoto A, Demetrick DJ, Spillare EA *et al* (1994) Mutations and altered expression of p16INK4 in human cancer. *Proceedings of the National Academy of Sciences of the USA* **91** 11045–11049

Pan HC and Griep AE (1994) Altered cell cycle regulation in the lens of HPV-16 E6 or E7 transgenic mice—implications for tumor suppressor gene function in development. *Genes and Development* **8** 1285–1299

Papadopoulos N, Nicolaides NC, Wei Y-F *et al* (1994) Mutation of a *mutL* homolog in hereditary colon cancer. *Science* **263** 1625–1629

Parker SB, Eichele G, Zhang P *et al* (1995) p53-independent expression of p21CIP1/WAF1Cip1 in muscle and other terminally defferentiating cells. *Science* **267** 1024–1027

Pollock PM, Pearson JV and Hayward NK (1996) Compilation of somatic mutations of somatic mutations of the CDKN2 gene in human cancers: non-random distribution of base substitutions. *Genes, Chromosomes and Cancer* **15** 77–88

Polyak K, Kato JY, Solomon MJ *et al* (1994) p27Kip1, a cyclin-Cdk inhibitor, links transforming growth factor-β and contact inhibition to cell cycle arrest. *Genes and Development* **8** 9–22

Prowse KR and Greider CW (1995) Developmental and tissue-specific regulation of mouse telomerase and telomere length. *Proceedings of the National Academy of Sciences of the USA* **92** 4818–4822

Purdie CA, Harrison DJ, Peter A *et al* (1994) Tumour incidence, spectrum and ploidy in mice with a large deletion in the p53 gene. *Oncogene* **9** 603–609

Quelle DE, Ashmun RA, Shurtleff SA *et al* (1993) Overexpression of mouse D-type cyclins accelerates G1 phase in rodent fibroblasts. *Genes and Development* **7** 1559–1571

Raycroft L, Wu H and Lozano G (1990) Transcriptional activation by wild-type but not transforming mutants of the p53 anti-oncogene. *Science* **249** 1049–1051

Reed MB, Woelker P, Wang Y, Wang ME anderson ME and Tegtmeyer P (1995) The C-terminal domain of p53 recognizes DNA damaged by ionizing radiation. *Proceedings of the National Academy of Sciences of the USA* **92** 9455–9459

Resnitzky D, Gossen M, Bujard H and Reed SI (1994) Acceleration of the G1/S phase transition by expression of cyclins D1 and E with an inducible system. *Molecular and Cellular Biology* **14** 1669–1679

Reznikoff CA, Belair C, Savelieva E *et al* (1994) Long-term genome stability and minimal genotypic and phenotypic alterations in HPV16 E7-, but not E6- immortalized human uroepiethelial cells. *Genes and Development* **8** 2227–2240

Sancar A (1995) Excision repair in mammalian cells. *Journal of Biological Chemistry* **270** 15915–15918

Sands AT, Abuin A, Sanchez A, Conti C and Bradley A (1995a) High susceptibility to ultraviolet-induced carcinogenesis in mice lacking XPC. *Nature* **377** 162–165

Sands AT, Suraokar MB, Sanchez A, Marth JE, Donehower LA and Bradley A (1995b) p53 deficiency does not affect the accumulation of point mutations in a transgene target. *Proceedings of the National Academy of Sciences of the USA* **92** 8517–8521

Serrano M, Hannon GJ and Beach D (1993) A new regulatory motif in cell cycle control causing specific inhibition of cyclin D/CDK4. *Nature* **366** 704–707

Serrano M, Lee H-W, Chin L, Cordon-Cardo C, Beach D and DePinho RA (1996) Role of the INK4a Locus in tumor suppression and cell mortality. *Cell* **85** 27–37

Shapiro GI and Rollins BJ (1996) p16(INK4A) as a human tumor suppressor. *Biochimica et Biophysica Acta Reviews on Cancer* **1242** 165–170

Sherr CJ, Kato J, Quelle DE, Matsuoka M and Roussel MF (1994) D-type cyclins and their cyclin-dependent kinases: G1 phase integrators of the mitogenic response. *Cold Spring Harbor Symposia on Quantitative Biology* **59** 11–17

Sherr CD and Roberts JM (1995) Inhibitors of mammalian G1 cyclin-dependent kinases. *Genes and Development* **9** 1149–1163

Smith ML and Fornace AJ (1995) Genomic instability and the role of p53 mutations in cancer cells. *Current Opinion in Oncology* **7** 69–75

Smith ML, Chen IT, Zhan Q, O'Connor PM and Fornace Jr AJ (1995) Involvement of the p53 tumor suppressor in repair of u.v. type damage. *Oncogene* **10** 1053–1059

Srivastava S, Zou Z, Pirollo K, Blattner W and Chang EH (1990) Germ-line transmission of a mutated p53 gene in a cancer-prone family with Li-Fraumeni syndrome. *Nature* **348** 747–749

Stark GR (1993) Regulation and mechanisms of mammalian gene amplification. *Advances in Cancer Research* **61** 87–113

Stewart N, Hicks GG, Paraskevas F and Mowat M (1995) Evidence for a second cell cycle block at G2/M by p53. *Oncogene* **10** 109–115

Symonds H, Krall, L, Remington L *et al* (1994) p53-dependent apoptosis suppresses tumor growth and progression in vivo. *Cell* **78** 703–711

Tlsty TD, Briot A, Gualberto A *et al* (1995) Genomic instability and cancer. *Mutation Research* **337** 1–7

Tsukada T, Tomooka Y, Takai S *et al* (1993) Enhanced proliferative potential in culture of cells from p53-deficient mice. *Oncogene* **8** 3313–3322

Waga S, Hannon GJ, Beach D and Stillman B (1994) The p21CIP1/WAF1 inhibitor of cyclin-dependent kinases controls DNA replication by interaction with PCNA. *Nature* **369** 574–578

Wang TC, Cardiff RD, Zukerberg L, Lees E, Arnold A and Schmidt EV (1994) Mammary hyperplasia and carcinoma in MMTV-cyclin D1 transgenic mice. *Nature* **369** 669–671

Wang XW, Forrester K, Yeh H, Feitelson MA, Gu JR and Harris CC (1994) Hepatitis B virus X protein inhibits p53 sequence-specific DNA binding, transcriptional activity and association with transcription factor ERCC3. *Proceedings of the National Academy of Sciences of the USA* **91** 2230–2234

Wang XW, Yeh H, Schaeffer L *et al* (1995) p53 modulation of TFIIH-associated nucleotide excision repair activity. *Nature Genetics* **10** 188–195

Weinberg RA (1991) Tumor suppressor genes. *Science* **254** 1138

Weinberg RA (1995) The retinoblastoma protein and cell cycle control. *Cell* **81** 323–330

White AE, Livanos EM and Tlsty TD (1994) Differential disruption of genomic integrity and cell cycle regulation in normal human fibroblasts by the HPV oncoproteins. *Genes and Development* **8** 666–677

Xiong Y, Hannon GJ, Zhang H, Casso D, Kobayashi R and Beach D (1993a) p21CIP1/WAF1 is a universal inhibitor of cyclin kinases. *Nature* **366** 701–704

Xiong Y, Zhang H and Beach D (1993b) Subunit rearrangement of the cyclin-dependent kinases is associated with cellular transformation. *Genes and Development* **7** 1572–1583

Yahanda A, Bruner J, Donehower LA and Morrison RS (1995) Astrocytes derived from p53-deficient mice provide a multistep in vitro model for development of malignant gliomas. *Molecular and Cellular Biology* **15** 4249–4259

Yin Y, Tainsky MA, Bischoff FZ, Strong LC and Wahl GM (1992) Wild type p53 restores cell cycle control and inhibits gene amplification in cells with mutant p53 alleles. *Cell* **70** 937–948

Yonish-Rouach E, Resnitzky D, Lotem J, Sachs L, Kimchi A and Oren M (1991) Wild-type p53 induces apoptosis of myeloid leukaemic cells that is inhibited by interleukin-6. *Nature* **352** 345–347

Zhou P, Jiang W, Weghorst CM and Weinstein IB (1996) Overexpression of cyclin D1 enhances gene amplification. *Cancer Research* **56** 36–39.

The author is responsible for the accuracy of the references.

Biographical Notes

J Julian Blow was a graduate student at Cambridge University and a postdoctoral fellow at Oxford University. In 1991 he joined ICRF, where he has continued his studies into the control of eukaryotic DNA replication using cell free extracts of *Xenopus* eggs.

Larry A Donehower obtained a PhD from George Washington University and received postdoctoral training at the University of California, San Francisco. Since 1985 he has been a faculty member at Baylor College of Medicine, where his research interests focus on the role of the *TP53* tumour suppressor gene in tumour progression.

Joyce L Hamlin is a professor in the department of biochemistry at the University of Virginia School of Medicine. She received her PhD in molecular biology from the University of California at Los Angeles and did postdoctoral work at Princeton University and Harvard Medical School. Her research interests are the control of mammalian chromosomal replication and genomic instability.

Calvin Harley is chief scientific officer of Geron Corporation. He received his PhD from McMaster University in 1980 and had postdoctoral training with John Maynard Smith (University of Sussex) and Herbert Boyer (University of California, San Francisco). From 1982 to 1993 he conducted research as a faculty member at McMaster on cell senescence and the role of telomeres and telomerase in cell ageing and immortalization. In 1993 he joined Geron as director of cell biology and led a group of scientists investigating the research and drug discovery opportunities in this field.

J Wade Harper obtained his PhD from the Georgia Institute of Technology and received postdoctoral training at Harvard Medical School. Since 1988 he has been a faculty member in the department of biochemistry at Baylor College of Medicine, Houston, where his research has focused on cell cycle control and its regulation by cyclin dependent kinases and their inhibitors.

Philip Hieter, PhD, is professor of molecular biology and genetics at the Johns Hopkins School of Medicine, Baltimore. He received his PhD degree in biochemistry from Johns Hopkins University in 1981, where he worked with Phil Leder on human immunoglobulin gene rearrangement. His postdoctoral training was in yeast genetics with Ron Davis at Stanford University. His laboratory works on aspects of the cell cycle and determinants of chromosome transmission and stability.

Li-chun Huang obtained her PhD from the University of Wisconsin-Madison in 1992. Her research at the Salk Institute has been devoted to analysing the effects of DNA damage on cell cycle progression and DNA repair in individual mammalian cells.

Satoshi Inoue is a visiting scientist at the Dana-Farber Cancer Institute from the National Institute of Health of Japan, Tokyo, where he is a senior research scientist. His research focuses on the understanding of radiation induced cell cycle checkpoints.

Mark Jackman was a graduate student at Cambridge University and is now a postdoctoral worker at the Wellcome/CRC Institute, Cambridge, working on the regulation of the human cell cycle.

Shengfang Jin is a research fellow in the department of microbiology and molecular genetics, Harvard Medical School, and the division of tumor immunology, Dana-Farber Cancer Institute, Boston. His research interests are the molecular roles of DNAPK components in DNA repair and nuclear signalling.

Michael Kastan graduated in chemistry from the University of North Carolina and received his MD and PhD in cellular biology from Washington University in St Louis in 1984. He is now an associate professor of oncology and paediatrics at Johns Hopkins University School of Medicine, Baltimore. His research focuses on the role and mechanisms of action of TP53 and other gene products in controlling cell cycle progression and cell survival following DNA damage.

James M Larner is an associate professor in the department of radiation oncology at the University of Virginia School of Medicine. He received his MD from the University of Virginia in 1980 and did his residency training in radiation oncology at Thomas Jefferson University. His research interests are the molecular aspects of DNA damage sensing pathways.

Hoyun Lee is a research assistant professor in the department of radiation oncology at the University of Virginia. He received his PhD in microbiology from the University of Guelph, Canada, in 1992. He did a postdoctoral fellowship in the department of biochemistry, University of Virginia. His research interest is DNA replication in the context of cell cycle controls.

Steven P Linke, PhD, received BS degrees in microbiology and biology with a minor in chemistry from North Dakota State University in 1991. He obtained his PhD from the University of California, San Diego, in 1996 and is currently a postdoctoral fellow at the Salk Institute for Biological Studies in La Jolla. His research has focused on the role of tumour suppressors in cell cycle control and the maintenance of genetic stability.

Patrick M O'Connor received his PhD from Manchester University and completed his postdoctoral training under the guidance of Dr Kurt Kohn in the laboratory of molecular pharmacology, National Cancer Institute, Bethesda, where today he is a senior investigator. His research focuses on cell cycle checkpoint control and chemosensitivity in mammalian cells.

Andrew M Page, BSci, is a second year PhD candidate in the biochemistry, cellular and molecular biology program at the Johns Hopkins School of Medicine, Baltimore. Before entering graduate school he worked as a research assistant in the laboratory of Dr Richard Kolodner at the Dana-Farber Cancer Institute in Boston, Massachusetts, where he conducted mutational studies of exoribonucleases in yeast. He presently studies cyclin ubiquitination and the cell cycle in the laboratory of Phil Hieter.

Huichin Pan obtained her PhD from the University of Wisconsin, then joined Dr Terry Van Dyke's laboratory as a postdoctoral fellow. Her research interest is the study of cell growth regulation in mouse models.

Thomas G Paulson received BSEd degrees in education and biology from Northwestern University in 1988 and is now in the biology department at the University of California, San Diego. His studies have analysed how genome structure and cell growth conditions impact on gene amplification competence, interrelationships between DNA repair and cell cycle control and correlations between microsatellite instability and clinical performance in breast cancer.

Jonathon Pines was an undergraduate and graduate student at Cambridge University and a postdoctoral worker at the Salk Institute, California. He is now at the Wellcome/CRC Institute, Cambridge, working on the regulation of the human cell cycle.

Steven I Reed is professor of molecular biology at the Scripps Research Institute in La Jolla, California. He received a BSc degree in molecular biophysics and biochemistry from Yale University then trained in biochemistry at Stanford University under George Stark, receiving his PhD in 1976. In 1979 he joined the biological sciences faculty of the University of California, Santa Barbara, where he established an independent research programme focused on the molecular genetics of the yeast cell cycle. In 1986 he moved to the Scripps Research Institute, where his current research interests include various aspects of regulation of yeast and mammalian cell cycles.

Galit Rotman graduated from the Hebrew University of Jerusalem and completed her graduate studies at Tufts University and the Hebrew University, where she received her PhD. Since 1989 she has been a senior member of the Tel Aviv University team that cloned the ataxia-telangiectasia gene, *ATM*. Her research has concentrated on the development, refinement and application of positional cloning methodology. She is currently studying the structure and regulation of the *ATM* gene.

Steven Sherwood is a section leader at Geron Corporation. He received his PhD from the University of California, Berkeley, in 1983 and was a senior research scientist at Stanford University from 1986 to 1994. His research interests are the regulation of cell growth, cell cycle controls and apoptosis and how these processes regulate tissue remodelling associated with ageing and disease. He currently leads a group of scientists investigating therapeutic opportunities based on changes in gene expression during cellular senescence.

Yosef Shiloh graduated from Israel's Institute of Technology in Haifa and received his PhD from the Hebrew University of Jerusalem. He trained at Harvard Medical School and carried out research at the University of Michigan. He joined the Tel Aviv University Sackler School of Medicine in 1985. His early research focused on the cellular phenotype of ataxia-telangiectasia, and the *ATM* gene was later identified and cloned in his laboratory. He is currently investigating the function and metabolism of the ATM protein.

Pia Thömmes was a graduate student at Konstanz University and has since worked at Zurich University and Imperial College, London. Her research has focused on the enzymology of eukaryotic DNA replication, including the mammalian DNA polymerases and helicases. Her recent work at the ICRF concerns the role of MCM/P1 proteins in DNA replication.

Terry Van Dyke is an associate professor in the department of biochemistry and biophysics at the University of North Carolina at Chapel Hill. She received her PhD at the University of Florida and postdoctoral training at the State University of New York at Stony Brook and at Princeton University. She was an assistant professor, then associate professor, at the University of Pittsburgh from 1986 to 1993. Her research focuses on the genetic, molecular and cellular basis of cancer development.

Geoffrey M Wahl is a professor in the gene expression laboratory at the Salk Institute. He trained in bacteriology at the University of California, Los Angeles, received a PhD in 1976 from the department of biological chemistry at Harvard University Medical School and studied mechanisms of gene amplification in cancer cells during postdoctoral training at Stanford University. His current research focuses on understanding how defects in cell cycle control destabilize the cancer cell genome, how the initiation of mammalian DNA replication is controlled and how chemotherapeutic strategies based on such knowledge can be developed.

David T Weaver is an associate professor of microbiology and molecular genetics at Harvard Medical School and in the division of tumor immunology, Dana-Farber Cancer Institute, Boston. His laboratory studies mechanisms of double strand break repairs in eukaryotes.

Ted Weinert obtained his PhD from Yale University in 1984, then did postdoctoral research with Lee Hartwell. Since 1990 he has been associate professor of molecular and cellular biology at the University of Arizona, Tucson. His research has focused on checkpoint controls in yeast and their relevance to cancer.

Chaoying Yin obtained her PhD from Pennsylvania State University and did postdoctoral research at Rutgers. Currently she is doing postdoctoral research in Terry Van Dyke's laboratory. Her research interest is the regulation of TP53 dependent apoptosis in tumorigenesis.

Index

357

LIST OF PREVIOUS ISSUES

VOLUME 21 1994

Palliative Medicine: Problem Areas in Pain and Symptom Management
Guest Editor: G W Hanks

VOLUME 22 1995

Molecular Mechanisms of the Immune Response
Guest Editors: W F Bodmer and M J Owen

VOLUME 23 1995

Preventing Prostate Cancer: Screening versus Chemoprevention
Guest Editors: R T D Oliver, A Belldegrun and P F M Wrigley

VOLUME 24 1995

Cell Adhesion and Cancer
Guest Editors: I Hart and N Hogg

VOLUME 25 1995

Genetics and Cancer: A Second Look
Guest Editors: B A J Ponder, W K Cavenee and E Solomon

VOLUME 26 1996

Skin Cancer
Guest Editors: I M Leigh, J A Newton Bishop and M L Kripke

VOLUME 27 1996

Cell Signalling
Guest Editors: P J Parker and T Pawson

VOLUME 28 1996

Genetic Instability in Cancer
Guest Editor: T Lindahl